Organelle Genes and Genomes

Organelle Genes and Genomes

NICHOLAS W. GILLHAM

Department of Zoology
Duke University

New York   Oxford
OXFORD UNIVERSITY PRESS
1994

Oxford University Press

Oxford New York Toronto
Delhi Bombay Calcutta Madras Karachi
Kuala Lumpur Singapore Hong Kong Tokyo
Nairobi Dar es Salaam Cape Town
Melbourne Auckland Madrid

and associated companies in
Berlin Ibadan

Copyright © 1994 by Oxford University Press, Inc.

Published by Oxford University Press, Inc.,
200 Madison Avenue, New York, New York 10016

Oxford is a registered trademark of Oxford University Press

All rights reserved. No part of this publication may be reproduced,
stored in a retrieval system, or transmitted, in any form or by any means,
electronic, mechanical, photocopying, recording, or otherwise,
without the prior permission of Oxford University Press.

Library of Congress Cataloging-in-Publication Data
Gillham, Nicholas W.
Organelle genes and genomes / Nicholas W. Gillham.
p. cm. Includes bibliographical references and index.
ISBN 0-19-508247-8.—ISBN 0-19-508248-6 (pbk.)
1. Extrachromosomal DNA. 2. Cytoplasmic inheritance.
3. Mitochondria. 4. Chloroplasts.
I. Title. QH452.G537 1993 574.87'34—dc20 93-33374

9 8 7 6 5 4 3 2 1

Printed in the United States of America
on acid-free paper

Preface

I first became interested in organelle genes before it was really thought that they existed. True, scientists, particularly in Germany, had for many years studied the inheritance and segregation of plastid phenotypes in flowering plants so there was good reason to believe that these organelles might have genes of their own. Yet in the late 1950s phenomena such as these were grouped together under the general rubric of cytoplasmic inheritance. This expression was used to encompass those bizarre genetic phenomena that were non-Mendelian in nature. The rigorous studies of scientists such as Sonneborn, working on *Paramecium,* Rhoades studying cytoplasmic male sterility and the *iojap* mutation in maize, and investigators such as Michaelis and Stubbe reporting on plastid inheritance had made the study of cytoplasmic inheritance respectable and even interesting. Nevertheless, the field was widely regarded as phenomenological.

The picture changed dramatically in the 1960s. Ephrussi and Slonimski studying petite mutants in yeast and Sager investigating the inheritance of antibiotic resistance in *Chlamydomonas* opened the way for detailed genetic investigations of what later proved to be the mitochondrial and chloroplast genomes. Also, a great rush of papers began to appear demonstrating that chloroplasts and mitochondria contained their own DNA and could make their own unique proteins. At first, heated arguments ensued about contamination, bacterial and otherwise. But in the end the evidence was overwhelming that chloroplasts and mitochondria contained their own genes, genomes, and protein-synthesizing systems. Much as Mendel's factors became respectable with the discovery of chromosomes, organelle genetics was accepted with discovery that chloroplasts and mitochondria possessed their own genomes.

In the 1970s the study of organelle heredity became a mature field. Much was learned about the structure of organelle DNA, the genes encoded in chloroplasts and mitochondria, and the protein-synthesizing systems of these organelles. More and more it began to look as if these organelles were degenerate bacterial symbionts that had enjoyed their association with the eukaryote nucleus for a billion years or so. This hypothesis, resurrected at the beginning of the decade by Margulis, was the subject of much debate in the 1970s, but is now generally accepted. Studies of organelle genomes also showed that chloroplasts and mitochondria were very much the junior, albeit essential, partners in this association, the corporate takeover by the nucleus being incomplete. By 1980 the genetic and physical maps of the mitochondrial genome of yeast had been aligned.

Earlier gains continued to be consolidated during the 1980s. Of greater import, however, were the contributions studies of organelle genes and genomes were beginning to make to our understanding of basic biological phenomena including intron structure, protein import, RNA editing, the evolution of cells, and the classification of animal and plant taxa. By the end of this decade transformation methodologies had been established for the chloroplast of *Chlamydomonas* and the mitochondrion of yeast. These were the ultimate technical tools required to make the genomes of the two organelles accessible to all the modern techniques of molecular biology. By the early 1990s transformation technology has also spread to flowering plants.

So what about the 1990s? This decade is seeing the continued contribution of organelle gene and genomic studies to the greater good of many fields in biology. However, students of the organelles themselves will still have much to keep them occupied. One unsolved problem concerns the high frequency of transmission of chloroplast and mitochondrial genomes by only one parent. While the mechanisms of uniparental inheritance remain to be elucidated at the molecular level, the overriding question is really evolutionary. Why has uniparental inheritance of organelle genomes arisen independently not only for chloroplasts and mitochondria, but at least several times for each organelle? A second problem is the copy number paradox. The problem of reconciling the copy number paradox can be likened to that of reconciling the quantum and wave theories of electromagnetic reasonance, although this is undoubtedly of lesser moment to all but organelle geneticists. The copy number paradox simply put is this. The physical number of organelle genomes per organelle and per cell is large. The genetic copy number is small. What is the reason? While no one really knows, despite some very useful theoretical modeling, this paradox does not affect empirically determined experimental results. One just lives with it and wonders why. Two other areas in which much progress is currently being realized are the elucidation of mechanisms involved in protein uptake into chloroplasts and mitochondria and the regulatory signaling that goes on between a chloroplast or mitochondrion and their chief executive, the nucleus.

Organelles other than chloroplasts and mitochondria are not discussed in this book. One might question why the controversy surrounding basal body DNA is not considered or why the origin of peroxisomes is not investigated. It is my belief that our picture concerning the origin and genetic capabilities of these organelles will have clarified considerably a few years hence. For instance, the current dispute concerning basal body DNA is reminiscent of the controversy that surrounded the existence of chloroplast and mitochondrial genomes in the early 1960s. Certainly, by the time another book like *Organelle Genes and Genomes* is written, basal bodies, peroxisomes, and other eukaryotic organelles will probably be included as well.

In writing this book I have had to rely largely on review articles, although I have also consulted many original papers as well. The terrain encompassed in *Organelle Genes and Genomes* is simply too vast to permit a single author to read and digest the multitudes of relevant and pertinent scientific papers that have been published on the diverse topics that are the subject of this book. I felt this was a safe approach as many of these reviews have been written by people that I know and whose opinions I respect. Original papers were chosen largely because they made points that I felt were especially relevant to a specific discussion or because they considered topics that had not yet been thoroughly reviewed. I have tried to cover the literature in the field approximately through the late spring of 1993. However, I am quite aware that I have missed important papers and some reviews. For this, I apologize to both author and reader in advance.

Durham, N.C.
October 1993   N.W.G.

Acknowledgments

I have been fortunate for 25 years to have enjoyed a fruitful and rewarding collaboration with Professor John Boynton of the Botany Department at Duke. Together with our students, postdoctoral fellows, and research associates we have tried to get to the bottom of a variety of problems in the genetics and molecular biology of chloroplasts and mitochondria using *Chlamydomonas* as a model system. I particularly wish to note the contributions of Dr. Elizabeth Harris, Director of the Chlamydomonas Genetics Center, Ms. Barbara Randolph-Anderson, and Ms. Anita Johnson to much of this work. Drs. Heriberto Cerutti, Charles Hauser, Amnon Lers, Paul Liu, Scott Newman, and Mr. Peter Heifetz have made particularly significant contributions to our most recent work.

I am greatly indebted to Dr. Jeffrey Palmer for reading the whole text of *Organelle Genes and Genomes.* Dr. Palmer made major suggestions concerning the organization of the book and suggested shortening many chapters. The result is that the text now contains 17 chapters instead of the original nine and is divided into four sections. Dr. Palmer also made substantive comments throughout the book, particularly with respect to the subjects of organelle genome organization and evolution. The text has benefitted markedly from his input. I also wish to thank Drs. C.W. Birky, Jr. and Wilhelm Gruissem who read Chapters 7–9 and 16–17, respectively. Their comments on such things as intracellular population genetics (Birky) and genetic regulation in organelles (Gruissem) were extremely useful and helped me to avoid pitfalls. Any errors of omission or interpretation that remain in the text are mine alone and I accept full responsibility for them.

It gives me great pleasure to thank the following people who kindly provided illustrations from their work for reproduction in this book: Drs. Giuseppe Attardi, Wesley Brown, Roberto Cattaneo, Thomas Cech, David Clayton, Paul Englund, Susan Forsburg, Benjamin Glick, Michel Goldschmidt-Clermont, Michael Gray, Wilhelm Gruissem, Leonard Guarente, Maureen Hanson, Reinhold Herrmann, Hans Kossel, Tsuneyoshi Kuroiwa, Thanda Manzara, Walter Neupert, Jeffrey Palmer, Ellen Prager, Jean-David Rochaix, Gottfried Schatz, Ronald Sederoff, James Siedow, Mark Stoneking, Masahiro Sugiura, and Thomas Bruns. In addition Drs. John Avise, Rob DeSalle, Bernard Dujon, David Lonsdale, Satoshi Horai, Douglas Wallace and Klaus Wolf gave me permission to reproduce illustrations from their papers. I am also greatly indebted to Dr. Richard Hallick for providing me with a preprint of his paper with Erhard Stutz and colleagues on the complete sequence of the *Euglena* chloroplast genome, to Dr. Michael Reith for sending a preprint of his paper describing the gene content of much of the chloroplast genome of *Porphyra,* and to Drs. Don Bryant and Hans Bohnert for providing their unpublished data on the almost completely sequenced cyanelle genome of *Cyanophora.*

Much of this book was written while I was on a half-year sabbatical at our house on Long Island. Drs. Benjamin and Frances Burr arranged a visiting investigator position for me at Brookhaven National Laboratory so that I could use the Brookhaven library and Dr. David Luck sponsored me as a Visiting Professor at Rockefeller University so that I could use the library there. I wish to acknowledge with thanks their invaluable help in this regard.

I also want to thank Mr. Kirk Jensen, Senior Editor, Oxford University Press for his excellent editorial guidance.

Last, but by no means least, I wish to acknowledge my wife Carol for her unstinting support. Without it this book never would have been written.

Notes on Terminology

In determining the functions of organelle genes through a variety of genetic and molecular approaches different systems of gene designation have arisen. Throughout this book I have tried to use a three letter designation for both chloroplast and mitochondrial genes. This convention is in general use for chloroplast genes (see Hallick and Bottomley, 1983, *Plant Molecular Biology Reporter* 1:38–43; Hallick, 1989, *Plant Molecular Biology Reporter* 7:266–275), but is much more variable for mitochondrial genes (see Chapter 7). Special problems arise in certain cases such as the genes encoding components of mitochondrial complex I *(ndh)* where both alphabetical *(ndhA)* and numerical *(ndh1)* systems of designation exist. I have chosen the numerical system since the number of *ndh* subunits in mammalian mitochondrial complex I is now in considerable excess of the numbers of letters in the alphabet. Also, numerical designations are in routine use for genes specifying proteins of other mitochondrial complexes (e.g., *cox1* and *cox3* are subunits of cytochrome oxidase) even though by tradition certain genes have always been designated by letters (e.g., *cytb*, cytochrome *b; cytc*, cytochrome *c*). In the standardized nomenclature for chloroplast genes, those genes involved in photosynthesis are designated by letters (e.g., components of photosystem I are symbolized by *psaA, psaB* etc. while those belonging to photosystem II are indicated by *psbA, psbB* etc.). Genes encoding chloroplast and mitochondrial ribosomal proteins are always numbered (e.g., *rps12*, protein S12 of the small subunit of the chloroplast ribosome or the small subunit of the mitochondrial ribosome of *Paramecium* on the basis of sequence similarity to the corresponding protein in *E. coli*). However, the *E. coli* genes themselves have alphabetical designations (e.g., *rpmB*). In yeast, genes encoding mitochondrial ribosomal proteins are designated *MRP1* etc. reflecting the fact that these genes are nuclear and that capital letters are used in yeast to designate wild-type nuclear genes. I have tried to follow this convention. I have chosen not to use the one letter code commonly in use for designating tRNA genes (e.g., *trnK*) in many cases because I have found it easier to discuss tRNAs and the genes encoding them using the older system (e.g., tRNA$^{Lys}_{UUU}$) because the anticodon is often referred to. Certain unavoidable problems arise because of the ways in which mutations are designated in different systems (see Chapter 7), but, in general, a three letter system has been used.

At the suggestion of Dr. Jeffrey Palmer, I have used "flowering plant" when referring to an angiosperm and "land plant" rather than "higher plant" in discussing nonalgal plants from Psilophytes and Lycophytes through Anthophytes (angiosperms).

He points out that plants such as waterlillies have become secondarily aquatic and that "higher" seems pejorative. I think these are reasonable suggestions and have done my best to excise "higher" throughout the text although it may have crept in inadvertantly from time to time. Similarly, Dr. Palmer recommends the use "animal" in place of "metazoa," an entirely reasonable suggestion.

Contents

Part I   Introduction to Chloroplasts and Mitochondria, 1

  1. Chloroplasts and Mitochondria, 3

Part II   Evolution and Organization of Organelle Genomes, 35

  2. Endosymbiotic Origin of Chloroplasts and Mitochondria, 37

  3. Organelle Genome Organization and Gene Content, 50

  4. Evolution of Organelle Genomes, 92

  5. Cytological Organization and Replication of Organelle Genomes, 110

  6. Phylogenetic Uses of Organelle DNA Variation, 127

Part III   Organelle Genetics, 147

  7. Model Systems and Gene Manipulation, 149

  8. Transmission and Compatibility of Organelle Genomes, 181

  9. Intracellular Genetics of Organelles, 195

  10. Introns, Mobile Elements, and Plasmids, 207

  11. Mitochondrial Diseases, 240

Part IV   Expression and Biogenesis, 269

  12. Transcription and mRNA Processing in Organelles, 271

  13. Protein Synthesis in Organelles, 292

14. Reading and Editing Messages in Chloroplasts and Mitochondria, 319

15. Protein Targeting and Import, 333

16. Control of Gene Expression in Chloroplast Biogenesis, 364

17. Control of Gene Expression in Mitochondrial Biogenesis, 396

Index, 409

PART I

INTRODUCTION TO CHLOROPLASTS AND MITOCHONDRIA

1

Chloroplasts and Mitochondria

One of the major features distinguishing eukaryotic cells and prokaryotes such as bacteria, including blue-green algae (Cyanobacteria), is that eukaryotic cells are populated with a variety of subcellular organelles. Of the many types of organelles found within the eukaryotic cell the chloroplast and mitochondrion are distinctive in possessing their own unique genetic systems. The genes present in the chloroplast and mitochondrial genomes rely for their expression on RNA and protein synthesizing systems unique to each organelle. Given the sophistication of the eukaryote nucleus, the economies of this division of labor are not obvious. A much more rational solution to the casual observer would have been to put all of the genes in one place, the nucleus. The reason this path was not followed probably is related to the fact that chloroplasts and mitochondria arose as endosymbionts, but why certain genes remain incarcerated in organelles while others were banished to the nucleus remains a mystery. Although the events leading to the evolution of modern chloroplasts and mitochondria occurred a billion or so years ago, one can reach several important conclusions by analyzing the chloroplast and mitochondrial genomes of living organisms. For one thing, organelle genes code for specific subsets of components required for chloroplast and mitochondrial function and these tend to be highly conserved phylogenetically. There are also striking similarities between the components coded by the two genomes. Thus, the rRNAs and usually the tRNAs and some ribosomal proteins of the organelle protein-synthesizing system are coded by most chloroplast and mitochondrial genomes as are specific proteins of the respiratory and photosynthetic electron transport chains.

We have known for some time that, with the exception of the large subunit of the CO_2-fixing enzyme, ribulose bisphosphate carboxylase (RUBISCO), a chloroplast gene product, the vast majority of envelope proteins and soluble enzymes of chloroplasts and mitochondria are nuclear gene products. However, this picture is beginning to change markedly as organelle genomes are being characterized in more and more algae and protists. What is now apparent is that the genomes of chloroplasts and mitochondria are much more plastic in gene content than would have seemed possible only 5 years ago (Chapter 3).

The purpose of this book is to provide the reader with an introduction to the fascinating world of organelle genomes and what they do. We will investigate their anatomy, expression, genetics, and evolution. How the expression of organelle and nuclear genes is coordinated in the construction of chloroplasts and mitochondria is a question whose answer is just now being unraveled. Along the way forays will be made into particularly interesting areas of study including RNA editing, splicing of intervening sequences, protein transport, and mitochondrial diseases. To enter into such an exploration it is appropriate to preface the discussion with a brief consideration of the organelles themselves. That is the aim of this chapter.

In terms of their bioenergetic function, chloroplasts and mitochondria complement one another. Chloroplasts derive energy from light that is employed for splitting of water and production of molecular oxygen. The electrons produced from the splitting of water are used via the photosynthetic electron transport chain to drive photosynthetic phosphorylation. Ultimately, molecular CO_2 is reduced by the protons and electrons derived from water and is converted into carbohydrate by the soluble enzymes of the chloroplast *stroma*. The mitochondrion, in contrast, catalyzes the aerobic oxidation of reduced carbon compounds via soluble enzymes of the tricarboxylic acid cycle found in its *matrix*. The electrons produced by the oxidation of reduced carbon compounds flow via the respiratory electron transport chain and drive oxidative phosphorylation. The electrons and protons derived from oxidation of reduced carbon compounds convert molecular oxygen to water, and CO_2 is released as an oxidation product of the tricarboxylic acid cycle. In summary, the chloroplast reduces CO_2 and splits water with the

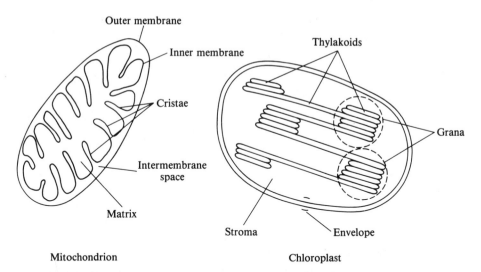

Fig. 1–1. Diagram comparing a typical mitochondrion and a land plant chloroplast. The mitochondrion is surrounded by an outer membrane. Separating the outer and inner membranes is the intermembrane space containing soluble enzymes. Localized in the inner membrane, which is infolded into cristae, is the respiratory electron transport chain. Within the inner membrane is the mitochondrial matrix containing many soluble enzymes. The chloroplast is surrounded by a double membrane envelope. Within the envelope is the fluid phase or stroma, which, like the mitochondrial matrix, contains many soluble enzymes. The thylakoid membranes are the sites of photosynthetic electron transport and are not connected to the envelope. Thylakoids are differentiated into areas where adjacent membranes become tightly appressed into stacks called grana, which are connected by unstacked stromal thylakoids.

release of O_2 while the mitochondrion oxidizes reduced carbon compounds with the formation of CO_2 and water. However, chloroplasts and mitochondria are not simply energy-generating and utilizing systems. A vast array of other metabolic processes go in within their confines as well. These are just as much key to the health and well-being of the cell as electron transport and energy generation. They are discussed briefly as the structure and function of chloroplasts and mitochondria are enumerated in more detail.

There are many similarities between chloroplasts and mitochondria in addition to the fact that both organelles contain their own DNA and protein-synthesizing systems. Both organelles have an outer bounding membrane, single in the case of the mitochondrion and double for chloroplasts of flowering plants, but triple or quadruple in certain algal groups (see following); an inner membrane possessing cytochromes, quinones, and other compounds required for electron transport; and a soluble phase containing enzymes required for carbon metabolism and a variety of other metabolic pathways (Fig. 1–1). Attached to the inner membranes of each organelle are proteins required for electron transport and coupled phosphorylation. There are obvious differences between chloroplasts and mitochondria too. Some of these have been alluded to in a general way in the last paragraph and will be detailed for each organelle separately in the remainder of this chapter.

MEMBRANE STRUCTURE

Because so much of the structure and function of chloroplasts and mitochondria is bound up with their membranes, a few general remarks on current notions of membrane structure are appropriate. Most membranes contain about 40% lipid and 60% protein on a weight basis, but there are distinct variations depending upon the membrane in question. For example, a unique sulfolipid is found in the internal or *thylakoid* membranes of chloroplasts which are also rich in galactolipids. In contrast, the inner mitochondrial membrane contains an abundance of phospholipids including cardiolipin. The most abundant lipids are *polar* or *amphipathic* species. Polar phospholipids predominate in the microsomal and mitochondrial membranes whereas glycolipids and chlorophylls are the principal polar lipids in thylakoid membranes.

The "fluid mosaic" model of Singer and Nicolson (1972) is consistent with most evidence derived from studies with different membranes (Fig. 1–2; Kleinfeld, 1987). In the fluid mosaic model *extrinsic* (peripheral) proteins attach to the outside of the membrane and are normally water soluble. In contrast, *intrinsic* (integral) proteins are usually amphipathic molecules in which the hydrophobic portions are buried inside the membrane and the hydrophilic portions are exposed at the membrane surface. Bitopic proteins cross the membrane only once while polytopic proteins cross the membrane more than once. Extrinsic proteins are more loosely associated with the membrane through electrostatic interactions with phospholipids and intrinsic proteins. Intrinsic proteins constitute 70–80% of all membrane proteins.

A variety of independent lines of evidence supports the Singer–Nicolson model, but one aspect of the Singer–Nicolson model that appears not to be correct is that the membrane constituents form a well-mixed fluid. In fact, numerous studies now indicate that membrane components exhibit varying degrees of inmiscibility (Kleinfeld, 1987).

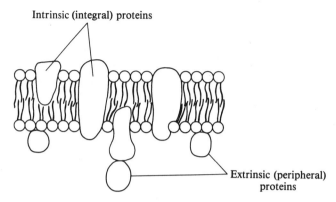

Fig. 1–2. The Singer–Nicholson model for membrane structure. The model assumes a lipid bilayer with extrinsic proteins on the outside of the membrane while intrinsic proteins with sequences of hydrophobic amino acids may partially or totally span the membrane. Modified from Tzagoloff (1982).

MITOCHONDRIA

Mitochondrial structure and function have been reviewed in detail in the excellent monograph by Tzagoloff (1982) from which this account is largely drawn. Mitochondria are most commonly observed to be oval structures about 1–2 μm long and 0.5–1.0 μm in diameter (Fig. 1–1). The outer membrane is smooth and continuous without obvious foldings or protuberances and has a diameter of 50 to 70 Å. The inner membrane is also continuous and its width varies from 75 to 100 Å, but this membrane is highly convoluted into folds that frequently appear almost as transverse septa. These folds were named *cristae mitochondriales* or *cristae* by Palade who published the first high-resolution electron micrographs of mitochondria in 1952. The lumen between the inner and outer membranes is called the *intermembrane space*, and the lumen contained within the inner membrane is the *matrix* (Fig. 1–1).

Mitochondria can assume different configurations depending on the physiological state of the cell or the medium in which the isolated organelles are suspended. Hackenbrock (1966) applied the terms *orthodox* and *condensed* to describe the reversible states of mitochondria that can be induced by altering their metabolism. These reversible states result from alterations in the conformation of the inner membrane, with a corresponding shift in the ratio of intermembrane space to matrix space. Exposure of mitochondria to a hypotonic medium, or to an isotonic medium containing readily penetrating cations and anions, etc., leads to a swelling of the mitochondria. Subsequent exposure to a hypertonic medium or to ATP, Mg^{2+}, Mn^{2+}, etc. causes a contraction. A characteristic of swollen mitochondria is a distended, often ruptured, outer membrane. The inner membrane unfolds, but retains its structural continuity. Contraction involves a refolding or reaggregation of the inner membrane without any restoration of the outer membrane, which tends to become detached from the inner membrane. Methods for mitochondrial isolation and fractionation have been reviewed by Tzagoloff (1982).

Mitochondrial Outer Membrane

The outer membrane is less dense (1.13 g/cm^3) than the inner membrane (1.21 g/cm^3), which aids in separation of the two membranes by differential centrifugation. This density difference is accounted for by the fact that on a protein basis the outer membrane contains two to three times the amount of phospholipid as the inner membrane. Not only do the inner and outer membranes differ in total amounts of lipids and proteins, but also in their composition of lipids and proteins. Cardiolipin, a major phospholipid component of the inner membrane, is present in only trace amounts in the outer membrane. Phosphatidylinositol is an important constituent of the outer membrane, but is much less conspicuous in the inner membrane. The neutral lipid cholesterol is six times more concentrated in the outer membrane than the inner membrane. Comparison of the proteins present in inner and outer membranes by sodium dodecyl sulfate (SDS)-gel electrophoresis, which separates proteins by molecular weight, reveals little correspondence between inner and outer membrane proteins.

The outer membrane is rich in enzyme activities and these are distinct from those associated with the inner membrane. Prominent among these are enzymes concerned with phospholipid biosynthesis (Ernster and Kuylenstierna, 1970). The two most frequently used marker enzymes for the outer membrane are monoamine oxidase and the rotenone-insensitive nicotinamide adenine dinucleotide (NADH) cytochrome c reductase. The latter activity is catalyzed by a flavoenzyme and the cytochrome b_5 complex and is distinct from the rotenone sensitive NADH–cytochrome c reductase of the respiratory chain. A flavoprotein–cytochrome b_5 reductase system is also present in the endoplasmic reticulum. So far as is known, no proteins of the outer membrane are coded by any mitochondrial genome.

Mitochondrial Intermembrane Space

Very few enzyme activities have been attributed unambiguously to the intermembrane space. Part of the problem is that most mitochondrial fractionation procedures described to date are not perfect. Rupture of the inner membrane leads to leakage of soluble components from the matrix into the soluble phase containing enzymes of the intermembrane space. Also, extrinsic proteins loosely bound to the inner and outer membranes may appear in the supernatant. However, if the fractionation is reasonably clean, outer membranes will be disrupted, but inner membranes will not. In this case, there should be an enrichment for specific enzymes in the supernatant phase associated with the intermembrane space following removal of inner and outer membranes by centrifugation. This is the basis for assigning several enzyme activities apparently coded by nuclear genes to the intermembrane space.

Mitochondrial Inner Membrane

The inner membrane is a highly specialized structure containing the respiratory electron transport chain and the coupling factor responsible for oxidative phosphorylation. Since this membrane is a patchwork of mitochondrial and nuclear gene products, it deserves more detailed description in this book. Coupling factor structure as well as the process of oxidative phosphorylation are discussed later in this chapter.

The electron transport chain includes four multimeric complexes that can be solubilized from the inner membrane using agents such as the bile acids cholate and deoxycholate. A complex is then purified employing a combination of other methods including centrifugation, ammonium sulfate precipitation and further solubilization. The composition of each complex has been studied in detail and the sequence of electron transfer established (Fig. 1–3).

NADH-ubiquinone reductase (complex I). The structure and function of complex I have been reviewed by Ragan (1987), Weiss and Friedrich (1991), and Weiss *et al.* (1991). Fearnley and Walker (1992) consider the degree of conservation of complex I subunits and how they are related to other proteins. Complex I catalyzes the oxidation of NADH (nicotinamide adenine dinucleotide) by ubiquinone (coenzyme Q) and is the entry point for electrons traveling from NADH into the mitochondrial electron transport chain. Complex I is inhibited by rotenone, and exhibits oxidative phosphorylation at site 1. The prosthetic group of the enzyme is flavin mononucleotide (FMN). Electron transfer is carried out by numerous redox groups whose sequence of operation and function is largely unknown. According to Weiss and Friedrich (1991) there are four iron–sulfur clusters associated with complex I; one, N-1, possesses two iron atoms (binuclear) and three, N-2, N-3, and N-4, have four iron atoms (tetranuclear). However, Weiss *et al.* (1991) caution that "the number of iron sulfur clusters in the enzyme are still under debate."

Complex I was first isolated from bovine heart mitochondria by Hatefi *et al.* (1962) and only more recently from another source, *Neurospora crassa* (Ise *et al.*, 1985). Complex I is dimeric, at least in the isolated state, and quite possibly in the native state as well. Beef heart complex I was thought to contain ca. 25 different polypeptides, but recently the number has been revised upward to ca. 41 (Arizmendi *et al.*, 1992; Fearnley and Walker, 1992). In describing bovine complex I Fearnley and Walker state that "it is the most complex enzyme that has been characterized, the total protein sequence in its subunits exceeding the combined sequences of all the constituent polypeptides of the *Escherichia coli* ribosome." Approximately 30 polypeptides are associated with the complex from *Neurospora* (Weiss *et al.*, 1991). A two subunit enzyme has been described from the bacterium *Paracoccus denitrificans*, which is likely part of a larger complex complex with at least 10 polypeptides (Ragan, 1987; Weiss *et al.*, 1991). The yeast *Candida albicans* has a respiratory chain similar to that of mammals and a high-molecular-weight NADH dehydrogenase has been partially purified from this organism. There is also evidence for an NADH-dehydrogenase similar to complex I in plant mitochondria, but the complex has not been purified. Site I phosphorylation is absent in the yeast *Saccharomyces* and in many bacteria in which the NADH dehydrogenase is much simpler and bears no obvious resemblance to complex I. In fact Weiss *et al.* (1991) point out that mitochondria of fungi and plants contain two additional NADH: ubiquinone reductases that are unrelated to the NADH dehydrogenase discussed here. *S. cerevisiae* possesses only these two enzymes.

Complex I can be degraded in several ways to yield flavoprotein (FP), iron–protein (IP), and hydrophobic (HP) fragments (Ragan, 1987). The FP fragment is soluble, accounts for all the FMN, six of the iron atoms and three of the polypeptides in a 1:1 molar ratio. The FP fragment appears to contain both a binuclear and a tetranuclear iron–sulfur cluster. The NADH-binding site has been localized to the 51-kDa protein

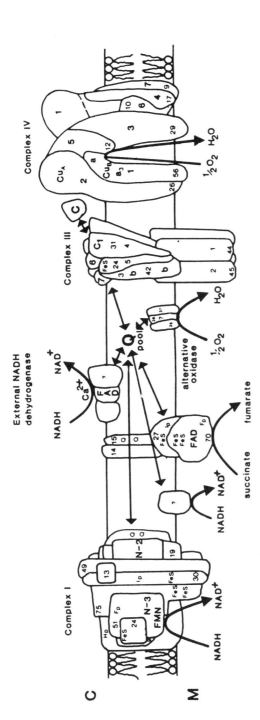

Fig. 1-3. Organization of the respiratory chain of plant mitochondria. FMN, flavin mononucleotide; FAD, flavin adenine dinucleotide; FeS, iron–sulfur center; C, cytochrome c; C_1, cytochrome c_1; b, cytochrome b; Cu_A and Cu_B, the atoms of Cu associated with cytochromes a and a_3, respectively; Q, proposed ubiquinone binding site; Q pool, ubiquinone pool; H_p, insoluble hydrophobic fraction; I_p, water-soluble iron–sulfur protein fraction; F_p, water-soluble iron–sulfur flavoprotein fraction; N-2, N-3, two iron–sulfur clusters; small numbers refer to the molecular weights of the respective polypeptides whereas large numbers refer to the subunit number; C and M refer to the cytoplasmic and matrix sides of the inner membrane, respectively. From Moore and Siedow (1991) with permission.

Table 1–1. Mitochondrial Gene Products in Human Complex I[a]

Gene Product	Molecular Weight (Da)
ND-1	35,666
ND-2	38,949
ND-3	13,188
ND-4	51,603
ND-4L	10,743
ND-5	66,937
ND-6	18,689

From Ragan (1987).

[a]Molecular weights were determined from the DNA sequences of each gene.

by photoaffinity labeling suggesting that FMN, the most likely oxidant of NADH, also resides in this subunit. Six polypeptides have been consistently identified with the IP fragment, which, like the FP fragment, is fully soluble. The IP fragment possesses at least three iron–sulfur clusters of which two copurify with a complex of 49-, 30-, and 13-kDa subunits. A 75-kDa subunit that is separable from this complex also possesses an iron–sulfur cluster. A 15-kDa Q-binding protein may be involved in the terminal electron transfer reaction. The HP fragment accounts for 70% of the total protein and six or seven of the enzyme's iron atoms and possesses at least two iron–sulfur clusters with one being binuclear and the other tetranuclear. The latter, tentatively identified as N-2, is thought to be the likely donor to ubiquinone.

The HP fragment contains all of the proteins in contact with phospholipid, in keeping with its operation in a hydrophobic environment, as well as all of seven of the mitochondrially encoded subunits of complex I found in vertebrates (Table 1–1).

Weiss and colleagues (Weiss and Friedrich, 1991; Weiss et al., 1991) review the assembly and evolution of complex I. They note that when mitochondrial protein synthesis is inhibited in *Neurospora*, a small NADH dehydrogenase complex (350 kDa) is synthesized containing ca. 30 nuclear-encoded polypeptides. These are identical to the nuclear encoded subunits of the large (700-kDa) enzyme. The small complex is hydrophilic and contains FMN and iron–sulfur centers N-1, N-3, and N-4, but lacks N-2. This complex has the same high affinity for NADH as the large complex. The hydrophobic (integral) portion contains the mitochondrially encoded subunits of the complex (Table 1–1) and is complementary to the hydrophilic (peripheral) part of the complex whose proteins are encoded by nuclear genes. This led Weiss and colleagues to propose a model for complex I that differs from that shown in Fig. 1–3 where complex I is completely embedded in the inner mitochondrial membrane. They suppose that complex I has both peripheral and integral portions.

Fearnley and Walker (1992) remark that sequence comparisons reveal that 19 complex I polypeptides are conserved in more than one species. They speculate that the minimal mitochondrial complex I probably requires these polypeptides at least.

Succinate-ubiquinone oxidoreductase (complex II). The properties of complex II, which catalyzes the oxidation of ubiquinone by succinate (Fig. 1–3), have been reviewed by Ohnishi (1987). The oxidoreductase is composed of hydrophilic succinate dehydrogenase (SDH) and hydrophobic proteins. The prosthetic group is flavin adenine dinucleotide (FAD) rather than FMN. Bovine heart SDH consists of a 70-kDa flavoprotein-binding (F_p) subunit that binds FAD and a 27-kDa subunit (I_p) possessing two binuclear and one tetranuclear iron–sulfur clusters. SDH catalyzes the oxidation of succinate to fumarate using an acceptor such as ferricyanide, but SDH, in contrast to intact complex II, is unable to reduce coenzyme Q. The hydrophobic portion of complex II contains two proteins of 15.5 and 17 kDa that copurify with a special form of cytochrome b_{558}. This form of cytochrome b is made in the cytoplasm and encoded by a nuclear gene unlike the cytochrome b in complex III, which is invariably a mitochondrial gene product. A specific pair of ubiquinone species is associated with the hydrophobic portion of complex II. No mitochondrial gene products are found in complex II.

Ubiquinol–cytochrome c reductase (complex III). Electron transfer chains of mitochondria, chloroplasts, and some photosynthetic bacteria all contain a multiprotein complex called $bc_1(b_6 f)$ with similar structural and functional properties (Fig. 1–4). Complexes from different sources possess two cytochrome b molecules, one cytochrome c_1, one high potential iron–sulfur cluster (Rieske FeS center), and probably a ubiquinone molecule. In the mitochondrion this assemblage is referred to as complex III. It catalyzes the oxidation of reduced coenzyme Q by cytochrome c, is inhibited by antimycin A and myxothiazol, and possesses the second of the three sites for oxidative phosphorylation.

The structure of complex III is reviewed by Weiss (1987). This complex contains 8 to 11 subunits depending on species. The five largest are highly conserved and include the core proteins (subunits I and II), cytochrome b (subunit III), cytochrome c_1 (subunit IV), and the Rieske iron–sulfur protein (subunit V) (Table 1–2). The core subunits have a peripheral location in the enzyme and are water soluble. Cytochrome b, the only

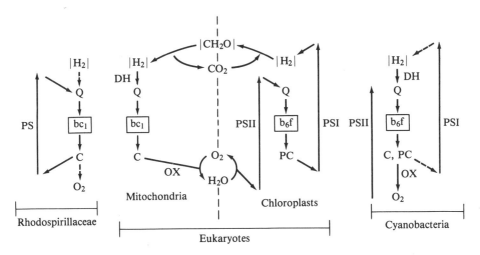

Fig. 1–4. Cytochrome bc complexes in different electron transport chains. DH, dehydrogenase; OX, oxidase; PS, photosystem; Q, quinone; PC, plastocyanin; C, cytochrome c; H_2, hydrogen bound to NAD(P)H or $FADH_2$. Modified from Hauska (1985).

Table 1–2. Subunit Composition of Complex III[a]

Subunit Number	Apparent Molecular Weight (kDa)	Name	Remarks
I	48–52	Core protein I	
II	42–45	Core protein II	
III	43–44*	Cytochrome b	
IV	27–28*	Cytochrome c_1	
V	22*	Fe–S subunit	
VI	13–15*		
VII	11–14		
VIII	9–17*		Associates with cytochrome c_1
IX	8*		Absent in yeast
X	7*		Associates with cytochrome c_1 Absent in yeast and *Neurospora*
XI	6*		Absent in yeast and *Neurospora*

Modified from Weiss (1987).

[a] A range of apparent molecular weights is given for each subunit based on SDS-gel electrophoresis of complex I subunits or determination of primary structure (*) from cow, *Neurospora,* and yeast.

mitochondrially encoded component of the complex, is a ca. 43-kDa protein and associates with two protoheme groups (b_{566}, low-potential and b_{562}, high-potential). This is a very hydrophobic protein with eight to nine membrane-spanning segments and can bind ubiquinone in photoaffinity labeling experiments. The 27- to 28-kDa cytochrome c_1 is an amphipathic protein that is associated with two smaller proteins of 9.2 and 7.2 kDa in preparations from bovine heart. The iron–sulfur subunit is the last of the five major subunits of complex III and has an average molecular weight of 21.6 kDa. Like cytochrome c_1, the iron–sulfur protein is also amphipathic. The primary structures of five of the six smaller subunits of bovine complex III have been determined and sequence similarity has been found between two of these and two of the three sequenced yeast subunits. A model of *Neurospora* complex III suggests that subunits I and II are peripheral proteins facing the matrix space, subunit III (cytochrome b) is an integral membrane protein, and subunits IV and V are peripheral proteins facing the intermembrane space.

While three of the five conserved subunits of complex III have documented electron transport functions, the two core proteins (subunits I and II) do not. Their equivalents are not found in isolated cytochrome $b_6 f$ from chloroplasts nor in bacterial bc complexes. However, the mitochondrial core proteins are essential since yeast mutants deficient in subunit II exhibit strongly reduced enzymatic activity and the complex is completely inactive in subunit I mutants (Weiss *et al.*, 1990). Weiss *et al.* (1990) suggest that subunits I and II function by interacting with other subunits of complex III and thus may contribute to the assembly and stabilization of the complex. Subunits I and II share striking amino acid sequence homology, respectively, with the processing-enhancing (PEP) and matrix-processing (MPP) proteins of yeast and *Neurospora,* which are involved in protein import (Chapter 15).

CHLOROPLASTS AND MITOCHONDRIA

Cytochrome c. This water-soluble protein mediates transfer of electrons between complexes III and IV. The enzyme from horse heart has a molecular weight of 12.4 kDa. The amino acid sequences of at least 70 different cytochrome *c* species have been determined and phylogenetic trees of the evolution of the protein have been constructed (Tzagoloff, 1982). The enzyme is encoded by nuclear genes. In the yeast *Saccharomyces cerevisiae* the *CYC1* and *CYC7* genes encode, respectively, iso-1-cytochrome *c* and iso-2-cytochrome *c*. *CYC2* encodes a protein required for the import of cytochrome *c* into the mitochondrion and the *CYC3* gene product is the heme lyase that catalyzes covalent attachment of the heme group to the isocytochromes *c* (see Chapter 15). The isocytochromes *c* of yeast and the genes encoding them have been intensively investigated by Sherman and his colleagues for many years. These studies are summarized in a short retrospective article by Sherman (1990).

Cytochrome c oxidase (complex IV). This complex catalyzes the oxidation of reduced cytochrome *c* by molecular O_2. Its activity is inhibited by azide, cyanide, and carbon monoxide. Complex IV contains the third oxidative phosphorylation site. The structure and evolution of cytochrome oxidase have been the subject of several reviews (e.g., Capaldi, 1990; Capaldi *et al.*, 1987; Kadenbach *et al.*, 1987).

The redox centers of both prokaryotic and eukaryotic aa_3-type cytochrome *c* oxidases are the two heme moieties and two copper atoms (Cu_A and Cu_B). Bacterial aa_3-type cytochrome *c* oxidases contain three subunits that are homologous with the generally mitochondrially encoded subunits (I, II, and III) found in eukaryotes (Table 1–3). Six nuclear-encoded subunits found in yeast are also found in mammals. Mammals contain an additional four subunits not found in yeast (Table 1–3). There is genetic

Table 1–3. Subunit Structure of Cytochrome *c* Oxidase in Bovine Heart, Yeast, and the Bacterium *Paracoccus denitrificans*

Bovine Heart[a]		Yeast		P. denitrificans	
Subunit Number	Apparent Molecular Weight (Da)	Subunit Number	Apparent Molecular Weight (Da)	Subunit Number	Apparent Molecular Weight (Da)
I	56,993	I	56,000	I	55,000
II	26,049	II	26,678	II	27,000
III	29,918	III	30,340	III	30,000
IV	17,153	V	14,858		
Va	12,434	VI	12,627		
Vb	10,670	IV	14,570		
VIa	9,418				
VIb	10,068				
VIc	8,480	VIIa	6,303		
VIIa	6,234	VII	6,603		
VIIb	6,350				
VIIc	5,541	VIII	5,364		
VIII	4,962				

From Capaldi (1990).

[a]The nomenclature of Kadenbach and Merle is given for the bovine heart subunits.

evidence that at least five of the six nuclear-encoded subunits in yeast are required either for the assembly or activity of cytochrome oxidase. The mitochondrially encoded subunits I, II, and III are the catalytic core of the enzyme in eukaryotes and the hemes and coppers have been localized to subunits I and II while subunit III functions in proton pumping by the enzyme (Fig. 1–3).

Subunits I–III are relatively hydrophobic proteins. Subunit I possesses 12 putative transmembrane regions while subunits II and III have a possible two and seven such regions, respectively. In contrast to the mitochondrially encoded subunits of cytochrome oxidase, the amino acid sequences of many of the subunits encoded in the nucleus are poorly conserved. Nevertheless, the overall features of those subunits conserved between different organisms remain the same. Bovine heart subunits IV, VIa, VIc, VIIa, VIIb, VIIc, and VIII all have tripartite structures with a hydrophilic N-terminus, a hydrophobic central domain, and a hydrophilic C-terminus. Subunits Va, Vb, and VIb do not have any obvious hydrophobic domains and are probably extrinsic proteins. Several of the nuclear-encoded proteins are present in two or more isoforms.

Alternative oxidase. The mitochondria of higher plants, the green alga *Chlamydomonas*, many filamentous fungi (e.g., *Neurospora*), yeasts (but not *Saccharomyces cerevisiae*), and various protista including several members of the *brucei* group of African trypanosomes possess a cyanide- and antimycin-insensitive alternative oxidase in addition to the usual cytochrome-mediated respiratory electron transfer chain (Fig. 1–3). However, the alternative oxidase, unlike the cytochrome-mediated electron transfer chain, is sensitive to salicyl hydroxamic acid (SHAM). Electron flux from reduced ubiquinone to O_2 via alternative oxidase is not associated with energy conservation and the free energy released is lost as heat. The pathway is nonphosphorylating when either succinate or external NADH is used as respiratory substrates, but site I phosphorylation occurs when NAD-linked substrates are used. Alternative oxidase activity is associated with a single 35- to 37-kDa protein and a cDNA encoding this protein has been cloned from the Voodoo Lily, *Sauromatum,* by Rhoads and McIntosh (1991). Lambowitz *et al.* (1989) demonstrated that an antibody made against the Voodoo Lily protein cross-reacts with two *Neurospora* mitochondrial proteins (36.5 and 37 kDa). These proteins

Table 1–4. Location of Genes Coding for the Four Respiratory Chain Complexes in the Cow[a]

Complex	Number of Genes	
	Mitochondrion	*Nucleus*
I	7	34
II	0	4
III	1	10
IV	3	10

[a]See nomenclature for seven mitochondrially encoded subunits of complex I in Table 1–1. The other mitochondrially encoded subunits of these complexes are *cytb* encoding cytochrome *b* and *cox1–3* encoding the three largest subunits of cytochrome oxidase (complex IV).

were made constitutively, but increased in amount under conditions that induced alternative oxidase activity. *Neurospora* mutations in the *aod-1* gene make apparently inactive polypeptides. This suggests that *aod-1* encodes the alternative oxidase and the two molecular weight forms seen on gels are the result of protein modifications. Mutations in the *aod-2* gene fail to produce these polypeptides, indicating that this gene encodes a component required for alternative oxidase induction. The characteristics of the alternative oxidase have been reviewed by Moore and Siedow (1991).

In summary, the mitochondrial respiratory electron transport chain is a mosaic of proteins coded in the nucleus and the mitochondrion (Table 1–4). Only in the case of complex II have polypeptides encoded in the mitochondrion so far not been reported. Why specific genes have been relegated to the mitochondrion and others to the nucleus is a mystery, although the mitochondrial gene products tend to be the most hydrophobic proteins in each complex.

Mitochondrial Matrix

Most of the many soluble enzymes of the mitochondrion are found in the matrix. Among the most important of these are the enzymes of the tricarboxylic acid cycle and fatty acid oxidation, the pyruvate dehydrogenase complex, carbamyl phosphate synthetase, and ornithine transcarbamylase. So far no genes encoding these enzymes have been found in any mitochondrial genome, but that does not mean that certain of these genes may not turn up as the mitochondrial genomes of more and more protists and algae are characterized. "Unexpected" genes involved in biosynthetic processes previously thought to be under strict nuclear control have been appearing with great regularity in algal chloroplast genomes. The mitochondrial matrix also contains mtDNA and the mitochondrial protein-synthesizing system.

CHLOROPLASTS

Chloroplast structure and function have been the subject of excellent monographs by Kirk and Tilney-Bassett (1978) and Hoober (1984). The latter book, in particular, has served as the basis for the introductory account given here to chloroplast structure and function. Chloroplasts vary in size, shape, and number per cell. A typical flowering plant chloroplast is a biconcave or lens-shaped structure 1–3 μm across by 5–7 μm in length (Fig. 1–1). The double membrane envelope of the chloroplast surrounds the fluid or soluble phase *(stroma)*. Chlorophyll and the entire photosynthetic apparatus are located in the *thylakoids*, an internal membrane system consisting of flattened vesicles. In a higher plant chloroplast the thylakoids are stacked in regular aggregates called *grana*. They are connected by thylakoids protruding beyond the ends of the grana *(stroma lamellae)*. There are many variations on this ground plan with respect to both color and structure. Thus, in 1885 Schimper classified the plastids into three groups based on their color. *Chloroplasts* were green, *leucoplasts* white, and *chromoplasts* yellow. Since that time a number of other plastid types have been described (Table 1–5).

The simplest photosynthetic systems are found in photosynthetic bacteria that lack a membrane-bounded structure equivalent to the chloroplast. In green, photosynthetic bacteria (Chlorobiaceae) the energy-transducing activities are restricted to localized

Table 1-5. Classification of Plastid Types

Plastid Type	Characteristics
Chloroplast	Typical mature green plastid
Chromoplast	Yellow plastid containing carotenoid pigment globules
Leucoplast	White plastid
Amyloplast	Starch-containing plastid
Elaioplast	Oil-containing plastid
Etioplast	Plastid of a dark-grown plant
Proplastid	Small, undifferentiated plastid of meristem
Proteinoplast	Plastid containing protein crystals

regions of the cell membrane specialized for this purpose. The light-harvesting complexes are found in nonmembranous particles called *chlorosomes* bound to these specialized regions. In most purple photosynthetic bacteria (Rhodospirillaceae, Chromatiaceae), the photosynthetic apparatus is located in pigment-rich specialized invaginations of the cell membrane that are highly differentiated chemically and functionally. These structures are called *chromatophores*. The Cyanobacteria contain free-standing thylakoids that extend through the cytoplasm usually near the periphery (Table 1–6). Attached to the surface of these membranes are structures called *phycobilisomes* that contain the accessory pigments phycocyanin (blue) and phycoerythrin (red) responsible for the colors of these organisms. These pigments are complexed with proteins called phycobiliproteins. The phycobiliproteins absorb wavelengths of light that are poorly absorbed by chlorophyll enhancing the photosynthetic capacity of cells containing them. The order of energy transfer within the phycobilisome is as follows: phycoerythrin → phycocyanin → allophycocyanin → photosystem II.

The thylakoids of the red algae (Rhodophyta) are contained within a double membrane chloroplast envelope, but beyond that they have important similarities to cyanobacterial thylakoids in that they exist individually in parallel to each other and have phycobilisomes on their surfaces. The Chrysophyte algae include the Bacillariophyta (diatoms), the Chrysophyta (yellow-green algae), and the Phaeophyta (brown algae). In these algae the thylakoids are usually grouped in bands of three with each thylakoid near, but not in contact with, its neighbor. The thylakoids usually extend from one pole of the chloroplast to the other and, in some organisms, there are several concentric rings of thylakoids inside the envelope called *girdle bands*. These chloroplasts are also surrounded by a double membrane structure in addition to the chloroplast envelope, which is called the *chloroplast endoplasmic reticulum*.

In Euglenophyte algae three membranes surround the chloroplast. This observation led to the notion that the outermost membrane is equivalent to the plasma membrane of a eukaryotic host cell that ingested the complete chloroplast from a green algal donor. In the Cryptophyta (cryptomonads) there is irrefutable evidence for endosymbiosis. In these algae the chloroplast and a vestigial nucleus, termed the *nucleomorph*, are enclosed within the chloroplast endoplasmic reticulum, which separates the putative endosymbiont from the cytoplasm and true nucleus of the cell. Green algal (Chlorophyta) chloroplasts have a double membrane envelope as do higher plants. The number of bounding membranes and their origin have been the subject of much speculation in studies of chloroplast evolution (Chapter 2). The higher classification of many algal

CHLOROPLASTS AND MITOCHONDRIA

Table 1-6. Some Structural Characteristics of the Photosynthetic Apparatus in the Cyanobacteria (Cyanophyta) and Different Algal Classes

Phylum and Class	Usual Number of Thylakoids Per Band	Girdle Bands Present	Grana Present	Degree of Apposition of Thylakoids	Number of Membranes Enclosing Chloroplast
Cyanophyta					
Cyanophyceae	1	No	No	None	No plastid
Rhodophyta					
Rhodophyceae	1	No	No	None	2
Cryptophyta					
Cryptophyceae	2	No	No	Loose	4
Dinophyta					
Dinophyceae (Dinoflagellata)[a]	3	No	No	Variable	3
Prymnesiophyta					
Prymnesiophyceae	3	No	No	Variable	4
Chrysophyta					
Raphidophyceae	3	Yes	No	—	4
Chrysophyceae	3	Yes	No	Variable	4
Bacillariophyceae[b]	3	Yes	No	Variable	4
Xanthophyceae[b]	3	Variable	No	Variable	4
Phaeophyta					
Phaeophyceae	3	Yes	No	Loose	4
Euglenophyta	3	No	No	Tight	3
Chlorophyta					
Prasinophyceae	2-4	No	Yes	Tight	2
Chlorophyceae	2-6	No	Yes	Tight	2

From Hoober (1984).

[a] Some dinoflagellate groups have plastids with only two membranes.

[b] These groups are referred to as the Bacillariophyta and Xanthophyta in some classifications.

groups is also in flux. For example, Cavalier-Smith (1986, 1989) recognized a kingdom Chromista, one of whose defining features is the existence of chloroplasts within the chloroplast endoplasmic reticulum.

In green algae the thylakoids are seen to be fused together into granal stacks as they are in land plants. The number of stacks per granum depends on species and growth condition. Fusion of membranes generally occurs over most of the thylakoid surface whereas in plants there are small discrete grana connected by unfused stromal thylakoids. An important feature characteristic of many algal chloroplasts is a granular structure called the *pyrenoid body*. In green algae such as *Chlamydomonas* the pyrenoid body is usually surrounded by starch, but in other algae the storage polysaccharides are found outside the chloroplast.

Chloroplast development is profoundly influenced by light (Fig. 1-5). Dividing meristematic cells of shoots or roots of higher plants contain small ameboid pale green or colorless structures called *proplastids*. Proplastids are double membrane structures about 1 μm in diameter. They have relatively little internal structure with the exception of a few apparently isolated vesicles and thylakoids. In plants exposed to light proplastids differentiate into chloroplasts. This process of chloroplast differentiation is

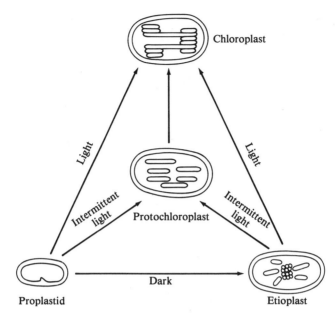

Fig. 1–5. Pathways of chloroplast development in higher plants. The proplastid is the youngest stage of development existing as a simple double membrane envelope. In continuous light the amount of thylakoid membranes and chlorophyll increases steadily to give a fully developed chloroplast. In the dark the proplastid develops into a colorless etioplast containing a prolamellar body. Exposure of etiolated plants to light results in transformation of etioplasts into chloroplasts. This transformation is accompanied by dispersal of the prolamellar body, thylakoid formation, and greening. Exposure of either proplastids or etioplasts to very short light pulses (2 min) separated by long dark periods (98 min) results in formation of protochloroplasts that contain few thylakoids, no chlorophyll *b*, and no grana. Exposure of protochloroplasts to continuous light results in rapid development of mature chloroplasts. Modified from Hoober (1984).

accompanied by an increase in chlorophyll levels and elaboration of thylakoid membranes. As cells move out from the meristematic regions, they cease to divide, but chloroplast division continues. In young wheat leaves the zone of mitotic divisions at the leaf base is spatially separated from a region in which constricted (dividing) chloroplasts are observed (Leech and Pyke, 1988). In the latter region plastid number is seen to increase from an average of 45 to 107 per cell. Further increases in plastid number subsequently occur yielding an average of 158 plastids per cell in the oldest cells near the leaf tip. In spinach leaves the number of plastids can change 10-fold from 10–15 plastids per cell in young leaves to 150–200 plastids per cell in fully expanded leaves (Possingham *et al.*, 1988).

If seeds are germinated in the dark and the young plants grown under these conditions, a different series of events takes place. The seedlings possess no chlorophyll and the plastids are referred to consequently as *etioplasts*. Within the etioplast an organized network of intersecting tubules, the *prolamellar body*, accumulates. Isolated prolamellar bodies are rich in lipid, but contain relatively little protein in contrast to the thylakoid membranes of light-grown plants. Exposure of etiolated plants to light results in the photoconversion of protochlorophyll(ide) with an *in vivo* absorption maximum at 650 nm to chlorophyll(ide) with an absorption maximum at 684 nm, following which the paracrystalline structure of the prolamellar body is lost, and the material is dispersed as primary lamellar layers. The protochlorophyll is attached to a lamellar protein called protochlorophyllide holochrome, which plays an essential role in the photoreduction of protochlorophyll to protochlorophyll(ide). During dispersal of the prolamellar body material into primary layers, the photochemically formed chlorophyll with an *in vivo* absorption maximum of 684 nm is converted to a form with an *in vivo* absorption maximum of 672 nm. In the final light-dependent step formation of new thylakoids and their aggregation occurs in strict correlation with chlorophyll synthesis and the *in vivo* absorption maximum shifts to 678 nm. The photochemical activity of the chloroplast and its capacity for electron transport and oxygen evolution develop simultaneously with the formation of grana structure. If etiolated seedlings are exposed to short (2 min) light pulses separated by long (98 min) dark periods, chlorophyll synthesis proceeds slowly. Under these conditions etioplasts are converted into *protochloroplasts* in which the prolamellar body disperses and primary thylakoids form. Parenthetically, it should be noted that a number of plants (e.g., pine, liverwort) and algae (e.g., *Chlamydomonas*) possess a light-independent protochlorophyllide reductase and are green in the dark.

The photoreduction of protochlorophyll to protochlorophyll(ide) is one of several light reactions implicated in chloroplast development. A second light reaction involves the plant regulatory photoreceptor *phytochrome* whose properties have been reviewed by Quail (1991). Phytochrome is reversibly interconvertible between its inactive (Pr) and active (Pfr) forms by sequential absorption of red (650–670 nm) and far red (730–750 nm) light according to the scheme shown below.

$$\text{Pr} \underset{\text{far red}}{\overset{\text{red}}{\longleftrightarrow}} \text{Pfr} \longrightarrow \underset{\substack{\text{gene} \\ \text{expression}}}{\text{altered}} \longrightarrow \underset{\text{response}}{\text{morphogenesis}}$$

Perception   transduction

Pfr formation (perception) initiates a transduction process that culminates in the altered expression of selected genes and a pattern of growth and development appropriate for the prevailing light environment. The phytochrome molecule consists of a specific polypeptide covalently linked to a linear tetrapyrrole chromophore. In *Arabidopsis* phytochrome is encoded by a small nuclear gene family.

The expression of many nuclear genes encoding components of the photosynthetic apparatus in response to Pfr formation has been documented (Thompson and White, 1991 and Chapter 16). These include genes specifying the small subunit of RUBISCO and the chlorophyll *a/b* binding proteins. Other less well characterized blue light receptors (blue/ultraviolet, UV-A and ultraviolet, UV-B) are also likely to be involved in chloroplast development (Thompson and White, 1991).

One final aspect of chloroplast structure deserves mention here. The classic Calvin–Benson or C_3 pathway of photosynthetic CO_2 involves carboxylation of the pentose sugar ribulose-1,5-bisphosphate (RuBP) by RUBISCO to form two molecules of the three carbon compound 3-phosphoglyceric acid (PGA). A number of hot-weather plants (e.g., corn, sugar cane, and crabgrass) fix CO_2 at roughly three times the rate of a C_3 plant such as spinach. In these plants the earliest products of CO_2 fixation are four carbon compounds such as aspartate, oxaloacetate, and malate. Such C_4 plants have two kinds of cells with structurally different chloroplasts arranged concentrically around vascular bundles of phloem tubules. Adjacent to the vascular bundle are the *bundle sheath* cells whose chloroplasts generally contain thylakoids that are primarily organized in single units and are not differentiated into grana. These chloroplasts carry out photosynthesis and make glucose via the C_3 pathway. Surrounding the bundle sheath cells is a layer of *mesophyll* cells, whose chloroplasts contain grana similar to typical C_3 plants. However, these cells do not carry out CO_2 fixation by the C_3 pathway, but rather by a different pathway called the C_4 pathway. In the C_4 pathway CO_2 in the cytoplasm combines with the three carbon compound phospho*enol*pyruvate to form the four carbon compound oxaloacetate. The reaction is catalyzed by phospho*enol*pyruvate carboxylase. The oxaloacetate synthesized is then converted to malate by reduction or to aspartate by transamination. The products of one or both of these reactions diffuse into adjacent bundle sheath cells where the four carbon compounds are decarboxylated and the CO_2 released is fixed again into PGA.

Chloroplast purification usually involves rupture of plant tissue and cells by grinding or blending. These rather extreme treatments are required to break open the plant cell wall although gentler methods involving enzymatic digestion of the cell wall have been developed. The homogenization is done in an isoosmotic medium containing sucrose or sorbitol. The chloroplast fraction is purified from the rest of the cellular debris by differential centrifugation. Intact and broken chloroplasts can then be separated by density gradient centrifugation using sucrose or silica sol since each particle class bands at position in the gradient equal to its own density. Chloroplasts, like mitochondria, can be subfractionated further into their constituent parts. The organelles can be lysed in dilute buffer and centrifuged. The supernatant contains the soluble stromal enzymes and the pellet the envelope plus thylakoids. The envelope and thylakoids can then be separated by density gradient centrifugation.

Chloroplast Envelope

The molecular aspects of plastid envelope biochemistry have been thoroughly reviewed (Douce *et al.*, 1984; Joyard *et al.*, 1991). The envelope membranes have been shown to have three major functions. First, the inner envelope membrane contains specific translocators and thereby regulates metabolite transport between the cytoplasm and the plastid stroma. Second, the envelope is involved in the biosynthesis of specific plastid components (e.g., glycerolipids, pigments, and prenylquinones). Third, the envelope membranes function in transport of plastid proteins encoded in the nuclear genome (Chapter 15).

Both plastid envelope membranes are very rich in glycerolipids. Like the cell envelopes of cyanobacteria, plastid envelopes contain large amounts of monogalactosyldiacylglycerol (MGDG), digalactosyldiacylglycerol (DGDG), and sulfolipid. The phospholipids found in plastid envelopes are phosphatidylglycerol (PG), phosphatidylcholine (PC), and phosphatidylinositol (PI). PC and PI are concentrated in the outer envelope membrane whereas the inner envelope membrane is virtually devoid of these compounds as are the thylakoids and cyanobacterial envelope. Similarly, MGDG accounts for over half the glycolipid content of the cyanobacterial envelope, the thylakoids and the plastid inner membrane, but the MGDG content of the outer membrane of the plastid is only 17%. The lipid matrix of the envelope membranes is highly conserved, being nearly identical for chloroplasts, etioplasts, and other nongreen plastids.

Glycerolipid biosynthesis requires the assembly of fatty acids, glycerol, and a polar head group (galactose for galactolipids, sulfoquinovose for sulfolipid; and phosphorylglycerol for phosphatidylglycerol). Plastids appear to be the major site of fatty acid synthesis in the plant cell and plastid envelope membranes play a central role in the biosynthesis of plastid glycerolipids. Thus, all of the enzymes required to synthesize plastid glycerolipids have been found to be associated with the plastid envelope. Carotenoids are also present in the plastid envelope membranes, but it is uncertain whether the enzymes of carotenoid biosynthesis, some of which are membrane bound, are associated with the plastid envelope. Although no chlorophyll is found in isolated envelopes, these membranes might contain some of the enzymes associated with chlorophyll biosynthesis.

Most proteins associated with the chloroplast envelope are coded by nuclear genes, but this is not true of all. Clemetson et al. (1992) report that at least four envelope proteins in the green alga *Chlamydomonas reinhardtii* may be chloroplast gene products and Sasaki et al. (1993) demonstrate that an open reading frame in the pea chloroplast genome specifies a chloroplast envelope protein.

Chloroplast Thylakoid Membrane

The thylakoid membrane is the site of photosynthetic electron transport and photophosphorylation. Since the thylakoid membrane, like the mitochondrial inner membrane, is a mosaic of organelle and nuclear gene products, this membrane deserves particularly careful scrutiny.

Like the mitochondrial inner membrane, the thylakoid membrane includes specific complexes, in this case three, that are responsible for electron transfer, plus the chloroplast coupling factor, which is discussed later in this chapter. Our current concept of how the three complexes of the photosynthetic electron transfer chain are organized depends on results obtained using a variety of techniques. For example, the pathway of electron transfer has been largely deduced from experiments with isolated chloroplasts or partially purified complexes using artificial electron donors and acceptors together with specific inhibitors. The protein components of the chlorophyll-containing light harvesting complexes (LHC) of photosystems I and II have been isolated together with chlorophyll by sodium dodecyl sulfate (SDS)-gel electrophoresis on so-called ''green gels'' without prior thermal denaturation. The proteins in each complex have been further characterized by gel electrophoresis. Active photosystem II particles have been obtained by high-pressure fragmentation of stacked thylakoids followed by partition in an aqueous polymer two phase system or by mild detergent treatment under controlled salt and pH conditions and differential centrifugation (see Andersson and Styring, 1991). Thylakoid membrane organization has also been the subject of electron microscopic investigation using freeze fracture techniques. These have provided information on the orientation of the photosynthetic complexes within the thylakoid membranes.

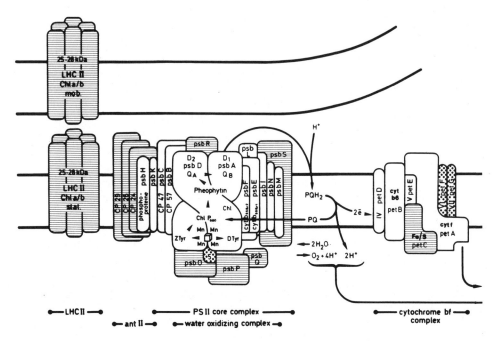

Fig. 1–6. Model for the thylakoid membrane of higher plants illustrating the three photosynthetic complexes and the CF_1/CF_0 ATP synthase. Most proteins are designated in terms of the genes that encode them (see Tables 1–7 and 1–8). PC, plastocyanin; Fd, ferredoxin; FNR, ferredoxin NADP$^+$ oxidoreductase; TR, thioredoxins; FTR, ferredoxin-thioredoxin oxidoreductase. Shaded components are nuclear-encoded in land plants, while unshaded components are coded in the chloroplast. From R. G. Herrmann with permission.

Photosystem II (PS II). The organization and function of the two photosystems have been the subject of several reviews (Andreasson and Vanngard, 1988; Anderson and Andersson, 1988; and Melis, 1991). Photosystem II specifically has been considered in reviews by Andersson and Styring (1991) and Ghanotakis and Yocum (1990). PS II is a multiprotein complex made up of at least 23 different polypeptides that include catalytic, regulatory, and structural subunits as well as several chlorophyll-binding proteins (Fig. 1–6, Table 1–7). At least seven proteins are involved in light harvesting, but the functions of the other subunits remain to be determined. Their roles probably concern the regulation of electron transfer; the structural organization of the PSII complex; and its assembly and turnover. Some proteins of the PSII complex are coded in the chloroplast and others in the nucleus (Table 1–7).

Photosynthetic electron transport initiates with absorption of light energy by chlorophyll *a/b*–protein complexes that serve as light harvesting "antennae." The light energy absorbed by the "antenna" chlorophyll is transferred to a pair of P_{680} reaction center chlorophyll *a* molecules. An excited P_{680} molecule then loses an electron to pheophytin. This electron reduces the primary quinone acceptor Q_a, oxidizing pheophytin followed by reduction of the secondary electron acceptor, Q_B, and oxidation of

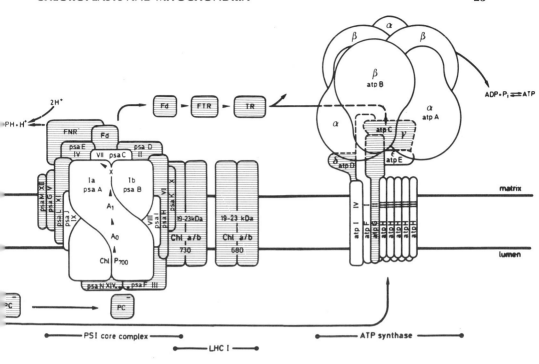

Q_A. The electron then proceeds to plastoquinone and the cytochrome b_6/f complex. Electron transport on the oxidizing side of PS II includes electron donation to $P680^+$ from Z, a tyrosine residue (Tyr_{161}) on the D-1 reaction center protein (see below), and electron donation from the ligand Mn to Z^+. The stepwise accumulation of four positive charges on the oxidizing side of PSII is sufficient for the oxidation of two H_2O molecules and the release of four electrons, four protons, and molecular O_2.

The core of the PSII reaction center (RCII) is a heterodimer of the 32-kDA D-1 protein and the 34-kDa D-2 protein plus cytochrome b_{559}. All of these proteins are encoded by chloroplast genes (Table 1–7). The D-1 and D-2 proteins are homologous to each other and to the L and M subunits of the reaction center of purple photosynthetic bacteria. The current models for D-1 and D-2, based on the X-ray crystal structures of bacterial reaction center, envision five hydrophobic membrane-spanning helices and two short regions of amphipathic solvent-exposed helix. The N-termini of these proteins would face the chloroplast stroma while the carboxy ends project into the thylakoid lumen (Marder and Barber, 1989). The stromally exposed loop connecting helices IV and V contains the ligands for Q_B, the second quinone acceptor in RCII, and is also the site of action of several important classes of herbicides. These herbicides, which include among others the triazine and phenylurea compounds, displace Q_B from its site thereby blocking electron transfer from Q_A to Q_B. Mutations to herbicide resistance in the chloroplast *psbA* gene encoding D-1 are all localized to the loop connecting helices IV and V. The D-1 protein undergoes a rapid cycle of synthesis and degradation, which is related to light intensity. At light intensities in excess of those required to saturate

Table 1-7. Proteins of Photosystem II and Its Light Harvesting Antenna

		Protein		
Gene[a]	Location[b]	Name	Apparent Molecular Weight (kDa)	Location
psbA	C	D1	32	Reaction center
psbB	C	CP 47	51	Core particles
psbC	C	CP 43	44	Core particles
psbD	C	D2	32	Reaction center
psbE	C	Cyt. b_{559}	9	Reaction center
psbF	C	Cyt. b_{559}	4	Reaction center
psbH	C	Phosphoprotein	9	BBY prep.[c]
psbI	C		4	Reaction Center
psbK	C		3.7	Core particles
psbL	C		4.8	Core particles?
psbM	C		3.5	?
psbN	C		4.4	?
psbO (psbI)	N	OEE1	33	O_2 evolving complex
psbP (psbII)	N	OEE2	23	O_2 evolving complex
psbQ (psbIII)	N	OEE3	16	O_2 evolving complex
psbR	N		10	Some core particles
psbS	N		22	Some core particles
	N		7	Some core particles
	N		5	Core particles
	N		3.2	Some core particles
cabII	N	CAB-II, type I	27	Antenna; BBY prep.
cabII	N	CAB-II, type II	27	Antenna; BBY prep.
	N	CP 24	24	Antenna; BBY prep.
cabII	N	CP 29	29	Antenna; some core particles

Compiled from reviews by Andersson and Styring (1991), Green et al. (1991), and Herrmann et al. (1991). Green et al. (1991) also indicate a minor CAB-II, type II protein associated with PSII.

[a] psbG is not shown since this gene is actually homologous to a gene encoding and ndh subunit and has been redesignated ndhk (Arizmendi et al., 1992).

[b] C, chloroplast; N, nucleus. Andersson and Styring (1991) list three polypeptides (7, 5, and 3.2 kDa) for which corresponding genes are not designated by them or by Herrmann et al. (1991).

[c] BBY prep., PSII prepared by the method of Berthold et al. (1981) in which PSII is still located in its natural lipid environment associated with its chlorophyll a/b antennae.

photosynthesis, the photosynthetic capacity of a plant can be severely lowered (*photoinhibition*). Photoinhibition probably results from an imbalance in the D-1 cycle such that the plant is unable to replace photodamaged D-1 molecules quickly enough.

Although cytochrome b_{559} is associated with RCII particles, this enzyme does not appear to be involved in the major electron transfer pathway of PSII. Cytochrome b_{559} is a heterodimer of α (9 kDa) and β (4 kDa) subunits encoded by two chloroplast genes (Table 1-7). Three hydrophilic, extrinsic proteins of 33, 23, and 16 kDa, present in equimolar amounts, form the O_2 evolving complex of PSII. A 10-kDa polypeptide seems to be responsible for anchoring the 23-kDa protein, which in turn is responsible for binding the 16-kDa protein. The 33-kDa protein appears to be closely associated with the D1/D2 heterodimer and was thought initially to be the catalytic manganoprotein of the water oxidation system. However, at present D-1 is the most likely candidate for binding the Mn^{2+} cluster. The 33-kDa protein may help to stabilize at least two of the 4 Mn^{2+} atoms. A 22-kDa protein may have a regulatory role on the acceptor side of PSII. All of these proteins are encoded by nuclear genes (Table 1–7).

Several classes of chlorophyll-binding proteins are associated with PSII. Two polypeptides of the PSII core CP 47 and CP 43 bind chlorophyll *a* and function in light harvesting and energy transfer to the reaction center. These proteins are encoded respectively by the chloroplast *psbB* and *psbC* genes (Table 1–7). PSI and PSII also contain numerous chlorophyll *a/b* (CAB) proteins encoded by several nuclear gene families (see Green, 1988; Green *et al.*, 1991; Chitnis and Thornber, 1988). The LHC II chlorophyll–protein complex is the major chlorophyll *a/b* antenna of PSII. This complex accounts for up to half of the chlorophyll and protein mass of the thylakoid membranes and is important in both short- and long-term adaptation of plants to different light conditions. Two major polypeptides of 27 and 25 kDa are the predominant subunits of LHC II in pea and spinach. They are the most closely related CAB proteins. A minor CAB protein (Type III) is also associated with LHC II. There are three additional CAB proteins, CP 29, CP 27, and CP 24, associated with PSII. These three proteins account for only 5–10% of the total chlorophyll associated with PSII and are encoded by nuclear genes. The CAB proteins of PSII are related to similar proteins in PSI.

Other small proteins are also associated with PSII including the 9-kDa phosphoprotein. Several of these are encoded by chloroplast genes (Table 1–7).

Electron transport between photosystem II and photosystem I. This topic has been reviewed by Cramer *et al.* (1991). Plastoquinone (PQ) is the mobile carrier that shuttles electrons from PSII to the the cytochrome b_6/f complex. The polypeptide composition of the cytochrome b_6/f complex is simpler than that of the comparable cytochrome b/c_1 complex (complex III) of mitochondrial membranes. The chloroplast b_6/f complex has only four polypeptides. None of these corresponds to the core polypeptides of complex III. They are cytochrome *f* (33 kDa, *petA*), cytochrome b_6 (23 kDa, *petB*), a Rieske iron–sulfur protein (20 kDa, *petC*), and "subunit IV (17 kDa, *petD*). All but the Rieske iron–sulfur protein are encoded by chloroplast genes. Subunit IV of the b_6/f complex may possess one or more of the functions specified by the C-terminal half of the larger mitochondrial and bacterial cytochromes including the quinone-binding function. Other genes that may encode proteins associated with the cytochrome b_6/f complex are *petE* (subunit V), *petF* (subunit VI), and *petG* (subunit VII) (Herrmann *et al.*, 1991).

The amino terminal 210–215 residues of cytochrome *b* proteins from cyanobacteria, photosynthetic bacteria, chloroplasts, and mitochondria possess highly conserved sequences containing the heme binding regions and the hydrophobic domains of these proteins. Mitochondrial cytochrome *b* has a binding site for the respiratory inhibitor antimycin A. However, this binding site appears to be absent on chloroplast cytochrome b_6 possibly due to the replacement of a Leu for a Gly at position 41 in the chloroplast enzyme. Plastocyanin, a copper-containing protein (ca. 12 kDa), is the second mobile electron carrier in the thylakoid membrane. Plastocyanin, which transfers electrons from the cytochrome b_6/f complex to photosystem I, is encoded by a nuclear gene.

Photosystem I (PSI). The structure and function of PSI are reviewed by Golbeck and Bryant (1991) and Golbeck (1992). The PSI reaction center (RCI) catalyzes the coupled photooxidation of plastocyanin and the reduction of ferredoxin. Electron transfer from plastocyanin to the reaction center chlorophyll (P_{700}, a chlorophyll *a* dimer) of PSI requires oxidation of P_{700}, the primary electron donor of PSI. Upon excitation P_{700} donates its electron, which migrates through a chain of electron acceptors in the order $A_0 \rightarrow A_1 \rightarrow F_x \rightarrow F_b \rightarrow F_a$. A_0, the primary electron acceptor, is probably a chlorophyll *a* monomer, but the secondary electron acceptor, A_1, is not. Most evidence currently suggests that A_1 may be the same molecule as phylloquinone, a species exclusively associated with PSI. F_x, F_b, and F_a are all iron–sulfur centers. Electrons are then transferred to a 1:1 complex composed of the nonheme [2Fe-2S] iron–sulfur protein, ferredoxin (12 kDa), and the FAD-flavoprotein ferredoxin $NADP^+$ oxidoreductase (40 kDa). This complex is tightly bound to the stromal side of the thylakoid membrane. The oxidoreductase catalyzes the transfer of electrons from ferredoxin, a one electron carrier, to $NADP^+$, a two electron carrier. Thus, the coenzyme FAD exists in a partially reduced intermediate form with one electron before it becomes fully reduced. This intermediate has been identified. NADPH then serves as the substrate for the reduction of CO_2 catalyzed by RUBISCO. Both ferredoxin and the oxidoreductase are encoded by nuclear genes. Electrons from PSI can also move cyclically from ferredoxin to the plastoquinone pool and hence back through the cytochrome b_6/f complex and plasto-

Table 1–8. Proteins of Photosystem I and Its Light Harvesting Antenna

		Protein		
Gene	Location[a]	Name	Apparent Molecular Weight (kDa)	Location[b]
psaA	C	P-700 apoprotein, 1a	83	Core
psaB	C	P-700 apoprotein, 1b	82.4	Core
psaC	C	PSI-7	8.9	Peripheral
psaD	N	PSI-2, ferredoxin docking psaC stability	17.9	Peripheral
psaE	N	PSI-4	9.7	Peripheral
psaF	N	PSI-3, plastocyanin docking	17.3	Peripheral
psaG	N	PSI-5	10.5	Peripheral
psaH	N	PSI-6	10.2	Peripheral
psaI	C	PSI-8	4	Core
psaJ	C	PSI-9	5.1	Core
psaK	N	PSI-10	8.4	Core
psaL[c]	N	PSI-11	18	Core
psaM[c]	C	PSI-12	3.5	?
cab-6A/B	N	CAB	22	Antenna
cab-7	N	CAB	24	Antenna
cab-8	N	CAB	24	Antenna

Compiled from reviews by Cramer *et al.* (1991), Golbeck and Bryant (1991), Golbeck (1992), and Herrmann *et al.* (1991).

[a]C, chloroplast; N, nucleus.

[b]The terms "core" and "peripheral" are used by Golbeck and Bryant (1991) in their classification of PSI proteins. According to Cramer *et al.* (1991) the *psaA* through *psaI* genes code for proteins of the PSI reaction center.

[c]Herrmann *et al.* (1991) reference papers describing the *psaL* and *psaM* genes. *psaM* is found in the liverwort mitochondrial genome, but not in rice or in any higher plant PSI complexes studied to date (Golbeck, 1992).

cyanin to PSI. This cyclic electron flow is used to power *cyclic photophosphorylation.* *Noncyclic photophosphorylation* is associated with passage of electrons down the main electron transport chain to $NADP^+$.

The PSI complex can be thought of as being comprised of two distinct components: the PSI core complex, which carries the P_{700} reaction center, and the light harvesting complex LHC1. The PSI core complex includes 11 polypeptides. The CP 1 apoproteins (82–83 kDa) are encoded by the chloroplast *psaA* and *psaB* genes. These proteins are thought to be bind most of the cofactors of PSI including the primary electron donor, P_{700}, and the primary electron acceptor, A_0. Both proteins are extremely hydrophobic. The PSI core also includes the products of the *psaI, psaJ, psaK* and *psaL* genes (Table 1–8). These genes encode integral membrane proteins. The *psaI* and *psaJ* genes are in the chloroplast genome. The iron–sulfur centers F_A and F_B are bound to the hydrophilic 8.9-kDa product of the chloroplast *psaC* gene. The other proteins of this extrinsic complex are the products of the *psaD, psaE,* and *psaH* genes, which are all nuclear in location (Golbeck, 1992). The psaD protein is involved in ferredoxin docking and stabilizing the psaA protein in the complex. The functions of the psaE and psaH proteins are not known. The psaF protein appears to play a role in electron transfer from plastocyanin to $P700^+$.

LHC 1 is comprised of three or four CAB proteins having molecular weights of 22 and 24 kDa, which are encoded by nuclear gene families like the CAB proteins of PSII (Chapter 16, Table 16–1). In tomato, four gene families have been identifed. These are closely related to each other and slightly more distantly to the PSII CAB proteins.

The PSI and PSII reaction centers are spatially separated with most PSI centers located within the unpaired stroma lamellae and PSII centers being found predominantly within the stacked membrane regions of the grana (Fig. 1–6).

Chlororespiration. Bennoun (1982) discovered an electron transfer pathway to O_2 in thylakoids of *Chlamydomonas* that he referred to as chlororespiration. Scherer (1990), in a short review, points out that in purple nonsulfur bacteria and cyanobacteria there is a close interaction between photosynthetic and respiratory electron transport chains that share several identical redox proteins. A membrane-bound NADH-plastoquinone oxidoreductase has been isolated and purified from *Chlamydomonas* and found to be a flavoprotein. NADPH and succinate have also been shown to act as electron donors in the chlororespiratory pathway of *Chlamydomonas*. Furthermore, the proteins encoded by the eleven *ndh* reading frames found in the completely sequenced chloroplast genomes of *Marchantia*, rice and tobacco share homology with comparable ND subunits in mitochondria including the seven found in vertebrate mitochondrial DNA (Table 1–1). There is recent indirect evidence for chlororespiration in tobacco and sugar beet chloroplasts. Thus, there appears to be a coupling between photosynthesis and chlororespiration that is reminiscent in some ways of that seen in photosynthetic bacteria.

Chloroplast Stroma

The mobile phase of the chloroplast, the stroma, contains the soluble enzymes involved in the Calvin–Benson cycle of C_3 photosynthesis. CO_2 fixation is catalyzed by RUBISCO.

RUBISCO is the principal protein component of the stroma and was discovered by Wildman and Bonner in 1947 who named it fraction I protein. In most photosynthetic organisms the intact enzyme has a molecular weight of ca. 560 kDa and is composed of eight large subunits of ca. 52 kDa and eight small subunits of ca. 14 kDa. In land plants and green algae, the large subunit is a chloroplast gene product and the small subunit is encoded by a nuclear gene family, but in all other algal groups both subunits are chloroplast gene products (Chapter 3). In the photosynthetic bacterium *Rhodospirillum rubrum* RUBISCO is a homodimer of two large subunits and in another photosynthetic bacterium, *Rhodopseudomonas sphaeroides*, there are two forms of the enzyme. Form one contains eight large and eight small subunits, but form two is a hexamer of large subunits with no small subunits being present. As these findings imply, the catalytic sites of RUBISCO are located in the large subunit, but in most organisms the small subunit is also required for function.

High concentrations of O_2 competitively inhibit CO_2 fixation because RUBISCO catalyzes a second reaction in which O_2 competes with CO_2 in the active site of the enzyme. This results in oxygenation rather than carboxylation of the substrate (Fig. 1–7). The products of the reaction are one molecule of phosphoglyceric acid (PGA) and one of phosphoglycollate instead of the two molecules of PGA produced when the enzyme carboxylates 1,5-ribulose bisphosphate (1,5-RuBP). *Photorespiration* causes the conversion of phosphoglycollate to CO_2 via a series of intermediates.

The chloroplast stroma is also home to a remarkable array of other metabolic pathways (see Hoober, 1984; Hrazdina and Jensen, 1992 for a discussion). They include enzymes involved in chlorophyll biosynthesis, aromatic amino acid metabolism, and fatty acid biosynthesis. Thus, in C_3 plants, the fatty acid-synthesizing system is located in the stroma and consists of several enzymatic components as do prokaryotic fatty-acid synthesizing systems. In yeast and animals, the steps in fatty acid biosynthesis are

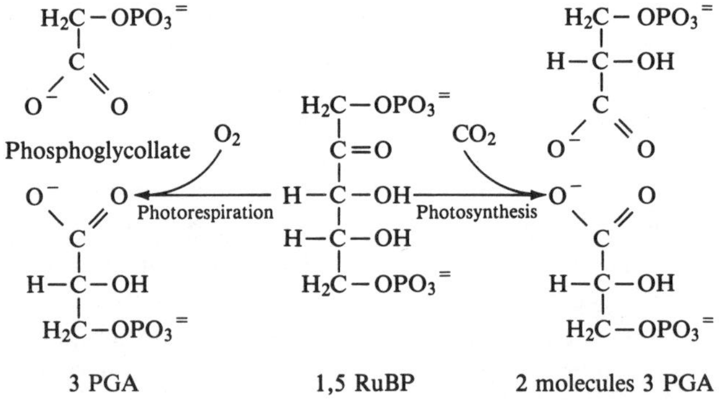

Fig. 1–7. Reactions carried out ribulose-1,5-bisphosphate carboxylase/oxygenase (RUBISCO). At low O_2 concentrations CO_2 is bound to the active site on RUBISCO and the enzyme catalyzes the condensation of CO_2 with the 5-carbon sugar ribulose 1,5-bisphosphate to form two molecules of 3-phosphoglyceric acid (PGA) and photosynthetic CO_2 fixation results. O_2 competitively inhibits CO_2 binding at high concentrations and RuBP is oxygenated rather than carboxylated as a result. The products of the reaction are one molecule of PGA and one of the two carbon compound phosphoglycollic acids, which is the substrate for photorespiration.

catalyzed by a large, soluble, polyfunctional protein. Certain enzymes and enzyme pathways include both chloroplast and cytoplasmic counterparts. For instance, distinct pyruvate dehydrogenase complexes are found in chloroplasts and mitochondria. Isoenzymes involved in aromatic amino acid synthesis are found in the chloroplast stroma and cytoplasm. Significantly, the systematic characterization of chloroplast genomes from diverse algal groups is beginning to reveal that certain proteins in these pathways are encoded by chloroplast rather than nuclear genes as in land plants (Chapter 3). This again exemplifies a degree of genetic plasticity in plastid genomes that would not have been suspected a few years ago.

Also found in the chloroplast stroma are chloroplast DNA and the chloroplast protein-synthesizing system.

ENERGY GENERATION IN MITOCHONDRIA AND CHLOROPLASTS

The process of oxidative phosphorylation, the concept of coupling, and the chemiosmotic hypothesis are nicely reviewed in Tzagoloff's (1982) monograph. Coupling of oxidative phosphorylation to the respiratory electron transport chain occurs at three sites. Site 1 is between NADH and coenzyme Q, site 2 between between coenzyme Q and cytochrome c, and site 3 occurs at the level of cytochrome oxidase. In tightly coupled mitochondria, mitochondria in which the energy of oxidation is used efficiently for ATP synthesis, the mole equivalents of phosphate esterified per oxygen consumed (P/O ratio) is three with NADH as the substrate and two if succinate is the substrate. Mitchell's (1961) chemiosmotic hypothesis is generally accepted as explaining the mechanism of coupling. The model has three major tenets: (1) The inner mitochondrial membrane has a low conductivity and is impermeable to ionic species including protonated water. (2) The electron transfer chain components are organized in the inner membrane in such a way that during oxidation–reduction an asymmetric translocation of electrons and protons occurs across the membrane. When electron transport takes place, a pH gradient and an electrical charge are built up across the membrane. The combined pH and electrical differential, the proton motive force, acts as the primary source of energy for ATP synthesis and other energy-coupled reactions. (3) The inner membrane contains a reversible proton-translocating ATPase that can either generate a proton-motive force from the energy of hydrolysis of ATP or, alternatively, can use the potential energy of the proton-motive force to synthesize ATP.

For each of the three phosphorylation sites there is an oxidation/reduction loop consisting of two circuits. One circuit carries protons and the other electrons. The proton-carrying circuit transfers two H^+ from the substrate SH_2 on the inside of the membrane to the outside while the electron circuit conveys electrons to the next electron carrier in the chain and ultimately to oxygen. The two flavoproteins and coenzyme Q are candidates for proton carriers and cytochromes, nonheme iron and copper for electron carriers. During electron transport a reversible ATPase discharges the proton gradient by translocating two H^+ in the direction opposite to proton translocation with the simultaneous esterification of ADP and inorganic phosphate (P_i) to form ATP. The chemiosmotic hypothesis of coupling applies to bacterial oxidative phosphorylation and to photophosphorylation as well.

The ATP synthases responsible for oxidative and photophosphorylation are highly conserved enzyme complexes (see reviews by Cox and Gibson, 1987; Nagley, 1988;

Nalin and Nelson, 1987; Glaser and Norling, 1991;). Each of these ATP synthases is composed of two domains (Fig. 1–6, Table 1–9): a coupling factor complex (F_1 bacteria, mitochondria; CF_1 chloroplasts) that is attached to an intrinsic membrane complex (F_0 bacteria, mitochondria; CF_0 chloroplasts). Bacterial F_0 is located in the cytoplasmic membrane while mitochondrial F_0 and chloroplast CF_0 are found in the inner and thylakoid membranes, respectively. Bacterial F_1 extends into the bacterial cytoplasm while mitochondrial F_1 protrudes into the mitochondrial matrix and CF_1 of the chloroplast into the stroma. F_1 and CF_1 contain catalytic sites where ADP, ATP, and P_i bind. F_0 and CF_0 have a transmembrane proton channel through which proton transport is coupled to ATP synthesis. The proton channel appears to be formed by a low-molecular-weight protein (8 kDa). This protein is called the proteolipid or subunit 9 in the case of the mitochondrion (Table 1–9).

F_1 and CF_1 are complexes of five subunits designated α, β, γ, δ, and ϵ with a stoichiometry of 3:3:1:1:1. β is a very highly conserved subunit and possesses the catalytic site. *E. coli* F_0 has three subunits a, b, and c, for which the stoichiometries 1:2:10–12 have been suggested. CF_0 possesses four subunits, but the stoichiometry has not been established. Mitochondrial F_0 possesses five polypeptides in yeast, seven in higher plants, and eight in beef heart. Despite the strong similarities between the ATP synthases of bacteria, mitochondria, and chloroplasts, there are some differences too. The mitochondrial enzyme possesses several additional polypeptides. These include mitochondrial ϵ that binds to F_1. This protein is unrelated to the ϵ subunits of bacterial F_1 and chloroplast CF_1, which are equivalent to the δ and δ' subunits of animal and plant mitochondrial F_1, respectively (Table 1–9). The δ subunits of plant and bacterial F_1 and of CF_1 are equivalent to the oligomycin sensitivity conferring proteins (OSCP)

Table 1–9. ATP Synthase Subunits in *E. coli*, Mitochondria, and Chloroplasts[a]

Complex	E. coli	Chloroplast	Gene Location[b]	Mitochondrion	Gene Location[b]
F_1/CF_0	α	α	C	α	N,M
	β	β	C	β	N
	γ	γ	N	γ	N
	δ	δ	N	OSCP[c]	N
	ϵ	ϵ	C	δ	N
	—	—	—	ϵ	N
	—	—	—	F_6	N
	—	—	—	ATPase-inhibitor	N
F_0/CF_0	a	CF_0-IV	C	ATPase-6	M
	b	CF_0-I	C	b	N
	c	CF_0-III	C	ATPase-9	N,M
	—	CF_0-II	N	—	—

Compiled from Glaser and Norling (1991) and Nagley (1988).

[a]In addition to the ATP synthase subunits shown here F_0 in yeast possesses two additional subunits and cow five additional subunits.

[b]C, chloroplast; M, mitochondrion; N, nucleus; N,M, coded in mitochondrion in one species, but not another (see Chapter 3).

[c]Higher plant mitochondria possesses a δ subunit equivalent to the δ subunit of bacterial F_1 and chloroplast CF_1 plus a δ' subunit equivalent to ϵ of bacterial F_1, chloroplast CF_1, and animal mitochondrial F_1 δ.

of yeast and mammals. The structure of the δ subunit is poorly conserved, which may account for the fact that bacterial and mitochondrial coupling factors, but not CF_1, are sensitive to oligomycin and related antibiotics. These antibiotics are thought to prevent the utilization of the oxidative energy for ATP synthesis by blocking some terminal phosphorylation step. The mitochondrial F_1 also interacts with two other loosely associated subunits F_6 and the ATPase inhibitor protein.

Like the other complexes of the electron transport chain discussed in this chapter, the ATPases are mosaics of organelle and nuclear gene products (Table 1–9). The chloroplast genes encoding ATP synthetase components are organized in much the same way as their homologs in cyanobacteria, except that certain of the chloroplast genes have been transferred to the nucleus (Chapter 3).

The intent of this chapter is not to present a detailed review of organelle biology. That is better left to monographs such as those by Hoober (1984) and Tzagoloff (1982). Rather, this chapter focuses on those aspects of chloroplast and mitochondrial structure and function that involve the interplay between organelle and nuclear products in the construction of the electron transport membranes of these organelles and their ATP synthases. The cooperative role of organelle and nuclear genomes in specifying the chloroplast and mitochondrial protein synthesizing systems is detailed later in this book.

REFERENCES

Anderson, J.M., and B. Andersson (1988). The dynamic photosynthetic membrane and regulation of solar energy conversion. *Trends Biochem. Sci.* 13: 351–355.

Andersson, B., and S. Styring (1991). Photosystem II: Molecular organization function, and acclimation. *Curr. Topics Bioenerget.*, 16: 1–81.

Andreasson, L.-E., and T. Vanngard (1988). Electron transport in photosystems I and II. *Annu. Rev. Plant Physiol. Mol. Biol.* 39: 379–411.

Arizmendi, J.M., Runswick, M.J., Skehel, J.M., and J.E. Walker (1992). NADH: Ubiquinone oxidoreductase from bovine heart mitochondria: A fourth nuclear encoded subunit with a homologue encoded in chloroplast genomes. *FEBS Lett.* 301: 237–242.

Bennoun, P. (1982). Evidence for a respiratory chain in the chloroplast. *Proc. Natl. Acad. Sci. U.S.A.* 79: 4352–4356.

Berthold, D.A., Babcock, G.T., and C.F. Yocum (1981). A highly resolved oxygen-evolving photosystem II preparation from spinach thylakoid membranes: EPR and electron transport properties. *FEBS Lett.* 134: 231–234.

Capaldi, R.A. (1990). Structure and function of cytochrome *c* oxidase. *Annu. Rev. Biochem.* 59: 569–596.

Capaldi, R.A., Takamiya, S., Zhang, Y.-Z., Gonzalez-Halphen, D., and W. Yanamura. (1987). Structure of cytochrome-*c* oxidase. *Curr. Topics Bioenerget.*, 15: 91–112.

Cavalier-Smith, T. (1986). The kingdom Chromista: Origin and systematics. *Prog. Phyc. Res.* 4: 309–347.

Cavalier-Smith, T. (1989). The kingdom Chromista. In *The Chromophyte Algae: Problems and Perspectives* (J.C. Green, B.S.C. Leadbeater, and W.L. Diver eds.), Systematics Association Special Volume 3*, pp. 381–407. Clarendon Press, Oxford.

Chitnis, P.R., and J.P. Thornber (1988). The major light-harvesting complex of photosystem II: aspects of its molecular and cell biology. *Photosynth. Res.* 16: 41–63.

Clemetson, J.M., Boschetti, A., and K.J. Clemetson (1992). Chloroplast envelope proteins are encoded by the chloroplast genome of *Chlamydomonas reinhardtii*. *J. Biol. Chem.* 267: 19773–19779.

Cox, G.B., and F. Gibson (1987). The assembly of F_1F_0-ATPase in *Escherichia coli*. *Curr. Topics Bioenerget.* 15: 163–175.

Cramer, W.A., Furbacher, N., Szczepaniak, A., and G.-S. Tae (1991). Electron transport between photosystem II and photosystem I. *Curr. Topics Bioenerget.* 16: 179–222.

Douce, R., Dorne, A.J., Block, M.A., and J. Joyard (1984). The chloroplast envelope and the origin of chloroplast lipids. In *Chloroplast Biogenesis* (R.J. Ellis ed.), *Soc. Exptl. Biol.* 21: 193–224.

Ernster, L., and B. Kuylenstierna (1970). Outer membranes of mitochondria. In *Membranes of Mitochondria and Chloroplasts* (E. Racker ed.), pp. 172–212. Van Nostrand Reinhold, New York.

Fearnley, I.M., and J.E. Walker (1992). Conservation of sequences of subunits of mitochondrial complex I and their relationships to other proteins. *Biochim. Biophys. Acta* 1140: 105–134.

Ghanotakis, D.F., and C.F. Yocum (1990). Photosystem II and the oxygen-evolving complex. *Annu. Rev. Plant Physiol. Plant Mol. Biol.* 41: 255–276.

Glaser, E., and B. Norling (1991). Chloroplast and plant mitochondrial ATP synthases. *Curr. Topics Bioenerget.* 16: 223–263.

Golbeck, J.H. (1992). Structure and function of photosystem I. *Annu. Rev. Plant Physiol. Plant Mol. Biol.* 43: 293–324.

Golbeck, J.H., and D.A. Bryant (1991). Photosystem I. *Curr. Topics Bioenerget. 16: 83–177.*

Green, B.R. (1988). The chlorophyll-protein complexes of higher plant photosynthetic membranes or just what green band is that? *Photosyn. Res.* 15: 3–32.

Green, B.R., Pichersky, E., and K. Kloppstech (1991). Chlorophyll *a/b* binding proteins: an extended family. *Trends Biochem. Sci.* 16: 181–186.

Hackenbrock, C.R. (1966). Ultrastructural bases for metabolically linked mechanical activity in mitochondria. I. Reversible ultrastructural changes with a change in metabolic steady state in isolated liver mitochondria. *J. Cell Biol.* 30: 269–297.

Hatefi, Y., Haavik, A.G., and D.E. Griffiths (1962). Studies of the electron transfer system. XL. Preparation and properties of mitochondrial DPNH-coenzyme Q reductase. *J. Biol. Chem.* 237: 1676–1680.

Hauska, G. 1985. Organization and function of cytochrome b_6f/bc_1 complexes. In *Molecular Biology of the Photosynthetic Apparatus* (K.E. Steinback, S. Bonitz, C.J. Arntzen, and L. Bogorad eds.), pp. 79–87. Cold Spring Harbor Laboratory, New York.

Herrmann, R.G., Oelmuller, R., Bichler, J., Schneiderbauer, A., Steppuhn, J. Wedel, N., Tyagi, A.K., and P. Westhoff (1991). The thylakoid membrane of higher plants: Genes, their expression and interaction. In *Plant Molecular Biology* (R.G. Herrmann and B. Larkins ed.), Vol. 2, pp. 411– 427. Plenum Press, New York.

Hoober, J.K. (1984). *Chloroplasts.* Plenum Press, New York.

Hrazdina, G., and R.A. Jensen (1992). Spatial organization of enzymes in plant metabolic pathways. *Annu. Rev. Plant Physiol. Plant Mol. Biol.* 43: 241–267.

Ise, W., Haiker, H., and H. Weiss (1985). Mitochondrial translation of subunits of the rotenone-insensitive NADH:ubiquinone reductase in *Neurospora crassa. EMBO J.* 4: 2075–2080.

Joyard, J., Block, M.A., and R. Douce (1991). Molecular aspects of plastid envelope biochemistry. *Eur. J. Biochem.* 199: 489–509.

Kadenbach, B., Kuhn-Nentwig, L., and U. Buge (1987). Evolution of a regulatory enzyme: cytochrome *c* oxidase (complex IV). *Curr. Topics Bioenerget.*, 15: 114–161.

Kirk, J.T.O., and R.A.E. Tilney-Bassett (1978). *The Plastids: Their Chemistry Growth and Inheritance,* 2nd ed. Elsevier/North Holland, Amsterdam.

Kleinfeld, A.M. (1987). Current views of membrane structure. *Curr. Topics Membranes Transport,* 29: 1–27.

Lambowitz, A.M., Sabourn, J.R., Bertrand, H., Nickels, R., and L. McIntosh (1989). Immunological identification of the alternative oxidase of *Neurospora crassa* mitochondria. *Mol. Cell. Biol.* 9: 1362–1364.

Leech, R.M., and K.A. Pyke (1988). Chloroplast division in higher plants with particular reference to wheat. In *Division and Segregation of Organelles* (S.A. Boffey and D. Lloyd ed.), *Soc. Exptl. Biol.* 35: 39–62.

Marder, J.B., and J. Barber (1989). The molecular anatomy and function of thylakoid proteins. *Plant Cell Environ.* 12: 595–614.

Melis, A. (1991). Dynamics of photosynthetic membrane composition and function. *Biochim. Biophys. Acta* 1058: 87–106.

Mitchell, P. (1961). Coupling of phosphorylation to electron and hydrogen transfer by a chemiosmotic type of mechanism. *Nature (London)* 191: 144–148.

Moore, A.L., and J.N. Siedow (1991). The regulation and nature of the cyanide-resistant alternative oxidase of plant mitochondria. *Biochim. Biophys. Acta* 1059: 121–140.

Nagley, P. (1988). Eukaryote membrane genetics: The F_0 sector of the mitochondrial ATP synthase. *Trends Genet.,* 4: 46–52.

Nalin, C.M., and N. Nelson (1987). Structure and biogenesis of chloroplast coupling factor CF_0/CF_1-ATPase. *Curr. Topics Bioenerget.* 15: 273–294.

Ohnishi, T. (1987). Structure of the succinate-ubiquinone oxidoreductase (complex II). *Curr. Topics Bioenerget.* 15: 37–65.

Palade, G. (1952). The fine structure of mitochondria. *Anat. Rec.* 114: 427–451.

Possingham, J.V., Hashimoto, H., and J. Oross (1988). Factors that influence plastid division in higher plants. In *Division and Segregation of Organelles* (S.A. Boffey and D. Lloyd eds.), *Soc. Exptl. Biol.* 35: 1–20.

Quail, P. (1991). Phytochrome: A light-activated molecular switch that regulates plant gene expression. *Annu. Rev. Genet.* 25: 389–409.

Ragan, C. I. (1987). Structure of NADH-ubiquinone reductase (complex I). *Curr. Topics Bioenerget.* 15: 1–36.

Rhoads, D.M., and L. McIntosh (1991). Isolation and characterization of a cDNA clone encoding an alternative oxidase protein of *Sauromatum guttatum* (Schott). *Proc. Natl. Acad. Sci. U.S.A,* 88: 2122–2126.

Sasaki, Y., Sekiguchi, K., Nagano, Y., and R. Matsuno (1993). Chloroplast envelope protein encoded by chloroplast genome. *FEBS Lett.* 316: 93–98.

Scherer, S. (1990). Do photosynthetic and respiratory electron transport chains share redox proteins? *Trends Biochem. Sci.* 15: 458–462.

Schimper, A.F.W. (1885). Untersuchungenuber die

Chlorophyllkorper und die ihnen Homologen Gebilde. *Jahrb. Wissenschaft. Bot. 16: 1–247.*

Sherman, F. (1990). Studies of yeast cytochrome *c*: How and why they started and why they continued. *Genetics* 125: 9–12.

Singer, S.J., and G.L. Nicolson (1972). The fluid mosaic model of the structure of cell membranes. *Science* 175: 720–731.

Thompson, W.F., and M.J. White (1991). Physiological and molecular studies of light-regulated nuclear genes in higher plants. *Annu. Rev. Plant Physiol. Plant Mol. Biol.* 42: 423–466.

Tzagoloff, A. (1982). *Mitochondria.* Plenum Press, New York.

Weiss, H. (1987). Structure of mitochondrial ubiquinol-cytochrome *c* reductase (complex III). *Curr. Topics Bioenerget.* 15: 67–90.

Weiss, H. and T. Friedrich (1991). Redox-linked proton translocation by NADH-ubiquinone reductase (Complex I). *J. Bioenerget. Biomembranes* 23: 743–754.

Weiss, H., Leonard, K., and W. Neupert (1990). Puzzling subunits of mitochondrial cytochrome *c* reductase. *Trends Biochem. Sci.* 15: 178–180.

Weiss, H., Friedrich, T., Hofhaus, G., and D. Preis (1991). The respiratory-chain NADH dehydrogenase (complex I) of mitochondria. *Eur. J. Biochem.* 197: 563–576.

Wildman, S.G., and J. Bonner (1947). The proteins of green beans. I. Isolation, enzymatic properties and auxin content of spinach cytoplasmic proteins. *Arch. Biochem.* 14: 381–413.

PART II

EVOLUTION AND ORGANIZATION OF ORGANELLE GENOMES

2

Endosymbiotic Origin of Chloroplasts and Mitochondria

Primarily on the basis of morphology Schimper in 1883 proposed that plastids were descended from cyanobacterial endosymbionts. This theory was further developed by Mereschkowsky (1905, 1910, 1920), but received scant attention until chloroplast DNA was discovered in the 1960s (Chapter 3). An endosymbiotic origin for mitochondria was proposed in 1922 by Wallin. In 1970 Margulis published an important book entitled *Origin of Eukaryotic Cells* that resurrected the notion that eukaryotic organelles had been acquired by the protoeukaryotic cell in a series of endosymbiotic events. A spirited debate ensued as to whether chloroplasts and mitochondria were indeed of endosymbiotic *(xenogenous)* origin or whether they arose *autogenously* in the process of cellular differentiation during the course of evolution (see Gray and Doolittle, 1982 and Taylor, 1987 for discussions). The debate took a distinct turn in favor of the endosymbiotic hypothesis in 1982 when Gray and Doolittle (see also Gray, 1989a, 1992) pointed out that the hypothesis would be proven if (1) the evolutionary histories of the nuclear and organelle genomes were known with certainty and were with certainty different, (2) the evolutionary history of the nuclear genome was demonstrably different from that of one or both organelle genomes even if its own history was not known with certainty, (3) it could be shown that mitochondria or chloroplasts arose from within distinctly different lineages of bacteria, and/or (4) one or both organelles originated in different eukaryotes from different subdivisions within a given bacterial phylum (i.e., that chloroplasts or mitochondria are of polyphyletic origin).

Although much information was available concerning plastids, mitochondria, and their component molecules in 1982, comparative data on single gene sequences that could be used to construct phylogenies and to test the proofs outlined above were still fragmentary. During the past decade such comparative analyses of single gene sequences either in the amino acid or nucleotide alphabets have proceeded apace (cf. Cedergren *et al.*, 1988). The molecules of choice have proven to be the rRNAs. These molecules possess highly conserved as well as variable regions (Gray and Schnare, 1989). They are components of all protein synthesizing systems. In fact Woese (1987) has called rRNAs "the ultimate chronometers." Collection of rRNA sequence data together with the development of sophisticated statistical testing techniques for assessing phylogenetic relatedness (cf. Cedergren *et al.*, 1988) have now provided convincing evidence that chloroplasts and mitochondria were derived from different eubacterial lines. This satisfies the third criterion of proof of the endosymbiont hypothesis described above. Also, while the ancestry of the nuclear genome remains uncertain, comparison of nuclear and organelle rRNA sequences indicates that they are derived from separate lineages satisfying the second proof (see Gray, 1989a). But as Gray (1992) notes in his review updating the evidence supporting the endosymbiont hypothesis "there are an increasing number of similarities that link archaebacterial and nuclear genomes."

This chapter considers theories of the evolution of organelle genomes, as a prelude to discussing their structure and organization and the forces contributing to their evolution.

THE ORIGIN OF CHLOROPLASTS AND MITOCHONDRIA

Woese and Fox (1977a) first hypothesized that there were three primary lineages of evolutionary descent from a common ancestor or "progenote" (Woese and Fox, 1977b; Fig. 2–1): eubacteria, archaebacteria, and eukaryotes. To propose separation of the bacteria into two separate "urkingdoms" based on catalogs of 16 S rRNA-derived oligonucleotides was a revolutionary step that has subsequently received strong support (see Brown *et al.,* 1989; Gray, 1992). As Gray (1992) points out, this view challenges the traditional eukaryote–prokaryote dichotomy and the more recent five-kingdom

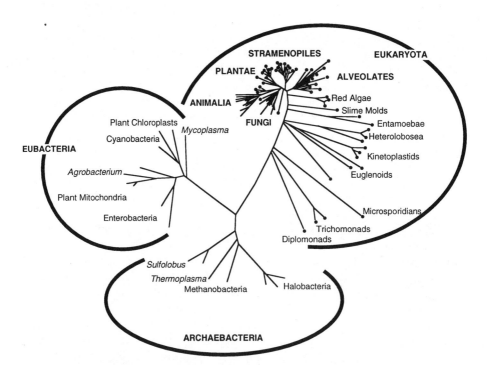

Fig. 2–1. An inferred phylogeny of eukaryotes based on a distance analysis of all positions that can be unambiguously aligned among complete 16S-like rRNA sequences from 75 taxa. The length of segments indicates the extent of molecular change. From Patterson and Sogin (1993).

scheme that divides the living organisms into four eukaryotic kingdoms (Animalia, Plantae, Fungi, and Protista) and a single kingdom of prokaryotes (Monera). Furthermore, the notion of protists as a unitary group embodied in the five-kingdom model is now obsolete since these organisms are cytologically, organizationally, and molecularly more diverse than plants, animals, and fungi (see following and Patterson and Sogin, 1993).

Whether the Archaebacteria themselves constitute a monophyletic group has been disputed by Lake (1987, 1988). He contends, based on rRNA sequence comparisons, that these bacteria are polyphyletic with the thermoacidophiles (eocytes) sharing common ancestry with eukaryotes (karyotes). In Lake's classification the extreme halophiles (photocytes) share ancestry with eubacteria while the methanogens are a sister group. These three groups comprise the parkaryotes. Nevertheless, subsequent analyses using rRNA phylogenetic trees have strongly reinforced the original assumption that the Archaebacteria form a unique monophyletic group (cf. Cedergren *et al.*, 1988; Gouy and Li, 1989; Gray, 1992).

A great variety of protists including the metamonads, microsporidians, and archamoebae appear primitively to lack chloroplasts and mitochondria. They have been grouped in the Superkingdom Archezoa by Cavalier-Smith (1987a). He supposes that the Archezoa includes the organisms that gave rise to the Eukaryota. However, based on small subunit rRNA phylogenies (see following) and ultrastructural criteria, Patterson and Sogin (1993) point out that the Archezoans, which they refer to as the Hypochondria, really constitute a paraphyletic lineage. That is, while they presumably shared a common ancestor at some time in the past, all descendants of this common ancestor cannot currently be identified. At the base of the tree are the diplomonads as represented by *Giardia* (Fig. 2–1) which is the only eukaryote known to have the "Shine-Dalgarno"-like mRNA binding sequence that characterizes prokaryotes (Chapter 12). The microsporidia, represented by *Vairimorpha,* possess a 23S-size large subunit rRNA that lacks a separate 5.8S rRNA typical of eukaryotic large ribosomal subunits (Vossbrinck and Woese, 1986). Patterson and Sogin (1993) also recognize three monophyletic assemblages (Fig. 2–1). The Euglenozoa include the photosynthetic and heterotrophic euglenids and the heterotrophic kinetoplastids. The alveolates are formed of dinoflagellates, ciliates and apicomplexan sporozoa. Finally, the stramenopiles are a very diverse group that includes oomycete fungi, diatoms and brown algae. Molecular phylogenetic studies also reveal that a much closer evolutionary link exists between animals and fungi than previously suspected (Fig. 2–1; Wainwright *et al.*, 1993).

It is important to keep these rapidly changing views of the proper classification of higher taxa in mind, particularly when reading Chapter 3, since they reveal previously unsuspected relationships between heterotrophic and photosynthetic groups of protists and algae. For instance, the malaria parasite *Plasmodium* possesses a degenerate chloroplast genome. This finding becomes less surprising in view of the new phylogenetic position of this organism, an apicomplexan that groups within the alveolates together with the photosynthetic dinoflagellates.

Unlike the eukaryote nucleus the endosymbiotic origin of plastids and mitochondria from different eubacterial lines now seems to be reasonably well established (Gray, 1992). Since the evidence supporting this conclusion also depends principally on rRNA phylogenies, a brief discussion of these is in order. Construction of rRNA phylogenies depends on aligning organelle rRNA sequences with their counterparts from eubacteria, archaebacteria, and eukaryotes. This cannot be done directly because of tremendous variation in the length and composition of homologous rRNAs (Chapter 3), but the problem is obviated because of the existence of a conserved core of secondary structure common to all small subunit and large subunit rRNAs. For example, in small subunit rRNAs eight noncontiguous, conserved regions of primary sequence (U1 to U8) are found to be interspersed with nonconserved regions (Fig. 2–2). These regions interact with one another by means of long range base pairing yielding a conserved central core of secondary structure common to all small subunit rRNAs (Stiegler et al., 1981). A similar core can be identified in the large subunit rRNA (cf. Gray, 1988).

These conserved sequences provide databases 500–600 nucleotides long present in all small and large subunit rRNAs so far examined with the exception of trypanosome mitochondrial rRNAs, which contain only a few elements of the universal cores (see Gray, 1988, for references). Figure 2–3 compares large and small subunit rRNA cores from chloroplasts of maize, mitochondria from several widely divergent organisms, and their homologs from archaebacteria, eubacteria, and the eukaryote cytoplasm. All sequences are at least 50% conserved. Utilization of these rRNA sequence databases together with sophisticated statistical methods for constructing phylogenies (e.g., Cedergren et al., 1988) provide the strongest evidence for the unique evolutionary origins of plastids and mitochondria.

Small subunit rRNA phylogenies argued that plastids were derived from the gram-positive cyanobacteria, as envisioned originally by Schimper (1883). Such phylogenies also pointed to eubacteria like *E. coli* and *Proteus vulgaris* as the progenitors of mitochondria (Gray et al., 1984). Subsequent investigations have confirmed the cyanobacterial derivation of plastids (Fig. 2–1). Comparisons of mitochondrial small subunit rRNA sequences from organisms as diverse as wheat, mouse, *Aspergillus,* and *Paramecium* now indicate that mitochondria originated from the α division of the purple bacteria (Yang et al., 1985; Fig. 2–1). This division also contains the rhizobacteria, agrobacteria, and rickettsia, all of which have established intracellular or other close relationships with eukaryotic cells (see Gray, 1989a). *Paracoccus denitrificans* is also a member of this division. This is significant because *P. denitrificans* "effectively assembles in a single organism those features of the mitochondrial respiratory chain and oxidative phosphorylation which are otherwise distributed at random among most aerobic bacteria" (John and Whatley, 1975).

Although this discussion has focused on rRNA phylogenies, many other pieces of evidence support the proposed eubacterial origins of organelle genomes. These are enumerated in various places throughout this text. For example, chloroplast genomes often retain the operon-like organization seen in bacteria. Many of the components of the transcriptional and translational apparatus, particularly those of plastids, have striking antecedents among the prokaryotes. Multigene phylogenies also bolster the evidence for origin of chloroplasts and mitochondria from specific eubacterial lineages. For instance, the cyanobacterial ancestry of chloroplasts is supported by phylogenies of several genes other than those encoding the rRNAs (Morden et al., 1992).

Since plastids and mitochondria have distinctly different phylogenetic ancestries among the eubacteria, they could not have originated autogenously within the same host cell (Doolittle, 1980). Therefore, the serial theory of endosymbiosis whereby mitochondria and plastids were acquired sequentially and in that order has become widely accepted (Taylor, 1974; Margulis, 1981). This view has, however, been disputed by Cavalier-Smith (1987b). He proposed the simultaneous acquisition of a cyanobacterium and a bacterium belonging to the α division of the purple bacteria by a primitive eukary-

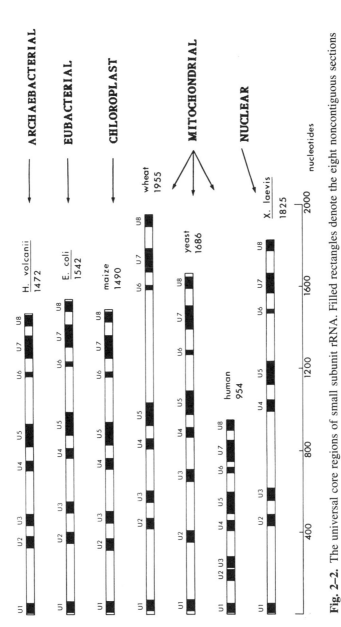

Fig. 2–2. The universal core regions of small subunit rRNA. Filled rectangles denote the eight noncontiguous sections of primary sequence (U1 to U8) that constitute the highly conserved central core of secondary structure. The length of each small subunit rRNA is indicated below the name of the organism. From Gray (1988).

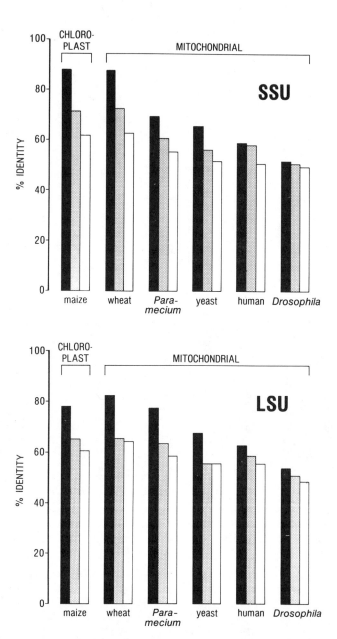

Fig. 2–3. Comparison of universal core regions in organelle small subunit (SSU) rRNA (upper) and large subunit (LSU) rRNA (lower). The bars indicate the percentage primary sequence identity in pairwise comparisons of individual organellar rRNAs with a eubacterial (*E. coli*, solid bar), archaebacterial (*Halobacterium halobium*, stippled bar), or eukaryotic cytosol (human, open bar) homologue. From Gray (1988).

ote such as an archezoan. He then posited that these endosymbiotic organisms evolved in concert into modern day plastids and mitochondria. Since the purple bacteria of the α division cannot photosynthesize in the presence of oxygen, Cavalier-Smith further supposed that the protomitochondrion lost the ability to photosynthesize. Conversely, the protoplastid would have retained its ability to perform oxygenic photosynthesis, but gradually lost its ability to respire. Evidently, the vast array of nonphotosynthetic eukaryotes would have had to lose their plastids as a consequence of a subsequent event(s).

Implicit in the simultaneous acquisition hypothesis is the notion that mitochondria and plastids were each of monophyletic origin. Monophyly is also consistent with the serial theory of endosymbiosis although not a necessary prerequisite. Certainly, a monophyletic origin for each organelle would most easily explain the striking amount of conservation of chloroplast and mitochondrial genes in different organisms even given that this apparent gene conservation partially reflects the inordinate past focus on chloroplast and mitochondrial genomes of a limited set eukaryotes (animals, mostly mammals, a few fungi, and flowering plants). What monophyly does not explain is the striking convergence in gene content between chloroplast and mitochondrial genomes. Both genomes always encode the rRNAs and usually most of the tRNAs of their protein-synthesizing systems. They also encode specific components of their respective electron transport chains and in all chloroplast genomes and most mitochondrial genomes, save those of animals, several (mitochondria) to many (chloroplasts) ribosomal proteins. Thus, while monophyly is a useful simplifying assumption, it may be unnecessary. Similar convergent evolution could have occurred for both plastids and mitochondria had each organelle been of polyphyletic origin. In fact, the diversity of plastid types in different phyla has long suggested that these organelles might be polyphyletic in origin (Table 2–1).

Table 2–1. Photosynthetic Pigments (chl, Chlorophyll; PB, Phycobilins) and Bounding Membranes of Plastids in Higher Plants and Different Groups of Eukaryotic Algae

Plastid Type	Number of Bounding Membranes		
	Two	Three	Four
Green (chl a/b)	Land plants Chlorophyta Gamophyta	Euglenophyta	
Red (chl a/PB)	Rhodophyta		
Yellow-brown (chl a/c)		Dinoflagellata[a]	Chrysophyta Haptophyta Cryptophyta[b] Xanthophyta Eustigmatophyta[b] Bacillariophyta Phaeophyta

Modified from Gray (1991).

[a] Certain dinoflagellate groups have plastids with only two membranes.

[b] Also contain phycobilins.

Two classes of observations are principally responsible for this notion (see Gray, 1991, 1992 for summaries). The first, and oldest, is that the composition of accessory photosynthetic pigments in eukaryotic algae divides them neatly into three groups. The second is that the number of bounding membranes ranges from two to four, which can also be interpreted as indicative of multiple endosymbiotic events.

The primary reaction center pigment in photosystems I and II in cyanobacteria, eukaryotic algae, and higher plants is chlorophyll a. However, the *Rhodophyta* and the photoautotrophic protist *Cyanophora paradoxa* share phycobilins with Cyanobacteria as accessory pigments. In the green algae *(Chlorophyta, Gamophyta)* and *Euglenophyta* the accessory pigment is chlorophyll b and in the chromophyte algae chlorophyll c. These differences in the reaction center pigment composition can be used to argue for at least three independent cyanobacterial acquisitions.

Alternatively, plastids may have been acquired only once with the original plastid-containing host being subject to a series of secondary endosymbiotic events (see Cavalier-Smith, 1982, 1987b; Taylor 1979, 1987). Following these secondary endosymbioses, pigment modifications would have occurred leading to the three major groups of algae classified on the basis of pigment composition (Table 2–1). Cavalier-Smith and Taylor point out that rather minor modifications are required to convert chlorophyll a to b (substitution of a methyl by a formyl group) or a to c [acquisition of one (c_1) or two (c_2) double bonds] with these reactions being secondary steps in chlorophyll biogenesis. Such accessory pigment changes coupled with the loss of phycobilins as secondary pigments in the lines leading to green algae and chromophytes could account for the pigment differences that distinguish these algal lines from each other and from the *Rhodophyta* and *Cyanobacteria*.

In summary, analyses of pigment composition in different algal groups can be used to support either monophyletic or polyphyletic theories of plastid origin.

Until recently prokaryotes containing chlorophylls b and c that might be precursors of green and chromophyte algae, respectively, had not been identified. The discovery of symbiotic (cf. Lewin, 1981) and free living (Burger-Wiersma et al., 1986) prochlorophytes suggested that these prokaryotes might have included organisms that became symbionts of the prototypical green algal cell (Whatley and Whatley, 1981). This conjecture appeared to be supported by a phylogeny based on a comparison of sequences of the *psbA* gene encoding the D-1 reaction center protein of photosystem II (Morden and Golden, 1989). Similarly the pigment composition of the brownish green photosynthetic bacterium *Heliobacterium chlorum* (Gest and Favinger, 1983) led to the supposition that a bacterium like this might have been the endosymbiont incorporated into the protoeukaryote ancestral to the chromophyte algae (Margulis and Obar, 1985). As Gray (1991) points out, neither of these conjectures has withstood tests based on 5S and 16S chloroplast rRNA trees (see Turner et al., 1989) or on a reassessment of the *psbA* phylogeny for prochlorophytes all of which show they are allied with the cyanobacteria (see Gray, 1989a, for a discussion). This relationship is also confirmed by other gene phylogenies (Morden et al., 1992). In fact prochlorophytes now appear to be polyphyletic within the cyanobacterial radiation and none of the known species specifically relates to chloroplasts (Gray, 1992). Bryant (1992) makes the interesting point that an economical hypothesis to explain the wide distribution of chlorophyll b-containing prochlorophytes among cyanobacteria would be to suppose that the ancestral oxygen-evolving prokaryotes synthesized both chlorophyll *a* and *b* in addition to the phycobilins and possessed binding proteins for these chlorophylls as well. Bryant imagines that "subsequent loss of one or the other antenna system during evolution has produced the distribution of species and capabilities observed today." Also, rRNA sequence comparisons do not support the proposal that an ancestor of *Heliobacterium* gave rise to the chromophyte plastid (Gray, 1991, 1992). Plastid rRNA trees group within the eubacterial lineage and specifically target the cyanobacteria.

The number of membranes bounding the plastid in different algal taxa (two, three, or four) can also be used to support the theory of secondary endosymbiosis (Table 2–1). The double membrane surrounding the plastids of green and red algae and higher plants is usually assumed to be a chimera. The inner membrane would form the plasma membrane of the endosymbiont (membrane 1) while the outer membrane (membrane 2) represents the phagosome (food vacuole) membrane of the host (host-1) (Fig. 2–4, Gray 1991). However, Cavalier-Smith (1987b) believes membrane 2 to be the outer membrane of the original endosymbiont. He supposes that the phagosome membrane of host-1 has been lost and argues that plastid and mitochondrial outer membranes have permeability properties as well as porin proteins reminiscent of the outer membranes of bacteria like *E. coli* (Benz, 1985).

"Complex" plastids with three and four bounding membranes are thought to have arisen via two successive endosymbioses (Fig. 2–4). First, host-1 was engulfed by a second eukaryote (host-2) following which reduction of host-1 ensued. In plastids bounded by three membranes this implies different origins for membrane 3 depending on the precursor of membrane 2. If membrane 2 is the phagosome membrane of host-1, membrane 3 could either be the host-1 outer membrane or the host-2 phagosome membrane (Fig. 2–4). In either case any remnants of host-1 would be found in the periplastid space between membranes 2 and 3. If membrane 2 is the outer membrane of the endosymbiont, membrane 3 could be the host-1 phagosome, the host-1 outer membrane or the host-2 phagosome (Fig. 2–4). The latter interpretation is favored by Cavalier-Smith. The periplastidial space could also contain host-1 remnants in this model unless membrane 3 is the host-1 phagosome. In this case the host-1 and host-2 cytoplasms and nuclei would have mixed with the presumptive loss during evolution of the host-1 components.

Plastids bounded with four membranes typify most chromophyte algae except dinoflagellates (Table 2–1). If membrane 2 is the host-1 phagosome, then membrane 3 becomes the host-1 outer membrane and membrane 4 the host-2 phagosome (Fig. 2–4). On the other hand, if membrane 2 is the outer membrane of the symbiont, three derivations of membranes 3 and 4 are theoretically possible. 1. Membrane 3 is the host-1 phagosome and membrane 4 the host-1 outer membrane. 2. Membrane 3 is the host-1 phagosome and membrane 4 the host-2 phagosome. 3. Membrane 3 is the host-1 outer membrane and membrane 4 the host-2 phagosome. The latter interpretation is in accord with the proposed loss of the host-1 phagosome by Cavalier-Smith.

Observations made on two algal groups are particularly interesting with respect to these models. The cyanelle of *Cyanophora paradoxa* is bounded by a rudimentary peptidoglycan wall suggesting that the cyanelle may represent a relatively recent acquisition (Hall and Claus, 1963). On the contrary, Cavalier-Smith (1982) argues that the original cyanobacterial progenitor of plastids possessed a peptidoglycan layer that sim-

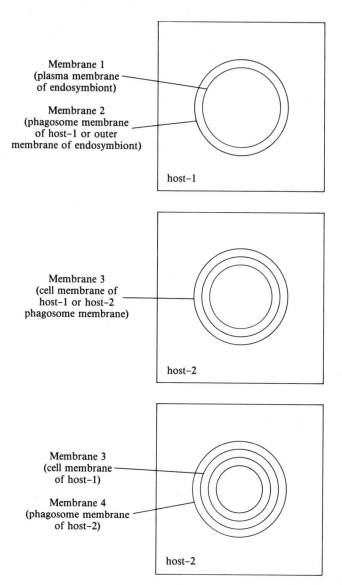

Fig. 2–4. Possible derivations of chloroplast bounding membranes. (Modified from Douglas (1991). Two bounding membranes typify land plants, Chlorophyta and Rhodophyta. Three bounding membranes are found in Euglenophyta and many Dinoflagellata although some dinoflagellates have only two bounding membranes. Four bounding membranes typify the chromophyte algae except the Dinoflagellata. Other possible interpretations are discussed in the text. The middle and lower panels of the figure interpreting the origins of membranes 3 and 4 assume that membrane 2 is the phagosome membrane of host-1. The alternate scenario for the origin of these membranes, which assumes that the host-2 membrane is the outer membrane of the endosymbiont, is discussed in the text.

ply has not been lost in this particular line of descent. The finding that cryptomonad algae possess a nucleomorph, presumably the vestigial nucleus of host-1 (Ludwig and Gibbs, 1985), together with ribosomes and rRNA whose phylogeny is distinct from both plastid and cryptomonad cytoplasmic rRNAs (McFadden, 1990; Douglas et al., 1991), provides particularly strong evidence favoring the secondary acquisition hypothesis. The nucleomorph and associated cytoplasm are located between membranes 2 and 3 and sequences from the cytoplasmic small subunit rRNA cluster specifically with nuclear 18S rRNA sequences of red algae (Douglas et al., 1991).

Morden et al. (1992) used multiple gene phylogenies to try and sort out relationships between different plant and algal lineages and their progenitors. Their analysis includes genes always found in the plastid (*atpB, psbA*, and *rbcL*) and genes that may be found in either the plastid or nuclear genome depending on taxon *(rbcS, tufA)*. All of these phylogenies plus that for the small subunit rRNA support a cyanobacterial origin for green plants and algae plus *Cyanophora*. In the other "nongreen" algae (cryptophytes, chromophytes, rhodophytes) four lineages (*atpB, psbA, tufA*, small subunit rRNA) plus the phylogeny for small subunit rRNA also support a single cyanobacterial of these algae. However, two lineages (*rbcL* and *rbcS*) suggest that the plastids of these nongreen algae derived their *rbcL* and *rbcS* genes from the purple bacteria. To get out of this conundrum Morden et al. (1992) propose a single endosymbiotic origin of plastids followed by a subsequent split of nongreen algae to give *Cyanophora* on the one hand and a second lineage comprising chromophytes, cryptophytes, and rhodophytes. After the second split the *rbcLS* operon (and potentially other genes as well) would have been laterally transferred from an α or β purple bacterium to the second lineage.

So where does all of this leave us with respect to the mono- or polyphyletic origin of plastids? In Gray's (1991, 1992) opinion the question cannot be answered presently despite much speculation. He points out that our current molecular database is simply too sparse and too biased toward plants and green algae to answer the question. Whether plastids were of monophyletic or polyphyletic origin, there are many modern examples of independent acquisition of algal and cyanobacterial endosymbionts by protists (ciliates, foraminiferans, radiolarians) and invertebrates (cnidarians, flatworms, sponges, ascidians) (see Douglas, 1991). However, none of these endosymbionts has undergone massive losses or transfers of genes to the host nucleus nor have mechanisms evolved to target proteins into these endosymbionts.

The question of whether mitochondria were of monophyletic or polyphyletic origin has also been considered (see Cavalier-Smith, 1992; Gray, 1988, 1989b). Mitochondrial rRNA sequences of the green alga *Chlamydomonas reinhardtii* do not branch with higher plants although nuclear and chloroplast rRNA phylogenies place *Chlamydomonas* and higher plants in the same branch (Cedergren et al. 1988). This anomaly has been examined in more detail by Gray et al. (1989). Although all mitochondrial small subunit rRNA sequences form a single tree that branches within the α subdivision of purple bacteria, higher plants cluster very near the root of this tree. *C. reinhardtii* branches separately and its small subunit rRNA sequences are more closely related to those of animals than to higher plants. Gray et al. (1989) originally suggested that the likely explanation of this anomaly was that the plant mitochondrial genome is a mosaic whose rRNA genes were acquired secondarily from a eubacterial endosymbiont existing in the same cell by lateral transfer. However, Gray (1992) now questions this hypothesis.

He points out that "a major issue is the validity of mitochondrial trees on which such a suggestion is based." The problem is one of establishing accurate branching patterns in a lineage where the compared sequences are diverging at radically different rates. As Gray points out, current methodology may be inadequate to deal with the rate difference problem. Perhaps multiple gene phylogenies such as those done by Morden *et al.* (1992) for several genes encoding plastid components may yield a more plausible phylogeny for the plant mitochondrial genome. On the other hand, the analysis of Morden *et al.* also runs into the problem of having to posit lateral gene transfer as discussed above. Although, the selection pressures promoting lateral gene transfer are difficult to visualize, the process clearly occurs in the case of plant mitochondria both from chloroplast and nucleus (Chapter 3). Furthermore, these genes, particularly those encoding tRNAs (Chapter 13), are often functional.

Nevertheless, in the absence of indisputable evidence to the contrary, it seems most prudent to pay heed to Cavalier-Smith's (1992) forceful arguments supporting monophyly. While he no longer seems to favor simultaneous acquisition to the same extent, Cavalier-Smith (1992) summarizes a body of evidence favoring monophyly for chloroplasts and mitochondria pointing out that lateral transfer clearly played a role in the acquisition of plastids in certain algal groups (e.g., the cryptomonads), but perhaps not others (e.g., the euglenoids).

REFERENCES

Benz, R. (1985). Porin from bacterial and mitochondrial outer membranes. *CRC Crit. Rev. Biochem.* 19: 145–190.

Brown, J.W., Daniels, C.J., and J.N. Reeve. 1989. Gene structure, organization, and expression in Archaebacteria. *CRC Rev. Microbiol.* 16: 287–338.

Bryant, D.A. (1992). Puzzles of chloroplast ancestry. *Curr. Biol.* 2: 240–242.

Burger-Wiersma, T., Veenhuis, M., Korthals, H.J., Van de Wiel, C.C.M., and L.R. Mur (1986). A new prokaryote containing chlorophylls *a* and *b*. *Nature (London)* 320: 262–264.

Cavalier-Smith, T. (1982). The origin of plastids. *Biol. J. Linn. Soc.* 17: 289–306.

Cavalier-Smith, T. (1987a). The origin of eukaryote and archaebacterial cells. *Ann. N.Y. Acad. Sci.* 503: 17–54.

Cavalier-Smith, T. (1987b). The simultaneous symbiotic origin of mitochondria chloroplasts, and microbodies. *Ann. N.Y. Acad. Sci.* 503: 55–71.

Cavalier-Smith, T. (1992). The number of symbiotic origins of organelles. *BioSystems* 28: 91–106.

Cedergren, R., Gray, M.W., Abel, Y., and Sankoff, D. (1988). The evolutionary relationships among known life forms. *J. Mol. Evol.* 28: 98–112.

Doolittle, W.F. (1980). Revolutionary concepts in evolutionary cell biology. *Trends Biochem. Sci.* 5: 146–149.

Douglas, A.E. (1991). Symbiosis in evolution. In *Oxford Surveys in Evolutionary Biology* (D. Futuyma and J. Antonovics, eds.), Vol. 8, pp. 347–382. Oxford University Press, New York.

Douglas, S.E., Murphy, C.A., Spencer, D.F., and M.W. Gray (1991). Cryptomonad algae are evolutionary chimeras of two phylogenetically distinct eukaryotes. *Nature (London)* 350: 148–151.

Gest, H., and J.L. Favinger (1983). *Heliobacterium chlorum*, an anoxygenic brownish green photosynthetic bacterium containing a "new" form of bacteriochlorophyll. *Arch. Microbiol.* 136: 11–16.

Gouy, M., and W.-H. Li. (1989). Phylogenetic analysis based on rRNA sequences supports the archaebacterial rather than the eocyte tree. *Nature (London)* 339: 145–147.

Gray, M.W. (1988). Organelle origins and ribosomal RNA. *Biochem Cell. Biol.* 66: 325–348.

Gray, M.W. (1989a). The evolutionary origins of organelles. *Trends Genet.*, 5: 294–299.

Gray, M.W. (1989b). Origin and evolution of mitochondrial DNA. *Annu. Rev. Cell Biol.* 5: 25–50.

Gray, M.W. (1991). Origin and evolution of plastid genomes and genes. In *The Molecular Biology of Plastids* (L. Bogorad and I. Vasil, eds.), pp. 303–330. Academic Press, San Diego.

Gray, M.W. (1992). The endosymbiont hypothesis revisited. *Int. Rev. Cytol.* 141: 233–357.

Gray, M.W., and W.F. Doolittle (1982). Has the endosymbiont hypothesis been proven? *Microbiol. Rev.* 46: 1–42.

Gray, M.W., and M.N. Schnare (1989). Evolution of the modular structure of rRNA. In *The Ribosome: Structure, Function, and Evolution* (W.E. Hill, A. Dahlberg, R. Garrett, P.B. Moore, D. Schlessinger, and J.R. Warner eds.), pp. 589–597. American Society of Microbiology, Washington, D.C.

Gray, M.W., Sankoff, D., and R. Cedergren (1984). On the evolutionary descent of organisms and organelles: A global phylogeny based on a highly conserved structural core in small subunit ribosomal RNA. *Nucl. Acids Res.* 12: 5837–5852.

Gray, M.W., Cedergren, R., Abel, Y., and D. Sankoff (1989). On the evolutionary origin of the plant mitochondrion and its genome. *Proc. Natl. Acad. Sci. U.S.A* 86: 2267–2271.

Hall, W.T., and G. Claus (1963). Ultrastructural studies on the blue-green algal symbiont in *Cyanophora paradoxa* Korschikoff. *J. Cell Biol.* 19: 551–563.

John, P., and F.R. Whatley (1975). *Paracoccus denitrificans* and the evolutionary origin of the mitochondrion. *Nature (London)* 254: 495–498.

Lake, J.A. (1987). Prokaryotes and archaebacteria are not monophyletic: rate invariant analysis of rRNA genes indicates that eukaryotes and eocytes form a monophyletic lineage. *Cold Spring Harbor Symp. Quant. Biol.* 52: 839–846.

Lake, J.A. (1988). Origin of the eukaryotic nucleus determined by rate-invariant analysis of rRNA sequences. *Nature (London)* 331: 184–186.

Lewin, R.A. (1981). *Prochloron* and the theory of symbiogenesis. *Ann. N.Y. Acad. Sci.* 361: 325–329.

Ludwig, M., and J.P. Gibbs (1985). DNA is present in the nucleomorph of *Cryptomonas* demonstrated by DAPI fluorescence. *Z. Naturforsch. Sect. C* 40: 933–935.

Margulis, L. (1970). *Origin of Eukaryotic Cells.* Yale University Press, New Haven.

Margulis, L. (1981). *Symbiosis and Cell Evolution.* Freeman, San Francisco.

Margulis, L., and R. Obar (1985). *Heliobacterium* and the origin of chrysoplasts. *BioSystems* 17: 317–325.

McFadden, G.I. (1990). Evidence that cryptomonad chloroplasts evolved from photosynthetic eukaryotic endosymbionts. *J. Cell Sci.* 95: 303–308.

Mereschkowsky, C. (1905). Uber natur and ursprung der chromatophoren im pflanzenreiche. *Biol. Zentr.* 25: 593–604.

Mereschkowsky, C. (1910). Theorie der zwei plasmaarten als grundlage der symbiogenesis, einer neuen lehre von der entstehung der organismen *Biol. Zentr.* 30: 278–303, 321–347, 353–367.

Mereschkowsky, C. (1920). La plante considérée comme une complex symbiotique. *Bull. Soc. Nat. Sci. Ouest France* 6: 17–21.

Morden, C.W., and S.S. Golden (1989). *psbA* genes indicate common ancestry of chlorophytes and chloroplasts. *Nature (London)* 337: 382–385.

Morden, C.W., Delwiche, C.F., Kuhsel, M., and J.D. Palmer (1992). Gene phylogenies and the endosymbiotic origin of plastids. *Biosystems* 28: 75–90.

Patterson, D.J., and M.L. Sogin (1992). Eukaryotic origins and protistan diversity. In *The Origin and Evolution of the Cell* (H. Hartman and K. Matsuno eds.), pp. 13–47. World Scientific Publishing Co. River Edge, N.J.

Schimper, A.F.W. (1883). Uber die entwicklung der chlorophyll korner und farbkorner. *Bot. Zeit.* 41: 105–114.

Stiegler, P., Carbon, P., Ebel, J.-P., and C. Ehresmann (1981). A general secondary-structure model for procaryotic and eukaryotic RNAs of the small ribosomal subunits. *Eur. J. Biochem.* 120: 487–495.

Taylor, F.J.R. (1974). Implications and extensions of the serial endosymbiosis theory of the origin of eukaryotes. *Taxon* 23: 229–258.

Taylor, F.J.R. (1979). Symbiontism revisted: A discussion of the evolutionary impact of intracellular symbioses. *Proc. R. Soc. London B* 204: 267–286.

Taylor, F.J.R. (1987). An overview of the status of evolutionary cell symbiosis theories. *Ann. N.Y. Acad. Sci.* 503: 1–16.

Turner, S., Burger-Wiersma, T., Giovannoni, S.J., Muir, L.R., and N.R. Pace (1989). The relationship of a prochlorophyte *Prochlorothrix hollandica* to green chloroplasts. *Nature (London)* 337: 380–382.

Vossbrinck, C.R., and C.R. Woese (1986). Eukaryotic ribosomes that lack a 5.8S RNA. *Nature (London)* 320: 287–288.

Vossbrinck, C.R., Maddox, M., Friedman, S., Debrunner-Vossbrinck, B.R., and C.R. Woese (1987). Ribosomal RNA sequence suggests microsporidia are extremely ancient eukaryotes. *Nature (London)* 326: 411–414.

Wainwright, P.O., Hinkle, G., Sogin, M.L., and S.K. Stickel (1993). Monophyletic origins of the mettzoa: An evolutionary link with fungi. *Science* 260: 340–342.

Whatley, J.M., and F.R. Whatley (1981). Chloroplast evolution. *New Phytol.* 87: 233–247.

Wallin, I.E. (1922). On the nature of mitochondria. *Am. J. Anat.* 30: 203–229, 451–471.

Woese, C.R. (1987). Bacterial evolution. *Microbiol. Rev.* 51: 221–271.

Woese, C.R., and G.E. Fox (1977a). Phylogenetic structure of the prokaryotic domain: The primary kingdoms. *Proc. Natl. Acad. Sci. U.S.A*, 74: 5088–5090.

Woese, C.R. and G.E. Fox (1977b). The concept of cellular evolution. *J. Mol. Evol.* 10: 1–6.

Yang, D., Oyaizu, Y., Oyaizu, H., Olsen, G.J. and C.R. Woese (1985). Mitochondrial origins. *Proc. Natl. Acad. Sci. U.S.A*, 82: 4443–4447.

3

Organelle Genome Organization and Gene Content

In 1951 Chiba, based on Feulgen staining results, suggested that chloroplasts of *Selaginella* and two flowering plants contained DNA. However, another decade passed before reports of DNA in chloroplasts and also mitochondria became common. Between 1960 and 1970 the existence of unique species of DNA in both organelles was established unequivocally. Initially, most of the evidence came from experiments in which DNA was isolated from chloroplasts and mitochondria and shown to be unique in base composition as judged by buoyant density centrifugation in cesium chloride. These kinds of experiments also suggested that organelle DNA was a minor cellular component compared to nuclear DNA. For instance, in 1963 Sager and Ishida observed that a unique species of DNA could be enriched to 40% of the total in a chloroplast preparation from *Chlamydomonas*. However, this chloroplast DNA (cpDNA) accounted for a mere 6% of the total DNA of the cell. In flowering plants the picture was at first confusing. Early reports suggested that cpDNA was denser (1.705 g/cm^3) than nuclear DNA (1.695 g/cm^3), but these chloroplast preparations also appeared to be "contaminated" with a lighter component. By the late 1960s reports began to appear suggesting that flowering plant cpDNAs had densities similar to nuclear DNA. The problem was finally laid to rest by Kirk in 1971. He concluded that cpDNAs from higher plants have buoyant densities in the range of 1.697 g/cm^3 and G+C contents of 37 to 38%. The earlier reports of dense satellites associated with chloroplasts were explained by bacterial or mitochondrial contamination of the isolated chloroplasts. Kirk also pointed out that plant nuclear DNA is relatively rich in 5-methylcytosine. This base is lacking in cpDNA, but would lower the density of nuclear DNA to bring it into the same range as cpDNA. Research leading to the identification of the "real" cpDNA has been reviewed by Kirk (1986).

By the late 1960s characterization of organelle DNA molecules by electron microscopy had begun. The most successful results were obtained with the circular mitochondrial DNA (mtDNA) molecules from animals. These molecules did not shear during isolation because of their small size (ca. 4.5–6.0 μm). In contrast, the larger DNA molecules from fungal and plant mitochondria and from chloroplasts were usually linear, but whether these molecules were linear or circular in the native state was unknown. In 1970 Hollenberg *et al.* demonstrated the presence of 25-μm circular DNA molecules from osmotically lysed mitochondrial preparations from the closely related yeasts *Saccharomyces cerevisieae* and *S. carlsbergensis*. Petes *et al.* (1973) confirmed the presence of circular molecules in yeast mitochondria. Neverthless, the vast majority of mtDNA molecules seen in these and other preparations were linear. These linear molecules may not result from the random fragmentation of circular molecules as generally assumed, but may be the products of rolling circle replication of yeast mtDNA (Maleszka *et al.*, 1991, Chapter 5). Manning *et al.* (1971) reported circular 40-μm cpDNA molecules from *Euglena*. The next year circular DNA molecules in the same size range were reported for higher plants (Manning *et al.*, 1972; Kolodner and Tewari, 1972). Through the use of appropriate standards the mass per unit length of these DNA molecules was computed. Thus, animal mtDNA molecules were 15- to 16-kb circles whereas cpDNA molecules were over 100 kb.

By combining observations on the molecular weights of individual organelle DNA molecules with data that yielded estimates of the total amount of DNA per cell, the conclusion was inescapable that there were several to many DNA molecules per organelle and per cell. Thus, a single mouse mitochondrion has, on average, five DNA molecules and a chloroplast of *Chlamydomonas* 80. The next question was whether these DNA molecules were genetically homogeneous or heterogeneous. Using renaturation kinetic analysis (Britten and Kohne, 1968; Wetmur and Davidson, 1968) investigators working on organelle DNAs were able to show that these molecules formed homogeneous collections whose molecular weights corresponded to those measured by electron microscopy.

Much of this early work has been reviewed by Gillham (1978).

In the mid-1970s restriction enzymes came into general use and restriction maps of organelle DNAs began to be published with a great regularity that continues to this day. At the same time specific genes began to be mapped by molecular methods to organelle genomes. The first such genes to be located were those coding for organelle rRNAs and tRNAs. Later these mapping techniques were extended to protein- coding genes using a wide variety of methods. Subsequently, DNA sequencing methods came into general use and were applied to organelle genomes. In fact human and bovine mitochondrial genomes were among the first genomes of any kind to be sequenced in their entirety (Anderson *et al.*, 1981, 1982). These sequencing studies uncovered not only genes corresponding to specific mitochondrially encoded proteins, tRNAs, and rRNAs, but also open reading frames (ORFs) specifying proteins (i.e., cytochrome *b*, three subunits of cytochrome oxidase, two subunits of the ATP synthase, and seven subunits of NADH dehydrogenase). Subsequently, sequencing of the much larger organelle genomes of land plant plastids (Hiratsuka *et al.*, 1989; Ohyama *et al.*, 1986; Shinozaki *et al.*, 1986; Wolfe *et al.*, 1992a) and mitochondria (Oda *et al.*, 1992) has been achieved.

This chapter considers the organization and gene content of chloroplast and mitochondrial genomes.

MITOCHONDRIAL GENOMES

Mitochondrial genomes vary enormously in size (ca. 400-fold), but less in gene content (ca. 20-fold). Animals, with the exception of certain cnidarians (e.g., hydra, sea anemones), possess small, circular molecules ranging in size from about 14 to 39 kb (Table

Table 3–1. Size and Conformation of Selected Mitochondrial Genomes[a]

Taxa	Number of Species Studied	Size (kb)	Conformation[b]
ANIMALS			
Cnidaria			
Cubazoa, Scyphozoa, Hydrazoa, Limnomedusae, Siphonophora	31	8–18	1L or 2L
Anthozoa, Corallimorpha, Ceriantharia	17[c]	17[d]	C
Platyhelminthes	1	15	C
Nematoda[d]	2	14	C
Echiurida	1	18	C
Arthropoda			
Drosophila[d]	39	15–19	C
Echinodermata[d]	3	15–16	C
Chordata			
Various	Many	16–19	C
Gallus (chicken)[d]	1	17	C
Xenopus (clawed toad)[d]	1	18	C
Rattus (rat)[d]	1	16	C
Mus (mouse)[d]	1	16	C
Balanoptera (fin whale)[d]	1	16	C
Bos (cow)[d]	1	16	C
Homo[d]	1	17	C
FUNGI			
Oomycetes	17	36–73	C,C(IV)
Hypochitridiomycetes	2	50,54	C,C(IV)
Ascomycetes			
Yeasts	27	17–101	25C, 1C(IV), 1L
Filamentous	13	26–115	C
Basidiomycetes			
Filamentous			
Suillus	15	36–121	C
Other species	9	34–176	8C, 1C(IV)
Yeasts	2	33–47	1C, 1C(IV)
PLANTS			
Bryophyta			
Marchantia[d]	1	187	C
Angiosperms			
Brassica (cabbage, etc.)	5	208–231	MC
Spinacia (spinach)	1	327	MC
Beta (sugar beet)	1	386	MC
Zea (maize)	1	570	MC
Cucumis (muskmelon)	1	2,400	Not known
GREEN ALGAE			
Chlorophyta			
Chlamydomonas[d]	4	16–20	C or L
Pandorina	1	20	L
Chlorella-like	1	76	

Taxa	Number of Species Studied	Size (kb)	Conformation[b]
OTHER ALGAE			
Chrysophyta	3	40	L
PROTISTS			
Sarcodina			
Acanthamoeba	1	40	C
Apicomplexa			
Plasmodium (malaria)	3	6	L
Ciliophora			
Paramecium[d]	1	40	L
Tetrahymena	3 strains	26–33	L
Zoomastigina			
Kinetoplastida			
Maxicircles	9	20–37	C
Minicircles	9	0.5–3	C

Data compiled from Clark-Walker (1992), Gray (1992), Lonsdale (1989), Sederoff (1984), Wolf and Del Giudice (1988), and Wolstenholme (1992). Data for cnidarians are from Bridge et al. (1992). Other references are cited in the text.

[a] Sizes in kilobase pairs (kb) are rounded to nearest whole number.

[b] C, circular; C(IV), circular with large inverted repeat; MC, master circle; L, linear.

[c] Seventeen species are known to have circular mitochondrial genomes, but the size of only one has been established with certainty.

[d] Completely sequenced mitochondrial genomes.

3–1). Cnidaria belonging to the classes Cubozoa, Scyphozoa, and Hydrozoa are unique among animals in having linear mitochondrial genomes of 8 to 18 kb (Bridge et al., 1992; Warrior and Gall, 1985). These are present as doublets in many *Hydra* species. In contrast, flowering plant mitochondrial genomes are the largest organelle genomes known, ranging in size from 208 to 2400 kb (Table 3–1). They are circular or at least their structures deduced from restriction enzyme mapping are circular. With one known exception they also recombine actively to yield smaller circular molecules carrying only a subset of the genes and restriction fragments present in the "master" molecule. Fungal mitochondrial genomes also exhibit great size variation and are usually circular. Thus, the mitochondrial genome of one yeast, *Schizosaccharomyces pombe* EF1, is only 17.3 kb whereas that of another yeast, *Brettanomyces custersii* is 101 kb (Tables 3–1 and 3–2). Information on the structure and gene content of protistan and algal mitochondrial genomes is still relatively meager, but it is clear already that these mitochondrial genomes are much more variable in gene content than those of animals, plants, and fungi (Tables 3–3 and 3–4).

Mammalian Mitochondrial Genomes

The structure and organization of mammalian mitochondrial genomes have been regularly reviewed (see Attardi, 1985; Attardi and Schatz, 1988; Cantatore and Saccone, 1987; Gray, 1989; Wolstenholme, 1992). The human, cow, and mouse mitochondrial genomes were the first organelle genomes to be sequenced completely. More recently

Table 3–2. Variability in Mitochondrial Genome Size in Ascomycota[a]

Organism	Mitochondrial Genome Size (kb)
Schizosaccharomyces pombe (4 strains)	17.3–24.6
Candida (Torulopsis glabrata) (2 strains)	18.9, 20.3
Saccharomyces exiguus[b]	23.7
Hansenula wingei	25.5
Kloeckeria africana	27.1
Saccharomyces unisporus[b]	27.4
Brettanomyces custersianus[b]	28.5
Aspergillus nidulans (3 strains)[c]	31–38
Eeniella (Brettanomyces) nana[b]	34.5
Saccharomyces telluris[b]	38.4
Kluyveromyces lactis	34.9–39.5
Brettanomyces naardenensis[b]	41.7
Pachytichospora transvaalensis	41.4
Saccharomyces lipolytica[b]	44–48.3
Hansenula mrakii	55
Brettanomyces anomalus[b]	57.7
Neurospora crassa[c]	60–73
Saccharomyces cerevisiae[b]	68–85
Dekkera intermedia[b]	73.2
Dekkera bruxellensis[b]	85
Podospora anserina[c]	87–100.3
Brettanomyces custersii[b]	101.1
Cochliobolus heterostrophus[c]	115

Data compiled from Clark-Walker (1992), Hoeben and Clark-Walker (1986), Sederoff (1984), and Wolf and Del Giudice (1988).

[a]All species indicated are yeasts, except as noted in foot c.

[b]*Saccharomyces* and *Bretannomyces/Dekkera* species.

[c]Filamentous species.

the entire sequences of the finwhale and rat mitochondrial genomes have been determined (Arnason et al., 1991; Gadeleta et al., 1989). Partial sequences have also been obtained for four other primates and two dolphins (see Wolstenholme, 1992). These 16.5-kb genomes all exhibit an invariant gene order (Fig. 3–1).

Apart from a short segment around the origin of replication, mammalian mtDNA is completely saturated with genes, all of which lack introns (Fig. 3–1). Most of the genes are transcribed from the heavy (H) strand, which is so named because it has a greater equilibrium density in cesium chloride gradients than its complement, the light (L) strand. These density differences are also found in the mtDNA molecules of other vertebrates and nematodes (Wolstenholme, 1992). The H strand coding sequences include the genes specifying the large (16S) and small (12S) mitochondrial rRNAs, 14 tRNAs, the three largest subunits of cytochrome oxidase, cytochrome *b*, two subunits of the ATP synthase, and seven subunits of NADH dehydrogenase (Fig. 3–1). The L strand includes coding sequences for eight tRNAs and one subunit of NADH dehydrogenase (ND6). The H strand genes are in most cases butt jointed to each other or

Table 3–3. Respiratory Chain Genes Encoded in Mitochondrial DNA[a]

Taxa	ndh										cytb	cox				atp		
	1	2	3	4	4L	5	6	7	11	"ORF169"		1	2	3	A	6	8	9
Animals[b]	+	+	+	+	+	+	+	−	−	−	+	+	+	+	−	+	+	−
Fungi																		
Aspergillus nidulans	+	+	+	+	+	+	+	−	−	−	+	+	+	+	−	+	+	+
Neurospora crassa	+	+	+	+	+	+	+	−	−	−	+	+	+	+	−	+	+	−
Podospora anserina	+	+	+	+	+	+	+	−	−	−	+	+	+	+	−	+	+	+
Saccharomyces cerevisiae	−	−	−	−	−	−	−	−	−	−	+	+	+	+	−	+	+	+
Schizosaccharomyces pombe	−	−	−	−	−	−	−	−	−	−	+	+	+	+	−	+	+	+
Allomyces macrogynus	+	+	+	+	+	+	+	−	−	−	+	+	+	+	−	+	+	+
Plants																		
Angiosperms	+	+	+	+	−	+	+	+	−	−	+	+	+	+	+	+	+	+
Bryophyte	+	+	+	+	+	+	+	P	−	+	+	+	+	+	+	+	−	+
Protists and algae																		
Phytophthora infestans	+	+	+	+	+	+	+	−	−	−	+	+	+	+	+	+	+	+
Prototheca wickerhamii	+	+	+	+	−	+	+	−	−	+	+	+	+	+	+	+	?	+
Chlamydomonas reinhardtii	+	+	−	+	−	+	+	−	−	+	+	+	−	−	−	−	−	−
Paramecium aurelia	+	+	+	+	−	+	−	+	+	+	+	+	+	+	−	−	−	+
Trypanosoma brucei	+	+	+	+	−	+	−	+	−	+	+	+	+	+	−	+	−	+

Modified from Gray (1992, Table II), which should be consulted for further details.

[a]Gene designations: *ndh*, subunit of NADH dehydrogenase (complex I); *cytb*, apocytochrome (component of complex III); *cox* (subunit of cytochrome oxidase (complex IV); *atp* (subunit of ATP synthase). Other ORFs exist in certain of these mitochondrial genomes, some of which could relate to respiration (see Gray, 1992). Symbols: +, gene present; −, gene absent; P, pseudogene. Where no gene symbol appears, insufficient data are available to decide.

[b]The *atp8* gene is absent in nematode mitochondrial genomes.

Table 3–4. Translation Apparatus Genes Encoded by Mitochondrial DNA[a]

Taxa	rrn			trn	rps												rpl				
	SSU	LSU	5S		1	2	3	4	7	8	10	11	12	13	14	19	2	5	6	14	16
Animals[b]	+	+	−	22	−	−	−	−	−	−	−	−	−	−	−	−	−	−	−	−	−
Fungi																					
Aspergillus nidulans	+	+	−	28	−	−	−	−	−	−	−	−	−	−	−	−	−	−	−	−	−
Neurospora crassa[c]	+	+	−	27	−	−	−	−	−	−	−	−	−	−	−	−	−	−	−	−	−
Podospora anserina	+	+	−	27	−	−	−	−	−	−	−	−	−	−	−	−	−	−	−	−	−
Saccharomyces cerevisiae[c]	+	+	−	24	−	−	−	−	−	−	−	−	−	−	−	−	−	−	−	−	−
Schizosaccharomyces pombe	+	+	−	25	−	−	−	−	−	−	−	−	−	−	−	−	−	−	−	−	−
Allomyces macrogynus	+	+	−	25(+?)	−	−	−	−	−	−	−	−	−	−	−	−	−	−	−	−	−
Plants																					
Angiosperms[d]	+	+	+	15–20	+	−	+	−	−	−	+	+	+	+	+	+	+	+	+	−	+
Bryophyte	+	+	+	27	+	+	+	+	+	+	+	+	+	+	+	+	+	+	+	+	+
Protists and algae																					
Phytophthora infestans	+	+	−	25(+?)	−	?	?	+	+	−	+	−	+	+	+	+	−	−	+	−	+
Prototheca wickerhamii	+	+	+	25(+?)	−	+	−	?	+	−	+	−	+	+	+	+	+	+	+	−	+
Chlamydomonas reinhardtii[d]	+	+	−	3	−	−	−	−	−	−	−	−	−	+	−	−	−	−	−	−	−
Paramecium aurelia[d]	+	+	−	4	−	−	−	−	−	−	−	−	−	−	+	−	+	−	−	+	−
Trypanosoma brucei[d]	+	+	−	0	−	−	−	−	−	−	−	−	−	−	−	−	−	−	−	−	−

Modified from Gray (1992, Table III), which should be consulted for further details. *Paramecium* data from Cummings (1992) and cnidarian data from Wolstenholme (1992).

[a]Gene designations: *rrn*, ribosomal RNA (SSU, small subunit; LSU, large subunit); *trn*, transfer RNA; *rps*, SSU ribosomal protein; *rpl*, LSU ribosomal protein. Symbols as in Table 3–3.

[b]Twenty-two tRNAs are coded by all animal mitochondrial genomes except cnidarians (represented by a sea anemone) and are sufficient to translate all codons. tRNA import must be invoked in the sea anemone since only three tRNAs are encoded by the mitochondrial genome.

[c]The *N. crassa* mitochondrial genome encodes a novel mitochondrial *rps* (S-5), which shares limited sequence similarity with a mitochondrially encoded *rps* of *S. cerevisiae* (*var-1*), but neither protein has any homology to bacterial ribosomal proteins. The *S. cerevisiae* mitochondrial genome also encodes an RNA component of mitochondrial RNAse P (Chapter 15).

[d]Taxa in which mitochondrial tRNA import must occur for protein synthesis to take place.

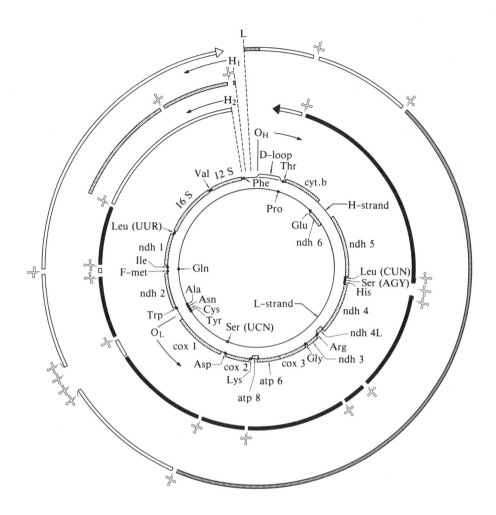

Fig. 3–1. Genetic and transcription maps of the human mitochondrial genome. The two inner circles show the positions of the two rRNA genes (12 S and 16 S), 14 tRNAs (black circles), and 12 reading frames transcribed from the heavy (H)-strand, and the positions of eight tRNA genes and one reading frame transcribed from the light (L)-strand. In the outer portion of the diagram, curved black bars represent the identified functional RNA species other than tRNAs resulting from processing of the two polycistronic primary transcripts of the H-strand starting at H_1 (rDNA transcription unit) and H_2 (total H-strand transcription unit). Cross-hatched bars represent the identified RNA species resulting from processing of the primary transcript of the L-strand. The white bars represent unstable presumably nonfunctional by products. Gene symbols: *cox1*, *cox2*, and *cox3* encode subunits I, II, and III of cytochrome oxidase, respectively; *cytb* encodes apocytochrome *b*; *ndh1, ndh2, ndh3, ndh4, ndh4L, ndh5,* and *ndh6* encode subunits of NADH dehydrogenase; *atp6* and *atp8* encode subunits of the F_0 portion of the ATP synthetase; O_H, O_L are the origins of H- and L-strand synthesis, respectively; tRNA genes are Ala (alanine), Arg (arginine), Asp (asparagine), Cys (cysteine), Gln (glutamine), Glu (glutamic acid), Gly (glycine), His (histidine), Ile (isoleucine), Leu (leucine), Lys (lysine), F-met (formylmethionine), Phe (phenylalanine), Pro (proline), Ser (serine), Thr (threonine), Try (tryptophan), Tyr (tyrosine), Val (valine). Modified after Attardi and Schatz (1988).

separated by only a few nucleotides. Most genes possess incomplete termination codons with a T or TA following the last sense codon. The termination codon, TAA, is completed posttranscriptionally by polyadenylation. Two pairs of H strand genes overlap. The 3' end of *atp6* overlaps the 5' end of *atp8* [by 46 base pairs (bp)in human mtDNA]. Similarly, *ndh4L* and *ndh4* overlap out of frame [by 7 bp in human mtDNA]. H strand and L strand genes overlap little if at all. The exceptions are *ndh5* and *ndh6*, which overlap 14 and 17 nucleotides, respectively, in mouse and cow. Another distinctive feature of the mammalian mitochondrial genome is the scattered distribution of the tRNA genes. These separate most of the rRNA and protein-coding genes. The tRNA structures in the primary transcripts appear to serve as signals for RNA processing.

In mammals a major fraction of the mtDNA molecules has a triple-stranded displacement loop (D-loop) region about 600 bp long found at a unique site on the molecule (Fig. 3–1). The D-loop or control region of the mammalian mitochondrial genome does not have a coding function, but the origin of H strand replication and the promoters for H and L strand transcription are in the D loop.

Mitochondrial Genomes of Other Animals

In addition to the five mammals described above, complete nucleotide sequences of the mitochondrial genomes of the chicken, the African Clawed Toad *(Xenopus laevis)*, two sea urchins, an insect *(Drosophila yakuba)*, three nematode worms, and a cnidarian have been reported (see Wolstenholme, 1992, for references). Partial mtDNA sequences have been obtained from a variety of other vertebrates and invertebrates. Each completely sequenced animal mitochondrial genome has the same gene content as mammalian mtDNA (Tables 3–3 and 3–4; Fig. 3–1) with the following exceptions (Wolstenholme, 1992). An *atp8* gene has not been found in the three sequenced nematode mitochondrial genomes. The mitochondrial genome of one nematode *(Meloidogyne javanica)* contains an ORF of 116 codons downstream of the *cox2* gene. Evolutionarily the most basal animal mitochondrial genome so far sequenced is also the most bizarre. The mitochondrial genome of a sea anemone *(Metridium senile)* contains but two tRNA genes. They specify tryptophan and formyl-methionine. Partial sequences of two other cnidarian mitochondrial genomes have also revealed the presence of only two tRNA genes. Thus, tRNA import must occur into cnidarian mitochondria for protein synthesis to take place as is also true of mitochondria of ciliates, kinetoplastid flagellates, flowering plants, and the green alga *Chlamydomonas*. Also, the *cox2* and *ndh5* genes of sea anemone each contains a group I intron. These are the only known introns in any animal mitochondrial genome. The *cox2* intron includes an ORF that may specify a protein required either for intron splicing or transposition.

Each animal mitochondrial genome also contains a region varying in length from 125 bp to 20 kbp that lacks genes and is equivalent to the control region of the mammalian mitochondrial genome. Variation in length of the control region accounts for most of the variability in animal mitochondrial genome size. In some cases this variation can be attributed to tandemly arranged repeats within the control region (see Wolstenholme, 1992 for references). For example, in the cricket *Gryllus firmus* there are one to seven copies of a 220 bp repeat whereas in lizards of the genus *Cnemidophorus* the repeat is 64 bp and is found in three to nine copies. In the root knot nematode *(Meloidogyne javanica)* three sets of tandemly arranged direct repeat sequences that differ in length (102, 63, and 8 bp) are found in a 7 kb region of the mitochondrial genome that is devoid of genes and may be the control region.

Mitochondrial gene rearrangements are apparent in comparisons of animals belonging to different phyla (Wolstenholme, 1992). Thus, Clary and Wolstenholme (1985) reported the sequence of the *Drosophila yakuba* mitochondrial genome and compared the gene arrangement to that of the mouse. They noted that five protein genes, the two rRNA genes, and 11 tRNA genes were aligned differently in the two genomes. One major difference in gene order may have resulted from translocation and inversion of a single segment containing the *ndh1*, small rRNA, tRNA^{val}, and large rRNA genes. A second rearrangement between the two genomes seems to have arisen from a single inversion of a segment containing the *ndh4L*, *ndh4* tRNA^{his}, and *ndh5* genes. Sequencing of the mitochondrial genomes of two nematodes (*C. elegans* and *A. suum*) reveals that their gene order is identical. Otherwise, gene arrangements among lower invertebrates are poorly conserved both with respect to each other and with regard to vertebrates, echinoderms, and insects (as represented by *Drosophila yakuba*).

Studies of animal mitochondrial genomes have revealed variations in the way the mitochondrial code is read and unusual features of tRNA structure (see Chapter 13 and 14).

The Mitochondrial Genomes of Plants and Algae

The largest, most spacious and complex mitochondrial genomes by far are those of angiosperms (see reviews by Fauron *et al.*, 1991; Hanson and Folkerts, 1992; Lonsdale, 1989; Palmer, 1992a; Newton, 1988). However, the great size of these circular mitochondrial genomes is not reflected by a proportional increase in gene content. In contrast, the linear mitochondrial genome of *Chlamydomonas reinhardtii* is only 15.8 kb and has one of the lowest gene contents of any mitochondrial genome (see Boynton *et al.*, 1992). Significantly, the completely sequenced mitochondrial genome of the liverwort (*Marchantia polymorpha*) is quite distinct in structure and gene content from both *Chlamydomonas* and flowering plants (Oda *et al.*, 1992).

The liverwort mitochondrial genome. The 94 possible genes sequenced in the liverwort mitochondrial genome make it richer in genes than any other known mitochondrial genome (Oda *et al.*, 1992). These genes encode three species of rRNA (26S, 18S, and 5S), 29 tRNA genes specifying 27 species of tRNA, and 30 genes encoding identifiable proteins. The latter include 16 ribosomal proteins, three subunits of the ATP synthase, the three largest subunits of cytochrome oxidase, apocytochrome *b*, and seven subunits of NADH dehydrogenase homologous to those encoded in the mitochondrial genomes of animals. There are also 42 additional open reading frames (ORFs). Four ORFs have homology to other ORFs, two of which are found in flowering plant mitochondrial genomes. Eleven ORFs may encode proteins involved in intron removal (maturases) or transposition.

Two features of liverwort mitochondrial genome structure are of particular note. The first is the presence of two clusters of genes encoding ribosomal proteins (Table 3–5). One encodes two proteins with homology to *E. coli* S7 (*rps7*) and S12 (*rps12*) and belonging to the *str* operon of that bacterium. The second cluster contains 12 genes encoding proteins with homology to *E. coli* genes in S10, *spc*, and α operons in the

Table 3-5. Ribosomal Protein Gene Clusters in *E. coli* and Several Organelle Genomes[a]

	str	S10		spc	α
	S12-S7-fus-tufA	S10-L3-L4-L23-L2-S19-L22-S3-L16-L29-S17	L14-L24-L5-S14-S8-L6-L18-S5-L30-L15-secY-L36		S13-S11-S4-rpoA-L17
Mitochondrion					
Liverwort	S12-S7-fus-tufA	S10 ——— S3-L16	L5-S14-S8-L6		S13-S11
Chloroplast					
Liverwort	S12[b]-S7	L23 — L2-S19-L22-S3-L16	L14 ——— S8 ——— L36		S11 —— rpoA
Tobacco	S12[b]-S7	L23 — L2-S19-L22-S3-L16	L14 ——— S8		S11 —— rpoA
Epifagus	S12[b]-S7	L23[c]-L2-S19 ——— S3-L16	L14[c] ——— S8		
Euglena	S12-S7	L23 — L2-S19-L22-S3-ORF-L16	L14 ——— S8 ——— L36 — S14-S2		
Porphyra	tufA	L2-S19-L22-S3-L16-L29-S17	L14-L24-L5 —— S8-L6 ———— secY		—— rpoA
Cyanophora	tufA	L3 — L2-S19-L22-S3-L16	L14 ——— L5 ——— S8-L6-L18-S5 ———— secY		

Data compiled from Bryant and Bohnert (1993), Fukuzawa *et al.* (1988), Hallick *et al.*, (1993), Oda *et al.* (1992), Reith and Munholland (1993), Sugiura (1989), Takemura *et al.* (1992), and Wolfe *et al.* (1992a).

[a] Genes set in italics are not found in the corresponding *E. coli* operons.

[b] *rps12* is a split gene whose mRNA is trans-spliced. This sequence represents the 3' half of the gene.

[c] Pseudogene.

same order that they are found in E. coli (Table 3-5; Takemura et al., 1992). Several genes in each operon are missing in the liverwort mitochondrial genome and are presumably located in the nucleus. The organization of these ribosomal protein-coding genes is strikingly like similar clusters seen in the chloroplast genomes of liverwort and flowering plants except that certain genes retained in the mitochondrial genome are not present in the chloroplast genome and vice versa. Second, a total of 32 group I and group II introns were found in 17 liverwort mitochondrial genes. Ten of these introns possessed ORFs. The ORFs found in these introns encode proteins with homology to proteins that catalyze movement of group I (endonucleases) and group II (reverse transcriptases) introns in other organisms (Chapter 10). Interestingly, the liverwort *cox1* gene, like its counterpart in yeast, is rich in introns, containing six belonging to group I and three of the group II type (Ohta et al., 1993). The insertion sites of one group I intron and four group II introns correspond to those of the respective fungal mitochondrial introns. The complete or fragmented ORFs found in these liverwort introns also relate to those found in yeast (Chapter 10).

Flowering plants. The mitochondrial genomes of angiosperms possess several features that distinguish them from the liverwort mitochondrial genome. First, although the liverwort mitochondrial genome possesses a number of repetitive sequences ranging up to 800 bp in size, they appear to be too short to catalyze recombination as is true of the longer "recombination repeats" found in angiosperms. Second, angiosperm mitochondrial genomes have incorporated plastid DNA sequences extensively as discussed in the next chapter, but foreign gene transfer to the mitochondrial genome of liverwort is not evident. Third, C → U RNA editing occurs in the mitochondria of flowering plants (Chapter 14), but probably not in liverwort since C residues that require editing in higher plants are T residues in liverwort.

Angiosperm plant mitochondrial genomes vary in size from 208 kb in *Brassica hirta* (white mustard) to 2400 kb in *Cucumis melo* (muskmelon) (Table 3-1). The mtDNA has a remarkably uniform buoyant density ranging from 1.705 to 1.707 g/cm^3 (Lonsdale, 1989). Initially, this uniformity in base composition seemed likely to reflect a corresponding physical uniformity since linear 30-μm (110-kb) molecules isolated from pea had molecular weights in good agreement with those obtained by renaturation kinetics (112 kb). However, a few years later it became apparent that this view was wrong and that the physical organization of plant mitochondrial genomes was far more complex. The main observations leading to this conclusion were these. First, depending on species and method of analysis, angiosperm mitochondria yielded both circular and linear forms of mtDNA that varied in both proportion and length. Second, kinetic complexity measurements of mtDNA from angiosperms gave widely different molecular weight estimates. Third, restriction endonuclease patterns of angiosperm mtDNAs were complex and fragments were often not stoichiometric in amount.

Resolution of this confusing situation began with a careful study by Ward et al. (1981). The mitochondrial genomes of pea, maize, and four species of cucurbits were compared using two independent methods of analysis (renaturation kinetics and summation of molecular weights of restriction fragments). These experiments yielded similar estimates of genome size except in the cases of cucumber and muskmelon where restriction enzyme analysis could not be used because of the extreme complexity of mtDNA restriction patterns. The results revealed a 7-fold range in mitochondrial

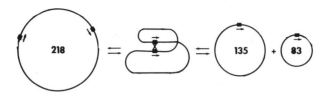

Fig. 3–2. Organization of the *Brassica campestris* mitochondrial genome into three circular molecules and their interconversion via recombination within the 2-kb repeat. Solid boxes indicate the position of the repeat and arrows its orientation. Adapted from Palmer and Shields (1984).

genome size among the cucurbits alone (340–2400 kb) indicating that changes in mitochondrial genome size can take place quite rapidly among related species of plants.

Because angiosperm mitochondrial genomes were so large they were among the last organelle genomes to be mapped physically. Palmer and Shields (1984) published the first restriction map of an angiosperm mitochondrial genome. They wisely chose to map the mitochondrial genome of *Brassica campestris* (Chinese cabbage, turnip), one of the smallest known. Palmer and Shields found that their map was consistent with a model that postulated the existence of a single "master circle" of 218 kb (Fig. 3–2). Within this molecule were a pair of 2-kb direct repeats (recombination repeats) separating unique sequence regions of 135 and 83 kb. The model proposed that reciprocal recombination between direct repeats yielded 135- and 83-kb circles in stoichiometric amounts. Similarly, black mustard *(B. nigra)* has a 231-kb master circle and subgenomic circles of 135 and 96 kb while radish *(Raphanus sativa)* possesses a master circle of 243 kb that interconverts into subgenomic circles of 139 and 103 kb (Palmer and Herbon, 1986). Unlike the mitochondrial genomes of the foregoing species, the 208-kb master circle of *Brassica hirta* does not recombine to form subgenomic cirles since the *B. hirta* mitochondrial genome lacks repeat elements (Palmer and Herbon, 1987). This is the simplest angiosperm mitochondrial genome known.

The maize mitochondrial genome has attracted particular attention because the determinants responsible for the cytoplasmic male sterile phenotype are located in its mitochondrial genome (Chapter 11). The various types of mitochondrial genomes of normal fertile (N) cytoplasm and cytoplasmic male sterile type T (cms T) maize have

Fig. 3–3. (A) Recombination between inverted and directly repeated sequences. Four unique sequences represented by A, B, C, and D flank a repetitive sequence (open box). The relative orientations of the repetitive elements are indicated (arrows). (I) recombination between inverted repeats leads to sequence inversion ("flip-flop"). (II) Recombination between direct repeats leads to the formation of subgenomic circles ("loop-out"). (B) Multicircular organization of the maize mitochondrial genome. The master circle of 570 kb has six pairs of repeats (open boxes), numbered 1 to 6. The relative orientation of the repeats is indicated (arrows). The master circle of 570 kb can exist in two isomeric forms, A and B. These result from recombination between two copies of repeat 4, which have inverted orientation. The resulting sequence inversion between the two copies of repeat 4 alters the relative orientation of repeat 5. Recombination between copies of repeats 1, 2, and 6 illustrates the formation of subgenomic circles. From Lonsdale (1989).

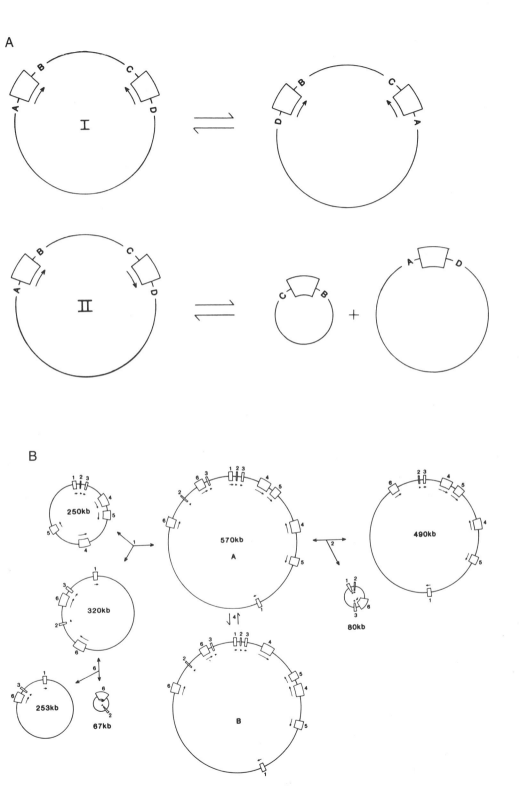

been characterized. Different forms of the maize mitochondrial genome arise as a consequence of (1) intramolecular recombination between directly repeated elements that results in sequence inversion (isomerization) of the master molecule and certain subgenomic circles or (2) recombination between direct repeats that yields subgenomic circles (Fig. 3–3). Both processes are reversible. The 570 kb N cytoplasm master circle has six sets of repeats (Lonsdale et al., 1984; Lonsdale, 1989). The cmsT cytoplasm mitochondrial genome is even more complex than that of N cytoplasm since six different permutations of the master circle alone are possible (Fauron et al., 1991). Small et al. (1989) proposed a three-stage general model for the formation of novel duplications in plant mtDNA. The process initiates by infrequent homologous recombination between short direct repeats in the progenitor master circle followed by further recombination between direct repeats in subgenomic circles to yield a new master circle. This model is based on comparisons of the N genome of maize and its ancestral (RU) form.

Some of the salient features of recombination repeats in angiosperm mitochondrial genomes are as follows (see Hanson and Folkerts, 1992). First, most recombination repeats are 1–10 kb, but not all repeats are recombination repeats. Second, a recombination repeat in one species may not serve this function in a second species. For instance, in *Brassica campestris* and *B. oleracea* the recombination repeats are 2 kb directly repeated sequences, but in *B. nigra* and the related *Raphanus sativa* this 2-kb sequence is present as a single copy. In *R. sativa* the recombination repeat is a 10-kb sequence. Embedded within this 10-kb sequence is a 7-kb sequence that serves as the *B. nigra* recombination repeat. Third, The orientation of the recombination repeats (direct or inverted) determines whether they will yield subgenomic circles or isomers on recombination as mentioned above. Fourth, two contrasting hypotheses have been proposed to explain the mechanism of recombination at the repeated sequence: simple homologous recombination versus recombination requiring sequence specificity. Fifth, it is not known whether recombination repeats are essential in those mitochondrial genomes where they are found and no other function than recombination has been ascribed to any recombination repeat.

The patterns of intramolecular recombination described here are not unique to flowering plant mitochondrial genomes. Recombination within the inverted repeat of the chloroplast genome results in its isomerization and recombination between direct repeats yields subgenomic mtDNA molecules in certain fungi that can result in mitochondrial disease (Chapter 11).

Tissue culture manipulations can promote extensive alterations in plant mtDNA (see Hanson and Folkerts, 1992 and Shirzadegan et al., 1989 for references). Experiments with *Brassica campestris* suggest that rapid structural alterations of mtDNA seen in tissue culture cells are a consequence of preferential amplification and reassortment of mitochondrial genomes ("sublimons") already present in low abundance in the intact plant (Shirzadegan et al., 1989, 1991). Cytoplasmic mixing resulting from protoplast fusion, either interspecific or intergeneric, also leads to formation of novel mitochondrial genomes as a result of mitochondrial fusion and mtDNA recombination (see Hanson and Folkerts, 1992; Lonsdale, 1989).

There are still many unanswered questions concerning angiosperm mitochondrial genome structure. As Hanson and Folkerts (1992) put it in their review:

> It is evident that a major inconsistency exists between present microscopic and electrophoretic characterization of mtDNAs and restriction mapping analysis. When a genome

map is assembled only from cosmid walking, a linkage map could possibly be constructed that does not correspond to the major configuration of the genome *in vivo*. Alternatively, it may be that the physical state of plant mtDNA *in vivo*—perhaps complexed with proteins or associated membranes—prevents isolation of intact molecules by current methods, and the predictions of molecular configuration made by restriction mapping methods may turn out to represent the actual genome organization *in vivo*.

Angiosperm mitochondrial genes are widely scattered and for the most part solitary (Gray, 1989; Hanson and Folkerts, 1992). The genes discovered so far include those encoding the rRNAs (including a 5S rRNA unique to plant mitochondria), numerous mitochondrial tRNAs, the three largest subunits of cytochrome oxidase, cytochrome *b*, several ATPase and NADH subunits, a few mitochondrial ribosomal proteins, and some ORFs (Tables 3-3 and 3-4). In several species ORFs resembling reverse transcriptase genes have been identified. The noncoding spacer DNA separating plant mitochondrial genes is not A+T rich as is true of fungal mtDNA, but instead has much the same base composition as the bulk mtDNA (Gray, 1989). Angiosperm mitochondrial genomes also commonly contain chloroplast DNA sequences and occasionally nuclear DNA sequences, as will be discussed in Chapter 4.

Chlamydomonas. The mitochondrial genome of *C. reinhardtii* has been completely sequenced and its structure reviewed by Gray and Boer (1988) and Boynton *et al.* (1992). This small (15.8-kb) linear mitochondrial genome possesses terminal inverted repeats of 531 to 532 bp in length including 39 to 41 nt single-stranded 3' noncomplementary extensions of the left and rights ends that are identical in sequence (Ma *et al.*, 1992; Vahrenholz *et al.*, 1993). The *C. reinhardtii* mitochondrial genome encodes cytochrome *b*, cytochrome oxidase subunit I, five subunits of NADH dehydrogenase, and a protein resembling a reverse transcriptase (Table 3-3). Thus cytochrome oxidase subunits II and III, which are encoded by virtually all other mitochondrial genomes, must be imported. None of these protein coding genes is interrupted in *C. reinhardtii*. However, the *cytb* gene of the interfertile strain, *C. smithii*, possesses a group I intron that encodes an endonuclease catalyzing its transposition into the intron-free *cytb* gene of *C. reinhardtii* (Chapter 10). Only three tRNAs are encoded by the *C. reinhardtii* mitochondrial genome, implying that most tRNAs are imported (Chapter 13). Undoubtedly, the most novel feature of this unusual mitochondrial genome is the structure of the rRNA genes (Boer and Gray, 1988). The small subunit rRNA appears to consist of four separately transcribed sequences that can be assembled on paper into a composite structure that conforms to the secondary structure model for *E. coli* 16S rRNA (Gutell *et al.*, 1985). Similarly, there are eight interacting segments of large subunit rRNA. The small subunit rRNAs are transcribed in the same order they occupy in the proposed assembled structure, but are separated by two unordered segments of large subunit rRNA, two tRNAs and two protein-coding genes. Interrupted rRNA genes have since been found in the mitochondrial genome of *Plasmodium* as discussed later.

The *C. eugametos/C. moewusii* species pair have circular mitochondrial genomes of 24 and 22 kb, respectively (Lee *et al.*, 1991; Denovan-Wright and Lee, 1992). Hybridization experiments suggest gene content is similar to *C. reinhardtii* and the rRNA genes are probably interrupted. The colonial volvocalean alga *Pandorina morum* is more closely related to *C. reinhardtii* than to the *C. eugametos/C. moewusii* species pair. The mitochondrial genome of *P. morum* is a 20-kb linear molecule (Moore and Coleman, 1989).

Other algae. Gray (1992) reviews what little is known about the structure of other algal mitochondrial genomes. A 76-kb mtDNA has been described from a *Chlorella*-like alga that contains several genes (*cox2, atpA, atp6, atp9*) not found in *C. reinhardtii* mtDNA. Mitochondrial rRNA genes have been characterized in *Prototheca zopfii*, a colorless relative of *Chlorella* and in *Scenedesmus obliquus*. In contrast to *Chlamydomonas*, the rRNAs encoded by these genes appear to be continuous. Coleman *et al.* (1991) have identified a 40 kb-linear mitochondrial genome with 1.6-kb terminal inverted repeats in the chrysophyte alga *Ochromonas danica*. Two related algae also have linear mitochondrial genomes similar in size.

Fungal Mitochondrial Genomes

The ascomycete fungi include some of the best model systems for investigating mitochondrial genome function (the yeasts *Saccharomyces cerevisiae* and *Schizosaccharomyces pombe* and the filamentous species *Aspergillus nidulans, Neurospora crassa*, and *Podospora anserina*). Ascomycetes possess mitochondrial genomes that are variable in size, often spacious, and usually circular (Tables 3–1, 3–2). The structures of fungal mitochondrial genomes in general and ascomycete mitochondrial genomes in particular are reviewed respectively by Clark-Walker (1992) and Wolf and Del Giudice (1988).

The mitochondrial genome of bakers' yeast (Saccharomyces cerevisiae). S. *cerevisiae* has long occupied a central position in investigations of mitochondrial genome function. The mitochondrial genome of bakers' yeast is much larger than the mammalian mitochondrial genome ranging from 68 to 85 kb in different strains (Tables 3–1 and Table 3–2), but its gene content is probably lower (Tables 3–3 and 3–4).

Like the mammalian mitochondrial genome, the mitochondrial genome of yeast encodes the two mitochondrial rRNAs, a complete set of tRNAs, the three large subunits of cytochrome oxidase, cytochrome *b*, and ATPase subunits 6 and 8 (Fig. 3–4; Tables 3–3 and 3–4). However, *atp9* is located in the mammalian nucleus whereas it is a mitochondrial gene in yeast. A gene encoding a mitochondrial ribosomal protein *(var1)* is also found in yeast and some other fungi. On the other hand, the *ndh* genes are absent from the yeast mitochondrial genome presumably because the complex containing these proteins is missing in yeast (Chapter 1). The larger size of the *S. cerevisiae* mitochondrial genome can be attributed to the presence of features lacking in most animal mitochondrial genomes such as lengthy intergenic spacers, introns, and intergenic ORFs. These ORFs are of questionable importance as mutations have not been assigned to them and they are missing from the mitochondrial genomes of the yeasts *Candida (Torulopsis) glabrata* and *Schizosaccharomyces pombe* (Clark-Walker, 1992). Introns are found in three "mosaic" genes of *S. cerevisiae* encoding cytochrome *b (cytb)*, cytochrome oxidase subunit I *(coxI)*, and the large rRNA molecule *(lg-rRNA)* (Chapter 10).

The introns present in the three mosaic genes of bakers' yeast are referred to as optional since their presence or absence generally has no effect on the phenotype of the cell. The short form of the *cytb* gene has two introns and three exons, but in the long form the first exon of the short form is fragmented into four exons separated by three introns. The short form of the *cox1* gene is almost 10 kbp in length, but the seven

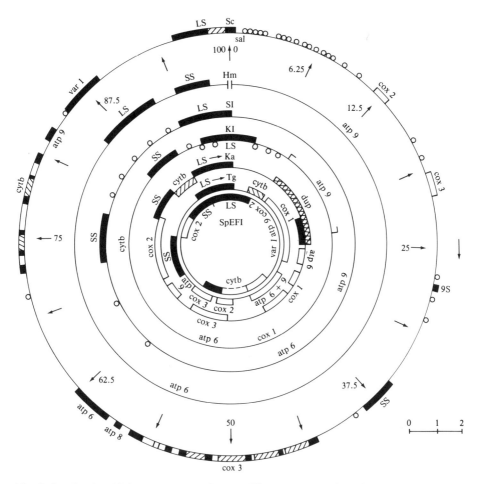

Fig. 3–4. Mitochondrial genome maps of seven different yeast species. The maps are presented as concentric circles drawn approximately to scale where the circumference of each circle is proportional to the length of the genome. The maps are scaled in 100 total units so that all map distances are percentages of the total map, indicated by arrows on the *Saccharomyces cerevisiae* map. The same scale may be used for all of the maps because the full map is always considered to be 100 units. For *Schizosaccharomyces pombe* the scale is reduced by 0.9 compared to the other genomes because *Candida (Torulopsis) glabrata* and *S. pombe* have mitochondrial genomes approximately equivalent in size (Table 3–2). Each species is oriented with a *Sal*I restriction site at the top of the map, with the exception of *Hansenula mrakii*, which has a linear map. The ends of the *H. mrakii* map are at the top. Species, in decreasing order of size, are (1) *Saccharomyces cerevisiae* KL14-4A (Sc), 77.8 kb; (2) *Hansenula mrakii* (Hm), 55 kb; (3) *Saccharomyces lipolytica* (Sl), 44 kb; (4) *Kluyveromyces lactis* (Kl), 37 kb; (5) *Kloeckeria africana* (Ka), 26.5 kb; (6) *Candida (Torulopsis) glabrata* CBS 138 (Tg), 18.9 kb; (7) *Schizosaccharomyces pombe* EF1 (Sp), 18.9 kb. tRNAs are indicated by circles on the map. Open regions within a block indicate intervening sequences. Gene abbreviations are given in the legend to Fig. 3–1 except for *atp9*, which encodes a subunit of the F_0 portion of the ATP synthetase 9, *var1*, which specifies a mitochondrial ribosomal protein, and 9S which encodes the RNA portion of RNase P (Chapter 13). The large and small subunit rRNAs are designated LS and SS, respectively. Modified after Sederoff (1984), which cites all of the original references. The *S. cerevisiae* map has been partially updated according to Attardi and Schatz (1988) with certain features including unidentified reading frames and origins of replication omitted.

exons constitute only 16% of the gene with the rest being occupied by six large introns. In the long form the fifth exon of the short form is split into three by two additional large introns. The shortest *cox1* gene is found in the closely related species *S. carlsbergensis,* which lacks introns 1 and 4. Many of these introns encode proteins required for splicing or intron transposition (Chapter 10). The 21S rRNA gene of ω^+ strains contains a 1143-bp group I intron encoding an endonuclease required for its transposition to ω^- strains lacking the intron (Chapter 10).

In contrast to animal mitochondrial genomes, which generally have G+C contents greater than 40%, the G+C content of yeast mtDNA is only 18%. Intensive study of this A+T-rich DNA, particularly by Bernardi and colleagues (see review by de Zamaroczy and Bernardi, 1985), led to the discovery of extreme compositional heterogeneity with G+C and A+T base pairs being tightly clustered. This finding led to the proposition that yeast mtDNA was organized into four distinct classes of elements. About 50% of the mtDNA is made up of AT-rich sequences containing less than 5% G+C ("spacers"). These spacers range in size from 150 to 1500 bp. About 2–3% of the genome consists of very short (50–80 bp) G+C rich (>50% G+C) sequences. These are of two types. Half, called "site clusters," contain a *Hae*III restriction site while the other half, named "GC clusters," possess an *Hpa*II restriction site. The rest of the mtDNA is composed of sequences with a size range similar to the spacers and a variable G+C content (average 26%). Bernardi and colleagues referred to the latter sequences as "genes." Because of the approximate equality in number of all four elements a regular arrangement of these elements was proposed: GC cluster—site cluster—gene—spacer.

With the advent of DNA sequencing, a more precise description of the structure and arrangement of these elements became possible. Bernardi's proposed genes indeed turned out to be coding sequences. However, the GC-rich clusters were not at the beginning of genes, but rather within the AT-rich intergenic or intragenic regions. The *var1* gene sequence discussed later (Chapter 10) more closely resembles "nongenic" DNA since it is approximately 90% AT.

Mitochondrial genomes of other ascomycetes. The mitochondrial genomes of the ascomycetes vary greatly in size. The yeast *Schizosaccharomyces pombe* EF1 has a small (17.3-kb) mitochondrial genome tightly packed with genes (Fig. 3–4), but in another yeast, *Brettanomyces custersii,* the mitochondrial genome is a 101-kb molecule (Table 3–2). In the genus *Saccharomyces* alone, mitochondrial genome sizes range from 23.7 to 85 kb in certain *S. cerevisiae* strains (Table 3–2). The yeasts *S. cerevisiae* and *C. glabrata* differ in size by roughly 60 kb, but their genic regions are more than 80% identical. The intergenic sequences in both species are A+T rich, but the spacers of *S. cerevisiae* are much larger than they are in *C. glabrata* accounting for at least 53 kb of the *S. cerevisiae* mitochondrial genome (Clark-Walker, 1985, 1992; de Zamaroczy and Bernardi, 1987). Yeasts belonging to the genus *Dekkera/Brettanomyces* have mitochondrial genomes ranging in size from 28.5 to 101 kb (Hoeben and Clark-Walker, 1986). This size variability again seems to result from differences in sizes of spacer regions. In *S. cerevisiae* variable numbers of GC clusters also contribute to differences in genome size (Sor and Fukuhara, 1982). Near perfect repeats of GC clusters have been reported (e.g., Sor and Fukuhara, 1982; Farelly *et al.,* 1982). The mitochondrial genome of the yeast *Kloeckeria africana* possesses a long inverted duplication containing a part of the large rRNA gene and some tRNA genes (Clark-Walker *et al.,* 1981).

Since at least 1 kb of the large rRNA gene is missing, this gene is nonfunctional. The presence or absence of certain genes is another source of variability. Thus, the *var1* gene, encoding a mitochondrial ribosomal protein, is present in the mitochondrial genomes of *S. cerevisiae* and *C. glabrata*, but absent from the mitochondrial genomes of *S. pombe* and *Aspergillus nidulans*. Two variant mitochondrial genome types have been found among laboratory strains of *Neurospora crassa* (Mannella et al., 1979). They are characterized by a 2.1-kb fragment that is tandemly repeated and inserted into the mtDNA. Additional structural variations were found in the related species *N. intermedia* and *N. sitophila* (Collins et al., 1981). Variant types usually contain insertions which may be as large as 8 kb. Many of the insertions are in the region of the *cytb* and *cox1* genes. Certain fungal mitochondrial genes contain optional introns while others do not. The underlying causes for size variation in organelle genomes are reviewed more extensively in the next chapter.

Mitochondrial gene organization also varies greatly in ascomycetes as is evident from comparing gene maps for different yeast species (Fig. 3–4). The rRNA genes may be adjacent as in *Schizosaccharomyces pombe* or separated by a region containing other genes, which is as much as 27 kb as in *S. cerevisiae* (Sederoff, 1984). Other species of yeasts fall between these two extremes and there is no evident proportionality between the degree of separation of the rRNA genes and genome size. The orientation of the rRNA genes varies in different species.

The mitochondrial genomes of the yeast genus *Dekkera/Brettanomyces* are particularly striking in size variation ranging from 28.5 kb (*B. custersianus*) to 101.1 kb (*B. custersii*) (Table 3–2). The organization of these mitochondrial genomes and their phylogeny has been studied by Clark-Walker and colleagues (Hoeben and Clark-Walker, 1986; Hoeben et al., 1993). The three smallest genomes (28–42 kb) have the same gene order, whereas the three larger mtDNAs (57–101 kb) are rearranged relative to each other and the smaller mitochondrial genomes. Comparisons of *cox2* sequences are consistent with a phylogeny in which the three smaller mitochondrial genomes are ancestral to the larger genomes. These latter mitochondrial genomes are more prone to rearrangement than the smaller molecules even though they seem to be more closely related phylogenetically.

The sizes of the mitochondrial genomes of three filamentous ascomycetes *Aspergillus nidulans, Neurospora crassa,* and *Podospora anserina* vary over 3-fold (Table 3–2). Gene orders also vary extensively so that no two marker genes are adjacent in all three species (Wolf and Del Giudice, 1988). As in the yeasts, large deletions and additions presumably account for size differences between the three genomes. In *N. crassa* comparison of wild-type strains reveals a size variation of 60–73 kb. As mentioned earlier, these variations can be accounted for by insertions that are sometimes as large as 8 kb and often in the region of the *cox1* and *cytb* genes. In both *Aspergillus* and *Neurospora* mitochondrial tRNA genes fall into two main clusters. The upstream cluster (between the rRNA coding sequences) contains nine genes with the second having ll tRNA genes being downstream of the large rRNA gene. With only three differences, the genes in this tRNA cluster have the same order in *Aspergillus* and *Neurospora*. In *S. cerevisiae* the main cluster of 19 tRNA genes is situated between the large rRNA and *cox2* genes. The *N. crassa* and *Aspergillus* mitochondrial genomes contain all of the *ndh* genes found in the mammalian mitochondrial genome (Table 3–3) The mitochondrial genome of *N. crassa* possesses 50–100 highly conserved GC-rich palindromes that flank most of its mitochondrial genes each containing two *PstI* sites (Yin et al., 1981). These do not serve as signals for RNA processing as proposed originally by Yin et al.(1981, see Wolf and Del Giudice, 1988, for references).

Oomycetes and hypochytridiomycetes. Although the *Oomycetes* have traditionally been classified as fungi (e.g., Clark-Walker, 1992) molecular phylogenetic data now indicate that they are better considered as protists (e.g., Gray, 1992) closely related to plants (Karlovsky and Fartmann, 1992). They are retained in this section together with the *Hypochytridiomycetes* following the treatment of Clark-Walker (1992) with the full realization that the *Oomycetes* and their allies will probably appear elsewhere in future classifications of organelle genomes.

The mitochondrial genomes of these related groups, which include the water molds *(Achyla)* and the potato blight organism *(Phytophora infestans)*, possess two unusual features (Clark-Walker, 1992). They normally contain sizable inverted repeats, like most chloroplast genomes, and the single copy length varies within a narrow range (36 to 45 kb). The smallest inverted repeat (0.5–0.9 kb) is found in *Phytophora megasperma,* but these repeats typically range from around 10 to 30 kb in size. Those fungi lacking the repeat structure are thought to have lost them secondarily as is the case for chloroplast genomes of certain legumes. Intramolecular recombination (isomerization or flip-flop) occurs within the inverted repeats of these mitochondrial genomes as it does within the inverted repeat of the chloroplast genome. The oomycete mitochondrial inverted repeat encodes the genes for the large and small rRNAs as is true of the chloroplast inverted repeat. The degree of parallelism between these mitochondrial genomes and most chloroplast genomes is eerie.

Basidiomycetes. Strains of *Agaricus bitorquis* possess the largest mitochondrial genomes yet found in fungi (176.3 kb), but the majority of filamentous basidiomycetes have mitochondrial genomes ranging from 35 to 76 kb in size (see Clark-Walker, 1992). The basidiomycetes characterized to date, with the exception of *Agaricus brunnescens,* lack an inverted repeat. In this species intramolecular recombination by isomerization occurs within the inverted repeat as it does in mitochondrial genomes of oomycetes and in chloroplast genomes.

The power of combining molecular phylogenetic analysis with mitochondrial genome mapping is illustrated by studies on the genus *Suillus* and its relatives (Bruns et al., 1988; Bruns and Palmer, 1989; Bruns et al., 1989). The size of the mitochondrial genome in 15 species of *Suillus* varies from 36 to 121 kb. Gene order is identical in all species examined except *S. luteus,* in which a single transposition of the small rRNA gene region accounts for the difference seen in gene order (Bruns and Palmer, 1989; regions 10–11, Fig. 3–5). Extension of this analysis to three related genera reveals additional rearrangements. The four different gene orders seen can be reconciled by postulating four rearrangements.

Phylogenetic comparisons in *Suillus* using restriction site differences in the mitochondrial ribosomal RNA gene regions alone or combined with two morphological characters yield a tree in which closely related species have similar-sized mtDNA molecules. This is consistent with the theory of Bruns et al. (1988) that large interspecific differences result from accretion of numerous smaller length mutations. The analysis also reveals that the related genera are more distant from *Suillus* than each of the *Suillus* species is from each other. Finally, there is a suggestion that one specific region of the

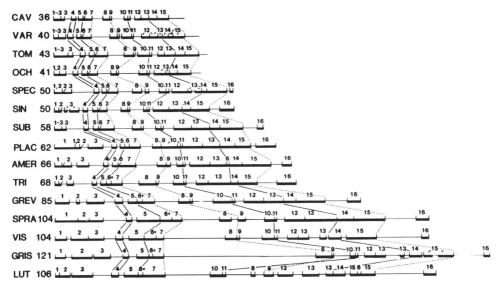

Fig. 3–5. Fragment order and size variation in the mitochondrial genomes of 15 *Suillus (Boletaceae)* species. The circular mtDNAs are linearized in approximate alignment to the positions of hybridization of 16 different probes. Since probe 16 overlaps and includes sequences that hybridize to probes 1 and 2, its position is shown only for regions to which it hybridizes uniquely. Genome sizes are given in kb after the species name abbreviation (*S. americanus*, AMER; *cavipes*, CAV; *grevillei*, GREV; *grisellus*, GRIS; *luteus*, LUT; *ochraceoroseus*, OCH; *placidus*, PLAC; *sinuspaulianus*, SIN; *spectabilis*, SPEC; *spraguei*, SPRA; *subalutaceus*, SUB; *tomentosus*, TOM; *tridentinus*, TRI; *variegatus*, VAR; *viscidus*, VIS). Solid lines connect restriction sites that border the same hybridization region and are shared by all *Suillus* species. Dotted lines connect hybridization borders that are defined by one of several closely spaced restriction sites. Note that distances between regions vary dramatically, but order differs only in *S. luteus* and only by single transposition of regions 10 and 11. From Bruns and Palmer (1989).

mitochondrial genome may be more susceptible to transposition or inversion than the rest of the mitochondrial genome.

An interesting corollary of these studies was the discovery that the mitochondrial genome of *Rhizopogon subcaerulescens* (Hymenogastraceae) bears a surprisingly close relationship to *Suillus* (Boletaceae) despite extensive morphological divergence (Bruns et al., 1989). These results imply a remarkably rapid rate of morphological change compared to the rate of molecular change during the evolution of these fungi.

These experiments, and similar investigations on chloroplast genome evolution in the genus *Chlamydomonas* (Chapter 4), illustrate the value of combining phylogenetic analysis with organelle genome characterization in less well characterized eukaryotes. The taxonomy of these organisms is often based on criteria that may not reflect true relationships with great precision. In the case of *Suillus* and *Rhizopogon* morphological differences are not a good guide, but in the case of *Chlamydomonas* morphological similarity is a poor indicator.

Protists

Probably because of an understandable interest in ourselves, our close relatives, and the plants we grow, scientists have tended to pay the most attention to the organelle genomes of the phylogenetically most advanced organisms. These tend to be highly conserved. This is certainly true of mammalian mitochondrial genomes and the chloroplast genomes of flowering plants. It is also true of the mitochondrial genomes of flowering plants in terms of gene content although the variability in size of these genomes, their ability to acquire foreign gene sequences, and the phenomenon of RNA editing (Chapter 14) certainly render them unconventional in many respects. To get some idea of the variability in gene content possible in organelle genomes and, perhaps, to glimpse some of the intermediates in organelle genome evolution one must study protists and algae. This section is devoted to the description of the mitochondrial genomes of three protistan groups, two of which have been studied in detail in part because they are important human and animal parasites (i.e., *Trypanosoma* and *Plasmodium*). *Plasmodium* also possesses a second organelle genome with striking similarities to the reduced chloroplast genomes of colorless plants and so provides an appropriate transition from mitochondrial to chloroplast genomes.

Paramecium. The structure and replication of mitochondrial genomes in ciliates have been reviewed by Cummings (1992). By far the most is known about the *Paramecium* mitochondrial genome, with *Tetrahymena* coming in second. As Cummings points out, phylogenetic tree construction using small cytoplasmic rRNA molecules reveals that ciliates constitute a loose phylogenetic assemblage with an ancient common ancestor. Although morphological criteria led to the grouping of *Paramecium* and *Tetrahymena* in the same order, rRNA phylogenies indicate that their relationships are not much closer than those of different classes of ciliates.

The linear 40.5-kb mitochondrial genome of *Paramecium aurelia* (species 4) has been sequenced completely (Pritchard *et al.*, 1990a, b) with the exception of a region containing the extreme termination sequence, which is probably only a few hundred base pairs in length (Cummings, 1992). The genome encodes the two mitochondrial rRNAs, cytochrome *b*, subunits I and II of cytochrome oxidase, one ATP synthase subunit (subunit 9), and five of the seven subunits of NADH dehydrogenase encoded by the mammalian mitochondrial genome (Tables 3–3 and 3–4). These are the main similarities to other mitochondrial genomes. The differences are more striking. Only three tRNAs are encoded in the *Paramecium* mitochondrial genome. The mtDNA of *Tetrahymena* is also deficient in tRNA genes and encodes only eight (Ziaie and Suyama, 1987). However, the *Paramecium* mitochondrial genome contains four genes with homology to genes encoding *E. coli* ribosomal proteins (*rpl2*, *rpl4*, *rps12*, and *rps14*). Three of these genes (*rpl2*, *rps12*, and *rps14*) are found in higher plant chloroplast genomes (see below). There is also a gene with homology with *ndhK* (*ndh11*) (originally misidentified in the chloroplast genome as *psbG*, see Arizmendi *et al.*, 1992). Another gene, ORF400, also encodes an NADH dehydrogenase subunit and is now designated *ndh7* (Gray, 1992). Although both of these genes are in the mammalian nucleus, they are chloroplast genes in land plants (Shimada and Sugiura, 1991). The three genes encoding cytochrome oxidase subunit III and ATP synthase subunits 6 and 8, which are commonly found in mtDNA, are lacking in the *Paramecium* mitochondrial genome.

Most *Paramecium* mitochondrial genes are highly diverged from their counterparts in other mitochondrial genomes (Cummings, 1992; Pritchard *et al.*, 1990a). Thus, the *ndh* complex shows only 16 to 35% identity to the same complex from other organisms. Sequence divergence may account for the large number of ORFs (17) in the *Paramecium* mitochondrial genome since identification of homologies is rendered difficult.

From the bizarre to the implausible: The Trypanosome kinetoplast. The *Paramecium* mitochondrial genome is bizarre, but the complex mitochondrial genome of the trypanosome kinetoplast and its edited transcripts (Chapter 14) are truly hard to imagine. Fortunately, for the bemused or intrigued there are numerous excellent recent reviews (e.g., Borst, 1991; Ryan *et al.*, 1988; Simpson, 1987; Simpson and Shaw, 1989; Stuart, 1991a, b; Stuart and Feagin, 1992).

The kinetoplastid protozoa, also knowns as trypanosomatids or trypanosomes, consist of a group of eight genera of flagellated protozoa belonging to the family Trypanosomatidae. Species of the monogenetic genera are parasitic on a single invertebrate host (e.g., *Crithidia*) whereas species belonging to the digenetic genera (e.g., *Leishmania, Trypanosoma*) are medically important human and animal parasites whose life cycles are usually divided between an invertebrate vector (e.g., insect, leech) and a vertebrate host. The diseases caused by these organisms include African trypanosomiasis (sleeping sickness), South American trypanosomiasis (Chagas' disease), and leishmaniasis (kala azar). Analysis of nuclear small rRNA sequences has led to the conclusion that trypanosomes represent a deep branch of the eukaryotic line (Sogin *et al.*, 1989).

The kinetoplast is found in the matrix of the cell's single, giant mitochondrion located at the base of the flagellum. It appears as a dark purple granule when stained with ordinary basic dyes used for DNA. Consistent with its prominence based on staining, kinetoplast DNA (kDNA) comprises ca. 7% of the total cellular DNA (Stuart and Feagin, 1992). kDNA is organized into networks of thousands of interlocking (catenated) DNA molecules. These are of two kinds, minicircles and maxicircles. There are approximately 5,000–10,000 minicircles and 20–50 maxicircles per network. The minicircles vary in size from 0.5 to 2.8 kb (Ryan *et al.*, 1988). There are 300–400 different minicircle sequence classes in one *Trypanosoma brucei*, 20 in *Crithidia fasciculata*, three in *Leishmania tarentolae*, and one each in *Trypanosoma evansi* and *T. equiperdum* (Stuart and Feagin, 1992). Although minicircles within a network may be heterogeneous in sequence, they always have a ca. 120-bp region that is conserved within different strains of the same species (Stuart and Feagin, 1992). A 13 nucleotide sequence embedded in the conserved sequences of all species examined may be required for minicircle replication. In different species, the conserved sequences may be present once per molecule, found as two direct repeats 180° apart, or four direct repeats 90° apart. Minicircles of most species also have a single major region of bent helix. The function of minicircles was a mystery until it was discovered that these molecules encoded guide RNAs (gRNAs) used for RNA editing (Chapter 14). Thus, the *T. brucei* minicircles contain three gRNA coding cassettes between 18 bp inverted repeats in addition to the conserved region (see Stuart, 1991a and Stuart and Feagin, 1992 for references). The 0.9-kb minicircles of *L. tarentolae* lack inverted repeats and may encode only one gRNA per molecule.

Maxicircles are the true mitochondrial genomes, but the existence of heavily edited transcripts of many genes (Chapter 14) has slowed the process of gene identification. Maxicircles range between 20 and 39 kb in size and contain genes encoding the rRNAs, cytochrome *b*, the three largest subunits of cytochrome oxidase, five subunits of NADH dehydrogenase, and possibly *atp6* (Stuart, 1991a, b; Stuart and Feagin, 1992). Of the *ndh* genes two (*ndh7* and *ndh8*) are encoded in vertebrate nuclear DNA, but in the chloroplast genomes of land plants. Apparently no tRNAs are encoded in either minicircle or maxicircle DNA although a complete set of tRNAs has been found in the kinetoplastid mitochondrion (Simpson and Shaw, 1989). There are also three conserved maxicircle unidentified reading frames (MURFs 1 and 2 and ORF 8/10). Additionally, six G vs C strand-biased sequences (CR1-6) exist whose positions, but not sequences, have been conserved among species (Stuart, 1991a; Stuart and Feagin, 1992). They are transcribed from the G biased strand and, with the exception of CR6, seem to specify small, hydrophobic proteins. CR6 may encode a heavily edited transcript for ribosomal protein S-12. The maxicircle also contains a large variable (VR) region that differs in sequence and size among different species and isolates. The VR is principally composed of A+T sequences and does not encode proteins, but does encode gRNAs. In *T. brucei* stocks, differences in A+T-rich repeats probably account for variations in size of the VR. Replication of the kDNA network is discussed in Chapter 5 and RNA editing in Chapter 14.

The strange story of the malaria parasite. The wonders that await the investigator inquisitive about the organization of organelle genomes in protists and algae are aptly illustrated by the story of the malaria parasite *(Plasmodium)*. This rapidly evolving tale has been chronicled in a number of recent papers (e.g., Aldritt *et al.*, 1989; Feagin *et al.*, 1991, 1992; Joseph *et al.*, 1989; Wilson *et al.* 1992) and two short reviews (Palmer, 1992b; Wilson *et al.*, 1991).

Plasmodium possesses two double-stranded extranuclear molecules. One of these is a multicopy linearly reiterated element having a 6-kb repeat size that is enriched in mitochondrial fractions (Wilson *et al.*, 1992). This genome has the smallest number of protein-encoding genes so far encountered in any mitochondrial genome (i.e., *cox1*, *cox3*, and *cytb*), the rRNA genes are very small and interrupted, and no tRNA-encoding genes have so far been reported (Feagin *et al.*, 1991, 1992). The sequences encoding the rRNA fragments correspond to about 50–60% of the conserved core sequenced in *E. coli* rRNA. Their fragmentation is reminiscent of that seen for the mitochondrial rRNA genes in *Chlamydomonas*.

The second species is a 35-kb circular molecule that is enriched in a cellular fraction less dense than the fraction containing mitochondria (Wilson *et al.*, 1992). This species possesses inverted repeats containing continuous rRNAs with secondary structures quite similar to those predicted for *E. coli* (Feagin *et al.*, 1992; Gardner *et al.*, 1993). This genome also encodes tRNAs, two subunits of a eubacterial-type RNA polymerase, and four ribosomal proteins (Feagin, 1994). Hence, the 35-kb genome bears some remarkable similarities to the chloroplast genome of the colorless, parasitic plant *Epifagus* (Beechdrops), which retains only chloroplast genes required for transcription and translation (see The Plastid Genomes of Colorless Plants and Algae: How Essential Is Chloroplast Protein Synthesis?). Phylogenetic comparisons with a gene encoding RNA polymerase subunit (*rpoC*) (Howe, 1992) and the large subunit rRNA gene (Gardner *et al.*, 1993) indicate that the 35-kb circular genome is more akin to plastid than to mitochondrial genomes. Palmer (1992b) lists this and the following five reasons for considering the 35-kb molecule as the remnant of a plastid genome of a photosynthetic eukaryote:

1. The only two organellar genomes known are those of plastids and mitochondria. The mere existence of a nonmitochondrial organelle genome is suggestive evidence for plastid origin.
2. Like most plastid DNAs, but only a limited number of mtDNAs, the 35-kb circle contains an rRNA-encoding inverted repeat. However, the arrangement and content of rRNA and tRNA genes within the *Plasmodium* repeat differ from those of other plastid genomes.
3. An operon of *rpo* genes encoding subunits of a eubacterial RNA polymerase is found in the *Plasmodium* 35-kb genome. An *rpoBC* operon is a universal feature of plastid genomes, but mtDNA is not known to contain any polymerase subunit encoding genes.
4. There are striking similarities between the 35-kb organelle genome of *Plasmodium* and those of colorless plants and algae in that virtually the only genes retained are those required for transcription and translation.
5. The 35-kb circle is conserved among malaria parasites and other apicomplexian protozoa (Wilson *et al.*, 1991). Molecular phylogenetic studies indicate that these protozoa may share a common ancestor with the dinoflagellates (e.g., Barta *et al.*, 1991).

CHLOROPLAST GENOMES

As Palmer (1991) presciently remarks in a review, one should keep two phylogenetic perspectives in mind in thinking about chloroplast genomes. "First, chloroplasts are undoubtedly of eubacterial, most likely cyanobacterial, origin" (Chapter 2). "Therefore, the inclusion of bacterial genomes in the comparisons drawn herein highlights both the primitive bacterial features of cpDNA as well as those derived subsequent to its marriage with the eukaryotic cell. Second, our knowledge of cpDNA is quite lopsided phylogenetically: The cpDNAs of a few land plants (mostly angiosperms) are extremely well characterized and those of >1000 other land plants moderately so, whereas cpDNAs of algae (particularly the great diversity of rhodophytes and chromophytes) are, with few exceptions, only poorly known. Thus, the present generalization that cpDNA evolution is in all respects conservative is based largely on consideration of land plants and may ultimately prove to be, at least in part, invalid for algal lineages." This bias is reflected in the following account, which begins with a discussion of the chloroplast genomes of two flowering plants and liverwort.

Land Plant Chloroplast Genomes

Two angiosperm chloroplast genomes have been completely sequenced: tobacco *(Nicotiana)*, a dicotyledon (Shinozaki *et al.*, 1986), and rice *(Oryza)*, a monocotyledon (Hiratsuka *et al.*, 1989). The liverwort chloroplast genome has also been sequenced in its entirety (Ohyama *et al.*, 1986). In striking contrast to the divergence seen between the mitochondrial genomes of liverwort and angiosperms discussed earlier the liverwort and angiosperm chloroplast genomes are very similar in organization and gene content. These complete sequences as well as restriction maps and partial sequences of a multitude of other land plants have yielded a detailed picture of these chloroplast genomes. The residual plastid genome of the colorless parasitic angiosperm *Epifagus* has also been sequenced (Wolfe *et al.*, 1992a, b) and is discussed later in this chapter (The Plastid Genomes of Colorless Plants and Algae: How Essential Is Chloroplast Protein Synthesis?).

Land plant chloroplast genomes normally possess two unique sequence regions and two large inverted repeats (typically 10–30 kb) containing the rDNA genes (Table 3–6 and Fig. 3–6). The two repeat sequences are always positioned asymmetrically dividing the genome into small (15–25 kb) and large (80–100 kb) unique sequence regions (Palmer, 1985). The 23 S rRNA gene is closest to the small single copy region and the 16 S rRNA gene closest to the large single copy region. The two repeats are identical in sequence as a consequence of an active copy correction system (Chapter 7 and Boynton *et al.*, 1992). Nearly two thirds of the 120–216 kb size variation in land plant chloroplast genomes is accounted for by the inverted repeat (Palmer, 1991). The smallest angiosperm chloroplast genomes, not counting those of colorless plants, are those of four separate groups of angiosperms lacking the inverted repeat (Table 3–6; Palmer, 1991). The best characterized of these are six tribes of legumes in the family Fabaceae in which the 18 genera examined all lack the inverted repeat. These plants have a chloroplast genome size of 120–130 kb. Other tribes in the Fabaceae retain the inverted repeat structure. At the other extreme is the geranium, *(Pelargonium hortorum)*, whose 216-kb chloroplast genome has a 76-kb inverted repeat that has swallowed up nearly half of the chloroplast genome resulting in duplication of many genes normally in the single copy region. While inversions occur very rarely in higher plant chloroplast genomes containing the inverted repeat (tobacco differs from rice by three inversions), there is a greater frequency of rearrangement in most legume chloroplast genomes that have lost the inverted repeat (see Chapters 4 and 7 for further discussion of the inverted repeat).

The completely sequenced chloroplast genomes from tobacco, rice, and liverwort contain about 110–120 genes (Fig. 3–6, Palmer, 1991; Sugiura, 1992). The majority of these genes are located in the large unique sequence region. The known products of most chloroplast genes function either to promote chloroplast gene expression (4 rRNAs, 30–31 tRNAs, 20 ribosomal proteins, 4 RNA polymerase subunits) or photosynthesis (28 thylakoid proteins plus one soluble protein, RUBISCO large subunit). Also, 11 subunits of bovine mitochondrial complex I have now been found to have homologs encoded by cpDNA providing support for the view that chloroplasts of land plants contain a complex similar to mitochondrial complex I that may be involved in chlororespiration (Chapter 1, Arizmendi *et al.*, 1992). Other chloroplast genes in one or more of the foregoing plants or in the colorless angiosperm *Epifagus* include a plastid homolog of the β subunit of the carboxyltransferase component of *E. coli* acetyl-CoA carboxylase *accD*, which catalyzes the first step in fatty acid synthesis in *E. coli* (Wolfe *et al.*, 1992a), *clpP*, the plastid homolog of the proteolytic subunit of the ATP-dependent Clp protease of *E. coli* (Wolfe *et al.*, 1992a); and a chloroplast envelope protein (Sasaki *et al.*, 1993). The *mbpX* and *Y* genes of *Marchantia* are homologous to cyanobacterial genes involved in sulfate uptake. Many land plants (e.g., gymnosperms, liverwort) and the green alga *Chlamydomonas* synthesize chlorophyll in the dark because they possess

ORGANELLE GENOME ORGANIZATION AND GENE CONTENT

Table 3–6. Size of Selected Chloroplast Genomes[a]

Taxa	Number of Species Studied	Size (kb) Genome	Size (kb) Inverted Repeat
Algae			
Chlorophyta			
Chlamydomonas	4	195–292	20–41
Chlorella	3	120–174	0–23
Codium	1	89	0
Spirogyra	1	130	0
Euglenophyta[b]			
Astasia	1	73	0
Euglena	1	130–152	0
Rhodophyta			
Griffithsia	1	178	0
Glaucophyta			
Cyanophora	1	127–136	10–11
Phaeophyta			
Dictyota	1	123	5
Pylaeiella	1	191	5
Xanthophyta			
Vaucheria	1	124	6
Bacillariophyta			
Coscinodiscus	1	118	9
Cyclotella	1	128	17
Odontella	1	118	9
Chrysophyta			
Ochromonas	1	121	15
Olisthodiscus	1	150	22
Cryptophyta			
Cryptomonas	1	118	6
Land plants			
Bryophyta			
Marchantia[c]	1	121	10
Physcomitrella	1	122	9
Ferns	50 in 6 families	140–160	10–24
Gymnosperms			
Gingko	1	158	17
Conifers	7 in 2 families	120–130	0
Angiosperms	2 genera, 1 family[d]	50–71	0–24
	20 genera, 3 families[e]	120–130	0
	470 genera, 96 families	130–160	10–30
	4 genera, 4 families[f]	160–180	25–40
	2 genera, 1 family[g]	200–220	70–80

Adapted from Palmer (1991).

[a]Sizes in kilobase pairs (kb) are rounded to the nearest whole number.

[b]*Astasia* is colorless. Both *Astasia* and *Euglena* have 6-kb tandem repeats.

[c]The chloroplast genome of *M. polymorpha* has been sequenced completely (Ohyama et al., 1986).

[d]Nonphotosynthetic plants belonging to the genera *Conophilus* (50 kb) and *Epifagus* (71 kb).

[e]Four groups of angiosperms lack the inverted repeat: (1) six tribes of legumes of the family Fabaceae and all 18 examined genera; (2) two genera of Geraniaceae; (3) one genus of Scrophulariaceae; (4) *Conophilus* (see above).

[f]Includes *Spirodela* (Lemnaceae), *Nicotiana* (Solanaceae), *Linum* (Linaceae), and *Begonia* (Begoniaceae).

[g]*Pelargonium* and *Geranium* (Geraniaceae) have duplications of half their genome.

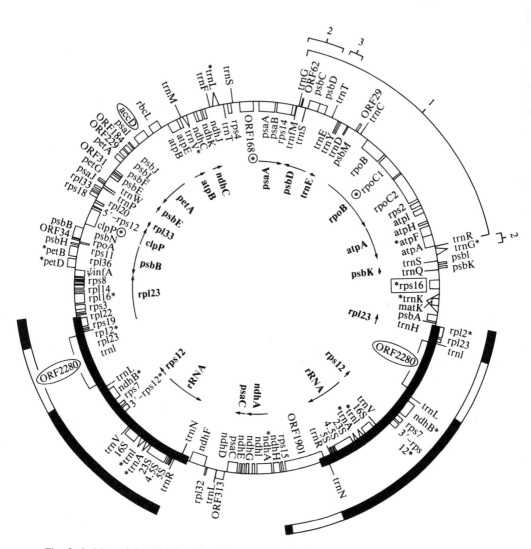

Fig. 3–6. Map of the 156-kb chloroplast genome of tobacco *(Nicotiana tabacum)*, which has been sequenced completely (Shinozaki et al., 1986). Arrows on the inside of the circle indicate sets of genes thought to constitute operons. In general, the operons are named according to the first transcribed gene. The *rbcL* operon consists of a single gene in higher plants and chlorophytes, but two genes *(rbcL* and *rbcS)* in other algae. Circled gene names indicate genes present in tobacco and liverwort *(Marchantia)*, but absent in rice. The solid box around *rps16* calls attention to the fact that this gene is also present in rice, but absent from liverwort. Asterisks denote genes that have the same intron(s) in all three sequenced genomes; circled asterisks denote genes that are split into only two of the three genomes. The thickened parts of the circle represent the 25.3-kb inverted repeat of tobacco. The thick lines outside the tobacco inverted repeat represent the extent of the inverted repeat in rice (20.8 kb), with regions deleted from rice indicated as open boxes. The numbered brackets outside the circle indicate regions that have been moved by three overlapping inversions in rice relative to tobacco. Gene nomenclature: rRNA genes are indicated by 16S, 23S, 4.5S, and 5S; tRNA genes by *trn* followed by the one letter amino acid code; RNA

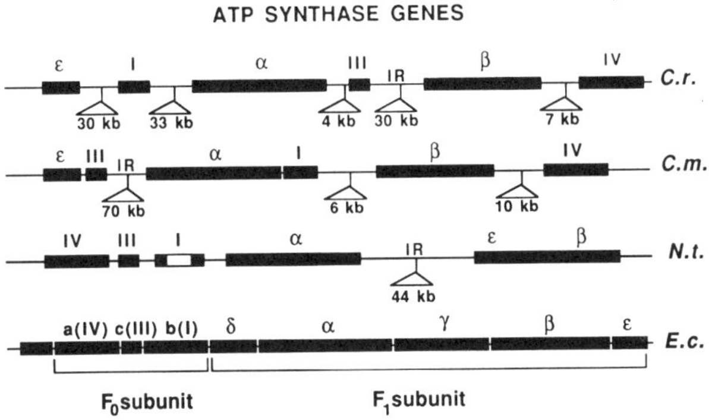

ATP SYNTHASE GENES

Fig. 3–7. Organization of genes encoding polypeptides of the ATP synthase complex in *Escherichia coli (E.c.)* and the chloroplast genomes of *Chlamydomonas reinhardtii (C.r.)*, *Chlamydomonas moewusii (C.m.)*, and *Nicotiana tabacum (N.t.)*. IR, inverted repeat region. From Harris et al. (1991). Protein subunits are designated in the figure and are encoded by the following genes: α, *atpA;* β, *atpB;* ε, *atpE;* I, *atpF;* III, *atpH;* IV, *atpI*.

a light-independent protochlorophyllide reductase. In the prokaryote *Rhodobacter capsulatus* this enzyme is comprised of subunits encoded by three genes (see Liu et al., 1993 for a summary). All three genes have now been identified in the chloroplast genomes of these plants (Liu et al., 1993; Suzuki and Bauer, 1992; Yamada et al., 1992), but are absent from angiosperm chloroplast genomes consistent with the inability of flowering plants to make chlorophyll in the dark. There are also 12 conserved ORFs of unknown function. Three genes (ORF2280, ORF1901 tobacco terminology, and *accD*) are present in tobacco, *Epifagus*, and liverwort, but not rice (Shimada and Sugiura, 1991; Wolfe et al., 1992a). Each of these plastid genomes also possesses unique ORFs that may or may not be functional (Shimada and Sugiura, 1991).

Chloroplast genes are generally close together with spacers usually only a few hundred base pairs in length. Chloroplast gene clusters tend to be cotranscribed into polycistronic mRNAs that are subsequently processed (Chapter 12). Many of these clusters bear striking resemblance to parts of bacterial operons. For example, a large cluster of genes encoding chloroplast ribosomal proteins contains some of the genes present in the S10, *spc*, and α operons of *E. coli* (Table 3–5). The gene order in this

---

polymerase genes by *rpo*, followed by a subunit-specific letter; genes for the 50S and 30S ribosomal proteins by *rpl* and *rps*, respectively, followed by a number corresponding to the cognate *E. coli* protein; genes for components of the thylakoid membrane complexes ATP synthase, photosystem I, photosystem II, and cytochrome b_6/f complex by *atp, psa, psb,* and *pet,* respectively, followed by a subunit-specific letter; genes for NADH dehydrogenase subunits by *ndh*, followed by a letter (the corresponding mitochondrial genes are followed by a number, e.g., *ndhA* equals *ndh1*); gene for the large subunit of RUBISCO, by *rbcL;* and open reading frames conserved in at least two of the three sequenced genomes are indicated by ORF, followed by their lengths in tobacco in codons. From Palmer (1991), who also cites all of the original references.

polycistronic transcription unit is identical to *E. coli*. The missing genes are the same in rice, tobacco, and liverwort and presumably transferred to the nucleus. Similarly, the chloroplast-encoded ATP synthase genes in flowering plants are arranged in two major clusters as they are in cyanobacteria rather than in a single cluster as in *E. coli*. (Fig. 3–7).

Numerous plastid genes contain group I or group II introns (see Palmer, 1991). Normally, the exons flank the intron and cis-splicing occurs, but in angiosperms the *rps12* gene transcript is trans-spliced. Its 5' exon is far away from the 3' segment, which is itself separated into two exons by a group II intron (see Sugiura, 1989).

Algal Chloroplast Genomes

Algal chloroplast genomes are much more variable in organization and gene content than those of land plants. This variability is readily illustrated by comparing the structure of several well-characterized algal chloroplast genomes.

Chlamydomonas. These algal chloroplast genomes are similar in structure to those of most land plants in containing an inverted repeat and two unique sequence regions (Boynton *et al.* 1992; Harris, 1989). However, the *Chlamydomonas* chloroplast genome is substantially larger than most land plant chloroplast genomes (*C. reinhardtii*, 196 kb; *C. eugametos* 243 kb). Unlike land plants the two unique sequence regions in *Chlamydomonas* are roughly the same size. Chloroplast genes in *Chlamydomonas* are also extensively rearranged with respect to land plants. For example, the *atpB* and *atpE* genes, which are adjacent and cotranscribed in cyanobacteria and land plants, are widely separated in different single copy regions in *Chlamydomonas* (Fig. 3–7). The remaining four chloroplast-encoded genes (*atpA, atpF, atpH,* and *atpI*) of land plants are clustered in a second operon. In *C. reinhardtii atpA* and *atpH* are close together in the same single copy region containing *atpE* and *atpF* while *atpB* and *atpI* are in the other single copy region. In *C. moewusii, atpE* and *atpH* are adjacent in one single copy region, and the remaining four genes map to a 22-kb segment in the other single copy region. Similarly, the ribosomal protein genes *rps7* and *rps12* are both continuous in *Chlamydomonas*, but they are in different unique sequence regions. In *E. coli* and cyanobacteria these genes are adjacent. In higher plants the two 3' exons of *rps12* are next to *rps7*.

Although the *rps12* gene is continuous in *Chlamydomonas* in contrast to higher plants, the *psaA* gene encoding the 83-kDa reaction center protein of PSI is split into three exons. These are widely separated on the chloroplast genome in different orientations (see Boynton *et al.*, 1992 and Rochaix, 1992 for reviews). The mRNA is assembled by trans-splicing. The genetic regulation of this process is an interesting story (Chapter 10).

Chloroplasts have active recombination systems (Chapters 4 and 7). Palmer (1991) points out that inversions frequently occur as the result of recombination between inversely oriented repeat sequences. No such repeats are found in the three sequenced higher plant chloroplast genomes or in most other higher plant chloroplast genomes that have been studied in any detail. However, short repeats are abundant in the chloroplast genomes of grasses, conifers, geranium, and subclover, which have highly rearranged chloroplast genomes (Palmer, 1991). Short dispersed repeat sequences (SDRs) are ubiquitous in intergenic sequences of the chloroplast genome of *C. reinhardtii* (Boynton et al., 1992). Individual SDRs are each composed of collections of shorter repeats that may appear in direct or inverted orientation and in different orders. Their distribution is also very different in the interfertile species *C. smithii* accounting for much of the variation in restriction fragment length polymorphisms between these two species. Probes for the *C. reinhardtii* SDRs do not hybridize with *C. eugametos* or *C. moewusii* cpDNA. The *Chlamydomonas* SDRs share no obvious homology with short repeat elements described in higher plants.

The unconventional chloroplast genome of Euglena. The 145-kb *Euglena* chloroplast genome, the subject of a comprehensive review by Hallick and Buetow (1989), has now been sequenced in its entirety (Hallick et al., 1993). The organization, structure, and gene content of this chloroplast genome depart markedly from the chloroplast genomes so far considered. The *Euglena* chloroplast genome lacks the inverted repeat and the rRNA genes are present as three tandem repeats plus an extra 16S rRNA gene. Variants have been reported with as many as five rRNA operons and two extra 16S rRNA genes and as few as one set of rRNA genes plus an extra 16S rRNA gene. Most of the tRNA genes are grouped in tight clusters of two to five genes whereas they tend to be scattered in land plant plastid genomes. In contrast to protein-coding chloroplast genes in land plants or *Chlamydomonas*, which contain one or two introns at most, comparable genes in *Euglena* are peppered with introns, a total of 149 being found so far in the chloroplast genome of this alga (Hallick et al., 1993). The gene encoding the large subunit of RUBISCO, for instance, has nine introns. Many intron-containing genes of *Euglena* are continuous in *Chlamydomonas* and land plants. Several genes encoding tRNAs have introns in land plants, but not in *Euglena* or *Chlamydomonas*. There are also differences in gene content between *Euglena* and land plants (Hallick et al., 1993). For instance, 11 *ndh* genes are absent from the chloroplast genome of *Euglena* as are some other genes (e.g., *rpoA*, *infA*, three to four tRNA genes, *psaI*, *psbM*, *petA*, *petD*, and four ribosomal protein genes. However, the *Euglena* chloroplast genome contains genes not found in land plants [e.g., *rpl5*, *rpl12*, *rpl15*, *tufA*, *psaM*, *ccsA* (chlorophyll biosynthesis), and several ORFs]. Ribosomal protein gene clusters are organized much as they are in land plants with a few additions (Table 3–5).

Further gene content differences: Cyanophora, Cryptomonas, and Porphyra. The plastid of *Cyanophora paradoxa*, usually referred to as the cyanelle, has a secondary peptidoglycan wall and its photosynthetic apparatus includes phycobiliproteins as is also true of cyanobacteria and red algae. Most of the 133-kb cyanelle genome has now been sequenced (Bryant and Bohnert, 1993). Although the gene content of the cyanelle genome closely resembles that of land plants, there are about 30% more genes. The additional genes include 15 involved in translation (*tufA* plus 14 ribosomal protein genes) and 10 photosynthesis genes (five encoding proteins specified by nuclear genes in flowering plants and six encoding phycobiliproteins). There are also reading frames with homology to genes involved in biosynthesis. They include the amino acid biosynthesis genes (*hisH* and *trpD*) and a gene, *acpA*, encoding an acyl carrier protein, a key cofactor in the synthesis and metabolism of fatty acids plus several chlorophyll biosynthesis genes. In addition, several genes encoding molecular chaperones (Chapter 15) are found in the cyanelle genome. Genes found in land plant chloroplast genomes, but so far not detected in the cyanelle genome, include the *ndh* genes, genes specifying three ribosomal proteins, and a number of ORFs. As in the red alga *Porphyra,* but unlike *Euglena,* introns are extremely rare. So far, only a single type I intron in a tRNA^{Leu} gene has been found. The same intron is found in cyanobacteria. The ribosomal protein gene clusters are similar to those of land plants with a few additions plus a deletion (L23, Table 3–5).

Cryptomonas Φ (see Douglas, 1992) contains a residual nucleus or nucleomorph in addition to the normal cell nucleus (Chapter 1). The plastid and nucleomorph are enclosed within the cytoplasmic endoplasmic reticulum that separates the putative endosymbiont from the cytoplasm and true nucleus of the cell. The nucleomorph 18S cytoplasmic rDNA genes are related to those of red algae while the nuclear rDNA clusters with the assemblage containing land plants and green algae (Douglas *et al.,* 1991). Partial sequencing of the plastid genome of this alga has revealed the presence of several novel genes (Douglas, 1992; Wang and Liu, 1991). They include *dnaK* encoding a protein of the hsp70 heat shock family (Chapter 15), *hlpA,* which specifies a polypeptide resembling bacterial histone-like proteins, an amino acid biosynthesis gene (*ilvB*), a gene involved in transcriptional regulation (*ompR*), and a protein export gene (*secY*). The *secY* gene as well as the *secA* and *hsp70* genes have been localized to the chloroplast genome of another chromophyte (see Scaramuzzi *et al.,* 1992, for a summary). These genes are homologs to bacterial genes involved in protein translocation across membranes raising "the unprecedented possibility of protein export from the chloroplast." The *Cryptomonas* Φ chloroplast genome, like the cyanelle genome, possesses the *acpA* gene encoding the acyl carrier protein (Douglas, 1992). This gene has also been found in the plastid genome of a marine diatom (Hwang and Tabita, 1991).

Reith and Munholland (1993) have sequenced ca. 60% of the chloroplast genome of the red alga *Porphyra.* They have identified over 125 genes and estimate that this chloroplast genome may contain as many as 200–220 genes or almost twice as many genes as land plant chloroplast genomes. Gene differences include the presence of at least seven photosynthesis and nine r-protein genes not found in land plants plus genes encoding all subunits of the ATP synthase except γ. The ribosomal protein gene clusters are similar to those of land plants with several additions, but the same deletion of L23 seen in *Cyanophora* (Table 3–5). The plastid genome of *Porphyra* also encodes eight genes required for phycobiliprotein synthesis and a variety of genes specifying proteins involved in biosynthesis and gene replication and expression. Introns have not been found in any of the 80 genes sequenced to date. *P. yezoensis* possesses an rRNA gene-containing inverted repeat (Shivji *et al.,* 1992), while the well-separated rRNA genes are directly repeated in *P. purpurea* (Reith and Munholland, 1993).

Other algae. Most algal lineages possess chloroplast genomes ranging in size from 120 to 160 kb (Table 3–6). In the majority of species the inverted repeat structure is retained although it is missing from some genera. Bicircular chloroplast genomes have been observed in brown algae. The best characterized of these is the chloroplast genome of *Pylaiella littoralis,* which exists as circles of 133 and 58 kb (Loiseaux-de-Goer *et al.,* 1988; Markowicz *et al.,* 1988). The large circle is about twice as abundant as the small circle. The two circles are dissimilar in sequence and do not recombine to form larger species. The 58-kb circle contains several pseudogenes, but functional genes have as yet not been identified (Palmer, 1991). Palmer (1991) points out one striking gene content difference between chlorophyll *a/b*-containing plants and algae and "nongreen'' algae. The gene encoding the small subunit of RUBISCO *(rbcS)* is nuclear in the former plants, but is chloroplast encoded in the nongreen algae. Interestingly, the Chrysophyte alga *Olisthodiscus luteus* possesses a 22-kb inverted repeat in its chloroplast genome that contains, in addition to the rRNA genes, cotranscribed genes for the large and small subunits of RUBISCO, plus three other photosynthesis genes (Shivji *et al.*, 1992).

Lastly, Wolfe *et al.* (1992a) point out that in contrast to land plants, no *ndh* genes have yet been reported from the chloroplast genomes of *Chlamydomonas, Euglena* (mentioned earlier), *Cyanophora,* and *Cryptomonas. Porphyra* can now be added to this list. The remarkable fact is that the process of chlororespiration, which is thought to involve these genes, was first described in *Chlamydomonas* (Chapter 1).

The Plastid Genomes of Colorless Plants and Algae: How Essential Is Chloroplast Protein Synthesis?

Chloroplast protein synthesis has long been known to be indispensable for survival of plants and algae that depend on CO_2 as a sole carbon source since numerous proteins required for photosynthesis are plastid gene products. However, as we learn more about the genes encoded in the plastid genomes of algae such as *Cyanophora, Cryptomonas,* and *Porphyra,* the likelihood is increasing that chloroplast protein synthesis is required for the production of one or more essential proteins not involved in photosynthesis. This viewpoint is supported by some, but not all, analyses of plastid genome function in colorless plants.

The colorless heterotroph, *Astasia longa,* is closely related to *Euglena gracilis* and possesses a circular 73-kb plastid genome. This genome is the homolog of the larger (145-kb) *Euglena* chloroplast genome except that the only chloroplast photosynthetic gene so far detected is the *rbcL* gene (Siemeister and Hachtel, 1989; 1990a, b; Siemeister *et al.*, 1990). In contrast, genes encoding the rRNAs, elongation factor tu *(tufA),* and several tRNAs and ribosomal proteins have been identified. Since the RUBISCO large subunit has been immunoprecipitated from *Astasia,* the *rbcL* gene must be transcribed and translated so the plastid protein synthesizing system of this colorless flagellate must be functional. Colorless, heterotrophic algae of the genus *Polytoma* are closely related to or derived from *Chlamydomonas. Polytoma* contains a spacious plastid genome (ca. 200 kb) similar in size to the *Chlamydomonas* chloroplast genome (Siu *et al.*, 1975; Vernon-Kipp *et al.*, 1989). Plastid rRNA genes are present and expressed in *Polytoma*; leucoplast ribosomes have been isolated; and the *tufA* gene identified. These results suggest that *Polytoma* too has a functioning plastid protein synthesizing system (Siu *et al.*, 1976 a, b; Vernon-Kipp *et al.*, 1989).

By far the most complete characterization of the plastid genome of a colorless plant has been done on *Epifagus virginiana* (beechdrops). This is a colorless, flowering plant belonging to a family of root-parasitic angiosperms. The 70-kb *Epifagus* plastid genome has been completely sequenced and contains only 42 genes (de Pamphilis and Palmer, 1989, 1990; Wolfe *et al.*, 1992 a, b). At least 38 of these genes encode components of the plastid gene expression system (rRNAs, tRNAs and ribosomal proteins). Photosynthesis genes and genes of the NADH dehydrogenase complex are absent, although several photosynthetic pseudogenes have been found. The *Epifagus* plastid genome contains only 17 tRNAs, suggesting that tRNAs must be imported if this plastid protein-synthesizing system is to function and lacks genes specifying the four RNA polymerase subunits encoded in the chloroplast genomes of flowering plants. Although experiments have not been reported in *Epifagus* similar to those in *Astasia*, which demonstrate the synthesis of a specific plastid-encoded protein, there are good reasons for believing that the plastid protein-synthesizing system is functional. Wolfe *et al.* (1992a) report transcription of all eight rRNA and protein-encoding genes so far examined and cite the following three evolutionary arguments in favor of function:

1. Plastid gene deletions in *Epifagus* are not random, but skewed. While only 5% of photosynthetic sequences have been retained with respect to tobacco, 80% of the ribosomal protein sequences are present.
2. Large open reading frames are retained in the *Epifagus* plastid genome. If these genes were nonfunctional, mutations, truncations and internal deletions would have been expected to occur as is true of pseudogenes in the *Epifagus* plastid genome.
3. *Conophilus* and *Epifagus*, both members of the nonphotosynthetic family *Orobranchaceae*, share the loss of the photosynthetic and *ndh* genes, but their rRNA genes are strongly conserved, suggesting their evolution is constrained by natural selection because they are functional.

These data strongly suggest that the *Epifagus* plastid protein-synthesizing system is essential. The question is why. Before the complete sequence of the *Epifagus* plastid genome was determined, Howe and Smith (1991) argued that the plastid protein synthesizing apparatus is retained in *Epifagus* for the sole purpose of making the chloroplast-encoded RNA polymerase subunits required for transcription of the tRNA^{Glu} gene. This tRNA is required for porphyrin synthesis in plants so its loss would be lethal since synthesis of heme and mitochondrial cytochromes could not occur. However, as Wolfe et al. (1992a) point out this cannot be the case for *Epifagus* because, while tRNA^{Glu} is encoded in the plastid genome of this plant, the RNA polymerase subunits are not, so transcription must rely on a nuclear-encoded enzyme. Does the *Epifagus* plastid genome encode proteins whose synthesis is required for survival? There are four possible candidates. Three are ORFs in the plastid genome of *Epifagus*, which are found in tobacco, but not in rice, so these can be excluded (Wolfe *et al.*, 1992a). The fourth gene, *clpP*, encodes one subunit of the plastid homolog of the ATP-dependent Clp protease of *E. coli*. Perhaps this protease is involved in processing protein precursors into an active form, or in protein degradation.

Whether the *Epifagus* results can be generalized remains to be seen. *Cuscuta reflexa* (Convolvulaceae) is an unrelated colorless, parasitic plant that contains residual thylakoids and traces of chlorophylls *a* and *b* (Machado and Zetsche, 1990). The plant also possesses very low levels of light-stimulated CO_2 fixation although RUBISCO large subunit is undetectable using immunological methods (Haberhausen *et al.*, 1992; Machado and Zetsche, 1990). Sequencing parts of the *C. reflexa* plastid genome (Bömmer *et al.*, 1993; Haberhausen *et al.*, 1992) revealed the presence of intact photosynthesis genes (e.g., *atpB, atpE, rbcL, psbA*), but a large deletion removed certain protein synthesis genes (i.e., *rpl2, rpl23* and several tRNA genes). Transcription analysis showed that *rbcL* was weakly transcribed while *psbA* was transcribed strongly. Bömmer *et al.* (1993) hypothesize that the translational apparatus of *C. reflexa* is nonfunctional based on the loss of specific protein synthesis genes from the plastid genome and their inability to detect RUBISCO large subunit using immunological techniques. However, an alternative interpretation is that the plastid protein synthesizing system of *C. reflexa* is functional and that photosynthetic proteins are made in very small quantities (accounting for the presence of trace levels of CO_2 fixation and chlorophylls *a* and *b* plus residual thylakoids).

In *Chlamydomonas* genetic evidence suggests that chloroplast protein synthesis is essential for survival of this alga. Hanson and Bogorad (1978) showed that a nuclear mutation conferring erythromycin resistance on chloroplast ribosomes underwent a marked reduction in chloroplast ribosome content when shifted from 25 to 15°C. Ribosome loss was accompanied by loss of the ability of the mutant to grow at 15°C under conditions permissive for photosynthetic function (see Chapter 7). Also, no *Chlamydomonas* mutant isolated to date with defective chloroplast ribosomes is completely deficient in chloroplast protein synthesis (Harris *et al.*, 1974, 1987; Harris, 1989). Lastly, mutations with symmetric deletions of the *psbA* gene encoded within the inverted repeat are frequently isolated, but deletion mutations in the rRNA gene region of the repeat only remove one set of rRNA genes (Palmer *et al.*, 1985). Many years ago Blamire *et al.* (1974) reported that treatment of wild-type cells with antibiotics blocking translation on chloroplast ribosomes inhibited replication of nuclear but not cpDNA. Inhibition did not occur in mutant strains having chloroplast ribosomes resistant to these antibiotics. These intriguing experiments have never been repeated.

The notion that chloroplast protein synthesis is indispensable is challenged by findings reported with plant tissue cultures. In *Nicotiana* antibacterial antibiotics such as streptomycin and lincomycin cause bleaching, but not death of callus in tissue culture. The bleached antibiotic-sensitive cells continue to divide at a reduced rate using sucrose as the carbon source (Maliga, 1984). Mutants resistant to these antibiotics are green in tissue culture. Streptomycin-resistant mutants result from specific base pair changes in the 16S rRNA or *rps12* genes while those resistant to lincomycin arise because of a specific base pair alteration in the gene encoding the 23S rRNA (Maliga *et al.*, 1990; Staub and Maliga, 1992). The ability of bleached antibiotic sensitive cells to continue to divide implies that chloroplast protein synthesis may be required only for the manufacture of photosynthetic and ribosomal proteins in *Nicotiana*. However, tobacco calli containing a chloroplast mutation to streptomycin resistance grow better in the dark on antibiotic than do sensitive calli (Cséplo *et al.* 1993; Malone *et al.*, 1992).

Experiments with calli cultured from roots of haploid rice plants derived from pollen grains provide convincing evidence that plastid protein synthesis is not essential in this system (Harada *et al.*, 1992). Albino plants obtained in this way from barley and wheat have long been known to contain large deletions in the plastid genome (Day and Ellis, 1984, 1985), but the deleted molecules form a heterogeneous collection. By inducing callus cultures from roots of albino rice plants Harada *et al.* (1992) obtained samples that were homoplasmic for different large deletions. Of five that were characterized, four lacked the inverted repeat and the plastid rRNA genes so none of these callus cultures can carry out chloroplast protein synthesis. Yet all five retained one region in common that contained the tRNA^{Glu} gene. Harada *et al.* (1992) suggest this gene has been retained because the tRNA^{Glu} encoded by the gene is essential for porphyrin biosynthesis. Obviously, expression of this gene would require functioning of a nuclear-encoded plastid RNA polymerase. Retention of small amounts of plastid DNA in bleached mutants of *Euglena* (Heizmann *et al.*, 1981) is probably not related to the requirement for tRNA^{Glu} in porphyrin synthesis since the mitochondrial heme in this flagellate is made via the animal-type δ aminolevulinic synthetase pathway, which does not require tRNA^{Glu}. In fact, the rRNA genes were the only ones detected in these deleted plastid genomes.

REFERENCES

Aldritt, S.M., Joseph, J.T., and D.F. Wirth (1989). Sequence identification of cytochrome *b* gene in *Plasmodium gallinaceum*. *Mol. Cell. Biol.* 9: 3614–3620.

Anderson, S., Bankier, A.T., Barrell, B.G., De Bruijn, M.H.L., Colson, A.R., Drouin, J., Eperon, I.C., Nierlich, D.P., Roe, B.A., Sanger, F., Schreier, P.H., Smith, A.J.H., Staden, R., and I.G. Young (1981). Sequence and organization of the human mitochondrial genome. *Nature (London)* 290: 457–465.

Anderson, S. de Bruijn, M.H.L., Coulson, A.R., Eperon, I.C., Sanger, F., and G. Young (1982). Complete sequence of bovine mitochondrial DNA: Conserved features of the mammalian mitochondrial genome. *J. Mol. Biol.* 157: 683–717.

Arizmendi, J.M., Runswick, M.J., Skehel, J.M., and J.E. Walker (1992). NADH: ubiquinone oxidoreductase from bovine heart mitochondria: A fourth nuclear encoded subunit with a homologue encoded in chloroplast genomes. *FEBS Lett.* 301: 237–242.

Arnason, U., Gullberg, A., and B. Widegren (1991). The complete nucleotide sequence of the mitochondrial DNA of the fin whale *Balaenoptera physalus*. *J. Mol. Evol.* 33: 556–568.

Attardi, G. (1985). Animal mitochondrial DNA: An extreme example of genetic economy. *Int. Rev. Cytol.* 93: 93–145.

Attardi, G., and G. Schatz (1988). Biogenesis of mitochondria. *Annu. Rev. Cell Biol.* 4: 289–333.

Barta, J.R., Jenkins, M.C., and H.D. Danforth (1991). Evolutionary relationships of avian *Eimeria* species among other apicomplexan protozoa: Monophyly of the apicomplexa is supported. *Mol. Biol. Evol.* 8: 345–355.

Blamire, J., Flechtner, V.R., and R. Sager (1974). Regulation of nuclear DNA replication by the chloroplast in *Chlamydomonas*. *Proc. Natl. Acad. Sci. U.S.A.* 71: 2867–2871.

Boer, P.H., and M.W. Gray (1988). Scrambled ribosomal RNA gene pieces in *Chlamydomonas reinhardtii* mitochondrial DNA. *Cell* 55: 309–411.

Bömmer, D., Haberhausen, G., and K. Zetsche (1993). A large deletion in the plastid DNA of the holoparasitic flowering plant *Cuscuta reflexa* concerning two ribosomal proteins *(rp12, rp123)*, one transfer RNA *(tRNAI)* and an ORF2280 homologue. *Curr. Genet.* 24: 171–176.

Borst, P. (1991). Why kinetoplast DNA networks? *Trends Genet.*, 7: 139–141.

Boynton, J.E., Gillham, N.W., Newman, S.M., and E.H. Harris (1992). Organelle Genetics and Transformation of *Chlamydomonas*. In *Plant Gene Research, Cell Organelles* (R.G. Herrmann, ed.), pp. 3–64. Springer-Verlag, Wien.

Bridge, D., Cunningham, C.W., Schierwater, B., DeSalle, R., and L.W. Buss (1992). Class-level relationships in the phylum Cnidaria: Evidence from mitochondrial genome structure. *Proc. Natl. Acad. Sci. U.S.A.* 89: 8750–8753.

Britten, R.J., and D.E. Kohne (1968). Repeated sequences in DNA. *Science* 161: 529–540.

Bruns, T.D., and J.D. Palmer (1989). Evolution of mushroom mitochondrial DNA: *Suillus* and related genera. *J. Mol. Evol.* 28: 349–362.

Bruns, T.D., Palmer, J.D., Shumard, D.S., Grossman, G.I., and M.E.S. Hudspeth (1988). Mitochondrial DNAs of *Suillus*: Three-fold size change in molecules that share a common gene order. *Curr. Genet.* 13: 49–56.

Bruns, T.D., Fogel, R., White, T.J., and J.D. Palmer (1989). Accelerated evolution of a false-truffle from a mushroom ancestor. *Nature (London)* 339: 140–142.

Bryant, D.A., and H. Bohnert (1993). Personal communication.

Cantatore, P., and C. Saccone (1987). Organization, structure, and evolution of mammalian mitochondrial genes. *Int. Rev. Cytol.* 108: 149–208.

Chiba, Y. (1951). Cytochemical studies on chloroplasts. I. Cytologic demonstration of nucleic acids in chloroplasts. *Cytologia (Tokyo)* 16: 259–264.

Clark-Walker, G.D. (1985). Basis of diversity in mitochondrial DNAs. In *The Evolution of Genome Size* (T. Cavalier-Smith, ed.), pp. 277–297. Wiley, Sussex.

Clark-Walker, G.D. (1992). Evolution of mitochondrial genomes in fungi. *Int. Rev. Cytol.* 141: 89–127.

Clark-Walker, G.D., McArthur, C.R., and K.S. Sriprakash (1981). Partial duplication of the large ribosomal RNA sequence in an inverted repeat in circular mitochondrial DNA from *Kloeckera africana*. Implications for the mechanism of the petite mutation. *J. Mol. Biol.* 147: 399–415.

Clary, D.O., and D.R. Wolstenholme (1985). The mitochondrial DNA molecule of *Drosophila yak-*

uba: Nucleotide sequence, gene organization and genetic code. *J. Mol. Evol.* 22: 252–271.

Coleman, A.W., Thompson, W.F., and L.J. Goff (1991). Identification of the mitochondrial genome in the chrysophyte alga *Ochromonas danica*. *J. Protozool.* 38: 129–135.

Collins, R.A., Stohl, L.L., Cole, M.D., and A.M. Lambowitz (1981). Characterization of a novel plasmid DNA found in mitochondria of *N. crassa*. *Cell* 24: 443–452.

Cséplo, A., Eigel, L., Horváth, G.V., Medgyesy, P., Herrmann, R.G., and H.-U. Koop (1993). Subcellular location of lincomycin resistance in *Nicotiana* mutants. *Mol. Gen. Genet.* 236: 163–170.

Cummings, D.J. (1992). Mitochondrial genomes of the ciliates. *Int. Rev. Cytol.* 141: 1–64.

Day, A., and T.H.N. Ellis (1984). Chloroplast DNA deletions associated with wheat plants regenerated from pollen. *Cell* 39: 359–368.

Day, A., and T.H.N. Ellis (1985). Deleted forms of plastid DNA in albino plants from cereal anther culture. *Curr. Genet.* 9: 671–678.

Denovan-Wright, E.M., and R.W. Lee (1992). Comparative analysis of the mitochondrial genomes of *Chlamydomonas eugametos* and *Chlamydomonas moewusii*. *Curr. Genet.* 21: 197–202.

de Pamphilis, C.W., and J.D. Palmer (1989). Evolution and function of plastid DNA: A review with special reference to nonphotosynthetic plants. In *Physiology, Biochemistry, and Genetics of Nongreen Plastids* (C.D. Boyer, J.C. Shannon, and R.C. Hardison, eds.), pp. 182–202. American Society for Plant Physiology, Rockville, MD.

de Pamphilis, C.W. and J.D. Palmer (1990). Loss of photosynthetic and chlororespiratory genes from the plastid genome of a parasitic flowering plant. *Nature (London)* 348: 337–339.

de Zamaroczy, M., and G. Bernardi (1985). Sequence organization of the mitochondrial genome of yeast—a review. *Gene* 37: 1–17.

de Zamaroczy, M., and G. Bernardi (1987). The AT spacers and the *var 1* genes from the mitochondrial genomes of *Saccharomyces cerevisiae* and *Torulopsis glabrata*: Evolutionary origin and mechanism of formation. *Gene* 54: 1–22.

Douglas, S.E. (1992). Eukaryote-eukaryote endosymbioses: Insights from studies of a cryptomonad alga. *BioSystems* 28: 57–68.

Douglas, S.E., Murphy, C.A., Spencer, D.F., and M.W. Gray (1991). Cryptomonad algae are evolutionary chimeras of two phylogenetically distinct eukaryotes. *Nature (London)* 350: 148–151.

Farelly, F., Zassenhaus, H.P., and R.A. Butow (1982). Characterization of transcripts from the *var1* region on mitochondrial DNA of *Saccharomyces cerevisiae*. *J. Biol. Chem.* 257: 6581–6587.

Fauron, C. M.-R., Hvlik, M., and M. Casper (1991). Organization and evolution of the maize mitochondrial genome. In *Plant Molecular Biology 2* (R.G. Herrmann and B. Larkins, eds.), pp. 345–363. Plenum Press, New York.

Feagin, J.E. (1994). The extrachromosomal DNAs of Apicomplexan parasites. *Annu. Rev. Microbiol.* 48: 81–104.

Feagin, J.E., Gardner, M.J., Williamson, D.H., and R.J.M. Wilson (1991). The putative mitochondrial genome of *Plasmodium falciparum*. *J. Protozool.* 38: 243–245.

Fukuzawa, H., Kohchi, T., Sano, T., Shirai, H., Umesono, K., Inokuchi, H. Ozeki, H., and K. Ohyama (1988). Structure and organization of *Marchantia polymorpha* chloroplast genome. III. Gene organization of the large single copy region from *rbcL* to *trnI*(CAU). *J. Mol. Biol.* 203: 333–351.

Gadeleta, G., Pepe, G., Decandia, G., Quagliariello, C., Sbisa, E., and C. Saccone (1989). The complete nucleotide sequence of the *Rattus norvegicus* mitochondrial genome: Cryptic signals revealed by comparative analysis between vertebrates. *J. Mol. Biol.* 28: 497–516.

Gardner, M.M., Feagin, J.E., Moore, D.J., Rangachari, K., Williamson, D.H., and R.J.M. Wilson (1993). Sequence and organization of large subunit rRNA genes from the extrachromosomal 35 kb circular DNA of the malaria parasite *Plasmodium falciparum*. *Nucl. Acids Res.* 21: 1067–1071.

Garey, J.R., and D.R. Wolstenholme (1989). Platyhelminth mitochondrial DNA: Evidence for early evolutionary origin of a tRNA^{ser}AGN that contains a dihydrouridine arm replacement loop, and of serine-specifying AGA and AGG codons. *J. Mol. Evol.* 28: 374–387.

Gillham, N.W. (1978). *Organelle Heredity*. Raven Press, New York.

Gray, M.W. (1989). Origin and evolution of mitochondrial DNA. *Annu. Rev. Cell Biol.* 5: 25–50.

Gray, M.W. (1992). The endosymbiont hypothesis revisted. *Int. Rev. Cytol.* 141: 233–357.

Gray, M.W., and P.H. Boer (1988). Organization and expression of algal (*Chlamydomonas reinhardtii*) mitochondrial DNA. *Phil. Trans. R. Soc. London B* 319: 135–147.

Gutell, R.R., Weiser, B., Woese, C.R., and H.F. Noller (1985). Comparative anatomy of 16-S-like ribosomal RNA. *Prog. Nucl. Acid. Res. Mol. Biol.* 32: 155–216.

Haberhausen, G., Valentin, K., and K. Zetsche (1992). Organization and sequence of photosynthetic genes from the plastid genome of the holoparasitic flowering plant *Cuscuta reflexa*. *Mol. Gen. Genet.* 232: 154–161.

Hallick, R.B., and D.E. Buetow (1989). Chloroplast DNA. In *The Biology of Euglena*, Vol. IV (D.E. Buetow, ed.), pp. 351–414. Academic Press, San Diego.

Hallick, B.R., Hong, L., Drager, R.G., Favreau, M.R., Monfort, A., Orsat, B., Spielmann, A., and E. Stutz (1993). Complete sequence of *Euglena* chloroplast DNA. *Nucl. Acids Res.* 21: 3537–3544.

Hanson, M.R., and L. Bogorad (1978). The *ery-M2* group of *Chlamydomonas reinhardtii*: Cold-sensitive, erythromycin-resistant mutants deficient in chloroplast ribosomes. *J. Gen. Microbiol.* 105: 253–262.

Hanson, M.R., and O.F. Folkerts (1992). Structure and function of the higher plant mitochondrial genome. *Int. Rev. Cytol.* 141: 129–172.

Harada, T., Ishikawa, R., Niizeki, M., and K.-i. Saito (1992). Pollen-derived rice calli that have large deletions in plastid DNA do not require protein synthesis in plastids for growth. *Mol. Gen. Genet.* 233: 145–150.

Harris, E.H. (1989). *The Chlamydomonas Sourcebook: A Comprehensive Guide to Biology and Laboratory Use.* Academic Press, San Diego.

Harris, E.H., Boynton, J.E., and N.W. Gillham (1974). Chloroplast ribosome biogenesis in *Chlamydomonas*. Selection and characterization of mutants blocked in ribosome formation. *J. Cell Biol.* 63: 160–179.

Harris, E.H., Boynton, J.E., and N.W. Gillham (1987). Interaction of nuclear and chloroplast mutations in biogenesis of chloroplast ribosomes in *Chlamydomonas*. In *Molecular and Cellular Aspects of Algal Development* (W. Wiessner, D.G. Robinson, and R.C. Starr, eds.), pp. 142–149. Springer-Verlag, Berlin.

Harris, E.H., Boynton, J.E., Gillham, N.W., Burkhart, B.D., and S.M. Newman (1991). Chloroplast genome organization in *Chlamydomonas*. *Arch. Protistenkd.* 139: 183–192.

Heizmann, P., Doly, Y., Hussein, Y., Nicolas, P., Nigon, V., and G. Bernardi (1981). The chloroplast genome of bleached mutants of *Euglena gracilis*. *Biochim. Biophys. Acta* 653: 412–415.

Hiratsuka, J., Shimada, H., Whittier, R., Ishibashi, T., Sakamoto, M., Mori, M., Kondo, C., Honji, Y., Sun, C.-R., Meng, B.-Y., Li, Y.-Q., Kanno, A., Nishizawa, Y., Hirai, A., Shinozaki, K., and M. Sugiura (1989). The complete sequence of the rice (*Oryza sativa*) chloroplast genome: Intermolecular recombination between distinct tRNA genes accounts for a major plastid DNA inversion during the evolution of the cereals. *Mol. Gen. Genet.* 217: 185–194.

Hoeben, P., and G.D. Clark-Walker (1986). An approach to yeast classification by mapping mitochondrial DNA from *Dekkera/Brettanomyces* and *Eeniella* genera. *Curr. Genet.* 10: 371–379.

Hoeben, P., Weiller, B., and G.D. Clark-Walker (1993). Larger rearranged mitochondrial genomes in *Dekkera/Brettanomyces* yeasts are more closely related than smaller genomes with conserved gene order. *J. Mol. Evol.* 36:263–269.

Hollenberg, C.P., Borst, P., and E.F.J. Van Bruggen (1970). Mitochondrial DNA V. A 25-µ closed circular duplex DNA molecule in wild-type yeast mitochondria. Structure and genetic complexity. *Biochim. Biophys. Acta*, 1–15.

Howe, C. (1992). Plastid origin of an extrachromosomal DNA molecule from *Plasmodium*, the causative agent of malaria. *J. Theor. Biol.* 158: 199–205.

Howe, C.J., and A.G. Smith (1991). Plants without chlorophyll. *Nature* (London) 349: 109.

Hwang, S.-R., and F.R. Tabita (1991). Acyl carrier protein derived sequence encoded by the chloroplast genome in the marine diatom *Cylindrotheca* sp. strain N4. *J. Biol. Chem.* 266: 13492–13494.

Joseph, J.T., Aldritt, S.M., Unnasch, T., Puijalon, O., and D.F. Wirth (1989). Characterization of a conserved extrachromosomal element isolated from the avian malarial parasite *Plasmodium gallinaceum*. *Mol. Cell. Biol.* 9: 3621–3629.

Karlovsky, P., and B. Fartmann (1992). Genetic code and phylogenetic origin of Oomycetous mitochondria. *J. Mol. Evol.* 14: 254–258.

Kirk, J.T.O. (1971). Will the real chloroplast DNA please stand up? In *Autonomy and Biogenesis of Chloroplasts and Mitochondria* (N.K. Boardman, A.W. Linnane, and R.M. Smillie, eds.), pp. 267–276. North Holland, Amsterdam.

Kirk, J.T.O. (1986). The discovery of chloroplast DNA. *Bioessays* 4: 36–38.

Kolodner, R., and K.K. Tewari (1972). Molecular size and conformation of chloroplast deoxyribonucleic acid from pea leaves. *J. Biol. Chem.* 247: 6355–6364.

Lee, R.W., Dumas, C., Lemieux, C., and M. Turmel (1991). Cloning and characterization of the *Chlamydomonas moewusii* mitochondrial genome. *Mol. Gen. Genet.* 231: 53–58.

Liu, X.-Q., Xu, H., and C. Huang (1993). Chloroplast *chlB* gene is required for light-independent chlorophyll accumulation in *Chlamydomonas reinhardtii*. *Plant Mol. Biol.* 23: 297–308.

Loiseaux-de Goer, A.L., Markowicz, Y., Dalmon, J., and H. Audren (1988). Physical maps of the two circular plastid DNA molecules of the brown alga *Pylaiella littoralis* (L.) Kjellm. *Curr. Genet.* 14: 155–162.

Lonsdale, D.M. (1989). The plant mitochondrial genome. In *The Biochemistry of Plants*, Vol. 15 (A. Marcus, ed.), pp. 229–295. Academic Press, San Diego.

Lonsdale, D.M., Hodge, T.P., and C. M.-R. Fauron (1984). The physical map and organisation of the mitochondrial genome from the fertile cytoplasm of maize. *Nucl. Acids Res.* 12: 9249–9261.

Ma, D.-P., King, Y.-T., Kim, Y., Luckett, W.S. Jr., Boyle, J.A., and Y.-F. Cheng (1992). Amplification and characterization of an inverted repeat from the *Chlamydomonas reinhardtii* mitochondrial genome. *Gene* 119: 253–257.

Machado, M.A., and K. Zetsche (1990). A structural, functional and molecular analysis of plastids of the holoparasites *Cuscuta reflexa* and *Cuscuta europaea*. *Planta* 181: 91–96.

Maleszka, R., Skelly, P.J., and G.D. Clark-Walker (1991). Rolling circle replication of DNA in yeast mitochondria. *The EMBO J.* 10: 3923–3929.

Maliga, P. (1984). Isolation and characterization of mutants in plant cell culture. *Annu. Rev. Plant Physiol.* 35: 519–542.

Maliga, P., Moll, B., and Z. Svab (1990). Towards manipulation of plastid genes in higher plants. In *Perspectives in Biochemical and Genetic Regulation of Photosynthesis* (I. Zelitch, ed.), pp. 133–143. Alan R. Liss, New York.

Malone, R., Horváth, G.V., Cséplo, A., Búzás, B., Dix, P.J., and P. Medgyesy (1992). Impact of the stringency of cell selection on plastid segregation in protoplast fusion-derived *Nicotiana* regenerates. *Theor. Appl. Genet.* 84: 866–873.

Mannella, C.A., Goewert, R.R., and A.M. Lambowitz (1979). Characterization of variant *Neurospora crassa* mitochondrial DNAs which contain tandem reiterations. *Cell* 18: 1197–1207.

Manning, J.E., Wolstenholme, D.R., Ryan, R.S., Hunter, J.A., and O.C. Richards (1971). Circular chloroplast DNA from *Euglena gracilis*. *Proc. Natl. Acad. Sci. U.S.A.* 68: 1169–1173.

Manning, J.E., Wolstenholme, D.R., and O.C. Richards (1972). Circular DNA molecules associated with chloroplasts of spinach, *Spinacia oleracea*. *J. Cell Biol.* 53: 594–601.

Markowicz, Y., Goer, S.L., and R. Mache (1988). Presence of a 16S rRNA pseudogene in the bimolecular plastid genome of the primitive brown alga *Pylaiella littoralis*. Evolutionary implications. *Curr. Genet.* 14: 599–608.

Moore, L.J., and A.W. Coleman (1989). The linear 20 kb mitochondrial genome of *Pandorina morum*, Volvacaceae, Chlorophyta. *Plant Mol. Biol.* 13: 459–465.

Newton, K.J. (1988). Plant mitochondrial genomes: Organization, expression, and variation. *Annu. Rev. Plant Physiol. Plant Mol. Biol.* 39: 503–532.

Oda, K., Yamato, K., Ohta, E., Nakamura, Y., Takemura, M., Nozato, N., Akashi, K., Kanegae, T., Ogura, Y., Kohchi, T., and K. Ohyama (1992). Gene organization deduced from the complete sequence of Liverwort *Marchantia polymorpha* mitochondrial DNA: A primitive form of plant mitochondrial genome. *J. Mol. Biol.* 223: 1–7.

Ohta, E., Oda, K., Yamato, K., Nakamura, Y., Takemura, M., Nozato, N., Akashi, K., Ohyama, K., and F. Michel (1993). Group I introns in the liverwort mitochondrial genome: the gene coding for subunit 1 of cytochrome oxidase shares five intron positions with its fungal counterparts. *Nucl. Acids Res.* 21: 1297–1305.

Ohyama, K., Fukuzawa, H., Kohchi, T., Sano, T., Sano, S., Umesono, K., Shiki, Y., Takeuchi, M., Chang, Z., Aota, S., Inokuchi, H., and H. Ozeki (1986). Chloroplast gene organization deduced from complete sequence of liverwort *Marchantia polymorpha* chloroplast DNA. *Nature (London)* 322: 572–574.

Palmer, J.D. (1985). Evolution of chloroplast and mitochondrial DNA in plants and algae. In *Molecular Evolutionary Genetics* (R.J. MacIntyre, ed.), pp. 131–240. Plenum, New York.

Palmer, J.D. (1991). Plastid chromosomes: Structure and Evolution. In *The Molecular Biology of Plastids* (L. Bogorad and I.K. Vasil, eds.), pp. 5–53. Academic Press, San Diego.

Palmer, J.D. (1992a). Comparison of chloroplast and mitochondrial genome evolution in plants. In *Plant Gene Research: Cell Organelles* (R.G. Herrmann, ed.), pp. 99–133. Springer Verlag, Wien.

Palmer, J.D. (1992b). Green ancestry of malarial parasites? *Curr. Biol.* 2: 318–320.

Palmer, J.D., and L.A. Herbon (1986). Tricircular mitochondrial genomes of *Brassica* and *Raphanus*: Reversal of repeat configurations by inversion. *Nucl. Acids Res.* 14: 9755–9765.

Palmer, J.D., and L.A. Herbon (1987). Unicircular structure of the *Brassica hirta* mitochondrial genome. *Curr. Genet.* 11: 565–570.

Palmer, J.D., and C.R. Shields (1984). Tripartite structure of the *Brassica campestris* mitochondrial genome. *Nature (London)* 307: 437–440.

Palmer, J.D., Boynton, J.E., Gillham, N.W., and E.H. Harris (1985). Evolution and recombination of the large inverted repeat in *Chlamydomonas* chloroplast DNA. In *Molecular Biology of the Photosynthetic Apparatus* (K.E. Steinbeck, S. Bonitz, C.J. Arntzen, and L. Bogorad, eds.), pp. 269–278. Cold Spring Harbor Press, Cold Spring Harbor, NY.

Petes, T.D., Byers, B., and W.I. Fangman (1973). Size and structure of yeast chromosomal DNA. *Proc. Natl. Acad. Sci. U.S.A.* 70: 3072–3076.

Pritchard, A.E., Sable, C.L., Venuti, S.E., and D.J. Cummings (1990a). Analysis of NADH dehydrogenase proteins, ATPase subunit 9, cytochrome *b*, and ribosomal protein L14 encoded in the mitochondrial DNA of *Paramecium*. *Nucl. Acids Res.* 18: 163–171.

Pritchard, A.E., Seilhamer, J.J., Mahalingam, R., Sable, C.L., Venuti, S.E., and D.J. Cummings (1990b). Nucleotide sequence of the mitochondrial genome of *Paramecium*. *Nucl. Acids Res.* 18: 173–180.

Reith, M., and J. Munholland (1993). A high resolution gene map of the chloroplast genome of the red alga *Porphyra purpurea*. *Plant Cell* 5: 465–475.

Ris, H., and W. Plaut (1962). Ultrastructure of DNA-containing areas in the chloroplast of *Chlamydomonas*. *J. Cell Biol.* 13: 383–391.

Rochaix, J.-D. (1992). Control of plastid gene expression in *Chlamydomonas reinhardtii*. In *Plant Gene Research, Cell Organelles* (R.G. Herrmann, ed.), pp. 249–274. Springer-Verlag, Wien.

Ryan, K.A., Shapiro, T.A., Rauch, C.A., and P.T. Englund (1988). Replication of kinetoplast DNA in trypanosomes. *Annu. Rev. Microbiol.* 42: 339–358.

Sager, R., and M. Ishida (1963). Chloroplast DNA in *Chlamydomonas*. *Proc. Natl. Acad. Sci. U.S.A.*, 50: 725–730.

Sasaki, Y., Sekiguchi, K., Nagano, Y., and R. Matsuno (1993). Chloroplast envelope protein encoded by chloroplast genome. *FEBS Lett.* 316: 93–98.

Scaramuzzi, C.D., Hiller, R.G., and H.W. Stokes (1992). Identification of a chloroplast-encoded *secM* gene homologue in a chromophyte alga: Possible role in chloroplast protein translocation. *Curr. Genet.* 22: 421–427.

Sederoff, R.R. (1984). Structural variation in mitochondrial DNA. *Adv. Genet.* 22: 1–108.

Shimada, H., and M. Sugiura (1991). Fine structural features of the chloroplast genome: Comparison of the sequenced chloroplast genomes. *Nucl. Acids Res.* 19: 983–995.

Shinozaki, K., Ohme, M., Tanaka, M., Wakasugi, T., Hayashida, N., Matsubayashi, T., Zaita, N., Chunwongse, J., Obokata, J., Yamaguchi- Shinozaki, K., Ohto, C., Torazawa, K., Meng, B.-Y., Sugita, M., Deno, H., Kamagoshira, T., Yamada, K., Kusuda, J., Takaiwa, F., Kato, A., Tohdoh, N., Shimada, H., and M. Sugiura (1986). The complete nucleotide sequence of the tobacco chloroplast genome: Its gene organization and expression. *EMBO J.* 5: 2043–2049.

Shivji, M.S., Li, N., and R.A. Cattolico (1992). Structure and organization of rhodophyte and chromophyte plastid genomes: Implications for the ancestry of plastids. *Mol. Gen. Genet.* 232: 65–73.

Shirzadegan, M., Christey, M., Earle, E.D., and J.D. Palmer (1989). Rearrangement, amplification, and assortment of mitochondrial DNA molecules in cultured cells of *Brassica campestris*. *Theor. Appl. Genet.* 77: 17–25.

Shirzadegan, M., Palmer, J.D., Christey, M., and E.D. Earle (1991). Patterns of mitochondrial DNA instability in *Brassica campestris* cultured cells. *Plant Mol. Biol.* 16: 21–37.

Siemeister, G., and W. Hachtel (1989). A circular 73 kb DNA from the colourless flagellate *Astasia longa* that resembles the chloroplast DNA of *Euglena*: restriction and gene map. *Curr. Genet.* 15: 435–441.

Siemeister, G., and W. Hachtel (1990a). Structure and expression of a gene encoding the large subunit of ribulose-1,5-bisphosphate carboxylase (*rcbL*) in the colourless euglenoid flagellate *Astasia longa*. *Plant Mol. Biol.* 14: 825–833.

Siemeister, G., and W. Hachtel (1990b). Organization and nucleotide sequence of ribosomal RNA genes on a circular 73 kbp DNA from the colourless flagellate *Astasia longa*. *Curr. Genet.* 17: 433–438.

Siemeister, G., Buchholz, C., and W. Hachtel (1990). Genes for ribosomal proteins are retained on the 73 kb DNA from *Astasia longa* that resembles *Euglena* chloroplast DNA. *Curr. Genet.* 18: 457–464.

Simpson, L. (1987). The mitochondrial genome of kinetoplastid protozoa: Genomic organization, transcription, replication, and evolution. *Annu. Rev. Microbiol.* 41: 363–382.

Simpson, L., and J. Shaw (1989). RNA editing and the mitochondrial cryptogenes of kinetoplastid protozoa. *Cell* 57: 355–366.

Siu, C.-H., Chiang, K.-S., and H. Swift (1975). Characterization of cytoplasmic and nuclear genomes in the colorless alga *Polytoma*. V. Molecular structure and heterogeneity of leucoplast DNA. *J. Mol. Biol.* 98:369–391.

Siu, C.-H., Chiang, K.-S., and H. Swift (1976a). Characterization of cytoplasmic and nuclear genomes in the colorless alga *Polytoma*. III. Ribosomal RNA cistrons of the nucleus and leucoplast. *J. Cell Biol.* 69: 383–392.

Siu, C.-H., Swift, H., and K.-S. Chiang (1976b). Characterization of cytoplasmic and nuclear genomes in the colorless alga *Polytoma*. II. General characterization of organelle nucleic acids. *J. Cell Biol.* 69: 371–382.

Small, I., Suffolk, R., and C.J. Leaver (1989). Evolution of plant mitochondrial genomes via substoichiometric intermediates. *Cell* 58: 69–76.

Sogin, M., Gunderson, J., Elwood, H., Alonso, R., and D. Peattie (1989). Phylogenetic meaning of the kingdom concept: An unusual ribosomal RNA from *Giardia lamblia*. *Science* 243: 75–77.

Sor, F., and H. Fukuhara (1982). Nature of an inserted sequence in the mitochondrial gene coding for the 15S ribosomal RNA of yeast. *Nucl. Acids Res.* 10: 1625–1633.

Staub, J.M., and P. Maliga (1992). Long regions of homologous DNA are incorporated into the tobacco plastid genome by transformation. *The Plant Cell* 4: 39–45.

Stuart, K. (1991a). RNA editing in trypanosomatid mitochondria. *Annu. Rev. Microbiol.* 45: 327–344.

Stuart, K. (1991b). RNA editing in mitochondrial mRNA of trypanosomatids. *Trends Biochem. Sci.* 16: 68–72.

Stuart, K., and J.E. Feagin (1992). Mitochondrial DNA of kinetoplastids. *Int. Rev. Cytol.* 141: 65–88.

Sugiura, M. (1989). The chloroplast chromosomes of land plants. *Annu. Rev. Cell Biol.* 5: 51–70.

Sugiura, M. (1992). The chloroplast genome. *Plant Mol. Biol.* 19: 149–168.

Suzuki, J.Y., and C.E. Bauer (1992). Light-independent chlorophyll biosynthesis: Involvement of the chloroplast gene *chlL* (*frxC*). *Plant Cell* 4: 929–940.

Takemura, M., Oda, K., Yamato, K., Ohta, E., Nakamura, Y., Nozato, N., Akashi, K., and K. Ohyama (1992). Gene clusters for ribosomal proteins in the mitochondrial genome of a liverwort. *Nucl. Acids Res.* 20: 3199–3205.

Vahrenholz, C., Riemen, G., Pratje, E., Dujon, B., and G. Michaelis (1993). Mitochondrial DNA of Chlamydomonas reinhardtii: the structure of the ends of the linear 15.8-kb genome suggests mechanisms for DNA replication. *Curr. Genet.* 24: 241–247.

Vernon-Kipp, D., Kuhl, S.A., and C.W. Birky, Jr. (1989). Molecular evolution of *Polytoma*, a nongreen chlorophyte. In *Physiology, Biochemistry and Genetics of Nongreen plastids* (C.D. Boyer, J.C. Shannon, and R.C. Hardison, eds.), pp. 284–286. American Society of Plant Physiologists, Rockville, MD.

Wang, S., and X.-Q. Liu (1991). The plastid genome of *Cryptomonas* Φ encodes an hsp70-like protein, a histone-like protein, and an acyl carrier protein. *Proc. Natl. Acad. Sci. U.S.A.*, 88: 10783–10787.

Ward, B.L., Anderson, R.S. and A.J. Bendich (1981). The mitochondrial genome is large and variable in a family of plants (Cucurbitaceae). *Cell* 25: 793–803.

Warrior, R., and J. Gall (1985). The mitochondrial DNA of *Hydra attenuata* and *Hydra littoralis* consists of two linear molecules. *Arch. Sc. Geneve* 38: 439–445.

Wetmur, J.G., and N. Davidson (1968). Kinetics of renaturation. *J. Mol. Biol.* 31: 349–370.

Wilson, R.J.M., Fry, M., Gardner, M.J., Feagin, J.E., and D.H. Williamson (1991). Have malaria parasites three genomes? *Parasitol. Today* 7: 136–138.

Wilson, R.J.M., Fry, M., Gardner, M.J., Feagin, J.E., and D. H. Williamson. (1992). Subcellular fractionation of the two organelle DNAs of malaria parasites. *Curr. Genet.* 21: 405–408.

Wolf, K., and L. Del Giudice (1988). The variable mitochondrial genome of ascomycetes: Organization, mutational alterations, and expression. *Adv. Genet.* 25: 185–308.

Wolfe, K.H., Morden, C.W., and J.D. Palmer (1991). Ins and outs of plastid genome evolution. *Curr. Opinion Genet. and Dev.*, 1: 523–529.

Wolfe, K.H., Morden, C.W., and J.D. Palmer (1992a). Function and evolution of a minimal plastid genome from a nonphotosynthetic parasitic plant. *Proc. Natl. Acad. Sci. U.S.A.* 89: 10648–10652.

Wolfe, K.H., Morden, C.W., Ems, S.C., and J.D. Palmer (1992b). Rapid evolution of the plastid translation apparatus in a nonphotosynthetic plant: Loss or accelerated sequence evolution of tRNA and ribosomal protein genes. *J. Mol. Evol.* 35: 304–317.

Wolstenholme, D.R. (1992). Animal mitochondrial DNA: Structure and evolution. *Int. Rev. Cytol.* 141: 173–216.

Wolstenholme, D.R., Goddard, J.M., and C.M.-R. Fauron (1983). Replication of *Drosophila* mitochondrial DNA. In *Developments in Molecular Virology*, Vol. 2 (Y. Becker, ed.), pp. 131–148l. Martinus Nijhoff, B.V. The Hague.

Yamada, K., Matsuda, M. Fujita, Y., Matsubara, H., and M. Sugai (1992). A *frxC* homolog exists in the chloroplast DNAs from various pteridophytes and gymnosperms. *Plant Cell Physiol.* 33: 325–327.

Yasuda, T., Kuroiwa, T., and T. Nagata (1988). Preferential synthesis of plastid DNA and increased replication of plastids in cultured tobacco cells following medium renewal. *Planta* 174: 235–241.

Yin, S., Heckman, J., and U.L. RajBhandary (1981). Highly conserved GC rich palindromic DNA sequences which flank tRNA genes in *Neurospora crassa* mitochondria. *Cell* 26: 326–332.

Ziaie, Z., and Y. Suyama (1987). The cytochrome oxidase subunit I gene of *Tetrahymena*: A 57 amino acid NH_2-terminal extension and a 108 amino acid insert. *Curr. Genet.* 12: 357–368.

4

Evolution of Organelle Genomes

Although we cannot currently trace a series of existing intermediates leading from primeval endosymbionts to the plastids and mitochondria of living organisms, we can make some conjectures concerning the steps that may have been involved in the evolution of these organelles as we know them today.

Most of the genes present in the endosymbiotic progenitors of chloroplasts and mitochondria were likely to have been duplicated in the nuclear genome of their protoeukaryotic(s) hosts. Many of these duplicated genes must simply have been lost from precursors of these organelles and replaced in a functional sense by their duplicates in the nucleus. However, this was *not* necessarily a straightforward process since these genes would have needed to acquire DNA sequences specifying protein presequences necessary to direct their uptake into mitochondria and plastids (Chapter 15). Furthermore, multitudes of gene products would have been involved (e.g., the enzymes involved in the tricarboxylic acid cycle of the mitochondrion or the dark reactions of photosynthesis in the plastid). Cavalier-Smith (1982, 1987) believes that development of mechanisms to fit these genes with appropriate targeting sequences and the organelles themselves with a responsive receptor machinery may have been the greatest barrier to organelle evolution. This is a principal reason why Cavalier-Smith argued that chloroplasts and mitochondria must each have been monophyletic in origin. However, the results of Baker and Schatz (1987) suggest that this need not have been such a great problem since 2.7% of *E. coli* DNA sequences can act to specify the presequences required for protein transport. Nuclear DNA rearrangements during evolution could have placed potential organelle targeting sequences in front of genes whose products were destined for import from the cytoplasm to the organelle. Cavalier-Smith (1992) acknowledges these results saying that "it is possible that this innovation was not quite as difficult as it first seemed" and that "one can envisage models in which elements of the protein export machinery of the bacterial symbiont were incorporated into the new organelle import mechanism." However, Cavalier-Smith then goes on to argue that mitochondrial import mechanisms may differ more radically from bacterial export mechanisms than previously thought.

Gene loss from the endosymbiotic precursors of chloroplasts and mitochondria must also have been accompanied by transfer of genes unique to these endosymbionts to the nuclei of the protoeukaryotes. One presumes, for instance, that some fraction of the genes required for respiratory electron transport and photosynthesis must have been transferred to the nucleus where they did not previously exist. For example, the gene encoding the small subunit of RUBISCO is nuclear in land plants and chlorophytes, but located in the plastid genomes of nongreen algae. Similarly, nuclear genes encoding many chloroplast ribosomal proteins show marked homology to *E. coli* ribosomal protein genes, but not to genes encoding cytoplasmic ribosomal proteins (Chapter 13).

The steps required for the introduction and successful expression of an organelle gene donated to the nucleus are as follows. First, a collision between two organelles or between organelle and nucleus must result in DNA transfer. Alternatively, lysis of protochloroplasts and protomitochondria might have been accompanied by uptake and integration of pieces of their DNA into the nuclear genome. Second, integration of donor DNA into the recipient genome has to take place. Third, the introduced gene either must acquire a presequence specifying a transit sequence plus the necessary cis-acting sequences for nuclear transcription and cytoplasmic translation. Fourth, codon usage in the mRNA transcribed from the introduced gene must be compatible with the cytoplasmic translation apparatus (i.e., tRNA anticodon–codon recognition). Evidence that acquisition of presequences required for protein transport may not have been such a great impediment has been discussed above. There is also abundant evidence for gene transfer from organelle to nucleus, which is discussed next.

Thorsness and Fox (1990) did a clever experiment with yeast that allows one to estimate the rate of successful gene transfer from mitochondrion to nucleus (at least in yeast). They constructed a plasmid having the mitochondrial *cox2* gene and nuclear *URA3* gene on a mitochondrial plasmid. Using techniques for mitochondrial transformation and gene identification discussed in Chapter 7, this plasmid was introduced into mitochondria of a yeast strain possessing a nonreverting *ura3* allele in its nuclear genome and lacking an endogenous mitochondrial genome Thorsness and Fox found that the rate of transfer of this plasmid from mitochondrion to nucleus was surprisingly high (ca. 2×10^{-5} Ura$^+$ prototrophs per cell generation). Thorsness and Fox (1993) have now identified 21 recessive mutations in six complementation groups in which the rate of DNA escape from the mitochondrion is increased. Some of these mutations apparently affect other mitochondrial functions.

Molecular evidence indicates that gene transfers are principally unidirectional from plastid or mitochondrion to the nucleus. A major transfer pathway from the flowering plant chloroplast to its mitochondrion also exists.

Transfers from plastid to nucleus. Palmer (1991) distinguishes what he refers to as "primordial" and "recent" transfers. The former class of transfers would presumably have occurred soon after the original symbiotic event prior to the diversification of plastid-containing lineages. The latter would have taken place in a specific plastid lineage. The genes encoding the metabolic enzymes glyceraldehyde-3-phosphate dehydrogenase and phosphoglucose isomerase are examples of primordial transfers. In both cases the nuclear gene encoding the chloroplast form of the enzyme is much more closely related to the equivalent eubacterial gene than to the gene encoding the cognate cytoplasmic enzyme. One example of a recent transfer involves the *tufA* gene, encoding elongation factor Tu (Baldauf and Palmer, 1990; Baldauf *et al.*, 1990). The three salient observations are that (1) the *tufA* gene is present in the nucleus of land plants and all Charophycean green algae [the presumptive ancestors of land plants, Mishler and Churchill (1985)], (2) the gene is absent from the chloroplast of land plants and certain Charophytes (i.e., *Zygnematales*), and (3) the gene is present in both nucleus and chloroplast of many Charophytes (i.e., *Charales, Coleochaetales*). These latter two groups are thought to include the Charophytes most closely related to land plants. The favored interpretation of these results is that an ancient gene transfer resulted in a duplication of the *tufA* gene in both chloroplast and nucleus in an ancestor of Charophytes and land plants. Subsequently, the chloroplast copy of the *tufA* gene was lost independently in the *Zygnematales* and land plants. A second example of a recent transfer involves the ribosomal protein gene, *rpl22*, whose transfer to the nucleus took place in the course of evolution of land plants (Gantt et al., 1991). This gene is present in all land plant lineages examined except legumes. The nuclear gene in pea has been identified and been shown to contain two exons separated by an intron. One of these encodes the putative presequence required to direct the CL22 protein encoded by *rpl22* into the chloroplast. Gantt et al. (1991) speculate that the *rpl22* structural gene, encoded by the second exon, might have acquired the exon of a different gene specifying the presequence.

While these experiments document the transfer of single functional plastid genes to the nucleus, other experiments reveal the presence of presumably nonfunctional cpDNA sequences in the nuclear genomes of land plants (Ayliffe and Timmis, 1992a, b). Although such sequences are usually short, long tracts of cpDNA in excess of 18 kb have now been identified in the tobacco nuclear genome. Five clones containing these cpDNA sequences were found to encompass one-third of the tobacco plastid genome. This raises the possibility that uninterrupted sequences as large as the entire plastid genome might reside in the tobacco nuclear genome.

Transfers from mitochondrion to nucleus. Schuster and Brennicke (1988) and Cantatore and Saccone (1987) summarized these transfers for yeast and animals. In locust, sea urchin, and humans apparently random fragments of mtDNA can be found in the nuclear genome by southern blotting. In *Saccharomyces* part of the ribosomal *var1* gene, the 3'-end of the *cytb* gene and a portion of an ori/rep sequence were reported to be integrated in single copies in the nuclear genome (Farelly and Butow, 1983). These sequences, which are noncontiguous in the mitochondrial genome, were adjacent in the nuclear genome. Farelly and Butow (1983) proposed that these "foreign" mitochondrial sequences may have originated from a petite mutant of yeast (Chapter 7) in which the mtDNA had been rearranged.

In each of the above cases the integrated mtDNA sequences were nonfunctional, but there are at least two examples of transfer that resulted in function. The first involves the fungal *atp9* gene (see Schuster and Brennicke, 1988, for references). This gene is found in the mitochondrion of yeast, but is encoded by a nuclear gene in *Aspergillus* and *Neurospora*. However, the two latter species contain a second copy of the gene in the mtDNA that is not expressed under normal growth conditions. In animals, whose closest relatives phylogenetically are the fungi (Chapter 2), the *atp9* gene is nuclear. The second example involves the *cox2* gene of plant mitochondria. This gene seems to have been transferred from mitochondrion to nucleus in the course of evolution of the legumes (Nugent and Palmer, 1991). This transfer was estimated to have taken place 60–200 million years ago, after the emergence of dicotyledons and monocotyledons, but before divergence of the legume subfamily Papilionoideae. Both mitochondrial and nuclear *cox2* genes are present in pea, soybean, and common bean, but the mitochondrial gene is operative in pea while the nuclear gene is expressed in soybean and common bean. In cowpea and mung bean only the nuclear gene is present and expressed. Of particular interest is the finding that the nuclear *cox2* sequence closely resembles edited mitochondrial transcripts. RNA editing converts C to U residues in plant mitochondrial mRNA (Chapter 14). The results suggest that transfer of *cox2* to the nucleus was mediated by an RNA intermediate that underwent reverse transcription to yield an edited *cox2* gene for nuclear integration. This possibility was first hypothesized by Schuster and Brennicke (1987, 1988) who discovered a reverse transcriptase-like sequence in the mitochondrial genome of the evening primrose (*Oenothera*).

Transfers from plastid to mitochondrion. The most massive gene transfers so far recorded have occurred from flowering plant plastids to their mitochondria (see Hanson and Folkerts, 1992; Lonsdale, 1989; Newton, 1988; Palmer, 1992; Schuster and Brennicke, 1988, for reviews). The best characterized examples come from maize and are summarized by Lonsdale (1989): (1) a 12-kb uninterrupted cpDNA sequence from the inverted repeat is the longest cpDNA sequence that has been found in any mitochondrial genome. This sequence includes the 16S rRNA gene, $tRNA^{Ile}$, $tRNA^{Val}$ *rps7*, 3' exon of *rps12*, *ndh2*, and $tRNA^{Leu}$; (2) the *rbcL* gene; (3) a 2-kb cpDNA fragment containing the chloroplast 5S rRNA, the 3'-end of the 23S rRNA, the 4.5S rRNA, and $tRNA^{Arg}$ and $tRNA^{Asp}$; and (4) genes with homology to $tRNA^{His}$, $tRNA^{Cys}$, and another $tRNA^{Arg}$ gene. One essential chloroplast-derived mitochondrial tRNA in maize is maintained on a plasmid (Chapter 10)

In a functional sense, the most important chloroplast genes in plant mtDNA encode tRNAs. Thus, flowering plant mitochondrial tRNAs have been divided into three groups termed ''chloroplast-like,'' ''cytosolic-like,'' and ''native'' (Chapter 13). Liverwort, once again, does not behave like flowering plants (Oda *et al.*, 1992). None of its 29 mitochondrially encoded tRNAs is ''chloroplast''-like, nor do any other foreign DNA sequences appear to be incorporated into this mitochondrial genome.

Nucleus to mitochondrion. There is no evidence for transfer of nuclear genes to the chloroplast in the three completely sequenced flowering plant chloroplast genomes. However, one good example exists for transfer of a nuclear DNA sequence to a plant mitochondrion. Integrated into the mitochondrial genome of *Oenothera* is a fragment of the nuclear gene encoding the 18S cytoplasmic rRNA of this plant (see Schuster and Brennicke, 1988).

ORGANELLE GENOME EVOLUTION

Changes in Organelle Genome Size and Composition

Palmer (1991, 1992) distinguishes four classes of factors affecting the evolution of organelle genome size: repeated sequences, gene and intron content, incorporation of foreign DNA sequences, and length mutations.

Repeated sequences. With the exception of the large inverted repeat, most land plant chloroplast genomes possess many fewer repeated sequences than do their mitochondrial genomes. Many land plant chloroplast genomes do not contain repeats larger than

50 bp including the completely sequenced genomes of rice, tobacco, and liverwort, but there are some exceptions. Thus, the chloroplast genomes of grasses, conifers, geranium, and subclover have a few short, dispersed repeats of (SDRs) 50–1000 bp. Certain of these are dispersed pseudogenes and could have arisen by reverse transcription and secondary integration of existing chloroplast genes.

Most of the restriction fragment polymorphism variation distinguishing the chloroplast genomes of the interfertile algal species *Chlamydomonas reinhardtii* and *C. smithii* can also be accounted for by insertions and deletions of 50–100 bp SDRs in the intergenic regions (Boynton et al., 1992). The individual SDRs are each composed of collections of shorter repeat motifs that may occur in direct or inverted orientation and in different orders. SDRs account for ca. 22% of the total chloroplast genome in *C. reinhardtii*. Whether SDRs are restricted to certain species of *Chlamydomonas* is unknown.

The large rRNA gene-containing inverted repeat is an important source of length variation in the chloroplast genome (Palmer, 1991, 1992). In land plants, with the exception of conifers, certain legumes, and a few other groups lacking the inverted repeat structure (Palmer, 1991), the inverted repeat is typically 20–30 kb in size, but ranges from 10 to 76 kb and always contains the rRNA genes. Most of this variation can be accounted for by spreading of the inverted repeat to encompass regions that are typically unique sequence or by contraction of the inverted repeat to expose new regions of unique sequence. This is nicely illustrated by comparing the inverted repeats of spinach, a typical flowering plant in this respect, with geranium and coriander. In two genera of the *Geraniaceae*, the large inverted repeat encompasses half of the chloroplast genome (Palmer, 1991). In contrast, in *Coriander* the repeat is less than half its normal size. Interestingly, nonphotosynthetic mutants of *Chlamydomonas* have been isolated with large duplications of the inverted repeat indicating that spreading of the repeat can occur in a single step (Boynton et al., 1992).

Although most algal chloroplast DNAs possess a large rRNA gene containing inverted repeat, the primitive or derivative nature of the inverted repeat in different lineages cannot be established for three reasons (Palmer, 1985, 1991). First, the repeat is absent from at least some members of three of nine algal phyla examined. Second, evidence of molecular homology of the repeats save for the rRNA genes is lacking. Third, rRNA genes are known to have been duplicated in parallel in several lineages of plastids and mitochondria.

Sizable inverted repeats featuring the rRNA genes have seemingly evolved independently in the mitochondrial genomes of certain fungi and protists (Clark-Walker, 1992; Chapter 3).

The large, complex mitochondrial genomes of flowering plants contain one or more small (2–3 copy) families of large (1–14 kb) repeats (Chapter 3). These are usually directly oriented and are implicated in high-frequency recombination that results in the multipartite organization of most of these organelle genomes (Hanson and Folkerts, 1992; Lonsdale, 1989; Palmer, 1992). SDRs of ca. 50–1000 bp are also common in these molecules. These repeats may serve as sites where recombination leading to inversion occurs. Tandemly repeated sequences have been found in the mtDNA molecules of a number of animals (Wolstenholme, 1992; Chapter 3). For instance, variability in the size of the mitochondrial genome in certain lizards can be accounted for by direct, tandem duplications (0.8-8 kb) of both coding and noncoding portions of the genome (Moritz et al., 1987).

Among fungi the well-characterized mitochondrial genome of bakers' yeast is known to contain eight *ori* or *rep* elements of 200–300 bp plus 200 G+C-rich sequences of 20–50 bp (G+C clusters, see Clark-Walker, 1992). The G+C rich clusters can be grouped into several families of repetitive sequences some of which include up to 30 members which are almost identical (Weiller et al., 1989). Analogous G+C-rich, repetitive sequences are also found in *Neurospora crassa* (Clark-Walker, 1992). These elements seem to be mobile (Chapter 10) and are missing from the mitochondrial genome of the yeast *Kluyveromyces lactis*, a relative of *S. cerevisiae*. This suggests that *K. lactis* never acquired G+C-rich repetitive sequences. Clark-Walker (1992) hypothesizes that in the past *S. cerevisiae* and *N. crassa*, at least, may have been separately infected with G+C-rich elements. The formation of tandem repeats in the *S. cerevisiae* mitochondrial genome is also thought to be of frequent occurrence (Clark-Walker, 1992). Thus, a spontaneously occurring mtDNA polymorphism arises from addition of two extra copies of a 14-bp A+T-rich sequence to a tandem repeat of this unit comprising six copies (Skelly and Clark-Walker, 1991).

Introns. mtDNA differs radically in structure among animals, plants, fungi, and protists (Chapter 3). With the exception of animals (Wolstenholme, 1992), this is largely the result of variations in lengths of intergenic spacers, the presence of introns, or both. Thus, variations in intron number and length profoundly affect the sizes of mitochondrial genes and genomes in fungi (Clark-Walker, 1992; Gray, 1989, 1992). These differences extend even to interfertile strains of the same species as in the case of the *cox1* and *cytb* genes of *Saccharomyces* (Chapters 3 and 10). Introns contribute little to the great size variation in mitochondrial genomes of higher plants although they are present in some genes (Gray, 1989, 1992; Palmer, 1992).

Although land plant cpDNAs contain about 20 introns ranging in size from 400 to 1000 bp, intron content contributes little to size variation of these organelle genomes (Palmer, 1992). Since all but one of these introns is shared by at least one angiosperm and the liverwort *Marchantia*, Palmer (1991) argues that they must have been present in the common ancestor of land plants. Two of these introns, found in the trn^{Ile} and trn^{Ala} genes located in the spacer between the rRNA genes (Chapter 3), have also been detected in the Charophyceae, the algal class from which land plants are thought to derive (see above). Evidence for an even earlier intron gain is the retention of a group I intron in the $tRNA^{Leu}_{UAA}$ genes of cyanobacteria, *Cyanophora*, many other algae, and land plants (Kuhsel et al., 1990; Palmer and Logsdon, 1991; Xu et al., 1990). As the $tRNA^{Leu}$ intron is not mobile, this observation argues for the monophyletic origin of plastids.

Intron content is much more variable in the plastid genomes of algae (Palmer, 1991). Although the aforementioned $tRNA^{Leu}$ intron is the only one so far discovered in the cyanelle genome (Bryant and Bohnert, 1993), 149 introns have been found in the *Euglena* plastid genome (Hallick et al., 1993). Palmer (1991) points out that no intron is shared among the four plant lineages so far examined for which substantial data on intron content are available (*Chlamydomonas reinhardtii*, *Euglena gracilis*, *Cyanophora paradoxa*, land plants), and only the cyanobacterially derived $tRNA^{Leu}$

intron is shared by even two of these lineages. The chloroplast genome of the red alga *Porphyra purpurea* is particularly notable in lacking introns in any of the >125 genes sequenced including tRNA^{Leu} (Reith and Munholland, 1993). If this intron represents an early gain as suggested above, it must have been lost from the tRNA^{Leu} gene of *Porphyra*. The almost completely sequenced cyanelle genome is also devoid of introns except for the aforementioned intron in the tRNA$^{Leu}_{UAA}$ gene (Bryant and Bohnert, 1993).

One of the most systematic comparisons of chloroplast intron distribution in related taxa comes from work done on the rRNA gene introns of the large subunit of the chloroplast ribosome of *Chlamydomonas* (Turmel et al. 1993). Careful phylogenetic reconstructions of this genus, more likely at least a family, using nuclear small subunit and chloroplast large subunit rRNA gene sequences, are congruent. They reveal the existence of two main lineages. A total of 39 group I introns representing 12 distinct insertion sites have been identified in domains II, IV, and V of the large subunit rRNA gene of the chloroplast ribosome. The regions where introns are inserted correspond to *E. coli* rRNA domains located on the ribosome surface. Turmel et al. (1993) speculate that the mechanism of intron insertion involved reversal of the self-splicing reaction. This permitted foreign group I introns to be spliced into rRNA followed by reverse transcription of the recombined rRNA and integration into the chloroplast genome by homologous recombination (see Woodson and Cech, 1989, Chapter 10). There is one intron in *C. eugametos* that does not use the proposed transfer mechanism. This intron encodes an endonuclease that catalyzes its transfer into the rRNA genes of strains lacking it (Chapter 10, Mobile Group I Introns).

Another thorough systematic study focuses on the intron in a chloroplast gene encoding a ribosomal protein (*rpl2*). Downie et al. (1991) determined the distribution of the *rpl2* intron for 390 species representing 116 angiosperm families and found that intron loss occurred independently six times.

Studies of chloroplast introns have provided fuel for the introns early or late debate. "Introns early" advocates believe introns were present in the common ancestor of eukaryotes, eubacteria, and archaebacteria with the general direction of evolution being toward intron loss (Darnell and Doolittle, 1986; Gilbert et al., 1986). Introns early proponents believe that the RNA self-splicing properties of certain introns are ancient. They also argue that the broad distribution of introns found among eukaryotic lineages is consistent with the idea that the original assembly of protein-coding genes in eukaryotes and eubacteria was determined by intron-mediated "exon shuffling." "Introns late" supporters assume that primeval genes lacked introns. These were instead acquired independently and more recently in the various lineages where they are found (Cavalier-Smith, 1985; Dibb and Newman, 1989; Palmer and Logsdon, 1991; Rogers, 1989). The presence of introns in chloroplast and mitochondrial genes is viewed by introns early advocates as evidence that intron elimination is under relaxed selection compared to the eubacteria from which these organelles derive. Introns late supporters believe organelles acquired their introns after endosymbiosis.

Palmer (1991) argued forcefully in favor of the introns late view for plastids since no one intron seems to be shared among all different plant lineages so far characterized. He also points out that in all but one case, when introns are found in the same gene in multiple lineages, they are not homologous, being different in location, sequence, and sometimes type. Thus, the *psbA* gene of *C. reinhardtii* possesses four group I introns

(Erickson *et al.*, 1984) located in entirely different positions than the four group II introns of the *Euglena gracilis psbA* gene (Karabin *et al.*, 1984). Furthermore, no introns have so far been found in the chloroplast genome of the red alga *Porphyra purpurea* and only the tRNA^{Leu} intron in the cyanelle genome as mentioned above. The discovery of the tRNA^{Leu} intron in Cyanobacteria (Kuhsel *et al.*, 1990; Xu *et al.*, 1990) discussed earlier raises the question of whether introns arose early or late in eubacteria. Palmer and Logsdon (1991) cite this as the singular exception to the late arrival rule that they promulgate not only for introns in organelle genes, but for nuclear gene introns as well. There is also abundant evidence for intron mobility in organelles (Chapter 5). The existence of mechanisms promoting intron mobility strengthens the argument that most introns are late arrivals.

The tRNA^{Leu} gene is no longer the sole example of a eubacterial intron. Ferat and Michel (1993) have discovered the first examples of group II introns outside of chloroplasts and mitochondria. In fact, they have found these introns in taxa containing the putative ancestors of chloroplasts and mitochondria, the cyanobacteria and purple bacteria (Chapter 2). These results are consistent with the early origin of group II introns in the organisms that gave rise to chloroplasts and mitochondria followed by intron loss in certain organellar lineages. In a sense the introns early versus introns late debate is to some degree semantic. Both views are probably correct. There is abundant evidence for intron mobility in organelles (Chapter 10). The existence of mechanisms promoting intron mobility strengthens the argument that many introns are late arrivals, but in no way negates the hypothesis that the first introns arose early. By casting a wider taxonomic net Ferat and Michel have shown that introns are more prevalent in eubacteria than previously suspected.

Spacers. Organelle genome size can also be varied by increasing or decreasing intergenic spacer regions (Clark-Walker, 1992; Gray, 1989, 1992). Animal mitochondrial genomes are extremely compact with genes being separated by at most a few nucleotides, but in fungi A+T-rich spacers together with introns account for the large variation in mitochondrial genome size with the smallest mitochondrial genomes having relatively short spacers. Clark-Walker (1992) discusses the possibility that the A+T-rich spaces of *S. cerevisiae* may be mobile elements and that they also can be built up as repeat elements through recombination or replication slippage (slipped strand mispairing). The large size as well as the great length variation of flowering plant mitochondrial genomes is also explained by spacer DNA. But in flowering plants the spacer DNA has much the same base composition as the rest of the mitochondrial genome (ca. 47% G+C). Chloroplast genes in land plants, in contrast, are fairly tightly packed with many being organized into operons (Chapter 12).

Length mutations. The length of spacer regions and introns can be affected by insertions and deletions (Palmer, 1991, 1992). Most length mutations in cpDNA are only 1–10 bp in size. They are often flanked by or close to short direct repeats, suggesting that they may arise by mispairing during DNA replication and repair. Larger length mutations (ca. 10–1200 bp) can occur as the result of addition or deletion of sequences from unequal recombination between misaligned tandem repeat sequences. Deletions in cpDNA are also produced by intramolecular recombination between short repeats (Boynton *et al.*, 1992; Palmer, 1991).

A fitting way to end this discussion of mechanisms affecting organelle genome size is to quote from Clark-Walker's (1992) discussion of how the roomy mitochondrial genome of *S. cerevisiae* might have been constructed:

> It would appear that intergenic regions in *S. cerevisiae* mtDNA have been built up by a variety of different processes. If the mitochondrial genome of *C. glabrata* is taken to resemble the ancestral molecule then expansion of small A+T-rich segments may first occur by slipped-strand mispairing. The resulting tandem repeats may be enlarged further by unequal crossing-over, and base substitutions, as well as additions or deletions, would alter their repetitive nature. Transposition and inversion of oligonucleotide segments probably occur as well as the acquisition of G+C-rich clusters. These sequences may have been built by similar processes to A+T segments or they could have an external origin. However, it appears that once acquired, G+C elements are mobile and have been transposed to target sites within intergenic regions and, where tolerated, into some genes as well.

Clark-Walker also points out that such a genome building process can result in a rather rickety structure in which deletion can lead to formation of mitochondrial mutants whose genomes become nonfunctional. Such mitochondrial petite mutants (Chapter 7) arise in many strains of *S. cerevisiae* at rates of 1% per generation. Clark-Walker puts it this way:

> There is nevertheless an unfortunate consequence arising from intergenic sequence that has been made from repetitive elements. The mitochondrial genome in *S. cerevisiae* is destabilized because frequent excisions occur at short directly repeated sequences, mainly G+C clusters, resulting in the formation of defective mtDNA that is manifested in petite mutants.

Changes in Organelle Genome Organization

Most vascular plants have the same chloroplast gene order as tobacco (Fig. 3-6). However, this order differs by a single large inversion (Palmer 1991, 1992) from nonvascular plants, represented by *Marchantia*, and what now seems likely to be the earliest diverging lineage of vascular plants, the lycopsids (Raubeson and Jansen, 1992). The ancient origin of this inversion in the vascular plant lineage is highlighted by its occurrence in the psilopsids, which have often been regarded as the earliest diverging lineage of vascular plants (Raubeson and Jensen, 1992). Other inversions with respect to the tobacco chloroplast gene order can also be found among flowering plants. Thus, the rice chloroplast genome differs from tobacco by three inversions. Even more complex cases of chloroplast genome rearrangement have, however, been described in two lineages of legumes, geranium, and conifers. Most of these seem to involve large numbers of inversions. In the two lineages of legumes and in conifers these rearrangements were preceded by loss of the inverted repeat suggesting that its absence may contribute to instability of the chloroplast genome (Palmer, 1991).

In contrast to most land plants the chloroplast genomes of the sibling species pairs *Chlamydomonas reinhardtii/smithii* and *C. moewusii/eugametos*, while colinear within species pairs, are enormously rearranged with respect to each other and to land plants (Lemieux and Lemieux, 1985; Lemieux *et al.*, 1985; Turmel *et al.*, 1993). As one might expect, this extensive molecular rearrangement is accompanied by rearrangement of the chloroplast genes themselves. This is particularly striking in the case of the chloroplast genes encoding components of the ATP synthase (Fig. 3-7). In higher plants these genes are organized in two operons similar to those of cyanobacteria, but in *Chlamydomonas* they are scattered about the chloroplast genome.

Euglena lacks the inverted repeat structure and its rRNA gene operons are arranged in tandem varying from one to five in number (Hallick and Buetow, 1989; Chapter 3). However, *Euglena* retains protein-encoding operons similar in organization to those of land plants although possessing a different overall arrangement. Similarly, operon arrangement again varies in *Cyanophora,* although many of the same operons found in land plants and *Euglena* are retained. Despite conservation of several operons, three distinct groups of Chromophytes (three diatoms, two brown algae and a Xanthophyte) differ from each other in plastid gene order and also with respect to other plastid genomes (Palmer, 1991).

Mitochondrial genomes in animals, fungi, and plants display contrasting evolutionary histories. Animal mitochondrial genomes are characterized by their compactness, but variations in mitochondrial gene order have been found in every phylum so far studied (see Gray, 1989, 1992; Wolstenholme, 1992). For instance, human mtDNA differs from that of sea urchin by two transpositions in the order of protein and rRNA genes and from that of *Drosophila* by three inversions.

In fungi, in contrast, unique mitochondrial gene orders have been found in every genus examined although conserved blocks of genes are evident in pairwise comparisons (Wolf and Del Giudice, 1988). Thus, in the yeasts *Candida glabrata* and *Saccharomyces exiguus*, which have relatively small mitochondrial genomes, the *atp6* and *atp9* genes are part of a five gene cluster. Their order is the same as in the *unc* operon of *E. coli* that encodes all of the ATPase genes of that bacterium. However, these genes are dispersed in larger mitochondrial genomes, such as that of *S. cerevisiae,* leading Clark-Walker *et al.* (1983) to hypothesize that smaller mitochondrial genomes may have been primitive. As discussed in the last chapter, this hypothesis is supported by studies on the genus *Dekkera/Brettanomyces* (Hoeben and Clark-Walker, 1986; Hoeben *et al.*, 1993) indicating not only that species with the smallest mitochondrial genomes are ancestral to those with larger mitochondrial genomes, but that the larger mitochondrial genomes are more prone to rearrangement. This model would be consistent with the formation of intergenic regions by acquisition and amplification of repeated sequences that could serve as sites for intramolecular recombination (Clark-Walker, 1992; Hoeben *et al.*, 1993).

The giant mitochondrial genomes of flowering plants with their generous intergenic spacer regions demonstrated recombinogenicity and possible mosaic origin (Gray, 1989; Gray *et al.*, 1989b; Chapter 3), exhibit distinctive gene orders in virtually every species so far examined (Palmer, 1992). Mitochondrial gene and sequence order is virtually randomized among angiosperms from different families with the only apparent conservations involving three pairs of cotranscribed genes: 18S/5S rRNA, *ndh3/rps12,* and *ndh1/rps13*. In some cases species in the same genus and even cultivars of the same crop species have extremely rearranged mitochondrial genomes.

Palmer (1992) considered the factors that may be responsible for the high rate of rearrangement of flowering plant mitochondrial genomes and identified three as potentially playing major roles. First, plant mtDNA has many SDRs that could act as sites of inversional recombination. This supposition is supported by sequencing studies showing that most inversions have SDRs present at the recombination breakpoints.

Second, plant mitochondrial genes are mostly monocistronic and separated by long spacers so rearrangement will not interrupt operon structure. Third, Palmer, in comparing chloroplast and mitochondrial genomes, points out that the inverted repeat in the former organelle genome may have a stabilizing influence.

In the foregoing discussion we have ignored protist mitochondrial DNAs, choosing to focus on those groups of organisms where the phylogeny is fairly well understood and about which there is much information. Nevertheless, it seems likely that genomic features similar to those described above are likely to contribute to rearrangement of organelle genomes in these organisms as well.

Role of Recombination in Organelle Genome Evolution

Although direct evidence for recombination of animal mtDNA is lacking (cf. Clayton, 1992; Wolstenholme, 1992), rearrangements have occurred during the evolution of animal mitochondrial genomes (see above). In contrast, intraorganelle recombination is a regular feature of flowering plant chloroplast and mitochondrial genomes and those of fungi. Recombination of molecules derived from different plastids or mitochondria also occurs following organelle fusion in the life cycles of *Chlamydomonas* and yeast (Chapter 7). Hence, it seems safe to say that intra- and intermolecular recombinations have probably played important roles in the evolution of most organelle genomes.

Four major types of recombination, namely, homologous, site-specific, transpositional, and irregular, have been identified in prokaryotes and eukaryotes (Low, 1988; Low and Porter, 1978). There is much evidence for homologous recombination in chloroplasts. The multimeric forms of cpDNA found in higher plants (cf. Deng *et al.*, 1989) probably result from single exchange events between monomers leading to cointegrate formation (Kolodner and Tewari, 1979). In *Chlamydomonas* homologous recombination between molecules carrying different genetic markers has been abundantly demonstrated among the progeny of biparental zygotes and also following chloroplast transformation (Boynton *et al.*, 1992; Chapter 7). One can infer that the 50- to 200-bp SDR sequences found in the plastid genome of *Chlamydomonas* are not required for cointegrate formation as these elements are absent from most plant chloroplast genomes. In flowering plant mitochondrial genomes, on the other hand, small (2–3 copy) families of large (1–14 kb) repeats, most of which are directly oriented, appear to be responsible for generating the master and subgenomic circles that characterize this organelle (Hanson and Folkerts, 1992; Lonsdale, 1989; Palmer, 1992; Chapter 3).

While homologous recombination between SDR elements is not responsible for cointegrate formation in plastid genomes, these elements are involved in generating deletion mutants in *Chlamydomonas* (Boynton *et al.*, 1992) and probably also in those plants possessing SDRs (Bowman and Dyer, 1986). Chloroplast DNA sequencing studies in grasses and conifers have shown that SDR sequences are located near the endpoints of inversions providing suggestive evidence that these sequences may be involved in the inversion process (Palmer, 1991). In fact, Palmer (1991) notes that SDR sequences are unusually abundant in those plant chloroplast genomes exhibiting the greatest degree of rearrangement.

Two other operationally distinguishable recombination pathways exist in the inverted repeat of the chloroplast genome. The first process (flip-flop recombination) yields 50–50 mixtures of the two isomeric forms of the chloroplast genome. The second

EVOLUTION OF ORGANELLE GENOMES

results in very efficient copy correction of point mutations, deletions, and inversions ensuring that both repeats are always identical (Boynton *et al.*, 1992; Palmer, 1991).

Recombination also plays an important role in rearrangement of fungal mitochondrial genomes. In *S. cerevisiae*, for example, short G+C-rich clusters within A+T-rich spacers are involved in recombination events (Gray, 1989). Generation of petite mutants involves intramolecular recombination between direct repeats followed by excision of the sequence in between, circularization, and amplification (Clark-Walker, 1992). The same mechanism is the cause of similar excision amplification events in other fungi as well. Flip-flop recombination is also observed in the inverted repeats found in the mitochondrial genomes of Oomycetes (Clark-Walker, 1992).

Despite extensive documentation on the role of recombination in the evolution of organelle genomes, virtually nothing is known of the genes and enzymes involved. A RecA-like protein that has been isolated from chloroplasts of pea and *Arabidopsis* appears to be coded in the nucleus. A cDNA clone possessing an open reading frame with 50 to 60% amino acid identity to a cyanobacterial RecA protein has also been identified from the latter organism (Cerutti *et al.*, 1992). Determination of the roles of proteins like RecA in recombination of organelle DNA will be important if we are ever to untangle the various pathways of recombination that exist in plastids and most mitochondria. In fact, the obvious question is whether these virtually ubiquitous recombination systems (animal mitochondria excepted) are there to promote recombination or repair. After all, recombination between organelle genomes from different parents is the exception and not the rule.

RATES OF EVOLUTION OF ORGANELLE DNA SEQUENCES

Nucleotide substitution rates have been studied in most detail in higher animals and plants (Table 4–1). Mammalian mtDNA has long been known to have rapid nucleotide substitution rates relative both to mammalian nuclear DNA and other organelle genomes (Brown, 1985; Cantatore and Saccone, 1987; Moritz *et al.*, 1987; Wilson *et al.*, 1985).

Table 4–1. Nucleotide Substitution Rates in Eukaryotic Genomes

Genome	*Synonymous Substitution Rate*[a]	*Relative Substitution Rate*[b]	*Nonsynonymous Substitution Rate*[a]
Angiosperm mitochondrial	0.5	1	0.1
Angiosperm chloroplast			
Both single copy regions	1.5	3	0.2
Inverted repeat	0.3	0.6	0.1
Angiosperm nuclear	5.4	12	0.4
Mammalian nuclear	2–8	4–16	0.5–1.3
Mammalian mitochondrial	20–50	40–100	2–3

From Palmer (1992).
[a]Estimated rate of substitutions/site/10^9 years. Rates are derived from mean values over all genes and from two estimates of divergence times (60 Ma for maize/wheat and 200 Ma for monocots/dicots). Ranges for mammalian rates are over different orders.

[b]Normalized relative to the angiosperm mitochondrial rate.

These high rates of nucleotide substitution are characterized by (1) very high frequencies of synonymous or relatively synonymous base pair substitutions; (2) a high transition/transversion ratio (ca. 10:1); and (3) a strong bias toward C ↔ T transitions in the L strand (see Gray, 1989). Rates of nucleotide substitution are highest at the control region where replication and transcription of mtDNA initiate and lowest for the tRNA and rRNA genes (Brown, 1985; Moritz et al., 1987). Each of the 13 mitochondrial protein-encoding genes with the exception of *cox3* exhibits its own characteristic rate of change with relative rates being similar among vertebrates. *cox3* undergoes a higher rate of change in the primate lineage than in rodents or artiodactyls. In contrast, nucleotide substitution rates in nuclear and mitochondrial protein-encoding genes of *Drosophila* are similar (Gray, 1989), except in the β lineage of Hawaiian *Drosophila* (Moritz et al., 1987). In this lineage the evolutionary rate for mtDNA is high relative to single copy nuclear DNA and three times faster than the rate calculated for the α lineage of Hawaiian *Drosophila*. Curves relating sequence differences to divergence time are biphasic for both mammalian and *Drosophila* mtDNA (see de Salle et al., 1987; de Salle and Hunt, 1987). There is an initial steep slope ascribable to the rapid accumulation of synonymous substitutions and a final shallow slope resulting from the slow accumulation of substitutions causing amino acid changes (Fig. 4–1).

Why does mammalian mtDNA evolve so rapidly? Initially the reason seemed to be that mitochondria possessed an error-prone DNA polymerase. Four distinct DNA polymerases have been identified in mammalian cells (Fry and Loeb, 1986). One of these, polymerase γ, has been isolated from mitochondria of amphibians, birds, mammals, echinoderms, and insects (Fry and Loeb, 1986; Wernette et al., 1988). Fidelity estimates made for this enzyme using a reversion assay with the single-stranded bacteriophage φX174 indicated an error rate of one mispairing per 8000 bases polymerized (Kunkel and Loeb, 1981). In contrast the error rate for the nuclear polymerase was 1–10/100,000 in the same assay (Grosse et al., 1983). However, subsequent experiments employing a forward mutation assay using the *lacZ* gene of *E. coli* carried on plasmid M13mp2 where the error rate is averaged over many positions showed that polymerase γ is very accurate with an error rate of ca. 1 in 500,000 (Kunkel, 1985). This is not surprising in light of the fact that polymerase γ has a 3' to 5' exonuclease activity that presumably confers a proof-reading capacity on the enzyme (Clayton, 1992). Also, the mispairings that polymerase γ permits are principally transversions, whereas transitions predominate in animal mtDNA.

A second reason why mammalian mtDNA might be mutation prone relates to the balance between DNA damage and repair. Thus the central role of the mitochondrion in oxidative metabolism might contribute to the high level of mitochondrial DNA damage because of its unavoidable proximity to oxygen radicals (Ames, 1988; Richter et al., 1988; Wallace, 1992). In fact evidence for differential chemical damage comes from observations indicating a 30-fold greater amount of oxidative damage in the mitochondrion compared to the nucleus. However, if oxidative damage is related to the increased mutation rate of animal mtDNA, the same should be true for plant mitochondria. This is clearly not the case (Table 4-1).

Third, although certain enzymes involved in DNA repair in other systems such as uracil-DNA glycosylase and AP endonuclease have been found in mammalian mitochondria (Gray, 1989), these mitochondria seem to be deficient in DNA repair since they appear to lack the enzymes necessary to remove thymine dimers from mtDNA.

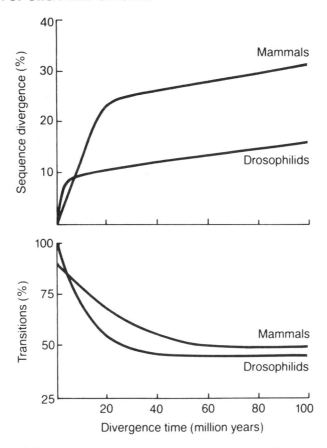

Fig. 4–1. Plots of divergence time versus total percentage sequence divergence (top) and percentage of total changes that were transitions (bottom). For *Drosophila* transitions were in a 900-bp segment of mtDNA (de Salle et al., 1987). Divergence times for mammals were based on the fossil record (Brown, 1983; Brown et al., 1982) and for drosophilids on biogeographical considerations and immunological calibrations. From de Salle and Hunt (1987).

This implies the absence of mechanisms for excision repair, photoreactivation, and possibly recombinational repair (Clayton et al., 1974; Lansman and Clayton, 1975) although certain types of repair activities have been reported in mammalian mitochondrial fractions (Clayton, 1992). Since transitions and length mutations are strongly favored over transversions in mammalian mitochondria, Wilson et al. (1985) speculated that mismatch repair might be defective in these mitochondria. Mismatch repair highly favors correction of transition mutations over transversions (Modrich, 1987).

Fourth, animal mtDNA may evolve rapidly because of relaxed selection (Brown et al.; 1982, Gray, 1989). Thus, an expanded codon recognition pattern in mammalian mitochondria (Chapter 14) means that a single tRNA is able to decode all four codons specifying a given amino acid in a given family. Hence, third position codon changes are effectively silent. In support of this observation is the high ratio of silent to replacement amino acid changes effected by mammalian mtDNA nucleotide substitutions

(Brown et al., 1982). However, Moritz et al. (1987) caution that "the possibility that the critical event is a slowdown in the rate of single copy nuclear DNA change among vertebrates—rather than an elevated rate of mtDNA divergence—places a rather different perspective on the matter. Perhaps more emphasis should be placed on investigating mechanisms that constrain the rate of vertebrate single copy nuclear DNA evolution."

Clark-Walker (1991) contrasted base substitution rates in yeasts and mammals in two mitochondrial genes (*cytb* and *cox2*) and the nuclear gene encoding cytochrome c. Clark-Walker found that in yeast the two mitochondrial genes had fewer changes than the nuclear gene, but the converse was true of the mammalian genes. Comparisons of the cytoplasmic small subunit rRNA genes were also done, which indicated that the yeasts studied were more divergent than the mammals. Clark-Walker concludes "that base-substitution rates are much slower in yeast mitochondrial genes than the equivalent genes from mammals."

Sequence evolution of plant mitochondrial and chloroplast genomes in higher plants proceeds much more slowly (Palmer, 1992; Table 4–1). The relative synonymous substitution rate for plant mtDNA is the lowest known for any organelle DNA. Chloroplast unique sequence DNA evolves at a 3-fold greater rate, but the rate of nucleotide substitution in the inverted repeat is actually slightly lower than for plant mtDNA and about four times lower than the rest of the plastid genome. This cannot simply attest to the efficiency of copy correction within the inverted repeat, since a priori one would expect new base pair substitutions to be fixed with equal probability to the original base pair that was already present. The low level of fixation of new mutations within the inverted repeat suggests that the copy correction machinery may be biased toward the resident base pair. The nonsynonymous amino acid substitution rate of plant chloroplast and mtDNA is also low. Angiosperm nuclear DNA evolves at a more rapid rate than either organelle genome (Table 4-1). Finally, the bias in the proportion of transition to transversion mutations is much more modest in plant chloroplast and mitochondrial DNA (ca. 2:1, Palmer, 1992). This may argue that these organelle DNAs are repaired more efficiently than mammalian mtDNA although it should be noted that in Hawaiian *Drosophila*, at least, transitions and transversions occur with similar frequency (de Salle and Hunt, 1987; de Salle *et al.*, 1987). Studies of enzymes involved in the recombination and repair of organelle DNA are still in their infancy. However, the discovery of a RecA type enzyme in chloroplasts of pea and *Arabidopsis* (Cerutti *et al.*, 1992) together with the observation that the *Chlamydomonas* chloroplast has its own photoreactivating enzyme (Cox and Small, 1985) suggest that repair systems in chloroplasts, at least, may be better developed than in mammalian mitochondria.

REFERENCES

Ames, B.N. (1988). Mutagenesis and carcinogenesis: Endogenous and exogenous factors. *Env. Mol. Mut.* 14(S16): 66

Ayliffe, M.A., and J.N. Timmis (1992a). Plastid DNA sequence homologies in the tobacco nuclear genome. *Mol. Gen. Genet.* 236: 105–112.

Ayliffe, M.A., and J.N. Timmis (1992b). Tobacco nuclear DNA contains long tracts of homology to chloroplast DNA. *Theor. Appl. Genet.* 85: 229–238.

Baker, A., and G. Schatz (1987). Sequences from a prokaryotic genome or the mouse dihydrofolate reductase gene can restore the import of a truncated precursor protein into yeast mitochondria. *Proc. Natl. Acad. Sci. U.S.A.*, 84: 3117–3121.

Baldauf, S.L., and J.D. Palmer (1990). Evolutionary

transfer of the chloroplast *tufA* gene to the nucleus. *Nature (London)* 344: 262–265.

Baldauf, S.L., Manhart, J.R., and J.D. Palmer (1990). Different fates of the chloroplast *tufA* gene following its transfer to the nucleus in green algae. *Proc. Natl. Acad. Sci. U.S.A.* 87: 5317–5321.

Bowman, C.M., and T.A. Dyer (1986). The locations and possible significance of small dispersed repeats in wheat ctDNA. *Curr. Genet.* 10: 931–941.

Boynton, J.E., Gillham, N.W., Newman, S.M., and E.H. Harris (1992). Organelle genetics and transformation of *Chlamydomonas*. In *Plant Gene Research: Cell Organelles* (R.G. Herrmann ed.), pp. 3–64. Springer-Verlag, Wien, New York.

Brown, W.M. (1983). Evolution of mitochondrial DNA. In *Evolution of Genes and Proteins* (M. Nei, and R.K. Koehn, eds.), pp. 62–88. Sinauer Associates, Sunderland, MA.

Brown, W.M. (1985). The mitochondrial genome of animals. In *Monographs in Evolutionary Biology: Molecular Evolutionary Genetics* (R.J. MacIntyre, ed.), pp. 95–129. Plenum Press, New York.

Brown, W.M., Prager, E.M., Wang, A., and A.C. Wilson (1982). Mitochondrial DNA sequences of primates: tempo and mode of evolution. *J. Mol. Evol.* 18: 225–239.

Bryant, D.A., and H. Bohnert (1993). *pers. comm.*

Cantatore, P., and C. Saccone (1987). Organization, structure, and evolution of mammalian mitochondrial genes. *Int. Rev. Cytol.* 108: 149–208.

Cavalier-Smith, T. (1982). The origin of plastids. *Biol. J. Linn. Soc.* 17: 289–306.

Cavalier-Smith, T. (1985). Selfish DNA and the origin of introns. *Nature (London)* 315: 283–284.

Cavalier-Smith, T. (1987). The origin of eukaryote and archaebacterial cells. *Ann. N.Y. Acad. Sci.* 503: 17–54.

Cavalier-Smith, T. (1992). The number of symbiotic origins of organelles. *BioSystems* 28: 91–106.

Cerutti, H., Osman, M., Grandoni, P., and A.T. Jagendorf (1992). A homolog of *Escherichia coli* RecA protein in plastids of higher plants. *Proc. Natl. Acad. Sci. U.S.A.* 89: 8068–8072.

Clark-Walker, G.D. (1991). Contrasting mutation rates in mitochondrial and nuclear genes of yeasts versus mammals. *Curr. Genet.* 20: 195–198.

Clark-Walker, G.D. (1992). Evolution of mitochondrial genomes in fungi. *Int. Rev. Cytol.* 141: 89–127.

Clark-Walker, G.D., McArthur, C.R., and K.S. Sriprakash (1983). Order and orientation of genic sequences in circular mitochondrial DNA from *Saccharomyces exiguus*: Implications for evolution of yeast mtDNAs. *J. Mol. Evol.* 19: 333–341

Clayton, D.A. (1992). Transcription and replication of animal mitochondrial DNAs. *Int. Rev. Cytol.* 141: 217–232.

Clayton, D.A., Doda, J.N., and E.C. Friedberg (1974). The absence of a pyrimidine dimer repair mechanism in mammalian mitochondria. *Proc. Natl. Acad. Sci. U.S.A.*, 71: 2777–2781.

Cox, J.L., and G.D. Small (1985). Isolation of a photorcactivation-deficient mutant of *Chlamydomonas*. *Mutat. Res.* 146: 249–255.

Darnell, J.E., and W.F. Doolittle (1986). Speculations on the early course of evolution. *Proc. Natl. Acad. Sci. U.S.A.*, 83: 1271–1275.

Deng, X.W., Wing, R.A., and W. Gruissem (1989). Chloroplast DNA exists in multimeric forms. *Proc. Natl. Acad. Sci. U.S.A.*, 86: 4156–4160.

de Salle, R., and J.A. Hunt (1987). Molecular evolution in Hawaiian Drosophilids. *Trends Ecol. Evol.* 2:212–216.

de Salle, R., Freedman, T., Prager, E.M., and A.C. Wilson (1987). Tempo and mode of sequence evolution in mitochondrial DNA of Hawaiian *Drosophila*. *J. Mol. Evol.* 26: 157–164.

Dibb, N.J., and A.J. Newman (1989). Evidence that introns arose at proto-splice sites. *EMBO J.* 8: 2015–2021.

Downie, S.R., Olmstead, R.G., Zurawski, G., Soltis, D.E., Soltis, P.S., Watson, J.C., and J.D. Palmer (1991). Six independent losses of the chloroplast DNA in dicotyledons: Molecular and phylogenetic implications. *Evolution* 45: 1245–1259.

Erickson, J.M., Rahire, M., and J.-D. Rochaix (1984). *Chlamydomonas reinhardtii* gene for the 32,000 mol. wt. protein of photosystem II contains four large introns and is located entirely within the chloroplast inverted repeat. *EMBO J.* 3: 2753–2762.

Farelly, F., and R.A. Butow (1983). Rearranged mitochondrial genes in the yeast nuclear genome. *Nature (London)* 301: 296–301.

Ferat, J.-L., and F. Michel (1993). Group II self-splicing introns in bacteria. *Nature (London)* 364: 358–361.

Fry, M., and L.A. Loeb (1986). *Animal Cell DNA Polymerases*. CRC, Boca Raton, FL.

Gantt, J.S., Baldauf, S.L., Calie, P.J., Weeden, N.F., and J.D. Palmer (1991). Transfer of *rpl22* to the nucleus greatly preceded its loss from the chloroplast and involved the gain of an intron. *EMBO J.* 10: 3073–3078.

Gilbert, W., Marchionni, M. and G. McKnight (1986). On the antiquity of introns. *Cell* 46: 151–154.

Grosse, F., Krauss, G., Knill-Jones, J.W., and A.R. Fersht (1983). Accuracy of DNA polymerase alpha in copying natural DNA. *EMBO J.* 2: 1515–1519.

Gray, M.W. (1989). Origin and Evolution of mitochondrial DNA. *Annu. Rev. CellBiol.* 5: 25–50.

Gray, M.W. (1992). The endosymbiont hypothesis revisited. *Int. Rev. Cytol.* 141: 233–357.

Hallick, R.B., and D.E. Buetow (1989). Chloroplast DNA. In *The Biology of Euglena* (D.E. Buetow, ed.), pp. 351–414. Academic Press, New York.

Hallick, R.B., Hong, L, Drager, R.G., Favreau, M.R., Monfort, A., Orsat, B., Spielman, A., and E. Stutz (1993). Complete sequence of *Euglena* chloroplast DNA. *Nucl. Acids Res.* 21: 3537–3544.

Hanson, M.R., and O.F. Folkerts (1992). Structure

and function of the higher plant mitochondrial genome. *Int. Rev. Cytol.* 141: 129–172.

Hoeben, P. and G.D. Clark-Walker (1986). An approach to yeast classification by mapping mitochondrial DNA from *Dekkera/Brettanomyces* and *Eeniella* genera. *Curr. Genet.* 10: 371–379.

Hoeben, P., Weiller, G. and G. D. Clark-Walker (1993). Larger rearranged mitochondrial genomes in *Dekkera/Brettanomyces* yeasts are more closely related than small genomes with a conserved gene order. *J. Mol. Evol.* 36: 263–269.

Karabin, G.D., Farley, M., and R.B. Hallick (1984). Chloroplast gene for M_r 32,000 polypeptide of photosystem II in *Euglena gracilis* is interrupted by four introns with conserved boundary sequences. *Nucl. Acids Res.* 12: 5801–5812.

Kolodner, R., and K.K. Tewari (1979). Inverted repeats in chloroplast DNA from higher plants. *Proc. Natl. Acad. Sci. U.S.A.*, 76: 41–45.

Kuhsel, M.G., Strickland, R., and J.D. Palmer (1990). An ancient group I intron is shared by eubacteria and chloroplasts. *Science* 250: 1570–1573.

Kunkel, T.A. (1985). The mutational specificity of DNA polymerases-alpha and gamma during in vitro DNA synthesis. *J. Biol. Chem.* 260: 12866–12874.

Kunkel, T.A. and L.A. Loeb (1981). Fidelity of mammalian DNA polymerases. *Science* 213: 765–767.

Lansman, R.A., and D.A. Clayton (1975). Selective nicking of mammalian mitochondrial DNA *in vivo*: Photosensitization by incorporation of 5-bromodeoxyuridine. *J. Mol. Biol.* 99: 761–776.

Lemieux, B., and C. Lemieux (1985). Extensive sequence rearrangements in the chloroplast genomes of the green algae *Chlamydomonas eugametos* and *Chlamydomonas reinhardtii*. *Curr. Genet.* 10: 213–219.

Lemieux, B., Turmel, M., and C. Lemieux (1985). Chloroplast DNA variation in *Chlamydomonas* and its potential application to systematics of this genus. *BioSystems* 18: 293–298.

Lonsdale, D.M. (1989). The plant mitochondrial genome. In *The Biochemistry of Plants*, Vol. 15 (A. Marcus, ed.), pp. 229–295. Academic Press, San Diego.

Low, K.B. (1988). Genetic recombination: A brief overview. In *The Recombination of Genetic Material* (K.B. Low, ed.), pp. 1–21. Academic Press, San Diego.

Low, K.B., and R.D. Porter (1978). Modes of gene transfers and recombination in bacteria. *Annu. Rev. Genet.* 12: 259–287.

Mishler, B.D., and S.P. Churchill (1985). Transition to a land flora: Phylogenetic relationship of the green algae and bryophytes. *Cladistics* 1: 305–328.

Modrich, P. (1987). DNA mismatch correction. *Annu. Rev. Biochem.* 56: 435–466.

Moritz, C., Dowling, T.E., and W.M. Brown (1987). Evolution of animal mitochondrial DNA: Relevance for population biology and systematics. *Annu. Rev. Ecol. Syst.* 18: 269–292.

Newton, K.J. (1988). Plant mitochondrial genomes: Organization, expression and variation. *Annu. Rev. Plant Physiol. Plant Mol. Biol.* 39: 503–532.

Nugent, J.M., and J.D. Palmer (1991). RNA-mediated transfer of the gene *coxII* from the mitochondrion to the nucleus during flowering plant evolution. *Cell* 66: 473–481.

Oda, K., Yamato, K., Ohta, E., Nakamura, Y., Takemura, M., Nozato, N., Akashi, K., Kanegae, T., Ogura, Y., Kohchi, T., and K. Ohyama (1992). Gene organization deduced from the complete sequence of liverwort *Marchantia polymorpha* mitochondrial DNA: A primitive form of plant mitochondrial genome. *J. Mol. Biol.* 223: 1–7.

Palmer, J.D. (1985). Evolution of chloroplast and mitochondrial DNA in plants and algae. In *Monographs in Evolutionary Biology: Molecular Evolutionary Genetics* (R.J. MacIntyre, ed.), pp. 131–240. Plenum Publishing, New York.

Palmer, J.D. (1991). Plastid chromosomes: Structure and Evolution. In *The Molecular Biology of Plastids* 1 (L. Bogorad and I. Vasil, eds.), pp. 5–53. Academic Press, San Diego.

Palmer, J.D. (1992). Chloroplast and mitochondrial genome evolution in land plants. In *Plant Gene Research: Cell Organelles* (R.G. Herrmann, ed.), pp. 99–133. Springer-Verlag, Wien.

Palmer, J.D., and J.M. Logsdon, Jr. (1991). The recent origins of introns. *Curr. Opinion Genet. and Development*, 1: 470–477.

Raubeson, L.A., and R.K. Jansen (1992). Chloroplast DNA evidence on the ancient evolutionary split in vascular land plants. *Science* 255: 1697–1699.

Reith, M., and J. Munholland (1993). A high-resolution map of the chloroplast genome of the red alga *Porphyra purpurea*. *Plant Cell* 5: 465–475.

Richter, C., Park, J.W., and Ames, B.N. (1988). Normal oxidative damage to mitochondrial and nuclear DNA is extensive. *Proc. Natl. Acad. Sci. U.S.A.*, 85: 6465–6467.

Rogers, J.H. (1989). How were introns inserted into nuclear genes? *TrendsGenet.*, 5: 213–216.

Schuster, W., and A. Brennicke (1987). Plastid, nuclear and reverse transcriptase sequences in the mitochondrial genome of *Oenothera:* is genetic information transferred between organelles via RNA? *EMBO J.* 6: 2857–2863.

Schuster, W., and A. Brennicke (1988). Interorganellar sequence transfer: Plant mitochondrial DNA is nuclear, is plastid, is mitochondrial. *Plant Sci.* 54: 1–10.

Skelly, P.J., and G.D. Clark-Walker (1991). Polymorphisms in tandemly repeated sequences of *Saccharomyces cerevisiae* mitochondrial DNA. *J. Mol. Evol.* 32: 396–404.

Thorsness, P.E., and T.D. Fox (1990). Escape of DNA from mitochondria to the nucleus in *Saccharomyces cerevisiae*. *Nature (London)* 346: 376–379.

Thorsness, P.E., and T.D. Fox (1993). Nuclear mutations in *Saccharomyces cerevisiae* that affect the escape of DNA from mitochondria to the nucleus. *Genetics* 134: 21–28.

Turmel, M., Gutell, R.R., Mercier, J.-P., Otis, C., and C. Lemieux (1993). Analysis of the chloroplast large subunit ribosomal RNA gene from 17 *Chlamydomonas* taxa. *J. Mol. Biol.* 232: 446–467.

Wallace, D.C. (1992). Mitochondrial genetics: A paradigm for aging and degenerative disease? *Science* 256: 628–632.

Wernette, C.M., Conway, C.M., and L.S. Kaguni (1988). Mitochondrial DNA polymerase from *Drosophila melanogaster* embryos: Kinetics, processivity, and fidelity of DNA polymerization. *Biochemistry* 27: 6046–6054.

Weiller, G., Schueller, C.M.S., and R.J. Schweyen (1989). Putative target sites for mobile G+C rich clusters in yeast mitochondrial DNA: Single elements and tandem arrays. *Mol. Gen. Genet.* 218: 272–283.

Wilson, A.C., Cann, R.L., Carr, S.M., George, M., Gyllensten, U.B., Helm-Bychowski, K.M., Higuchi, R.G., Palumbi, S.R., Prager, E.M., Sage, R.D., and M. Stoneking (1985). Mitochondrial DNA and two perspectives on evolutionary genetics. *Biol. J. Linn. Soc.* 26: 375–400.

Wolf, K., and L. Del Giudice (1988). The variable mitochondrial genome of Ascomycetes: Organization, mutational alterations, and expression. *Adv. Genet.*, 25: 185–308.

Woodson, S.A., and T.R. Cech (1989). Reverse self-splicing of the Tetrahymena group I intron: Implication for the directionality of splicing and for intron transposition. *Cell* 57: 335–345.

Wolstenholme, D.R. (1992). Animal mitochondrial DNA: Structure and evolution. *Int. Rev. Cytol.* 141: 173–232.

Xu, M.-Q., Kathe, S.D., Goodrich-Blair, H., Nierzwicki-Bauer, S.A., and D.A. Shub (1990). Bacterial origin of a chloroplast intron: Conserved self- splicing group I introns in cyanobacteria. *Science* 250: 1566–1570.

5

Cytological Organization and Replication of Organelle Genomes

Organelle genomes do not exist in isolation as naked DNA molecules. Rather electron microscopy as well as fluorescence microscopy with the fluorochrome 4′diamidino-2-phenylindole (DAPI) reveal that organelle DNA molecules are bundled together in nucleoids or nuclei (e.g., the mitochondrial nucleus). In fact, the organelle nucleoid is probably the unit of division and segregation of organelle DNA molecules. Preparatory to these processes organelle DNA molecules must replicate. Currently, the process of organelle DNA replication is understood in a descriptive sense in a very few organisms and not at all from the biochemical viewpoint in any organism. Similarly, our understanding of how the replication and segregation of organelle genomes relates to the division and partitioning of nucleoids is virtually nonexistent, although provocative theories have been proposed. Ultimately, these relationships must be understood if we are ever to make sense of the copy number paradox (Chapter 9). The latter expression refers to the fact that numbers of organelle genomes measured genetically are much lower than would be expected based on physical estimates of number of organelle DNA molecules.

THE ORGANIZATION OF ORGANELLE DNA MOLECULES

Kuroiwa has published comprehensive reviews of the structure and organization of the mitochondrial (1982) and chloroplast (1991) nucleoids. Based on electron microscopic studies, mitochondrial nucleoids have been classified into three groups: electron transparent (ET), semielectron transparent (SE), and electron dense (ED) (Kuroiwa, 1982). DAPI staining experiments reveal that mitochondria of most organisms contain a single nucleoid with the large mitochondria of animal cells being exceptional in this respect. Particularly detailed observations have been made by Kuroiwa on the ED nucleoid of the slime mold *Physarum*. These studies form the basis on which comparisons to other species can be made.

Kuroiwa (1982) recognizes four division patterns for mitochondrial nucleoids (Fig. 5–1). In the first, typified by *Physarum*, the mitochondrial nucleoid divides at the time of mitochondriokinesis as if the nucleoids were pinched by constriction of the mitochondrial bounding membranes. The second, exemplified by *Paramecium*, is similar to

CYTOLOGICAL ORGANIZATION/REPLICATION OF ORGANELLE GENOMES 111

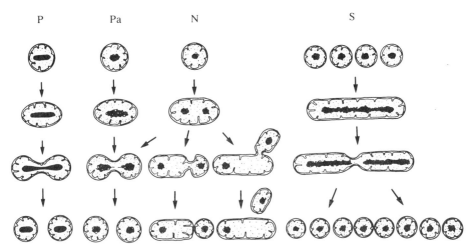

Fig. 5–1. Four types of mitochondrial division accompanied by division of mitochondrial nucleoids: *Physarum* type (P type), *Paramecium* type (Pa type); *Nitella* type (N type), and *Saccharomyces* type (S type). When the mitochondrion divides, at least one mitochondrial nucleoid is distributed to each daughter mitochondrion. From Kuroiwa (1982).

the first except that division of the nucleoid immediately precedes mitochondriokinesis. The third type, observed in the giant alga *Nitella*, is characterized by division of the nucleoid before mitochondrial division commences. This pattern of division is seen frequently in plant cells. The fourth type is found in yeast. Here 30 or so small spherical mitochondria fuse to form one large, long mitochondrion during the cellular G_1 period and after mtDNA synthesis has occurred. After fusion of the mitochondria, the mitochondrial nucleoids coalesce to form a single, large structure. This large mitochondrion now divides yielding two daughters, each containing an elongate nucleoid. The daughter mitochondria subsequently fragment to yield smaller mitochondria, each of which possesses a single spherical nucleoid. Fusion of mitochondrial nucleoids to form a "string-like network" also occurs during the first meiotic division in yeast (Miyakawa *et al.*, 1984). After the second meiotic division, a ring of mitochondrial nucleoids is observed surrounding each daughter nucleus. Respiratory inhibitors were found to block the formation of the string-like structure, but inhibitors of RNA synthesis and cytoplasmic or mitochondrial protein synthesis did not (Miyakawa *et al.*, 1988). Thymidine labeling experiments with both *Physarum* and *Tetrahymena* indicate that newly synthesized mtDNA is partitioned symmetrically to daughter nucleoids.

The mitochondrial nucleoid of *Physarum* has been isolated and characterized (Kuroiwa, 1982). This structure is estimated to contain ca. 32 linear mitochondrial genomes. Transmission electron microscopy revealed the presence of chromatin-like fibrils ranging from 10 to 50 nm in diameter. The *Physarum* nucleoids also contain specific proteins certain of which are required for maintainance of a normal conformation. At least one of these proteins is a basic histone-like polypeptide. Morphologically intact mitochondrial nucleoids have also been isolated from yeast (Miyakawa *et al.*, 1987). Each isolated nucleoid contained an average of 3.1 mtDNA molecules.

Twenty species of polypeptides were also detected in the isolated *S. cerevisiae* nucleoids ranging in molecular weight from 10 to 70 kDa. One of these proteins is a histone-like polypeptide with a molecular weight of 20 kDa.

Kuroiwa (1991; Kuroiwa *et al.*, 1981) classifies plants and algae into five types based on the shape, size, and distribution of chloroplast nucleoids (Fig. 5–2). The SN type (scattered plastid nucleoids) is characterized by chloroplasts having small, uniformly dispersed nucleoids in the matrix between the thylakoid membranes and/or the grana stacks. Land plants and many algal groups (Chlorophyceae, Prasinophyceae, Euglenophyceae, Cryptophyceae, Eustigmatophyceae, Dinophyceae) are of the SN type. Chloroplasts of the second or CN type (centrally located plastid nucleoids) possess one or a few nucleoids located centrally in the plastid and surrounded by thylakoids. *Cyanidium caldarium* RK-1, a probable red alga, typifies this pattern of nucleoid organization as do undifferentiated proplastids of higher plants. Chloroplasts of the third or CL type (circular plastid nucleoids) have a large, ring-shaped chloroplast nucleoid within the girdle lamellae. Circular nucleoids of the CL type have been isolated from

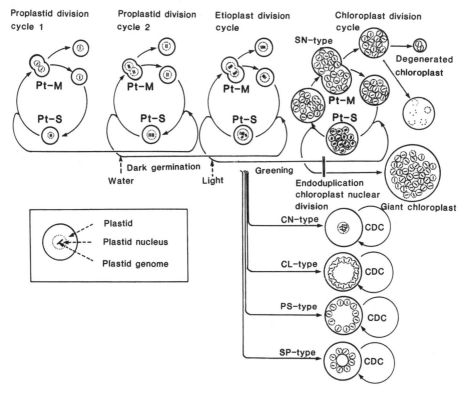

Fig. 5–2. Diagram of plastid nucleoid division during chloroplast development and the five types of chloroplasts that differentiate from proplastids: scattered nucleoids (SN type), centrally located nucleoids (CN type), circular nucleoids (CL type), spread pyrenoid nucleoid (SP type), and peripherally scattered nucleoids (PS type). The abbreviations Pt-M, Pt-S, and CDC are plastid division stage, plastid DNA synthesis stage, and chloroplast division cycle, respectively. From Kuroiwa (1991).

the brown alga *Ectocarpus indicus* and appear to be chains of small, spherical particles that may correspond to the small, individual nucleoids seen in SN type plants. The PS type (peripherally scattered plastid nucleoids) is a modification of the CL type in which the nucleoids are scattered peripherally around the chloroplast adjacent to the inner envelope. The SP type (spread pyrenoid plastid nucleoid) represents the final pattern of nucleoid organization and is a variant on the SN type. In SP plastids the nucleoids form a shell around the pyrenoid (e.g., the green alga, *Bryopsis plumosa*).

The nucleoid division cycle has been followed in plastids of the CL, SN, and CN types (Fig. 5-2). In CL type plants, the ring shaped nucleoid divides into two rings, each of which is transmitted to a daughter chloroplast. In SN type plants the 20–30 small nucleoids divide, and subsequently equal numbers of nucleoids appear to be distributed to the daughter chloroplasts. However, plants with these patterns of nucleoid division are less than ideal material for study because there are usually many plastids per cell and plastid divisions are not synchronized. By contrast, the red alga *Cyanidium caldarium* RK-1, which possesses the CN type of nucleoid division, is particularly well-suited for such investigations. This unicellular alga has a single mitochondrion and a single chloroplast and can be cultured synchronously. Under these conditions each mother cell increases in size for about 50 hr followed by two cell divisions to yield four endospores. The chloroplast, the cell nucleus, the mitochondrion, and the cell itself divide in that order. Chloroplast genomes appear to be partitioned essentially equally between daughter cells.

DAPI staining and epifluorescence microscopy have been used to follow plastid nucleoid differentiation and genome copy number during chloroplast development in early embryonic cells of *Brassica juncea* and cultured cells of *Nicotiana tabacum* (Kuroiwa *et al.*, 1981; Kuroiwa, 1991; Yasuda *et al.*, 1988). Proplastids can be conveniently classified into at least two types. The first possesses a single nucleoid (0.2 μm diameter) and one to two copies of the chloroplast genome. The second proplastid type contains two to five nucleoids of 2–3 μm diameter and several copies of the chloroplast genome per proplastid. The first proplastid type has been referred to as a "proplastid precursor" by Kuroiwa (1991). Often these two proplastid types are mixed in cultured cells of *N. tabacum* (Yasuda *et al.*, 1988). However, the corresponding proplastid types appear to be temporally separated in early embryonic and "dormant" embryonic cells of *B. juncea* where they measure 0.2 and 0.5–1.0 μm, respectively.

Kuroiwa (1991; Kuroiwa *et al.*, 1981) recognizes two cycles of proplastid division (Fig. 5–2). The first cycle involves the division of proplastid precursors to form new proplastid precursors while the second involves proplastid division. According to this scheme, proplastid precursors can also divide into proplastids and, in the dark, proplastids differentiate into etioplasts. On illumination, etioplasts metamorphose into green chloroplasts. This process involves an increase in nucleoid number of 2- to 3-fold without a marked increase in chloroplast genome number per nucleoid.

The process of change in nucleoid structure and number and the accompanying fluctuations in chloroplast genome number have received detailed attention in wheat using quantitative fluorescence microscopy (Miyamura *et al.*, 1986). On imbibition of water the small proplastid nucleoids of the leaf primordia in dry seeds increase in size and differentiate into larger ring-shaped structures in the developing leaf. These proplastid nucleoids contain ca. 120 copies of the chloroplast genome. In the dark these ring-shaped structures divide into ca. 10 circular nucleoids within 3 days. This morphological change is accompanied by a 7.5-fold increase in the chloroplast genome copy number and a 4.6-fold increase in etioplast number so the total increase in plastid genome copy number is 34-fold. Illumination results in differentiation of etioplasts into chloroplasts and is accompanied by further division of the nucleoids to yield 20–30 per plastid. However, there is little increase in the DNA copy number per plastid. An 8-fold increase in plastid genome number was also observed during the plastid division cycle following illumination.

Leech and Pyke (1988) have reviewed the process of chloroplast division in flowering plants with particular reference to wheat and their comments are relevant to the foregoing discussion (see also Possingham and Lawrence, 1983). Leech and Pyke distinguish between the processes of proplastid and chloroplast division, pointing out that proplastid division is restricted to leaf and root meristems, in which the process keeps pace with, or only slightly exceeds, the rate of cell division. In contrast, in expanding leaf cells, cell division ceases, but rapid chloroplast division continues. Thus, in wheat the youngest mesophyll cells that have just ceased to divide contain 10–15 plastids while a mature mesophyll cell possesses 130–150 plastids. In differentiating mesophyll cells most plastid division occurs in a zone just distal to mitotically dividing meristematic cells. Dividing plastids are easily recognized by their dumbbell shapes. They peak in number slightly before cells are observed in which the chloroplast number has increased 2- to 3-fold. Boffey and Leech (1982) estimate that before plastid replication takes place there are over 1000 genome copies per plastid. This number subsequently drops to 200 to 300 per plastid following plastid replication. Although these estimates are somewhat different than those obtained by Kuroiwa and colleagues (Kuroiwa, 1991), they lead to the same conclusion. Major increases in plastid DNA precede plastid division and little additional plastid DNA synthesis accompanies this process (also see Possingham *et al.*, 1988).

Whatley (1988) observes that proplastids and chloroplasts seem to follow the same sequential steps during the division process (1) elongation in the direction of one major axis, (2) progressive annular constriction in a plane at right angles to the major axis and usually near the midpoint, (3) assumption of a dumbbell shape with a narrow connecting isthmus, and (4) separation of the two daughter plastids following their twisting in orientation around the connecting neck. A structure called the "plastid-dividing ring" (PD ring), first identified in *C. caldarium* (Mita *et al.*, 1986; Mita and Kuroiwa, 1988) and later in other plant species (Kuroiwa, 1991), also appears to be involved. In synchronous cultures of *C. caldarium* the spherical mother cell chloroplast increases ca. 3.5-fold in size following which it changes in shape to resemble an American football. At the same time, the concentric, circular thylakoids of the chloroplast begin to separate into two parts. At this point the PD ring makes its appearance and subsequently contracts with the concurrent constriction of the chloroplast. A variety of experimental evidence suggests that the PD ring is made up of actin-like filaments (Kuroiwa, 1991).

Nucleoids have been isolated from proplastids of cultured cells of *Nicotiana tabacum* and from chloroplasts isolated from mesophyll protoplasts of this same species (Nemoto *et al.*, 1988, 1989, 1990, 1991). Isolated proplastid nucleoids were found to contain numerous polypeptides, only four of which (69, 31, 30, and 14 kDa) were tightly bound to cpDNA. In chloroplasts from mesophyll cells these same cpDNA binding proteins were absent and were replaced by a different set of polypeptides (35,

28, and 26 kDa). These results suggest that the population of DNA-binding proteins changes during plastid differentiation.

Rose (1988) reviews evidence indicating that chloroplast genomes may be membrane bound. He points out that nucleoid arrangements shown by DAPI staining can be readily interpreted in terms of membrane attachment. Thus, SN-type chloroplasts show thylakoid attachment of nucleoids, PS-type chloroplasts exhibit envelope attachment of nucleoids, and in CL-type chloroplasts the nucleoids are bound to girdle thylakoids. Rose also proposes a model showing how attachment of nucleoids to membranes could lead to segregation of cpDNA at plastid division. Direct evidence for membrane attachment of cpDNA molecules is now starting to emerge, but the results are still contradictory. Liu and Rose (1992) report that spinach chloroplast genomes are thylakoid-bound in the inverted repeat region near the rRNA genes. In contrast, Sato *et al.* (1993) find that a 130-kDa protein of the inner chloroplast envelope of pea binds specific cpDNA sequences, but these do not include the rRNA genes (pea has no inverted repeat). Sato *et al.* (1993) hypothesize that the protein may function to anchor cpDNA to the envelope during replication. So while it seems increasingly likely that chloroplast genomes are membrane bound, at least some of the time, the membrane(s) to which they are bound remain to be determined unequivocally. This point has to be resolved before satisfactory models for segregation and assortment of chloroplast genomes and nucleoids can be formulated.

Calculations of organelle DNA molecules per organelle and per nucleoid reveal that the numbers are quite variable and stage dependent (Table 5–1). Thus, the size of the *Physarum* mitochondrial nucleoid in a resting spore or a sclerotium is around 20% of that in the mitochondrion of an actively proliferating plasmodium. In the case of land plant proplastid precursors there is a single nucleoid with, sometimes, a single cpDNA molecule. Within a fully differentiated green chloroplast there are ca. 60 DNA molecules per organelle and these are distributed among 12 to 25 nucleoids. Thus, metabolically active organelles have greater numbers of DNA molecules than inactive

Table 5–1. Average Number of Organelle DNA Molecules per Organelle and Nucleoid

Organelle	Source	DNA Molecules per Organelle	Nucleoids per Organelle	DNA Molecules per Nucleoid
Mitochondrion	*Physarum*			
	Spore	6–8	1	6–8
	Plasmodium	32	1	32
	Tetrahymena	6	1	6
	Mouse	5–6	1–3	2–6
Chloroplast	*Chlamydomonas*	80	5–6[a]	13–16
	Euglena	100–300	20–34	3–15
	Flowering plant			
	Proplastid precursor	1	1	1
	Mature plastid	ca. 60	12–25	2–5

Data from Kuroiwa *et al.* (1981), Kuroiwa (1982, 1991), and Gillham (1978)

[a]From Birky *et al.* (1984).

or undifferentiated organelles. Furthermore, while "ploidy" on a per organelle basis may be high, it is usually not great on a per nucleoid basis. Although the ploidy of the *Physarum* mitochondrial nucleoid is quite high, there is but one nucleoid per mitochondrion.

The number of nucleoids per organelle can also vary. Thus, Birky et al. (1984) determined nucleoid numbers in plastids of vegetative cells, gametes, and zygotes of the green alga *Chlamydomonas reinhardtii* using the DAPI staining method. Means for vegetative cells ranged between 5 and 6 per plastid, with nucleoid numbers in different cells being distributed approximately normally around this mean.

REPLICATION OF ORGANELLE GENOMES

Replication of organelle DNA molecules is autonomous of the nucleus in that organelle DNA molecules act directly as templates for their own synthesis (i.e., a master copy in the nucleus is not involved), but not in terms of the enzymatic machinery required for replication. This conclusion is most compelling in the case of mammalian mtDNA where we know the function of every gene, not one of which is involved in DNA replication. However, this is unlikely to be the case for all organelle genomes. Thus, sequencing of the chloroplast genome of *Porphyra* reveals the presence of several genes whose products are involved in replication of prokaryotic genomes (Reith and Munholland, 1993).

Animal mtDNA. The replication of organelle DNA has been most thoroughly researched in the case of mammalian mtDNA molecules. The replication of these genomes has been comprehensively reviewed by Clayton (1982, 1991a, b; 1992). In addition to the basic monomeric genome, animal cells maintain a proportion of their mitochondrial genomes as catenated circles in which the monomers are linked together in a chain. Another topological variation is encountered in malignant human cells and some tissue culture lines, in which two monomer-size circles are linked together in head-to-tail configuration to form a simple double size circle.

The mtDNA copy number in mammalian cells does not appear to be tightly controlled. For example, in a mouse L cell line in which the entire mtDNA population consists of head-to-tail dimer molecules, the number of unit genomes is approximately two-thirds more than in a line where all molecules are monomers. Thus, neither the number of molecules nor the number of genome equivalents is subject to rigorous control. Cell size differences between mouse and human cells do not explain why the latter cells contain many more mitochondrial genomes than the former (estimated per cell as > 1000 in L cells and > 8000 in cultured human cells). Also, there is no apparent requirement in mammalian cells for all mtDNA molecules to be duplicated at each cell doubling. Some may replicate twice before others replicate at all. This finding has important consequences *if* it applies generally to organelle DNA molecules for it provides a rationale by which organelle gene mutations might segregate and be expressed (Chapter 11). Finally, in mouse L cells at least, mtDNA molecules can replicate at any and all phases of the cell cycle.

Most information relating to the replication of animal mtDNA molecules comes from mouse L cells (Fig. 5–3). In L cells the predominant form of mtDNA is a covalently closed circle containing a D-loop at the origin of replication. The triplex D-loop structure is formed by synthesis of a short daughter H strand that remains stably associated with the parental closed circle (D-mtDNA, Fig. 5–3). Replication intermediates are molecules in which H strand synthesis has proceeded past the D-loop region (ExpD). H-strand synthesis continues in a unidirectional manner until completion. When H strand synthesis is 67% complete, the L strand origin becomes exposed and synthesis of this strand commences (ExpD, 1, Fig. 5–3). This delay in initiation of L-strand

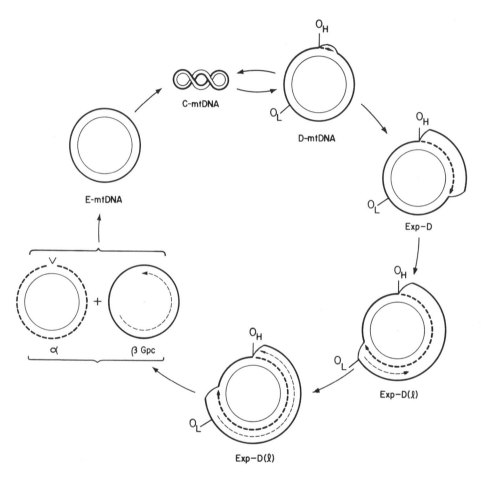

Fig. 5–3. Replication of monomeric mouse mtDNA molecules. Thick solid lines: parental heavy (H)-strands. Thin solid lines: parental light (L)-strands. Thick dashed lines: daughter H strands. Thin dashed lines: daughter L strands. The order of replication is clockwise starting with at D-mtDNA (see text). O_H and O_L, origins of H- and L-strand synthesis, respectively. The double arrows reflect the metabolic instability of D-loop strands and consequent equilibrium between D-mtDNA and C-mtDNA. Expanded D-loop replicative intermediates are termed Exp-D prior to initiation of L-strand synthesis and Exp-D (ℓ) after initiation of L-strand synthesis. Carat, interruption of at least one phosphodiester bond in the H-strand of the daughter molecules; β Gpc, gapped circular daughter molecule. See text for discussion. From Clayton (1982).

synthesis results in segregation of two classes of daughter molecules. One possesses a parental L-strand and a complete daughter H-strand (α). The other has a parental H-strand and partially synthesized daughter L-strand (β gpc). The α molecule is then converted into a closed circle with few if any superhelical turns as is the β molecule following completion of L-strand synthesis. Approximately 100 negative superhelical turns are now introduced into these daughter E-mtDNA molecules (C-mtDNA) and these DNA molecules serve as templates for synthesis of D-mtDNA. The entire cycle is completed in 2 hrs with a polymerization rate of ca. 270 nucleotides per minute per strand. This is 200 times slower than *E. coli* chromosome replication.

In the case of the mouse D-loop at least four major H strand forms ranging from ca. 520 to 690 nucleotides in length are associated with the parental L strand. Length differences are largely due to variations at the 3′-end. In human mtDNA three nascent H strand species are found from ca. 570 to 655 nucleotides in length with the major contribution to length differences being sequences at the 5′-ends. Synthesis of nascent H-strands is ribonucleotide primed. In mouse mtDNA the nascent H-strands have a half-life of ca. 1 h and greater than 95% of them turn over and cannot serve as primers. The finding that there are discrete sizes of nascent H-strands appears to relate to the existence of distinct stop points for synthesis (termination associated sequences, TAS) that are recognized in the L-strand template (Fig. 5–4). One such element upstream of the bovine D-loop specifically binds a 48 kDa protein that may be involved in D-loop termination (Madsden *et al.*, 1993).

Because of the rapid turnover of nascent H-strands in the D-loop, maintenance of D loop structure may be related to transcription rather than replication. This view is supported by the fact that the promoters for H- and L-strand transcription are located in the D loop (Chapter 12). Thus, turnover of D loop strands could be related to transcriptional control.

The origin for H-strand synthesis (O_H) has been mapped precisely in the D-loop region (Fig. 5–4A). H-strand synthesis initiates by RNA priming. Mapping of the 5′-end of the RNA primer reveals that the primer initiates at the promoter for L strand transcription. Therefore, replication priming for H strand synthesis depends on transcriptional function of the L-strand promoter (LSP). Since there are no known differences between transcription initiation and replication priming at this promoter, the discussion of transcription initiation in Chapter 12 is presumed to be applicable to replication priming as well. Transitions from RNA to DNA synthesis occur over a region of three short conserved sequence elements or blocks (CSBs) I, II, and III. A ribonucleoprotein called RNAse MRP (for mitochondrial RNA processing) recognizes features of the CSBs and cleaves mitochondrial H-strand mRNA between CSBII and III (Chapter 12). Its exact role in RNA primer metabolism remains to be established definitively although several possibilities exist (Clayton, 1991b). RNase MRP is a ribonucleoprotein containing an RNA component and protein components that are all encoded by nuclear genes. The 265 (human) to 275 (mouse) RNA component is the only other RNA besides tRNAs that has been found to be imported into mitochondria.

The origin of L-strand synthesis (O_L) is approximately 30 nucleotides in length and is surrounded by several tRNA genes (Fig. 5–4B). O_L has the potential to form a stem-loop structure. Replication is initiated by an mtDNA primase which recognizes O_L to initiate RNA priming and DNA synthesis. L-strand synthesis is primed by RNA, which is complementary to a T-rich loop structure in O_L. The transition from RNA to

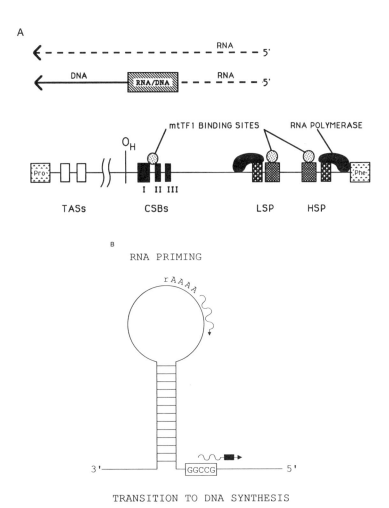

Fig. 5–4. (A) Schematic representation of several features of the vertebrate mtDNA D-loop. The dashed line represents RNA synthesis from the L-strand promoter (LSP) and the dashed/solid line represents an RNA-primed DNA strand initiated at the promoter. Both molecules are produced using the L-strand as a template and therefore have the sequence of the H-strand. The transitions from RNA to DNA synthesis occur within the boxed region; there are three conserved sequence blocks in this region (CSBs I, II, and III). O_H is the origin of H-strand synthesis. The entire D-loop region is bounded by the genes for proline tRNA (Pro) and phenylalanine tRNA (Phe). Mitochondrial RNA polymerase is shown at the transcriptional start sites at the bipartite LSP and H-strand promoter. The known binding sites for the mitochondrial transcription factor mtTF1 (Chapter 12) are indicated. The 3′-ends of D-loop DNA strands map near conserved sequence elements that have been termination associated sequences (TASs). (B) Schematic of O_L. RNA priming occurs in a thymidine-rich template sequence shown as a loop. The transition to L-strand synthesis in human mtDNA is at the template GGCCG shown. This site is critical for DNA synthesis. From Clayton (1991a).

DNA synthesis occurs near the base of the stem at a site that is actually part of a tRNA gene.

The mtDNA polymerase (polymerase γ) was originally thought to have a high error rate (Kunkel and Loeb, 1981), but subsequent experiments showed that animal polymerase γ is in a class with the least error prone of all DNA synthesizing enzymes (Chapter 4). Single-strand binding proteins have also been identified in mitochondrial preparations. These proteins are likely to be involved either in transcription or replication of animal mtDNA (see Clayton, 1992). Hehman and Hauswirth (1992) reported a DNA helicase from bovine mitochondria that they suppose precedes DNA polymerase γ during mtDNA replication. Although mammalian mitochondria lack a repair mechanism for removing pyrimidine dimers from the mitochondrial genome (Clayton, 1982), certain kinds of repair enzymes have been reported in mitochondrial fractions (Tomkinson et al., 1990).

Drosophila is the only other animal in which mtDNA replication has been extensively studied and there are important differences with respect to mammals (Wolstenholme et al., 1983). The two strands are not referred to as H and L since they are similar in buoyant density. Replication of the first strand in *Drosophila* begins in the A+T-rich region, but a class of D-loop-containing molecules is not seen. First strand synthesis proceeds in a direction opposite to that in mammals (i.e., toward the rRNA genes). Replication is highly asymmetrical and synthesis of the first strand can be up to 100% complete before second strand synthesis commences. Second strand synthesis can, apparently, initiate at various positions on the molecule.

Yeast. Yeast has unmatched advantages for the genetic dissection of mtDNA replication although these remain to be exploited fully. The basic observations on yeast mtDNA replication have been reviewed by Wolf and del Giudice (1988). There are approximately 50 molecules of mtDNA per haploid cell and twice that in diploid cells. The mtDNA molecules are organized in nucleoids, as discussed earlier, with the number of molecules per nucleoid averaging 3.9 in stationary phase cells (Miyakawa et al., 1987). Replication of mtDNA takes place continuously throughout the cell cycle. In a provocative paper Maleszka et al. (1991) report that the majority of mtDNA molecules from the yeast *Candida(Torulopsis) glabrata* are not the expected 19-kb circular monomers, but exist as linear molecules ranging in size from 50 to 150 kb or two to seven genome units. Their observation of circular molecules with single- or double-stranded tails (lariats) led them to propose that replication of mtDNA in *C. glabrata* occurs by a rolling circle mechanism. They suggest that *S. cerevisiae* molecules may also replicate by the same mechanism, pointing out that the majority of mtDNA molecules isolated from *S. cerevisiae* are linear and not circular. Maleszka and Clark-Walker (1992) extended these results to filamentous ascomycetes and oomycetes with the same results. Again, significant numbers of heterogeneous linear molecules were observed containing from one to seven genome lengths of mtDNA. These results suggest that rolling circle replication may be a general feature of fungal mitochondrial genomes.

Altogether eight sequences have been identified in yeast petite mutants (Chapter 7) that appear to promote replication of their defective mitochondrial genomes. Five of these *rep* or *ori* regions (*ori1–5*) are active. The *ori* regions consist of two GC clusters (A and B) separated by a short AT stretch, a central 200-bp AT stretch (*l*), and a third GC cluster (C) flanked by short AT stretches (*r* and *r'*) that contain sites for transcription initiation. Bidirectional RNA-primed DNA replication has been demonstrated at *ori1* and *ori5*. Two pairs of *ori*s, *ori2–ori7* and *ori3–ori4*, are close together and oriented in tandem. Stohl and Clayton (1992) identified an RNase MRP in yeast that cleaves the yeast *ori5* mitochondrial RNA sequence in a site-specific manner immediately adjacent to the vertebrate CSB II sequence homolog, GC cluster C. The exact location of the cleavage site corresponds to the mapped position of the transition from RNA to DNA synthesis. Foury (1989) cloned and sequenced the nuclear gene encoding the catalytic subunit of yeast mtDNA polymerase (*MIP1*). Two other nuclear genes whose products are necessary for mtDNA metabolism in yeast have also been identified (see Van Dyck et al., 1992, for a summary). *PIF1* encodes an ATP-dependent DNA helicase involved in mtDNA recombination, repair, and stability. *RIM1* specifies a single-stranded DNA-binding protein required for mtDNA replication in *S. cerevisiae* with homology to the SSB protein from *E. coli*.

Protists. Replication of the linear mitochondrial genomes of *Paramecium* and *Tetrahymena* proceeds by two distinct modes (Cummings, 1992). In *Paramecium* replication initiates at an origin (I), at one end of which is the site of a covalent crosslink between the two parental strands. The origin contains a series of A+T-rich repeats flanked by palindromic sequences. Replication is unidirectional from the origin first generating a lariat and subsequently a dimer that is processed to yield two daughter molecules. To explain how replication and processing could reconstitute a crosslinked end, a model was proposed that supposes that the dimer is cut in such a way that staggered ends result. Ligation of the single strands would then yield the desired cross-linked monomer molecules.

While initiation of mtDNA replication in *Paramecium* is understood, termination is not (Cummings, 1992). Exactly the converse situation applies in *Tetrahymena*. The *Tetrahymena* mitochondrial genome is a linear duplex molecule of about 50 kb. The DNAs of all strains differ in size with the size variations being dependent on the size of a duplication–inversion present at both ends. *Tetrahymena* mtDNA replication begins in the middle with the formation of an eye form and proceeds bidirectionally to the ends of the parental strands. Electron microscopic characterization of partially denatured mtDNA from *T. pyriformis* revealed the presence of "panhandle" structures. The single-stranded "pan" contained the entire mitochondrial genome except for the terminal duplication-inversion sequence that constituted the double stranded "handle." The first 24 kb of the handle consists of the inverted and duplicated large rRNA gene followed by a bubble representing nonhomologous sequence and then a region at the terminus of the molecule made up of direct repeats. Goldbach et al. (1979) proposed that replication of the *T. pyriformis* mitochondrial genome is completed by the introduction of a specific endonuclease initiated single-strand nick between two repeating units at a variable distance from the replication terminus. Displacement DNA synthesis would then lead to a molecule with a single-stranded tail. DNA ligase, possibly in combination with DNA polymerase, would then complete replication of the 3' tail. The model will also explain the variable length of the duplicated inversion in terms of increases and decreases in numbers of repeat sequences. Consistent with the model variable lengths of G+C-rich repeats have actually been demonstrated (see Cummings, 1992, for details).

Kinetoplast DNA. The replication of this extraordinary complex of minicircles and maxicircles has been dissected with remarkable success (see Fig. 5–5 and Englund *et*

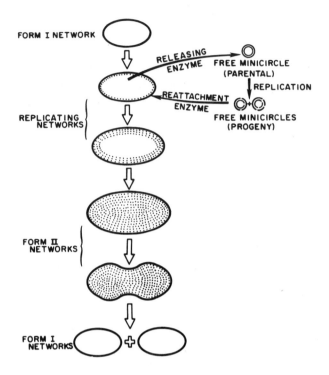

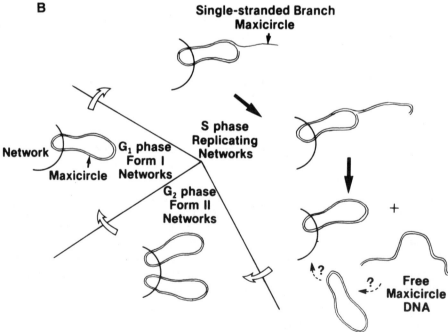

Fig. 5–5. Scheme for the replication of Trypanosome kinetoplast DNA. (A) Replication of the kinetoplast DNA network. Elliptical structures represent the outline of the entire network and the minicircles. Dots indicate the location of nicked or gapped minicircles that have undergone replication. (B) Rolling circle replication of maxicircles. From Ryan *et al.* (1988).

al., 1982; Ryan et al., 1988). Prior to replication in the G_1 phase of the cell cycle, the kDNA network consists of about 5000 covalently closed minicircles (network form I). Individual minicircles are released from the network at the beginning of S phase by a releasing enzyme, presumably a topoisomerase, to form covalently closed minicircles. These are replicated as theta structures on average exactly once. The two daughter minicircles are nicked or gapped. Following replication each minicircle returns to the progeny network. A reattachment enzyme, presumably also a topoisomerase, is thought to catalyze this process. Reattachment occurs at the network periphery. Therefore, the replicating network consists of a peripheral zone with replicated minicircles and a central zone containing unreplicated minicircles. As replication proceeds, the peripheral zone enlarges and the central zone shrinks. The size of the entire network increases because two replicated minicircles reattach for each minicircle removed from the network. When S phase ends all minicircles in the network have replicated and are nicked or gapped. Each of these form II networks, then splits to yield two form I networks in which the minicircles are covalently closed. Rauch et al. (1993) find that minicircle catenation does not involve supercoiling. In fact, there is no minicircle supercoiling at all. Rauch et al. speculate that supercoiling might prevent extensive catenation and, hence, network formation. Xu and Ray (1993) report isolation of proteins associated with kinetoplast DNA networks. These polypeptides have presequences and so are probably imported and two are lysine-rich basic proteins.

Maxicircle replication occurs at the same time in cell division as minicircle replication and each maxicircle probably replicates but once per cell generation. The maxicircles replicate as network-bound rolling circles (Fig. 5–5B). The current model for maxicircle replication imagines that rolling circle replication is initiated at a specific site on a network-bound maxicircle. The tail, initially a single strand, is converted to a double-stranded structure when lagging-strand synthesis begins. Once the replication fork has moved around the circle and a few kilobases beyond, the tail is cut off, yielding a free, linear maxicircle. This molecule has terminal repeats that permit its recircularization. Whether recircularization precedes or follows reattachment to the network is not known. In the latter case a topoisomerase could catalyze the process.

Borst (1991) set forth a novel hypothesis to explain this elaborate replication process. He points out that kDNA networks are heteroplasmic (i.e., contain more than one molecular species) and that organelle DNA heteroplasmons normally segregate during cell division (Chapter 9). This would be lethal in the case of the kDNA network because minicircles encode gRNAs necessary for editing the mRNAs encoded by the maxicircle (Chapter 14). Therefore, it is crucial that the maxicircle plus all minicircle species be retained in the kDNA network and its elaborate replication process is designed to ensure that this happens.

Chloroplast DNA replication. Bogorad (1991), in a review of cpDNA replication, summarizes the current situation in the last sentence of his article: "At present, we know little about cpDNA replication, but the tools are available for greatly advancing our understanding of the process."

Some of the most informative experiments are also the oldest. Kolodner and Tewari (1975a, b) analyzed cpDNA replication by electron microscopy in corn and pea. The replication of cpDNA in these plants initiates with the formation of D-loops located 7.1 kb apart. These D-loops expand toward each other to yield a single, looped out Cairns type intermediate. Synthesis of the two strands proceeds until two progeny cpDNA molecules are formed. At this point the nicked daughter circles may be sealed to form closed circles, but more probably the 3'-hydroxyl end of each nicked progeny molecule is extended by DNA polymerase. This displaces a single-stranded tail from the molecule that can be filled in by discontinuous duplex synthesis to yield a molecule with a double-stranded tail (rolling circle replication). The tails may then be converted to circular molecules by intrastrand recombination.

The experiments just described were done prior to the existence of physical maps of cpDNA so the D-loops could not be pinpointed on the cpDNA molecule. Since then D-loop regions have been mapped in cpDNA from *Euglena, Chlamydomonas,* and pea. Wu and colleagues (Waddel *et al.*, 1984; Wang *et al.*, 1984) identified two D-loop regions approximately 6.4 kb apart in the *Chlamydomonas* chloroplast genome. The maize cpDNA replication origin has also been sought using cloned cpDNA fragments and a partially purified preparation of pea cpDNA polymerase (Gold *et al.*, 1987). Two regions were found to be particularly active, one of which corresponds in sequence to the sequence identified in the *Chlamydomonas* D-loop. Later analyses delimited the maize replication origin to within 455 bp (Carillo and Bogorad, 1988).

REFERENCES

Birky, C.W., Jr., Katko, P., and M. Lorenz (1984). Cytological demonstration of chloroplast DNA behavior during gametogenesis and zygote formation in *Chlamydomonas reinhardtii. Curr. Genet.* 9: 1–7.

Boffey, S.A., and R.M. Leech (1982). Chloroplast DNA levels and the control of chloroplast division in light-grown wheat leaves. *Plant Physiol.* 69: 1387–1391.

Bogorad, L. (1991). Replication and transcription of plastid DNA. In *The Molecular Biology of Plastids* (L. Bogorad and I.K. Vasil, eds.), Chapter 4. Academic Press, San Diego.

Borst, P. (1991). Why kinetoplast DNA networks? *Trends Genet.*, 7: 139–141.

Carillo, N., and L. Bogorad (1988). Chloroplast DNA replication *in vitro*: Site-specific initiation from preferred templates. *Nucl. Acids Res.* 16: 5603–5620.

Clayton, D.A. (1982). Replication of animal mitochondrial DNA. *Cell* 28: 693–705.

Clayton, D.A. (1991a). Replication and transcription of vertebrate mitochondrial DNA. *Annu. Rev. Cell Biol.* 7: 453–478.

Clayton, D.A. (1991b). Nuclear gadgets in mitochondrial DNA replication and transcription. *Trends Biochem. Sciences* 16: 107–111.

Clayton, D.A. (1992). Transcription and replication of animal mitochondrial DNAs. *Int. Rev. Cytol.* 141: 217–232.

Cummings, D.J. (1992). Mitochondrial genomes of the ciliates. *Int. Rev. Cytol.* 141: 1–64.

Englund, P.T., Hajduk, S.L. and J.C. Marini (1982). The molecular biology of trypanosomes. *Annu. Rev. Biochem.* 51: 695–726.

Foury, F. (1989). Cloning and sequencing of the nuclear gene *MIP1* encoding the catalytic subunit of the yeast mitochondrial DNA polymerase. *J. Biol. Chem.* 264: 20552–20560.

Gillham, N.W. (1978). *Organelle Heredity.* Raven Press, New York.

Gold, B., Carillo, N., Tewari, K.K., and L. Bogorad (1987). Nucleotide sequence of a preferred maize chloroplast genome template for *in vitro* DNA synthesis. *Proc. Natl. Acad. Sci. U.S.A.,* 84: 194–198.

Goldbach, R.W., Bollen-DeBoer, J.F., Van Bruggen, E.F.J., and P. Borst (1979). Replication of the linear mitochondrial DNA of *Tetrahymena pyriformis. Biochim. Biophys. Acta* 562: 400–417.

Hehman, G.L., and W.W. Hauswirth (1992) DNA helicase from mammalian mitochondria. *Proc. Natl. Acad. Sci. U.S.A.,* 89: 8562–8566.

Kolodner, R.D., and K.K. Tewari (1975a). Chloroplast DNA from higher plants replicates by both the Cairns and the rolling circle mechanism. *Nature (London)* 256: 707–711.

Kolodner, R.D., and K.K. Tewari (1975b). Presence of displacement loops in the covalently closed circular chloroplast deoxyribonucleic acid from higher plants. *J. Biol. Chem.* 250: 8840–8847.

Kuroiwa, T. (1982). Mitochondrial nuclei. *Int. Rev. Cytol.* 75: 1–59.

Kuroiwa, T. (1991). The replication, differentiation, and inheritance of plastids with emphasis on the concept of organelle nuclei. *Int. Rev. Cytol.* 128: 1–62.

Kuroiwa, T., Suzuki, T., Ogawa, K., and S. Kawano (1981). The chloroplast nucleus: Distribution, number, size, and shape, and a model for the multiplication of the chloroplast genome during development. *Plant Cell Physiol.* 22: 381–396.

Leech, R.M., and K.A. Pyke (1988). Chloroplast division in higher plants with particular reference to wheat. In *Division and Segregation of Organelles* (S.A. Boffey and D. Lloyd, eds.), *Soc. Exp. Biol.* 35: 39–62.

Liu, J.-W., and R.J. Rose (1992). The spinach chloroplast chromosome is bound to the thylakoid membrane in the region of the inverted repeat. *Biochem. Biophys. Res. Commun.* 184: 993–1000.

Madsden, C.S., Ghivizzani, S.C., and W.W. Hauswirth (1993). Protein binding to a single termination-associated sequence in the mitochondrial DNA D-loop region. *Mol. Cell. Biol.* 13: 2162–2171.

Maleszka, R., and G. D. Clark-Walker (1992). In vivo conformation of mitochondrial DNA in fungi and zoosporic moulds. *Curr. Genet.* 22: 341–344.

Maleszka, R., Skelly, P.J., and G.D. Clark-Walker (1991). Rolling circle replication of DNA in yeast mitochondria. *The EMBO J.* 10: 3923–3929.

Mita, T., and T. Kuroiwa (1988). Division of plastids by a plastid-dividing ring in *Cyanidium caldarium*. *Protoplasma (Suppl 1)*: 133–152.

Mita, T., Kanbe, T., Tanaka, K., and T. Kuroiwa (1986). A ring structure around the dividing plane of the *Cyanidium caldarium* chloroplast. *Protoplasma* 130: 211–213.

Miyakawa, I., Aoi, H., Sando, N., and T. Kuroiwa (1984). Fluorescence microscopic studies of mitochondrial nucleoids during meiosis and sporulation in the yeast, *Saccharomyces cerevisiae*. *J. Cell Sci.* 66: 21–38.

Miyakawa, I., Sando, N., Kawano, S., Nakamura, S., and T. Kuroiwa (1987). Isolation of morphologically intact mitochondrial nucleoids from the yeast *Saccharomyces cerevisiae*. *J. Cell Sci.* 88: 431–439.

Miyakawa, I., Tsukamoto, T., Sakoda, M., Kuroiwa, T., and N. Sando (1988). Inhibition of yeast mitochondrial nucleoid fusion by ethidium bromide and respiration inhibitors. *J. Gen. Appl. Microbiol.* 34: 485–492.

Miyamura, S., Nagata, T., and T. Kuroiwa (1986). Quantitative fluorescence microscopy on dynamic changes of plastid nucleoids during wheat development. *Protoplasma* 133: 66–72.

Nemoto, Y., Kawano, S., Nakamura, S., Mita, T., Nagata, T., and T. Kuroiwa (1988). Studies on plastid-nuclei (nucleoids) in *Nicotiana tabacum* L. I. Isolation of proplastid-nuclei from cultured cells and identification of proplastid-nuclear proteins. *Plant Cell Physiol.* 29: 167–177.

Nemoto, Y., Nagata,, T. and T. Kuroiwa (1989). Studies on plastid-nuclei (nucleoids) in *Nicotiana tabacum* L. II. Disassembly and reassembly of proplastid-nuclei isolated from culture cells. *Plant Cell Physiol.* 30: 445–454.

Nemoto, Y., Kawano, S., Kondoh, K., Nagata, T., and T. Kuroiwa (1990). Studies on plastid-nuclei (nucleoids) in *Nicotiana tabacum* L. III. Isolation of chloroplast-nuclei from mesophyll protoplasts and identification of chloroplast DNA-binding proteins. *Plant Cell Physiol.* 31: 767–776.

Nemoto, Y., Kawano, S., Nagata, T., and T. Kuroiwa (1991). Studies on plastid-nuclei (nucleoids) in *Nicotiana tabacum* L. IV. Association of chloroplast-DNA with proteins at several specific sites in isolated chloroplast-nuclei. *Plant Cell Physiol.* 32: 131–141.

Possingham, J.V. and M.E. Lawrence (1983). Controls to plastid division. *Int. Rev. Cytol.* 84: 1–56.

Possingham, J.V., Hashimoto, H., and J. Oross (1988). Factors that influence plastid division in higher plants. In *Division and Segregation of Organelles* (S.A. Boffey and D. Lloyd, eds.). *Soc. Exp. Biol.* 35: 39–62.

Rauch, C.A., Perez-Morga, D., Cozzarelli, N.R., and P.T. Englund (1993). The absence of supercoiling in kinetoplast DNA minicircles. *EMBO J.* 12: 403–411.

Reith, M., and J. Munholland (1993). A high-resolution gene map of the chloroplast genome of the red alga *Porphyra purpurea*. *Plant Cell*, 5: 465–475.

Rose, R.J. (1988). The role of membranes in the segregation of plastid DNA. In *Division and Segregation of Organelles* (S.A. Boffey and D. Lloyd, eds.), *Soc. Exp. Biol.* 35: 171–195.

Ryan, K.A., Shapiro, T.A., Rauch, C.A., and P.T. Englund (1988). Replication of kinetoplast DNA in trypanosomes. *Annu. Rev. Microbiol.* 42: 339–358.

Sato, N., Albrieux, C., Joyard, J., Douce, R., and T. Kuroiwa (1993). Detection and characterization of a plastid envelope DNA-binding protein. *EMBO J.* 12: 555–561.

Stohl, L.L., and D.A. Clayton (1992). *Saccharomyces cerevisiae* contains an RNase MRP that cleaves at a conserved mitochondrial RNA sequence implicated in replication priming. *Mol. Cell. Biol.* 12: 2561–2569.

Tomkinson, A.E., Bonk, R.T., Kim, J., Bartfield, N., and S. Linn (1990). Mammalian nitochondrial endonuclease activities specific for ultraviolet-irradiated DNA. *Nucl. Acids Res.* 18: 929–935.

Van Dyck, E., Foury, F., Stillman, B., and S.J. Brill (1992). A single- stranded DNA binding protein required for mitochondrial DNA replication in *S. cerevisiae* is homologous to *E. coli* SSB. *EMBO J.* 11: 3421–3420.

Waddell, J., Wang, X.-M., and M. Wu (1984). Electron microscopic localization of chloroplast DNA replicative origins in *Chlamydomonas reinhardtii*. *Nucl. Acids Res.* 12: 3843–3856.

Wang, X.-M., Chang, C.H., Waddell, J., and M. Wu (1984). Cloning and delimiting one chloroplast DNA replication origin of *Chlamydomonas reinhardtii*. *Nucl. Acids. Res.* 12: 3857–3872.

Whatley, J. M. (1988). Mechanisms of plastid division. In *Division and Segregation of Organelles* (S.A. Boffey and D. Lloyd, eds.), *Soc. Exp. Biol.* 35: 63–83.

Wolf, K. and L. del Giudice (1988). The variable mitochondrial genome of ascomycetes: Organization, mutational alterations, and expression. *Adv. Genet.*, 25: 185–308.

Wolstenholme, D.R., Goddard, J.M., and C.M.-R. Fauron (1983). Replication of *Drosophila* mitochondrial DNA. In *Developments in Molecular Virology* Vol. 2 (Y. Becker, ed.), pp. 131–148. Martinus Nijhoff, B.V., The Hague.

Xu, C., and D.S. Ray (1993). Isolation of proteins associated with kinetoplast DNA networks *in vivo*. *Proc. Natl. Acad. Sci. U.S.A.*, 90: 1786–1789.

Yasuda, T., Kuroiwa, T., and T. Nagata (1988). Preferential synthesis of plastid DNA and increased replication of plastids in cultured tobacco cells following medium renewal. *Planta* 174: 235–241.

6

Phylogenetic Uses of Organelle DNA Variation

Organelle genomes possess the unique attribute for phylogenetic studies of being maternally or uniparentally inherited in many taxa (Chapter 8). Hence, analyses of mutational changes and rearrangement patterns are largely uncomplicated by recombination with other organelle genomes. This is in marked contrast to nuclear chromosomes that normally recombine vigorously at meiosis. Different organelle genomes also have specific attributes that make them particularly useful for one kind of study or another (see Chapters 3 and 4). Thus, animal mtDNA molecules have relatively rapid mutation rates and generally conservative gene order and genome size. These characteristics make these molecules particularly useful for phylogenetic analyses from the genus down to the population level. The chloroplast genomes of land plants have much lower mutation rates and also tend to have conservative gene order and genome size. Phylogenetic analyses conducted with these organelle genomes tend to be generally less useful for populations, but have more resolution at the generic level and above. Plant mtDNA molecules have very low mutation rates, but gene order and genome size are so variable that phylogenetic analyses employing restriction site polymorphisms with these molecules are largely useless. On the other hand, comparisons of specific gene sequences in these molecules can provide useful phylogenetic information at even higher taxonomic levels because of the very slow mutation rates found in these molecules. Fungal and protistan mtDNA molecules are also proving useful in phylogenetic analyses as discussed earlier in Chapters 3 and 4. However, the propensity of these genomes to rearrange (e.g., the *Brettanomyces/Dekkera* yeasts) means that gene sequence comparisons rather than restriction site analyses are the most useful taxonomic tools.

Methodologies currently used in molecular systematics are comprehensively reviewed in a book edited by Hillis and Moritz (1990) while the molecular systematics of plants is considered in a recent volume edited by Soltis *et al.* (1992a). The latter volume, in particular, is rife with articles on the uses of cpDNA to investigate the phylogeny of different plant groups. This chapter begins with a discussion of the use of animal mtDNA molecules to dissect phylogeny at the species level and below and then turns to the use of chloroplast DNA analyses to establish phylogeny at higher taxonomic levels. Unfortunately, in a book of this type it is possible to discuss only a few examples among hundreds in this burgeoning and productive field of molecular

ANIMAL mtDNA: EVOLUTION AT THE SPECIES LEVEL AND BELOW

The majority of mtDNA genotypic variants segregating in populations probably have no differential effect on organismal fitness since they commonly include base pair substitutions at silent positions in protein-coding genes and substitutions and small additions/deletions in the D-loop (control) region (see Avise et al., 1987; Wilson et al., 1985). However, some mtDNA mutations in protein-encoding genes do cause nucleotide pair substitutions that are not selectively neutral since they effect an amino acid change in the gene product. These mutations will be selected for or against in combination with neutral variants located elsewhere in the molecule since there is no recombination of animal mtDNA. The frequencies of these selectively neutral "hitch hikers" will depend on the frequency of the variant that is subject to natural selection.

The major methods employed in tha analysis of animal mtDNA evolution listed in historical order and in terms relative resolving power are restriction fragment pattern comparisons, restriction site mapping (usually with four base pair recognition sites), and amplification and sequencing of specific genes or noncoding regions using appropriate primers, and the polymerase chain reaction (PCR).

Evolution at the Species Level: The Hominoids

The evolution of primate, particularly hominoid, mtDNA has been investigated intensively. The results are instructive in that they illustrate how mtDNA comparisons can be used to construct phylogenies and estimate divergence times (see reviews by Brown, 1983, 1985; Wilson et al., 1985; Horai et al., 1992; Moritz et al., 1987). Certain of the results are also highly controversial, especially as they relate to the place of origin of the human race (the "Eve" hypothesis, see below).

Based on estimates from the fossil record concerning divergence times, the mean rate of divergence of primate mtDNA averaged over the entire molecule is about 2% per million years (Ma) (Wilson et al., 1985). However, this rate is not uniform across the molecule, with rates of nucleotide substitution being highest in the control region and lowest in the tRNA and rRNA genes. Also, the curve relating percent divergence to divergence time is biphasic, so rapid divergence rates are confined to divergence times of ca. 15 Ma or less (see Fig. 4–1). This means that species pairs that have diverged within the last 15 Ma should give the most informative comparisons. The hominoid primates, human, chimpanzee, gorilla, orangutan, and gibbon form such a group. Pairwise divergence times estimated for this group range from 5 to 10 Ma and parsimony analysis yields a phylogeny in agreement with that established using other methods (Brown, 1983). Horai et al. (1992) have reexamined this phylogeny (Fig. 6–1). using complete sequence data for a common 4938 bp region of mtDNA containing four complete and three partial gene sequences from five hominoids (pygmy and common chimpanzee, human, orangutan, and siamang). Their best estimates for divergence times for gorilla from hominoid, hominoid from chimpanzee, and the two chimpanzees from each other are 7.7 ± 0.7, 4.7 ± 0.5, and 2.5 ± 0.5 Ma respectively.

This kind of analysis can be used to predict the time and origin of the first female

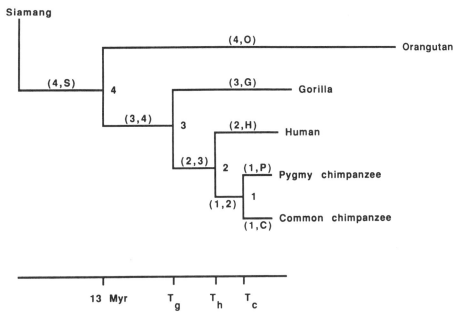

Fig. 6–1. Genealogy of six hominoid mtDNAs. All tree-making methods best support this mtDNA genealogy, irrespective of genes or data sets. Nodes are numbered 1–4. T_g is the estimated divergence time of the gorilla (7.7 ± 0.7 Ma), T_h is the estimated human divergence time (4.7 ± 0.5 Ma), and T_c is the divergence time of common and pygmy chimpanzees (2.5 ± 0.5 Ma). Modified from Horai et al. (1992).

in a given mtDNA lineage. This is possible because mtDNA is maternally inherited. In the case of humans, examination of mtDNA lineages has led to the conjecture that the common human mtDNA ancestor, Eve, lived in Africa between 166,000 and 249,000 years ago (Fig. 6–2; Cann et al., 1987; Vigilant et al., 1991). However, Eve would not have been the only woman in existence at the time. As Wilson and Cann (1992) put it,

> one might refer to the lucky woman whose lineage survives as Eve. Bear in mind, however, that other women were living in Eve's generation and that Eve did not occupy a specially favored place in the breeding pattern. She is purely the beneficiary of chance. Moreover, if we were to construct the ordinary lineages for the population, they would trace back to many of the men and women who lived at the same time as Eve. Population geneticists Daniel L. Hartl of Washington University School of Medicine and Andrew G. Clark of Pennsylvania State University estimate that as many as 10,000 people could have lived then. The name "Eve" can therefore be misleading—she is not the ultimate source of all the ordinary lineages, as the biblical Eve was.

The Eve hypothesis is the kind of scientific notion that gets everybody excited. The ensuing controversy involved broadsides from both evolutionary biologists and anthropologists. As Gibbons (1992) points out in a concise review in *Science* magazine "ironically, Eve's case never looked stronger than it did a year ago, just before she fell

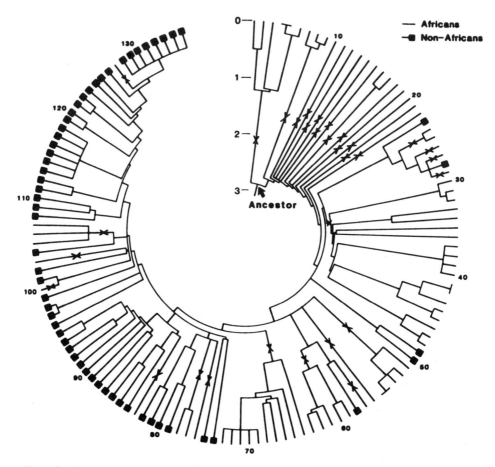

Fig. 6–2. Phylogenetic tree relating 135 mitochondrial DNA types found among 189 individuals. Markings on the branches indicate the 31 African clusters of mitochondrial DNA types; the remaining 24 non-African clusters are not labeled. Of the nucleotide positions examined 119 were informative and these were used construct 100 trees with a minimal length of 528 steps. Since the 100 trees examined differed only in the arrangement of some of the terminal twigs, conclusions about features common to all, such as the presence of deep African branches on both sides of the root, could be drawn. One of the 100 trees of length 528, chosen at random, is shown here. From Vigilant et al. (1991).

from grace.'' That was at the time of publication of the paper by Vigilant *et al.* (1991). The evolutionary biologists criticized the Eve hypothesis on statistical grounds. These investigators realized that something was wrong with the analysis (see Maddison, 1991; Templeton, 1991). As Gibbons puts it,

> when Maddison, then a postdoc at Harvard University, took a look at the phylogenetic tree, he realized right away that something was wrong—the 25 !Kung bushmen of Africa were split on the deepest branches of the tree, even though the !Kung are closely related.

So he contacted Wilson's co-authors on the *Science* paper, Stoneking and Linda Vigilant, now at Pennsylvania State University, and got their data. After 4500 computer runs, Maddison ended up with thousands of trees that were even more parsimonious—and many showed non-African roots.

At about the same time, Templeton was doing his own PAUP run (the method used by Vigilant *et al.*, 1991, which means Phylogenetic Analysis Using Parsimony) and coming to a similar conclusion. And an analysis by molecular systematist Blair Hedges and colleagues in the laboratory of Masatoshi Nei at Penn State showed that the order in which the data were entered into the PAUP program influenced whether the best tree was rooted in Africa or somewhere else. Hedges showed the data to Stoneking who agreed to sign a letter admitting to the error in *Science*'' [Hedges *et al.*, 1992].

Hedges *et al.* concluded that "the absence of a strong association between mtDNA sequence and geography, expecially among the non-Africans, suggests that the same multiple mtDNA types have been maintained in widely separated populations since those populations diverged, thus confounding an evolutionary interpretation of the data."

The anthropologists also weighed in. *Scientific American* published back-to-back articles by Wilson and Cann (1992) and Thorne and Wolpoff (1992). The first promulgated the Eve hypothesis and the second disputed the hypothesis. As Thorne and Wolpert (1992) succinctly put it. "On one side stand some researchers, such as ourselves, who maintain there is no single home for modern humanity—humans originated in Africa and then slowly developed their modern forms in every area of the Old World. On the other side are workers who claim that Africa alone gave birth to modern humans within the past 200,000 years." The popular press jumped into the fray. The *Economist* (February 15, 1992, p. 100) had this to say:

> Revenge, the Italians say, is a dish best tasted cold. In Chicago *insalata vendetta* provoked a certain amount of lip-smacking from that small but dedicated band, the Neanderthal-watchers. These fossil-hunters have looked on with increasing concern over the past few years as their hard-won interpretations of human origins have been elbowed aside by newcomers armed with blood samples and computers. Now one of the most cherished of these new notions, that of the "African" Eve, looks close to collapse.

Or is it? As Gibbons (1992) points out

> while Stoneking and Vigilant admit they made mistakes using PAUP, they maintain that other lines of genetic and fossil evidence still support putting Eve in Africa. The best evidence says Cann, is the diversity of Africans' DNA, which has been found, not just by their group, but by others in both mitochondrial and nuclear DNA. "The tree is only one part of the argument," says Cann. "A tree is an abstraction from the sequences, and the sequences themselves are not disputed. The diversity of sub-Saharan African lineages is still there."

At the end Gibbons (1992) turns to David Hillis, a molecular evolutionary biologist at the University of Texas who has been editor for articles on the Eve hypothesis for *Systematic Biology*. She quotes him as saying: "The data are simply ambiguous. They don't argue that there wasn't an African origin, and they don't argue that there was one. It's like saying you can't solve a mystery after reading one page of the book."

So while the jury is still out, this very healthy scientific controversy has focused attention on the methodology used in the analysis itself.

INTRASPECIFIC PHYLOGEOGRAPHY

Avise *et al.* (1987) have reviewed extensively the use of animal mtDNA differences to study microevolution within species. Their thesis is that animal mtDNA by virtue of its maternal pattern of inheritance, absence of recombination, rapid pace of evolution, and extensive intraspecific polymorphism "permits and even demands an extension of phylogenetic thinking to the microevolutionary level." This is, of course, the same

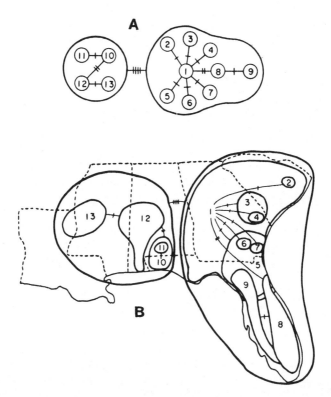

Fig. 6–3. Phylogenetic networks and phenograms summarizing evolutionary relationships among 13 mitochondrial DNA genotypes observed in a sample of 75 bowfin fish *(Amia calva)*. (A) Hand-drawn parsimony network. Slashes crossing branches indicate restriction site changes along a path; heavier lines encompass two major arrays of mitochondrial DNA genotypes distinguishable by at least three restriction site changes. (B) The parsimony network in A superimposed over the geographic sources of collections. (C) Wagner parsimony network computer generated from a presence–absence site matrix. Inferred restriction site changes are indicated, and numbers in the network represent levels of statistical support (by bootstrapping) for various clades. (D) UPGMA phenogram, where p is estimated nucleotide sequence divergence. This figure is taken from Avise et al. (1987) and the reader is referred to this paper for a detailed discussion of each method used.

PHYLOGENETIC USES OF ORGANELLE DNA VARIATION

rationale advanced for equivalent studies above the species level. The approach is to construct population phylogenies, this time within a given species, using tree building techniques such as parsimony analysis. Additionally, knowledge about the biogeographical distribution of these populations is employed to deduce microevolutionary trends.

This approach can be illustrated by comparative studies done on bowfin fish (*Amia calva*) in the Southeastern United States (Bermingham and Avise, 1986). A total of 13 distinct mtDNA genotypes or "clones" were observed. These could be grouped into two major phylogenetic and geographical assemblages (Fig. 6–3). The eastern assemblage included nine related clones from South Carolina, Georgia, and Florida. The western assemblage encompassed four related bowfin clones found from Alabama to

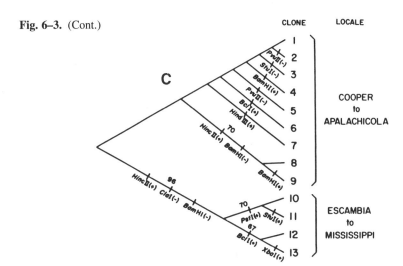

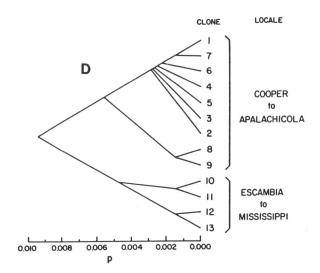

Fig. 6–3. (Cont.)

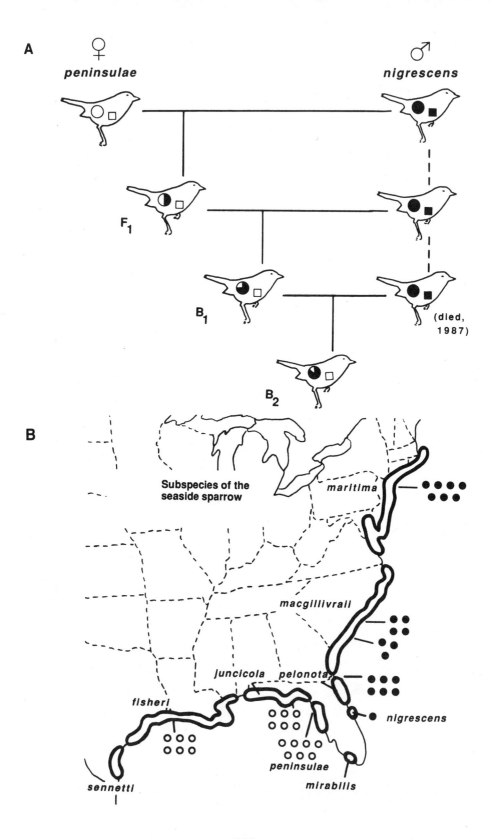

Mississippi. Clone 1 appeared to be the likely ancestor of the eastern assemblage because it was the most common and widely distributed eastern genotype (9 out of 10 river drainages). This clone formed the hub of a network whose spokes connected separately to seven other mtDNA genotypes (Fig. 6–3). Clone 1 is also at least one mutational step nearer the western clones than any other eastern clone and connects most closely with clones 10 and 12 occupying the western river drainages most proximate to the East.

The distribution of the bowfin into two assemblages in the Southeast fits the first of five categories recognized by Avise et al. (1987) to explain the "phylogeographic" data they have compiled for many different species, principally in this part of the United States. Category I is the most commonly encountered category and includes species

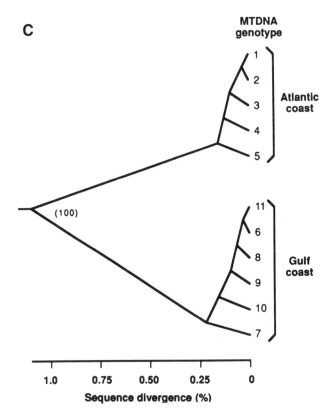

Fig. 6–4. Use of mitochondrial DNA analysis to establish the relationship of the extinct dusky seaside sparrow *(Ammodramus maritimus nigrescens)* to other populations of this species. (A) A pedigree in the captive breeding program for the dusky seaside sparrow. Three *nigrescens* males, the last of which died in 1987, were involved in crosses to *A. m. peninsulae*. The darkened areas within circles and squares represent the expected proportions of nuclear and mitochondrial genes, respectively, of dusky origin. (B) Geographic distributions of the taxonomically recognized subspecies of the seaside sparrow. Open and closed circles represent birds exhibiting, respectively, the distinctive Gulf coast and Atlantic coast mitochondrial genotypes. (C) UPGMA dendrogram showing the distinction between mitochondrial DNA genotypes of Atlantic Coast versus Gulf coast populations (numbered) of the seaside sparrow. From Avise and Nelson (1989).

where there is both a geographic and a phylogenetic discontinuity (e.g., two main assemblages differentiable by at least four assayed mutational steps). The redear sunfish, sampled from the same drainages as the bowfin, exhibits a phylogenetic dichotomy of even greater magnitude with the eastern and western assemblages differing by 17 or more assayed steps (Bermingham and Avise, 1986). In category II, phylogenetic discontinuities occur without spatial separation. Avise et al. (1987) had no good examples in this category. Hence, there was no evidence for sympatric differentiation of populations or secondary sympatry following allopatric population differentiation in the species they studied. In categories III–V the phylogenetic divergence pattern is continuous. In category III spatial separation is present, but mtDNA parsimony networks are more or less continuous with small numbers of mutational steps between phylogenetically adjacent clones, each of which is confined to a portion of the geographic range of the species. The marine oyster toadfish is an example of category III species in which limited gene flow between populations occurs in the absence of long-term zoogeographic barriers. Lack of spatial separation and phylogenetic continuity are assumed in category IV species (e.g., red-winged blackbird). In category V species there is an intermediate level of gene flow coupled with partial spatial separation.

Phylogeographic analysis using animal mtDNA provides a powerful methodology for establishing the genetic relatedness of different groups of populations within a species and has potentially important applications as two distinctly different examples—preservation of the dusky seaside sparrow and racial differentiation in human beings—will illustrate. The dusky seaside sparrow *(Ammodramus maritimus nigrescens)* is a recently extinct subspecies of the seaside sparrow found along the Atlantic and Gulf Coasts of the United States. In 1966 the dusky seaside sparrow, confined in its range to Brevard County, Florida, on the Atlantic Coast was listed as "endangered" by the U.S. Fish and Wildlife Service (Avise and Nelson, 1989). By 1980 only six male birds could be found. Five of these were brought into captivity and a backcrossing program initiated with Scott's seaside sparrow *(A.m. peninsulae)*, which resides on the Florida Gulf coast. Three dusky males were involved in the backcrossing program, the last one of which died in 1987 (Fig. 6–4A). By this time hybrids ranging from 50 to 87.5% dusky had been produced. They will constitute a core population for planned reintroduction of dusky-like birds into the wild.

In the meantime Avise and Nelson (1989) performed a phylogeographic analysis of seaside sparrow populations. Based on mtDNA restriction pattern differences, they found that the Atlantic and Gulf coast populations formed two distinct assemblages (Fig. 6–4B,C). Their results indicated long-term separation of these two assemblages estimated at 250,000–500,000 years. Furthermore, using the mtDNA criterion the dusky appears to be indistinguishable from the nominate subspecies *A. maritimus*. Avise and Nelson (1989) point to three practical consequences that might have resulted had phylogeographic analysis using mtDNA been available when loss of the dusky seaside sparrow appeared imminent. First, in the absence of a formal species or subspecies designation for the dusky seaside sparrow, the preservation efforts mandated by the Endangered Species Act would not have applied. Second, the backcrosses with the surviving dusky males would have involved the more closely related nominate subspecies rather than the distantly related Scott's seaside sparrow of the Florida Gulf coast. Third, conservation efforts would have focused on the preservation of the two major phylogenetic population groups within the species rather than on saving the Brevard County population.

Phylogenetic analysis of human mtDNA lineages is being employed as a tool to identify ancient migration routes. Thus, using linguistic comparisons Bellwood (1985) proposed two major prehistoric migrations into Southeast Asia. The first was an ancient "Australoid" migration from the Indo-Malaysian Archipelago that led to the colonization of Australia and New Guinea about 40,000 years ago. The second migration occurred 4000 to 6000 years ago. This migration was of "Southern Mongoloids" from Fujian or Zhejian provinces of modern China into much of island and mainland Southeast Asia. Ballinger *et al.* (1992) using mtDNA analysis concur that all Southeast Asian populations examined appear to have common origins consistent with southern Mongoloid origin hypothesized on by Bellwood linguistic grounds. The mtDNA data further suggest that Southern China is the center of Asian mtDNA radiation. Other analyses have confirmed the Mongoloid affinities of Philippine Negritos and other Asians (Japanese, Ainu, and Koreans) (Harihara *et al.*, 1988). Stoneking *et al.* (1990) examined geographic variation in mtDNA from 119 people from 25 locations in Papua New Guinea and reached three conclusions: (1) external and internal isolation of Papua New Guinea populations has taken place with respect to mtDNA lineages, (2) there is a likely Southeast Asian origin of these lineages, with no particularly close phylogenetic relationship between Australian and Papua New Guinea populations, and (3) an unusual relationship exists between mtDNA genetic distance and geographic distance separating populations in Papua New Guinea that probably reflects ancestral colonization routes in this gigantic, mountainous, and densely forested island.

Analysis of mtDNA polymorphisms has also been used to trace migration of Native American populations. North, Central, and South American populations exhibit the same set of rare Asian mtDNA variants, suggesting that they derived from a common ancestral population (Wallace *et al.*, 1985; Schurr *et al.*, 1990). There are three Native American languages, Amerind, Nadene, and Eskaleut (Greenberg, 1987). It has been hypothesized that three waves of migration populated the Americas corresponding to these three Native American languages. Torroni *et al.* (1992) tested this hypothesis for Amerind and Nadene populations using mtDNA analysis. Their results support the hypothesis and indicate that these two populations were founded by two independent migrations.

BOTTLENECKS AND LINEAGE EXTINCTION

Since mtDNA transmission in animals is from mother to daughter generation after generation, the question arises as to how far back in time specific maternal lineages can be traced. This problem has been addressed theoretically by Avise *et al.* (1984; see also Avise, 1986; Avise *et al.*, 1987) using a Poisson model that assumes that adult females in a population produce daughters according to a Poisson distribution with mean μ. Thus, if mothers give birth on the average to one surviving daughter, 37% of all maternal lineages will go extinct after one generation and less than 2% of the original mothers will have contributed mtDNA molecules to the population 100 generations later. Avise *et al.* (1984) have used this approach to estimate probabilities of survival of two or more independent mitochondrial lineages through time. The striking result of this analysis is that stochastic mtDNA lineage extinction within a population is expected to occur at a rapid pace with the result that mtDNA evolutionary trees are continuously "self-pruning."

The propensity for mitochondrial lineages to undergo extinction rapidly for stochastic reasons must be kept in mind in considering whether bottlenecks in absolute population size need to be invoked to account for certain observations made on the basis of mtDNA sequence differences. For instance, the observation has been made that the human species has an anomalously low level of mtDNA variability in spite of an apparently normal level of nuclear variability (cf. Wilson et al., 1985). This suggests that human populations might have passed through a transient bottleneck at some time during the course of evolution. Avise (1986) points out that his theoretical models suggest that a bottleneck in population size need not necessarily have to be invoked to account for such mtDNA data even if all living human beings trace to the primeval Eve. It seems prudent to remember this point in considering population bottlenecks that have been inferred to explain reduced mtDNA variability in other species of vertebrates (cf. Wilson et al., 1985).

HYBRIDIZATION AND INTROGRESSION

Animal mtDNAs have been very useful in tracing past hybridizations that have given rise to modern species (see Avise, 1986; Wilson et al., 1985). Two recent accounts of hybridization having potentially broad significance will be summarized here.

Mitochondrial DNA analysis has been used to study the spread of African honeybees in South and Central America (Page, 1989; Hall and Muralidharan, 1989; Smith et al., 1989). European honeybees are not very productive in Brazil's tropical climate so in 1956 South African queen honey bees were imported into southern Brazil by Warwick Kerr. His objective was to produce a strain of honey bees adapted to tropical climates through hybridization. Kerr hoped this strain could be easily managed like the European subspecies since the African bees have the nasty habit of severe stinging. In 1957 26 of the 47 African queens escaped and by 1963 African bees and their hybrids comprised an estimated 70% of the total honey bee population of the state of São Paulo. These bees have spread over South and Central America at a rate of 500 km per year. They have now arrived in Texas and are expected to begin to populate the southern tier of the United States. Mitochondrial DNA analysis reveals that these Africanized bees retain almost exclusively African mtDNA. This could mean that they either are descended from hybrids between European drones and the original escaped African queens or that these queens mated with escaped African drones. Isozyme data indicate that long-established neotropical honeybee populations do not differ significantly from African bees (Smith et al., 1989). These results suggest that the bees that have spread through Central and South America also have the nuclear gene phenotypes of African bees, but they do not yet prove that these bees are pure African bees rather than Africanized hybrids. Although the data appear to show that these bees share an unbroken maternal mtDNA lineage with the African bees introduced into Brazil in 1956, Meusel and Moritz (1992) sound a cautionary note. They are concerned about the use of mtDNA as an indicator of the "African" honey bee until racial variation of mtDNA has been documented in detail in honeybees.

The story of the red wolf has been nicely told in a *New York Times Magazine* article by Jan DeBlieu (1992). The red wolf was nearly driven to extinction, but a captive breeding program made possible its successful reintroduction into a swamp forest in coastal North Carolina. Based on morphological criteria biologists in the U.S. Fish and Wildlife Service during the 1970s had concluded that the red wolf was a full species distinct from coyotes and gray wolves. Their studies were used to justify the extensive captive breeding program that led to reintroduction of the red wolf to the wild.

Since little of this research had been published in scientific journals, some scientists questioned its validity, and so the service turned to two scientists, Robert Wayne and Susan Jenks, who were using mtDNA lineages to establish relationships between animal species. As DeBlieu puts it, "Was the red wolf a true species, the service asked? And if not, what was it? A subspecies? A cross of coyote and gray wolf? The question was pivotal, since it was against Federal policy to spend Endangered Species Act funds on the preservation of hybrid animals and plants."

Wayne and Jenks (1991) analyzed mtDNA from 94 red wolves, including some material that had been preserved in the early 1900s. DeBlieu says, "To their surprise, Wayne and Jenks failed to uncover any consistent genetic difference between the red wolves being released in North Carolina and coyote-gray wolf hybrids. 'We simply didn't find any evidence of a unique red wolf genotype', Wayne says. 'Maybe it's there and we just missed it; we're in the process of doing more tests. Right now, though, I can't say there's any good supporting evidence that the red wolf is a species.'"

"If Wayne and Jenks's findings are verified, it means that the Federal Government may have spent 20 years and hundreds of thousands of dollars to preserve a wild mutt." But according to DeBlieu, Wayne also avers, "'These tests are all very preliminary, and it's just too soon to draw firm conclusions. It's certainly too soon to talk about dropping protection for the red wolf." This case underlines the ever increasing importance mtDNA phylogenies are assuming in sorting out the dynamic processes involved in animal speciation and hybridization. Wayne (1993) summarizes this and other cases of hybridization in a recent review on the molecular evolution of the dog family.

CHLOROPLAST DNA PHYLOGENY: THE EVOLUTION OF HIGHER TAXA

Chloroplast DNA analyses have been particularly useful in untangling phylogenies at the level of genus and above because of the relatively low mutation rates of these molecules (for reviews see Downie and Palmer, 1992a; Palmer *et al.*, 1988; Clegg and Zurawski, 1992). Three principal methods have been used for phylogeny construction. In the first specific chloroplast gene sequences are compared. In the second method restriction site mapping over a well-defined region is used. Lastly, major structural rearrangements can be employed for phylogeny construction. These sorts of rearrangements include rare inversions, losses of genes and introns, the loss of the inverted repeat, expansion and contraction of the inverted repeat, and occasional length mutations (Chapter 4). Downie and Palmer (1992a) note that the analysis of major structural rearrangements is "a complementary approach to comparative sequencing for studying the higher-level relationships among angiosperms."

The method that yields the most unambiguous results and is applicable to the largest number of taxa involves phylogenies and makes use of specific chloroplast gene sequences. The most sequence comparisons are available for the *rbcL* genes, but a smaller number of sequences have also been obtained for a few other genes including *psbA* and *atpBE*. Although such standardized comparisons are relatively precise and can be applied broadly for phylogenetic reconstruction, the applications of the principal methods used for phylogenetic constructions themselves (parsimony, distance matrix methods, and maximum likelihood) can lead to controversy (Clegg and Zurawski, 1992). Hence, all three methods are often used in phylogenetic reconstructions. As Clegg and Zurawski (1992) put it, "this is done to establish that each method is reasonably consistent for a given data set and to placate proponents of each method." Felsentein (1988) has reviewed the properties of the three methods and Clegg and Zurawski (1992) discuss their application to *rbcL* sequence data. They conclude from the published literature that *rbcL* sequence data have the greatest phylogenetic resolution at the levels of family, order, and subclass.

One example of the application of *rbcL* sequence data to discriminate between advanced taxa of flowering plants comes from a paper by Albert *et al.* (1992) on the phylogeny and structural evolution of carnivorous plants. Their molecular phylogenetic data indicate "that both carnivory and stereotyped trap forms have arisen independently in different lineages of angiosperms." Albert *et al.* distinguish two basic trapping mechanisms, which they refer to as flypaper and pitcher traps. They find that flypaper traps have five separate origins among angiosperms whereas pitcher traps have three. Both trap types are found in two lineages. The first, the American pitcher plants (Sarraceniaceae), prove to be related by *rbcL* sequence data to a flypaper trapper, the South African fly bush *(Roridula),* while the second embraces the Old World pitcher plants *Nepenthes* and the sundews (Droseraceae) plus the Venus flytrap *(Dionaea).* In another study, Olmstead *et al.* (1992) used parsimony analysis and compared 57 *rbcL* sequences in the Asteridae, the second largest subclass of dicots, which is thought to be of relatively recent origin. In this particular study Olmstead *et al.* sought to test the monophyly of the Asteridae and to identify the major lineages in this group of flowering plants. They conclude that the Asteridae are indeed monophyletic and that *rbcL* sequence data can be used to infer the major phylogenetic lineages within this group of plants. This study is particularly notable because it is accompanied by a companion paper making use of restriction site mapping of the highly conserved inverted repeat to test independently the molecular phyogeny of the Asteridae (Downie and Palmer, 1992b). This approach was driven by the finding that rates of nucleotide substitution at silent and noncoding sites are four to six times lower in the inverted repeat than the single copy regions (Table 4–1). The results obtained by this method proved to be generally consistent with those derived from morpological data and from the *rbcL* sequence data discussed above. An important conclusion of this study is that by focusing on the highly conserved inverted repeat sequence, restriction site mapping can be extended to higher taxonomic levels than was previously thought possible.

Although phylogenetic analysis of cpDNA molecules has proved most useful for cladistic analysis of higher taxa, Soltis *et al.* (1992b) point out that cpDNA variation can also be used for analyses at the intraspecific level. They believe that in many studies sampling schemes were used "that were grossly inadequate for detecting intraspecific variation," which they note was not the goal in most of these cases in any event. Soltis *et al.* tabulate many instances of intraspecific variation of chloroplast genomes and show how this variation can be applied to analyze problems similar to those being studied with animal mtDNA. In a practical sense, one of the more important of these applications involves the analysis of cpDNA variation in helping to reconstruct the origins of certain crop plants. This topic is considered in a review by Doebley (1992). For example, he discusses the use of cpDNA variation to examine the relationship between maize (*Zea mays* ssp. *mays*) and its wild relatives. Three teosinte species (*Z. diploperennis, Z. perennis*, and *Z. luxurians*) differ from *Z. mays* ssp. *mays* by 19 mutations, but the four cpDNA genotypes found in *Z. mays* ssp. *mays* are also found in two other teosintes (*Z. mays* ssp. *mexicana* and ssp. *parviglumis*). Doebley says that "these cpDNA results are consistent with previous cytologic and allozymic data that suggest one of these two wild subspecies was the progenitor of maize. The fact that four cpDNA types were found in both teosinte and maize implies either multiple domestications or introgression of wild type cytoplasms into cultivated maize."

ANALYSIS OF ORGANELLE DNA FROM DEAD AND FOSSILIZED MATERIAL

All of the phylogenies discussed in this chapter so far deduce past events by analysis of organelle DNA in living species. Thus, it is fitting to conclude by considering the potential that exists for analysis of fossil organelle DNA. Animal mtDNA as well as highly conserved gene sequences in cpDNA (e.g., the *rbcL* gene) are particularly well suited for this endeavor since the PCR now makes possible the amplification of tiny DNA samples for sequencing as long as appropriate oligonucleotide primers are available. In practice this means that museum or herbarium samples of both extant and extinct species can be used for organelle DNA analysis, greatly increasing the potential uses of this tool for phylogenetic reconstructions.

One of the first taxonomic studies making use of organelle DNA from dead individuals involved determining the relationship between the mountain zebra and its extinct relative the quagga (Higuchi *et al.*, 1984; Wilson *et al.*, 1985). The analysis involved comparing *cox1* gene sequences in living zebras and a 140-year-old quagga museum specimen. The quagga and mountain zebra *cox1* genes were found to differ by 6% in nucleotide substitutions. This agrees with the percentage divergence of the mountain zebra and Burchell's zebra mtDNA estimated by restriction map comparisons. The significnance of the latter observation is that the mtDNA of Burchell's zebra is nearly identical to quagga mtDNA.

Paabo (1989) considered the methodologies and problems involved in the extraction of ancient DNA using dried specimens ranging in age from four (pork) to 13,000 (extinct ground sloth *Mylodon*) years. The DNA samples were invariably degraded to an average size of 100–200 bp and cloning efficiencies were poor. Analysis of nitrogen bases in total DNA revealed that the two pyrimidines, thymine and cytosine, were present in greatly reduced amounts. This is probably explained by the fact the pyrimidines, particularly thymine, are known to be substantially more sensitive to oxidative damage than purines. However, when oligonucleotide primers to mtDNA sequences and PCR were used, amplification of sequences up to 140 bp could be achieved. Two amplified sequences from the liver of a 4000-year-old Egyptian mummy showed only three base changes from the reference mtDNA. Only one of these has not been observed in other human mtDNAs so PCR amplification of mtDNA sequences from dried tissues produces material that is clearly suitable for phylogenetic analyses.

Bone mtDNA sequences are being used both for forensic and archeological analyses. Thus, Hagelberg *et al.* (1991a) report the use of bone mtDNA typing to establish the identity of a murder victim. Hagelberg et al. (1991b) have also analyzed ancient bone DNA recovery in relation to gross and microscopic preservation of bone. For this work they chose bone samples from excavations by the Oxford Archaeological Unit done in two cemeteries within the former precinct of Abingdon Abbey. One of these was a medieval cemetary used from 1000 to 1540 A.D. and the second a cemetery employed for burial of English Civil War dead from 1644 to 1663 A.D. Hagelberg et al. make this amusing aside. "Interestingly, the Civil War graves are all oriented north-south, in contrast to the medieval graves which all lie east-west. During the English Civil War and rule of Oliver Cromwell much of Anglican ritual was considered Papist, hence the unconventional orientation of the graves." Another smaller medieval cemetery about a kilometer northwest was also excavated by the Abingdon Archaeological and Historical Society.

Hagelberg et al. (1991b) examined five human femur samples from the Civil War Cemetery and one femur from the small medieval graveyard. To assess the level of degradation of the mtDNA samples four pairs of primers were then used specifying mtDNA lengths of 121 to 800 bp. Interestingly, while the largest fragment that could be amplified from the Civil War skeletons was one 375 bp in length, the 800-bp fragment was obtained from medieval bone DNA. This bone was also better preserved than the Civil War samples having extremely good micromorphology. This important analysis shows that mtDNA samples can be extracted from ancient bone and that the degree of preservation rather than age is the determining factor with respect to quality of mtDNA.

A truly astonishing paper was published in *Nature* by Golenberg et al. in 1990. Using methods of DNA isolation worked out for ancient plants by Rogers and Bendich (1985), Golenberg et al. reported amplifying an 820-bp sequence from the chloroplast *rbcL* of an extinct species of *Magnolia (M. latahensis)* (17–20 Ma) found in the Miocene Clarkia fossil beds of Northern Idaho. These are mainly well-preserved leaf compression fossils, in which the chloroplasts can be found in over 90% of the cells examined in some species (Niklas et al., 1985). Approximately one in 10 samples extracted yielded high molecular mass (HMM) DNA that could be visualized on gels (Golenberg, 1991). This DNA was used for PCR amplification of an *rbcL* sequence. Golenberg et al. (1990) reported 17 bp substitutions between the fossil *M. latahensis* and a recent species *(M. macrophylla)*. Twelve of these were transitions and only four of these were nonsynonymous first and second position substitutions. Phylogenetic statistical analysis revealed that the fossil species clustered well within the Magnoliidae.

Naturally, the claim that Miocene cpDNA could be amplified did not go unchallenged. Paabo and Wilson (1991) reviewed this work and were skeptical for two reasons. First, *in vitro* rates of spontaneous depurination are estimated to be 4×10^9 per second at 70°C at pH 7.4. Following depurination rapid breakage of the DNA chain is expected. Paabo and Wilson (1991) used these data to calculate a depurination rate of 6.33×10^{-6} sites per year. Based on this rate and a reasonable estimate of cpDNA copy number, no 800-bp fragments should be intact after 5000 years and no purines should remain at all after 4 Ma. Second, Paabo and Wilson (1991) were unable to amplify sequences from the Clarkia DNA extractions using the primers of Golenberg et al. (1990). These amplification experiments, reported in more detail by Sidow et al. (1991), showed that primers to conserved parts of genes encoding cytoplasmic rRNA also failed to yield a DNA product. However, primers designed to amplify a 370-bp piece of bacterial DNA did yield a product from the HMM DNA. Hence, Paabo and colleagues concluded that HMM DNA from the Clarkia fossils was contaminated with bacterial DNA that was probably of recent origin. This is reminiscent of the early days of chloroplast and mitochondrial DNA isolation when bacterial DNA contamination was frequently a problem (Chapter 3).

Golenberg (1991) rebutted the arguments made by Paabo and colleagues. Golenberg first pointed out that rates of depurination could not be estimated by simple linear extrapolation as done by Paabo and Wilson (1991). Also, the leaves themselves underwent a series of transformations that may have been responsible for their extraordinary preservation. "Thus, it is likely that the conditions in which the DNA was preserved are not similar to normal physiological conditions, and, therefore, depurination may have been slower than estimated." Golenberg (1991) then enters into a lengthy technical discussion of why Paabo and colleagues were unable to reproduce the results reported by Golenberg *et al.* (1990). It goes without saying that it would be very hard to confuse an *rbcL* sequence with a bacterial rDNA sequence. Since then, Soltis *et al.* (1992c) reported amplifying a 1320-bp sequence from the *rbcL* gene of bald cypress *(Taxodium)* from the Clarkia deposit. Phylogenetic analysis groups the fossil *Taxodium* with the extant *T. distichum*. These results coupled with those of Golenberg *et al.* make it extremely unlikely that artifacts are involved even given the notorious sensitivity of PCR techniques to contamination. In both *Magnolia* and *Taxodium*, the phylogenetic analyses reveal that the fossil species exhibit nucleotide substitutions in *rbcL* gene distinguishing them from their modern counterparts, yet they group with them. Therefore, the remote possibility of contamination with cpDNA either from living congeners of the fossil species or unrelated species is eliminated.

The significance of being able to sequence organelle DNA from fossils as old as 17–20 Ma years may be great. In the words of Golenberg *et al.* (1990):

> This should open up possibilities of research in several new areas. First, mutation rates to outgroup species may be estimated by using the calibrated time difference between pairs of closely related extant and fossil species and their last shared ancestor. Second, sequence data and molecular phylogenetic analyses may be used independently to test purported phylogenetic relationships of extinct taxa. Third, biogeographic trends may be investigated by comparing individual relationships of species within fossil assemblages to those of existing plant communities.

Despite these hopeful words, sequencing of fossil organelle DNAs is still in its infancy and we will not know how useful this approach is going to be for some time. In contrast, there is no question that comparisons of organelle DNAs from living organisms will continue to play a major role in phylogeny extending all the way from the intraspecific level to the highest levels of eukaryotic classification.

REFERENCES

Albert, V.A., Williams, S.E., and M.W. Chase (1992). Carnivorous plants: Phylogeny and structural evolution. *Science* 257: 1491–1495.

Avise, J. (1986). Mitochondrial DNA and the evolutionary genetics of higher animals. *Phil. Trans. R. Soc. London. B* 312: 325–342

Avise, J.C., and W.S. Nelson (1989). Molecular genetic relationships of the extinct dusky seaside sparrow. *Science* 243: 646–648.

Avise, J.C., Neigel, J.E., and J. Arnold (1984). Demographic influences on mitochondrial DNA lineage survivorship in animal populations. *J. Mol. Evol.* 20: 99–105.

Avise, J.C., Arnold, J., Ball, R.M., Bermingham, E.,

Lamb, T., Neigel, J.E., Reeb, C.A., and N.C. Saunders (1987). Intraspecific phylogeography: The mitochondrial DNA bridge between population genetics and systematics. *Annu. Rev. Ecol. Syst.* 18: 489–522.

Ballinger, S.W., Schurr, T.G., Torroni, A., Gan, Y.Y., Hodge, J.A., Hassan, K., Chen, K.-H., and D.C. Wallace (1992). Southeast asian mitochondrial DNA analysis reveals genetic continuity of ancient mongoloid migrations. *Genetics* 130: 139–152.

Bellwood, P. (1985). *Prehistory of the Indo-Malaysian Archipelago.* Academic Press, Sydney.

Bermingham, E., and J.C. Avise (1986). Molecular zoogeography of freshwater fishes in the southeastern United States. *Genetics* 113: 939–965.

Brown, W.M. (1983). Evolution of mitochondrial DNA. In *Evolution of Genes and Proteins* (M. Nei and R.K. Koehn, eds.), pp. 62–88. Sinauer Associates, Sunderland, MA.

Brown, W.M. (1985). The mitochondrial genome of animals. In *Monographs in Evolutionary Biology: Molecular Evolutionary Genetics* (R.J. MacIntyre, ed.), pp. 95–129. Plenum Press, New York.

Cann, R.L., Stoneking, M., and A.C. Wilson (1987). Mitochondrial DNA and human evolution. *Nature (London)* 325: 31–36.

Clegg, M.T., and G. Zurawski (1992). Chloroplast DNA and the study of phylogeny: Status and future prospects. In *Molecular Systematics of Plants* (P.S. Soltis, D.E. Soltis, and J.F. Doyle, eds.), pp. 1–13. Chapman and Hall, New York.

DeBlieu, J. (1992). Could the red wolf be a mutt? *New York Times Magazine* June 14: 30.

Doebley, J. (1992). Molecular systematics and crop evolution. In *Molecular Systematics of Plants* (P.S. Soltis, D.E. Soltis, and J.J. Doyle, eds.), pp. 202–222. Chapman and Hall, New York.

Downie, S.R., and J.D. Palmer (1992a). Use of chloroplast DNA rearrangements in reconstructing plant phylogeny. In *Molecular Systematics of Plants* (P.S. Soltis, D.E. Soltis, and J.J. Doyle, eds.), pp. 14–35. Chapman and Hall, New York.

Downie, S.R., and J.D. Palmer (1992b). Restriction site mapping of the chloroplast DNA inverted repeat: A molecular phylogeny of the Asteridae. *Ann. Missouri Bot. Gard.* 79: 266–283.

Felsenstein, J. (1988). Phylogenies from molecular sequences: Inferences and reliability. *Annu. Rev. Genet.* 22: 521–565.

Gibbons, A. (1992). Mitochondrial Eve: Wounded, but not dead yet. *Science* 257: 873–875.

Golenberg, E.M. (1991). Amplification and analysis of Miocene plant fossil DNA. *Phil. Trans. R. Soc. London.* B 333: 419–427.

Golenberg, E.M., Giannasi, D.E., Clegg, M.T., Smiley, C.J., Durbin, M., Henderson, D., and G. Zurawski (1990). Chloroplast DNA sequence from a Miocene *Magnolia* species. *Nature (London)* 344: 656–658.

Greenberg, J. (1987). *Language in the Americas.* Stanford University Press, Stanford, CA.

Hagelberg, E., Gray, I.C., and A.C. Jeffreys, (1991a). Identification of the skeletal remains of a murder victim by DNA analysis. *Nature (London)* 352: 427–429.

Hagelberg, E., Bell, L.S., Allen, T., Boyde, A., Jones, S.J., and J.B. Clegg (1991b). Analysis of ancient bone DNA: techniques and applications. *Phil. Trans. R. Soc. London.* B 333: 399–407.

Hall, H.G., and K. Muralidharan (1989). Evidence from mitochondrial DNA that African honey bees spread as continuous maternal lineages. *Nature (London)* 339: 211–213.

Harihara, S., Saitou, N., Hirai, M., Gojobori, T., Park, K.S., Misawa, S., Ellepola, S.B., Ishida, T., and K. Omoto (1988). Mitochondrial DNA polymorphism among five Asian populations. *Am. J. Hum. Genet.* 43: 134–143.

Hedges, S.B., Kuman, S., Tamura, K., and M. Stoneking (1992). Human origins and analysis of mitochondrial DNA sequences. *Science* 255: 737–739.

Higuchi, R., Bowman, B., Freiberger, M., Ryder, O.A., and A.C. Wilson (1984). DNA sequences from the quagga, an extinct member of the horse family. *Nature (London)* 312: 282–284.

Hillis, D.M., and C. Moritz (1990). *Molecular Systematics.* Sinauer Associates, Sunderland, MA.

Horai, S., Satta, Y., Hayasaka, K., Kondo, R., Inoue, T., Ishida, T., Hayashi, S., and N. Takahata (1992). Man's place in Hominoidea revealed by mitochondrial DNA geneaology. *J. Mol. Evol.* 35: 32–43.

Maddison, D.R. (1991). African origin of human mitochondrial DNA reexamined. *Syst. Zool.* 40: 355.

Meusel, M.S., and R.F.A. Moritz (1992). Mitochondrial DNA length variation in the cytochrome oxidase region of hhoney bees. *Apidologie* 23: 1–4.

Moritz, C., Dowling, T.E., and W.M. Brown (1987). Evolution of animal mitochondrial DNA: Relevance for population biology and systematics. *Annu. Rev. Ecol. Syst.* 18: 269–292.

Niklas, K.J., Brown, R.M. Jr, and R. Santos (1985). Ultrastructural states of preservation in Clarkia angiosperm leaf tissues: Implications on modes of fossilization. In *Late Cenezoic history of the Pacific Northwest* (C.J. Smiley, ed.), pp. 143–160. Pacific Division of the American Association for the Advancement of Science, San Francisco.

Olmstead, T.G., Michaels, H.J., Scott, K.M., and J.D. Palmer (1992). Monophyly of the Asteridae and identification of their major lineages inferred from DNA sequences of *rbc*L. *Ann. Missouri Bot. Gard.* 79: 249–265.

Paabo, S. (1989). Ancient DNA: Extraction, characterization, molecular cloning and enzymatic amplification. *Proc. Natl. Acad. Sci. U.S.A.* 86: 1939–1943.

Paabo, S., and A.C. Wilson (1991). Miocene DNA sequences—a dream come true? *Curr. Biol.* 1: 45–46.

Page, R.E. (1989). Neotropical African bees. *Nature (London)* 339: 181–182.

Palmer, J.D., Jansen, R.K., Michaels, H.J., Chase, M.W., and J.R. Manhart (1988). Chloroplast DNA variation and plant phylogeny. *Ann. Missouri Bot. Gard.* 75: 1180–1206.

Rogers, S.O., and A.J. Bendich (1985). Extraction of DNA from milligram amounts of fresh, herbarium and mummified plant tissues. *Plant Mol. Biol.* 5: 69–76.

Schurr, T.G., Ballinger, S.W., Gan, Y.Y., Hodge, J.A., Merriwether, D.A., Lawrence, D.N., Knowler, W.C., Weiss, K.M., and D.C. Wallace (1990). Amerindian mitochondrial DNAs have rare Asian mutations at high frequencies, suggesting they derived from four primary maternal lineages. *Am. J. Hum. Genet.* 46: 613–623.

Sidow, A., Wilson, A.C., and S. Paabo (1991). Bacterial DNA in Clarkia fossils. *Phil. Trans. R. Soc. London* B 333: 429–433.

Smith, D.R., Taylor, O.R., and W.M. Brown (1989). Neotropical Africanized honey bees have African mitochondrial DNA. *Nature (London)* 339: 213–215.

Soltis, P.S., Soltis, D.E., and J.J. Doyle eds. (1992a). *Molecular Systematics of Plants*. Chapman and Hall, New York.

Soltis, D.E., Soltis, P.S., and B.G. Milligan (1992b). Intraspecific chloroplast DNA variation: Systematic and phylogenetic implications. In *Molecular Systematics of Plants* (P.S. Soltis, D.E. Soltis, and J.J. Doyle, eds.), pp. 177–201. Chapman and Hall, New York.

Soltis, P.S., Soltis, D.E., and C.J. Smiley (1992c). An *rbcL* sequence from a Miocene *Taxodium* (bald cypress). *Proc. Natl. Acad. Sci. U.S.A.*, 89: 449–451.

Stoneking, M., Jorde, L.B., Bhatia, K., and A.C. Wilson (1990). Geographic variation in human mitochondrial DNA from Papua New Guinea. *Genetics* 124: 717–733.

Templeton, A.R. (1991). Human origins and analysis of mitochondrial DNA sequence. *Science* 255: 737.

Thorne, A.G., and M.H. Wolpoff (1992). The multiregional evolution of humans. *Sci. Am.* 266: 76–83.

Torroni, A., Schurr, T.G., Yang, C-C., Szathmary, E.J.E., Williams, R.C. Shanfield, M.S., Troup, G.A., Knowler, W.C., Lawrence, D.N., Weiss, K.M., and D.C. Wallace (1992). Native american mitochondrial DNA analysis indicates that the Amerind and the Nadene populations were founded by two independent migrations. *Genetics* 130: 153–162.

Vigilant, L., Stoneking, M., Harpending, H., Hawkes, K., and A.C. Wilson. (1991). African populations and the evolution of human mitochondrial DNA. *Science* 253: 1503–1507.

Wallace, D.C., Garrison, K. and W.C. Knowler (1985). Dramatic founder effects in Amerindian mitochondrial DNAs. *Am. J. Phys. Anthropol.* 68: 149–155.

Wayne, R.K. (1993). Molecular evolution of the dog family. *Trends Genet.*, 9: 218–225.

Wayne, R.K., and S.M. Jenks (1991). Mitochondrial DNA analysis implying extensive hybridization of the endangered red wolf, *Canis rufus*. *Nature (London)* 351: 565–570.

Wilson, A.C., and R.L. Cann (1992). The recent African genesis of humans. *Sci. Am.* 266: 68–73.

Wilson, A.C., Cann, R.L., Carr, S.M., George, M., Gyllensten, U.B., Helm-Bychowski, K.M., Higuchi, R.G., Palumbi, S.R., Prager, E.M., Sage, R.D., and M. Stoneking (1985). Mitochondrial DNA and two perspectives on evolutionary genetics. *Biol. J. Linn. Soc.* 26: 375–400.

PART III

ORGANELLE GENETICS

7

Model Systems and Gene Manipulation

In 1909, just 9 years after the rediscovery of Mendel's laws, Correns and Baur independently observed that different plastid phenotypes could be inherited in a non-Mendelian fashion. Correns followed the inheritance of green, variegated, and white color patterns in reciprocal crosses of the four-o'clock *(Mirabilis jalapa)* and observed that the seedlings resembled the maternal parent in phenotype. In contrast, Baur, studying the garden geranium *(Pelargonium zonale)*, found green, white, and variegated F_1 progeny in reciprocal crosses of green and white parents, but not in Mendelian ratios. Thus, in the same year Baur and Correns discovered the two most frequent modes of organelle gene transmission, *uniparental–maternal* and *biparental,* except they did not know that they were observing the transmission patterns of organelle genes. In angiosperms, plastid transmission has now been examined in crosses of >50 genera of dicotyledons and monocotyledons (Chapter 8; Smith, 1988). Although plastid transmission is most often maternal, biparental inheritance has been demonstrated in about 20 genera (Chapter 8; Gillham *et al.,* 1991). In conifers plastid transmission is largely uniparental–paternal. Non-Mendelian transmission in crosses differentiates organelle genes from nuclear genes with their characteristic Mendelian segregation patterns at meiosis.

A second important observation also derives from the work of Baur, Correns, and others on plastid or *plastome* inheritance as it is often called. In plants exhibiting uniparental–maternal inheritance, a variegated maternal parent will yield green, white, and variegated offspring. This pattern of transmission arises because the maternal parent is *heteroplasmic* for green and white plastid phenotypes. These sort out somatically in the F_1 progeny. Crosses between plants having green and white phenotypes, on the other hand, yield green or white progeny depending on the maternal parent, since these plants are *homoplasmic* for the plastids they contain. Sorting out, or somatic (vegetative) segregation, characterizes organelle genomes as well as the organelles themselves. For example, in yeast, where mitochondrial fusion seems to occur in the zygote, sorting out of mitochondrial genetic markers occurs rapidly during subsequent vegetative divisions.

During the two or three decades following the discoveries of Baur and Correns, plastome inheritance was extensively investigated, particularly in Germany, and a vast literature developed. One of the central questions at issue was whether sorting out variegation reflected the cytoplasmic assortment of mutant plastids rather than effects of different cytoplasms on plastid development. For example, somatic mutation in a nuclear gene could affect the cellular environment or the plastids themselves in such a way that proper plastid development failed to occur. Thus a white leaf sector might result from somatic mutation of a nuclear gene rather than sorting out of plastids. Demonstration that the plastids themselves were the culprits in sorting out variegation predicted the discovery of mixed cells containing both green and white plastid types. Such cells were sought and found and, until the advent of molecular biology, provided the best evidence for plastome mutation in flowering plants.

The theoretical basis of sorting out variegation was established by Michaelis long before the advent of the computer. Investigations, also in Germany, of plastome inheritance in interspecific crosses of the *Oenothera* by Renner and his students Stubbe, Schotz, and Schwemmler led to the concepts of plastid–genome compatibility and plastid competition. In a sense, the theoretical and experimental investigations of Michaelis on the basis of sorting out variegation in Fireweed, *Epilobium*, and the plastid competition studies of Schotz in *Oenothera* were initial ventures into organelle population genetics.

Most of this early work has been lucently reviewed by Kirk and Tilney-Bassett (1978).

During the early 1970s Wildman and his colleagues took advantage of maternal inheritance of presumed plastid mutations in *Nicotiana* together with interspecific hybridization experiments to provide the first definitive evidence as to where the genes for individual plastid components are located (reviewed by Gillham, 1978). These studies, particularly with respect to the localization of the large and small subunits of RUBISCO, are classic and deserve mention here. The genus *Nicotiana* contains more than 60 species, many of which can be hybridized. By examination of large and small subunits of RUBISCO from several species of *Nicotiana*, Wildman and his colleagues were able to detect specific peptide differences. These differences were used as markers in interspecific crosses to show that the large subunit was maternally inherited whereas the small subunit exhibited biparental Mendelian inheritance.

Mitochondrial genetics begins in 1949 with the discovery in France by Ephrussi and collaborators, notably Slonimski, of petite mutations in bakers' yeast (see Ephrussi, 1953 for a summary). These mutations formed small colonies under the growth conditions used (hence the name). They were respiration deficient and could be induced quantitatively with acriflavine and related dyes in diploid as well as haploid strains. The vast majority of such spontaneous and induced petites exhibit non-Mendelian inheritance and are called *vegetative petites*. The cytoplasmic factor responsible was given the name ρ by Marquardt (1952). Ephrussi and colleagues also found a small fraction of petite mutations that were Mendelian in inheritance. These were termed *segregational petites*.

Unfortunately all vegetative petites proved to be pleiotropic and similar in their effects on respiratory phenotype. Hence, until new mitochondrial mutations having unique phenotypes could be isolated, genetic analysis of the mitochondrial genome was stalled. The isolation of such mitochondrial mutants became possible in the late 1960s when Linnane and colleagues discovered that antibiotics blocking mitochondrial protein synthesis (Chapter 13) prevented yeast cells from growing on a medium containing a respirable carbon source such as glycerol (e.g., Clark-Walker and Linnane, 1966). However, this effect was conditional since yeast cells can grow in the presence of antibiotics on a medium with a fermentable carbon source like glucose. Subsequently, antibiotic-resistant mutations exhibiting non-Mendelian inheritance were isolated that were able to grow on glycerol medium in the presence of antibiotics (Linnane *et al.,* 1968; Thomas and Wilkie, 1968a; Wilkie *et al.,* 1967). Mutations having different phenotypes were then shown to recombine (Thomas and Wilkie, 1968b). Elegant genetic methods were then developed for the analysis of mitochondrial gene recombination by Slonimski and colleagues (e.g., Coen *et al.,* 1970). Sophisticated genetic experiments allowed Slonimski and colleagues to correlate the vegetative petite and antibiotic resistance phenotypes with the same ρ factor (Deutsch *et al.,* 1974).

In parallel with these genetic investigations, structural studies on petite mitochondrial genomes were proceeding (e.g., Faye *et al.,* 1973; Lazowska *et al.,* 1974; Locker *et al.,* 1974). The majority of vegetative petites were found to contain mtDNA, but the molecules differed strikingly from wild-type mtDNA molecules. These petite mtDNAs had undergone extensive deletion with the concomitant amplification of the short, retained sequences. Since the retained sequences were unique to each independently arising petite, a novel cloning system became available to yeast mitochondrial geneticists in which the petite mutation, as it were, cloned parts of its own mtDNA. By combining genetic and molecular characterization of the mtDNA molecules of different vegetative petites, the genetic and physical maps of the yeast mitochondrial genome were correlated (see *Physical mapping* of *mitochondrial genes in yeast*).

By the mid-1970s methods had been developed for the isolation of mitochondrial point mutations causing respiration deficiency (e.g., Flury *et al.,* 1974; Tzagoloff *et al.,* 1975a,b). Also, a mutagen, manganese, had been discovered that would specifically induce mitochondrial gene mutations (Putrament *et al.,* 1973). A method for complementing yeast mitochondrial mutations was also developed (Foury and Tzagoloff, 1978; Slonimski *et al.,* 1978). Now a full range of mitochondrial phenotypes was available together with a battery of techniques for their study. Hence, mitochondrial mutations were rapidly localized on the mitochondrial genome of yeast, leading to the detailed map we have today (see *Physical mapping of mitochondrial genes in yeast*). By the end of the 1970s yeast mitochondrial geneticists were, with one exception, able to combine the full array of "classical" genetic tools with the rapidly advancing technologies disseminating from molecular biology to study the structure and expression of yeast mitochondrial genes. That exception was the absence of a method for transforming mitochondrial genes. Transformation was accomplished almost a decade later (Johnston *et al.,* 1988).

Much of this early work on yeast has been reviewed by Dujon (1981), Gillham (1978), and Gingold (1988).

During the 1950s non-Mendelian inheritance in filamentous fungi was also discovered and later many of the phenotypes were related to specific mitochondrial diseases (Chapter 11). The *poky* mutation in *Neurospora,* so named because it grew slowly, was described by the Mitchells in 1952. This mutation exhibited a uniparental–maternal pattern of inheritance and had a respiratory-deficient phenotype. However, unlike vegetative petites, *poky* was not completely lacking in respiratory chain cytochromes. Tracing the primary defect in this mitochondrial mutant proved a tortuous undertaking (see Gillham, 1978 and Chapter 11). Other *Neurospora* mitochondrial mutants with respiratory defects were subsequently discovered, some of which showed stop-start growth and were called *stoppers* (e.g., Bertrand and Pittenger, 1972). In *Aspergillus* a variety of cytoplasmic variants were described and studied during the 1950s (e.g., Jinks, 1964). Later on, in the 1970s, mitochondrial genetics in *Aspergillus* became the subject of more systematic exploration (e.g., Rowlands and Turner, 1974). Indefinite asexual propagation is not possible in a number of fungi because the mycelium eventually sickens and dies. This phenomenon has been long investigated in *Podospora anserina* where it is called senescence. These studies begin in the 1950s at a descriptive level (e.g., Rizet, 1953; Marcou, 1954) and continue to this day at the molecular level (Chapter 11).

The green alga *Chlamydomonas reinhardtii* occupies a central position in the study of chloroplast genetics. The chloroplast genetic system of *C. reinhardtii* was discovered by Sager who, in 1954, showed that mutations to streptomycin resistance were of two kinds. Low level resistance mutations were Mendelian in inheritance, but high-level resistance mutations were transmitted uniparentally by the mating type plus (mt^+) parent in crosses. Subsequently, rare transmission of uniparentally inherited markers from the mating type minus (mt^-) parent was discovered. In most cases chloroplast markers from the mt^+ parent were transmitted too, and recombination of phenotypically different resistance markers could be detected among the progeny of these *biparental* zygotes (Gillham, 1965; Sager and Ramanis, 1965). In 1967 Sager and Ramanis reported that UV irradiation of the mt^+ parent increased the frequency of biparental zygotes dramatically. This greatly simplified the analysis of chloroplast gene recombination and led to the construction of genetic maps (e.g., Boynton *et al.*, 1976; Harris *et al.*, 1977; Sager and Ramanis, 1970, 1976,b; Singer *et al.*, 1976). Sager and Ramanis felt their results were consistent with a circular genetic map that might encompass the entire chloroplast genome, while the genetic map of Boynton, Gillham, and colleagues was linear. Today we know that the principal markers used in these studies, mutations to antibiotic resistance, form a linear array localized in the rRNA genes in the inverted repeat (Harris *et al.*, 1989 see also *Mapping chloroplast genes in Chlamydomonas*).

In the late 1970s methods were developed for the specific induction of chloroplast gene mutations in *Chlamydomonas* using the thymidine analog 5-fluorodeoxyuridine. This resulted in isolation of the first nonleaky nonphotosynthetic mutations in the chloroplast genome (Shepherd *et al.*, 1979). The latter mutants were able to survive because *Chlamydomonas reinhardtii* will use acetate or CO_2 as carbon sources. A method for assaying complementation of chloroplast genes was then developed by Bennoun *et al.* (1980). However, correlating the genetic and physical maps proved a more difficult task in *Chlamydomonas* than in yeast since mutations analogous to vegetative petites do not exist in *Chlamydomonas*. Nevertheless, mutations have now been mapped in many chloroplast genes using molecular methods (see Boynton *et al.*, 1992). A highly effective method for chloroplast gene transformation in *Chlamydomonas* was developed in the 1980s (Boynton *et al.*, 1988), which has since been extended to higher plants (Svab *et al.*, 1990). Organelle transformation methods are described later in this chapter.

MODEL SYSTEMS FOR THE STUDY OF ORGANELLE GENETICS

Since the topic of organelle genetics was explored in detail by Gillham (1978) the discussion in this chapter is greatly condensed. Mitochondrial genetics in yeast has also been reviewed by Dujon (1981), Gingold (1988), Tzagoloff (1982), and Wolf and Del Giudice (1988). Chloroplast genetics in *Chlamydomonas* has been reviewed by Boynton *et al.* (1992) and Harris (1989).

The Mitochondrial Genetics of Bakers' Yeast *(Saccharomyces cerevisiae)*

Life cycle. When haploid cells of opposite mating type (a and α) are mixed, they fuse pairwise to form vegetative diploids. Both haploid and diploid cells reproduce vegetatively by budding. Diploid cells will undergo meiosis (sporulation) if their growth medium becomes exhausted or if placed in a medium containing a carbon source such as acetate. Four haploid ascospores result and Mendelian alleles segregate 2:2.

When *S. cerevisiae* is grown aerobically on carbon sources such as ethanol or glycerol the respiratory and morphological phenotypes of the mitochondria are fully developed (Fig. 7–1) and oxidative phosphorylation supplies energy. If *S. cerevisiae* is grown on a carbon source such as glucose, respiration becomes *glucose repressed* and the cells ferment glucose with the production of ethanol. Under these conditions there is a tremendous depletion of respiratory chain components and the mitochondria themselves are fewer in number and larger in size (Fig. 7–1). Anaerobic growth of *S. cerevisiae* not only causes great reductions in respiratory chain components, but also prevents formation of two important lipid components of mitochondrial membranes, unsaturated fatty acids and ergosterol. Thus, healthy anaerobic cultures of cells can be obtained only if the growth media are supplemented with ergosterol and a source of unsaturated fatty acids. The mitochondria of anaerobically grown cells are called *prom-*

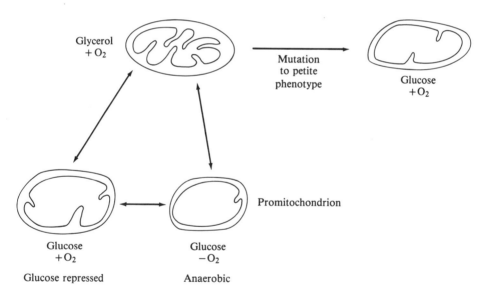

Fig. 7–1. Diagram of yeast mitochondria from cells growing under different conditions. In aerobically grown cells using glycerol as a carbon source the respiratory phenotype is fully expressed and the cristae are well differentiated. In cells growing aerobically on glucose medium respiration is repressed and glucose is fermented to produce ethanol and the mitochondrial inner membrane shows correspondingly less differentiation. This phenotype is even more pronounced in the *promitochondria* of yeast cells growing anaerobically on glucose. The vegetative petite mutation whether ρ^- or ρ^0 leads to the loss of functional mtDNA and as a consequence the ability of the cells to respire. This change is reflected in the ultrastructure of the mitochondrial inner membrane.

itochondria. Exposure of anaerobically grown cells to oxygen results in differentiation of promitochondria into mitochondria. Morphologically, promitochondria of lipid-supplemented cells resemble normal mitochondria; they are the same size, have cristae, and the inner membrane is visible. Promitochondria of lipid-depleted cells are less well developed.

Petite mutants. Vegetative petite mutations are all phenotypically similar because they either lack mtDNA or possess very large mtDNA deletions. As a result, genes essential for mitochondrial protein synthesis (i.e., rRNA or tRNA genes) are lost and all of those respiratory enzyme subunits made on mitochondrial ribosomes are missing. Petite mutations arise spontaneously at a high rate (ca. 1–2%/cell/generation).

Vegetative petite mutations are subdivided into two classes. The *neutral petite* phenotype is not transmitted in crosses to wild type, either among the vegetative diploid progeny of the resulting zygotes or among the haploid progeny of diploids induced to undergo meiosis (Fig. 7–2). The *suppressive petite* phenotype is transmitted to a fraction of the vegetative diploid progeny in a cross ranging from 1% or less in some strains to over 99% in others (hypersuppressive petites). Aerobic metabolism is required for sporulation in *S. cerevisiae* so suppressive petite diploids will not sporulate. However, respiratory-competent diploids issuing from a cross of a wild-type strain by a suppressive petite do sporulate and segregate 4 wild type:0 petite progeny (Fig. 7–2). Only when the immediate zygotic progeny of the cross are sporulated can transmission of the petite phenotype be seen. Under these conditions some zygotes will segregate 0 wild type:4 petite. *Segregational petites* segregate 2:2 in crosses as expected for nuclear gene mutations (Fig. 7–2). These PET gene mutations fall into >215 complementation groups (Tzagoloff and Dieckmann, 1990).

The term ρ, first used by Marquardt (1952) to designate the cytoplasmic factor responsible for the petite phenotype prior to the discovery of mtDNA, has been used ever since in the petite literature. Thus, wild-type strains are ρ^+, petite strains retaining mtDNA ρ^- and petite strains that have lost mtDNA are ρ^0. Suppressive petites are always ρ^-, but neutral petites may be ρ^0 or ρ^-. The wild-type allele of a segregational petite is designated *PET* and the mutant allele *pet*. Vegetative petite mutations are induced by a wide variety of agents, of which acridine dyes and ethidium bromide are the most important (reviewed by Gillham, 1978).

The organization of mtDNA in ρ^- mutants has been extensively investigated (reviewed by Clark-Walker, 1992; Dujon, 1981; Gillham, 1978; Wolf and Del Giudice, 1988). These mutations result from very large deletions of the ρ^+ mitochondrial genome, with the mtDNA sequences retained usually being less than one-third of the genome and often as little as 1%. Nevertheless, diminution of the total amount of mtDNA does not occur. The retained sequences are amplified in the ρ^- mutant to yield as much mtDNA as was present in the ρ^+ cell from which it derived. Amplification in ρ^- mutants is achieved by repetition of the conserved sequences along the molecule. Two major classes of sequence arrangements are found. In the first the petite mtDNA molecules consist of tandem repeats (A–B A–B) with the retained sequence being identical to wild type. In the second the repeat unit in the petite mtDNA is an inverted duplication (palindrome) of the conserved sequence (A–B B–A). Other more complex ρ^- petites, are also found in which the conserved sequences show internal rearrangement or are even "scrambled" with different segments of the wild-type mitochondrial

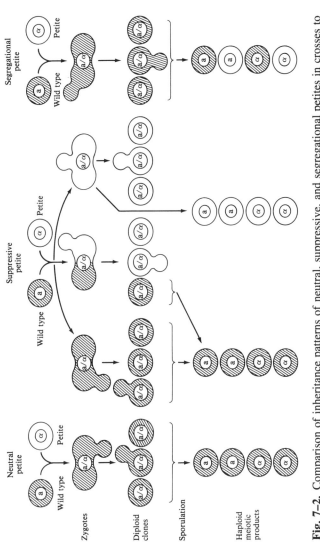

Fig. 7–2. Comparison of inheritance patterns of neutral, suppressive, and segregational petites in crosses to wild type. Neutral petites produce only wild-type diploid progeny in such crosses that segregate 4:0 wild-type:petite progeny following sporulation. Zygotes issuing from a cross between a suppressive petite and wild-type may yield wild-type progeny, petite progeny, or a mixture, with the ratio depending on the level of suppressivity of the mutation. Petite vegetative diploids will not sporulate, but zygotes destined to yield petite progeny will do so, and they segregate 0:4 wild-type:petite progeny. Phenotypically wild-type progeny result from crosses between recessive segregational petites and wild-type. On sporulation of these diploids, the petite phenotype is found to segregate 2:2 wild-type:petite progeny in a typical mendelian fashion. After Gillham (1978).

genome that were not contiguous originally being juxtaposed. A significant fraction of ρ^- mtDNA molecules is relatively long and probably similar in size to the mtDNA molecules of ρ^+ cells. However, smaller circular molecules have also been seen, with monomers of the repeat unit being more frequent than dimers, dimers more frequent than trimers, etc. In fact the rule seems to be that the frequency of a given oligomer is inversely proportional to the number of repeat units it contains.

Clark-Walker (1992) has reviewed current thinking on petite formation. The mitochondrial genome of *S. cerevisiae* can be destabilized by excision events occurring at short directly repeated sequences, usually G+C clusters. This explanation for petite formation in *S. cerevisiae* is consistent with the observation that spontaneous petites are formed with a frequency four orders of magnitude less in *Candida glabrata,* which lacks G+C-rich repetitive elements.

The mitochondrial genomes of hypersuppressive (HS) petite mutants have have very short repeat units of 400–900 bp (Wolf and Del Giudice, 1988). All HS mutants share common sequences of about 300 bp with the wild-type mitochondrial genome called *rep* or *ori.* An *ori* region consists of three G+C clusters separated by two A+T stretches. Two G+C clusters (A and B) are separated by an A+T stretch while the third (C) is separated from B by a sequence of 200 A+T bp (1). Cluster C is flanked by A+T sequences called r and r*, which contain sites for transcript initiation. A total of eight different *ori* regions have been identified in the mitochondrial genomes of 20 different wild-type yeast strains. Two of these, *ori6* and *ori7,* have not been found at all in spontaneous petites, suggesting that they may be inactive. Partial deletion of an *ori* sequence in an HS mutant decreases suppressivity.

The *ori/rep* sequences can be thought of as mobile elements that recently infected the mitochondrial genome and were transposed to different sites (Clark-Walker, 1992). This view would be consistent with the idea that the three G+C clusters associated with each *ori/rep* region are themselves mobile elements (Weiller *et al.,* 1989; Chapter 10). Alternatively, *ori/rep* elements may have evolved from mitochondrial promoters in association with G+C rich DNA (De Zamaroczy and Bernardi, 1987).

Other mitochondrial mutants. Many other mitochondrial mutations have been isolated in *S. cerevisieae.* Such mutations segregate vegetatively (somatically) in both haploids and diploids. The resulting meiotic progeny segregate 4 wild type:0 mutant or 4 mutant: 0 wild type depending on whether the diploid being sporulated was wild type or mutant. These mutations are broadly divisible into three classes:

1. Antibiotic-resistant mutants (ant^R) can grow in the presence of inhibitor on a nonfermentable carbon source. Such mutations have been assigned to the genes encoding the two rRNAs, *cytb* and the two mitochondrially encoded subunits of the F_0 portion of the ATP synthase (Table 7–1).
2. The *mit*⁻ mutants, like petites, are unable to grow on nonfermentable carbon sources. However, *mit*⁻ mutations are point or deletion mutations affecting specific mitochondrial genes encoding components of the ATP synthase or respiratory chain (Table 7–1).
3. *syn*⁻ mutants are also like *mit*⁻ mutants, but affect genes encoding components required for mitochondrial protein synthesis. Mitochondrial suppressors of *mit*⁻ mutations are designated *mim.*

MODEL SYSTEMS AND GENE MANIPULATION

Table 7–1. Mitochondrial Genes and Gene Mutations in Yeast

Protein and Gene	Gene Symbol Yeast	Mutant Symbol Yeast	Mutant Phenotype
Cytochrome oxidase			
Subunit I *(cox1)*	*oxi3*	*oxi3$^-$*	Respiration-deficient
Subunit II *(cox2)*	*oxi1*	*oxi1$^-$*	Respiration-deficient
Subunit III *(cox3)*	*oxi2*	*oxi2$^-$*	Respiration-deficient
Apocytochrome *b (cytb)*	*cob-box*	*cob1$^-$, cob2$^-$*	Respiration-deficient
		box1$^-$ to *box10$^-$*	Respiration-deficient
		diu1^R, diu2^R	Diuron-resistant
		muc1^R to *muc3^R*	Mucidin-resistant
		ana1^R, ana2^R	Antimycin-resistant
		fun1^R	Funiculosin-resistant
ATPase subunit 6 *(atp6)*	*pho1*	*pho1$^-$*	Phosphorylation-deficient
		oli2^R, oli4^R	Oligomycin-resistant
		oss1^R	Ossamycin-resistant
ATPase subunit 9 *(atp9)*	*pho2*	*pho2$^-$*	Phosphorylation-deficient
		oli1^R, oli3^R	Oligomycin-resistant
		oss2^R	Ossamycin-resistant
		ven1^R	Venturicidin-resistant
Mitoribosomal protein	*var1*		
21 S rRNA		*rib1*	Chloramphenicol-resistant *cap^R*)
		rib2, rib3	Erythromycin-resistant *(ery^R)* or spiramycin-resistant *(spi^R)* polarity of recomination between flanking markers *rib1, rib2, rib3*)
15 S rRNA		*par1^R*	Paramomycin-resistant
tRNA		*syn$^-$*	Mitochondrial protein synthesis-deficient

See Dujon (1981)

The gene designations for the three mitochondrially encoded subunits of cytochrome oxidase are confusing for the historical reason that they received genetic designations before their products were defined precisely (Table 7–1). Thus, the *oxi1* gene codes for cytochrome oxidase subunit II, *oxi2* for subunit III, and *oxi3* for subunit I. However, the conventional nomenclature will be used here (i.e., *cox1*, cytochrome oxidase subunit I; *cox2*, cytochrome oxidase subunit II; *cox3*, cytochrome oxidase subunit III).

Isolation of *mit* and *syn* mutants posed a problem for many years until Putrament et al. (1973) discovered that Mn^{2+} selectively induces mitochondrial gene mutations, and methods were developed to distinguish or eliminate vegetative petite mutations from consideration (see Tzagoloff, 1982).

The mitochondrial cross. Mitochondrial genes in *S. cerevisieae* were first mapped by genetic recombination (Dujon, 1981; Gillham, 1978). The interpretation of the results of such crosses is complicated by the fact that mitochondria are parts of cells and not distinct organisms. One cannot score the phenotype of a specific mtDNA molecule or even the organelle within which that molecule resides. Only cellular phenotype can be determined. Initially, genotypic purity for a given mitochondrial allele in a cell line could only be inferred, not rigorously proved despite strongly suggestive genetic evidence (e.g., Lewis and Birky, 1984). Fortunately, this is no longer the case since one can use molecular techniques (e.g., restriction site and length polymorphisms) to demonstrate physical purity. Another problem is that the input and output of mitochondrial genomes in a cross can be controlled to a limited degree, for example, only by inhibiting synthesis of mtDNA in one parent in a cross. One is tempted to make an analogy to a phage cross, in which the average multiplicity of infection, the average burst size, and the frequency of phenotypically different progeny phages cannot be measured directly (see Chapter 9).

Fortunately, the foregoing problems are more apparent than real and mitochondrial gene recombination can be rationally examined in yeast. Three analytical methods have been used. First, pedigrees of the diploid buds produced by zygotes and their vegetative diploid progeny are studied. Pedigree analysis has the advantage of yielding detailed information about the pattern of segregation and recombination in the early cell divisions following zygote formation. The disadvantages are that pedigree analysis is time consuming, so only limited numbers of pedigrees are easily done. Hence, quantitative information concerning recombination frequencies or output ratios of mitochondrial alleles from the two parents is difficult to obtain.

Pedigree studies revealed that during the first few cell divisions following zygote formation, a large fraction of the cells becomes homoplasmic for one allele or another. Rapid segregation is unexpected because, with ca. 100 copies of mtDNA in the zygote, random partitioning during cell division would rarely produce a homoplasmic cell. Rapid segregation is not unique to mitochondrial genes, but also occurs for chloroplast genes in *Chlamydomonas*.

In the second method, zygote clone analysis, the progeny of individual zygotes are analyzed after ca. 20 cell generations by plating appropriate dilutions on nonselective media, and replica-plating the resulting colonies onto appropriate selective media for genotype identification. Quantitative zygote clone analysis of this kind is much faster than pedigree analysis, and it yields more information. However, zygote clone analysis can give only indirect information about the timing of recombinational events and segregation patterns.

The third method, referred to as the standard cross, has been used most extensively in the analysis of mitochondrial gene recombination. In the standard cross a large population of zygotes, formed by random mass mating between two parents, is plated on nonselective medium. The zygotes and diploid vegetative cell progeny are allowed to grow for about 20 generations until a confluent lawn of cells is formed. The resulting population is harvested, diluted, and plated on nonselective medium. The colonies formed by the diluted cells are then replica-plated to appropriate selective media. The standard cross yields quantitative information on the genotypic composition of the population as a whole irrespective of zygote of origin. This is a characteristic and reproducible figure for a specific mitochondrial cross (i.e., between two given parental strains defined by their nuclear and mitochondrial genetic backgrounds).

Since the most comprehensive theoretical model of mitochondrial gene recombination in yeast likens the process to phage recombination (see Chapter 9), it is worth noting that zygote clone analysis and the standard cross, have their analogs in the single burst experiment and phage cross, respectively. That is, each diploid zygote can be compared to a single-phage infected bacterium in which many virus genomes capable of recombination are present with their composition being analyzed following lysis of that bacterium. In the yeast zygote, phage are replaced by mitochondrial genomes and the burst consists of progeny buds. In contrast, in the standard cross the population of mitochondrial genomes is analyzed irrespective of the zygote of origin, in the same way that progeny phage are analyzed in a phage cross irrespective of the bacterium of origin. Again, by analogy to phage, the upper limit for recombination is predicted to be 20–25% rather than 50% because pairings between mitochondrial genomes are not restricted to genotypically different mtDNA molecules, but may occur between genotypically identical molecules as well. Recombination mapping of mitochondrial genes in yeast has proved useful only over short distances (ca. 1 kb) because recombination is very frequent and because theory predicts multiple rounds of mating of mtDNA molecules (Dujon, 1981).

Despite these limitations, recombination frequencies have been used for mapping closely linked markers, in allele testing of mit^- and syn^- mutants in pairwise crosses, and in fine structure mapping of mitochondrial genes. Also, data gathered from such crosses led to the realization that yeast mitochondrial genes could be classified in two groups with respect to polarity of recombination. Mutations exhibiting polar behavior map in the 21 S rRNA gene and are resistant to antibiotics such as chloramphenicol (cap^r) and erythromycin (ery^r). They are found in the vicinity of a locus called ω (see Chapter 10 for a discussion of polarity).

Physical Mapping of mitochondrial genes in yeast. The fact that independently isolated $ρ^-$ petites retain different segments of the mitochondrial genome provides a unique and powerful tool for correlating the genetic and physical maps of the yeast mitochondrial genome. Two general mapping methods have been used, both of which depend on characterizing markers retained by different $ρ^-$ mutants. The principle involved in measuring marker retention in mutants can be simply illustrated for an ant^r mutant. The ant^r cells are mutagenized with ethidium bromide and the resulting petite mutants isolated (Table 7–2). Some petites will be $ρ^0$ and lack mtDNA whereas others will be $ρ^-$. Some of the $ρ^-$ petites will have retained segments of the mitochondrial genome containing the ant^r mutation, but many will not. Since antibiotic resistance can be scored

Table 7–2. Classification of $ρ^-$ Stocks for Retention of an Antibiobic-Resistant (ant^r) Marker in Crosses to a Wild-Type Antibiotic-Sensitive Stock $(ρ^+ ant^s)$

Petite genotype	Phenotypes of Respiration-Sufficient Diploids Produced in Crosses to Wild Type	
	ant^r	ant^s
$ρ^- ant^r$	+	+
$ρ^- ant^0$	−	+
$ρ^0 ant^0$	−	+

only on a nonfermentable carbon source on which the petites would die, the different petites must be distinguished by mating them to wild-type antibiotic sensitive (ant^s) cells. Recombination between the retained ant^r allele in the petite and the ant^s allele in the ρ^+ strain will yield respiratory-competent, antibiotic-resistant progeny able to grow under restrictive conditions (Table 7–2). The other petites will yield only antibiotic-sensitive diploid progeny (Table 7–2). The same principle can be applied to mit and syn mutants except that here one induces petite formation in a wild-type strain, makes diploids with a mit^- or syn^- strain, and determines whether the diploids can grow on a nonfermentable carbon source. Diploids derived from some ρ^- mutants will grow under these conditions and these must have retained the wild-type allele of the mit^- or syn^- mutant.

The first mapping method simply extends petite marker retention analysis. The genetic markers retained by each ρ^- mutant are determined after their rescue into a ρ^+ mitochondrial genome. The order and map distances of the markers are obtained on the basis of two criteria. First, if markers are considered pairwise, the observed frequency with which a ρ^- mutant retains one marker or the other (a^+b° or $a^\circ b^+$) is compared to the expected frequency calculated assuming that each marker is lost independently of the other. A ratio of 1 means the markers are behaving independently of one another while a ratio of <1 means that they are not. That is, they are being deleted or retained together more frequently than expected by chance. The lower the value the tighter the linkage. Second, markers are considered four at a time. Three possible orders exist for four markers on a circular genetic map (Fig. 7–3). Given the order ABCD, the retention or loss of the pairs AB, BC, CD, and DA requires only one continuous deletion whereas pairs AC and DB require two independent deletions. Since ρ^- mutants having two deletions are less frequent than mutants with a single continuous deletion, one need only ask which marker order yields the two least abundant genotypes (Fig. 7–3).

The second method, petite deletion mapping, is also based on rescue of the genetic markers carried by the ρ^- genome by recombination into a ρ^+ genome. Crosses are made between a series of point mutants and a set of ρ^- mutants that can discriminate between the point mutants. By determining which combinations yield wild-type progeny, a unique order of the point mutations can be deduced (Fig. 7–4). At the same time restriction enzyme analysis of the mtDNA sequences retained in the ρ^- mutations and comparison to ρ^+ mtDNA make possible the alignment of the genetic and physical maps of the mitochondrial genome.

Complementation tests. A priori one might suppose that complementation testing of mitochondrial gene mutations would prove difficult because of rapid mitochondrial gene segregation, of recombination to yield wild-type mtDNA molecules, and of compartmentalization of genetically distinct mtDNA molecules in different mitochondria. In fact complementation testing has been used quite effectively in assigning mit^- mutants to complementation groups (see Dujon, 1981). The technique involves measuring respiratory activity in a synchronously formed population of zygotes. For some combinations of mit^- mutants, full restoration of respiration is achieved in 5–8 hr. Other combinations show no significant increase in respiration for about 15–20 hr, the time required for the formation of wild-type recombinants. Complementation is assumed to occur for the former combinations because respiration is restored long before recombination occurs. Mutations in different loci as defined by recombination may or may

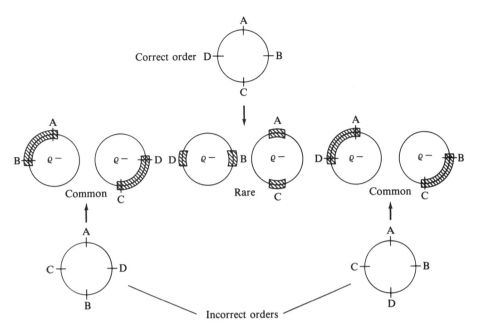

Fig. 7–3. Diagram illustrating the principles involved in double-deletion/double-retention mapping of mitochondrial markers using mutants. The correct order for the genes A, B, C, and D in ρ^+ mtDNA is shown at the top of the figure. Below, three possible pairs of double-deletion/double-retention ρ^- genotypes are shown. Each member of two of these pairs can be generated by a single deletion, but in the case of the final pair, two deletions are required to generate the two complementary double-retention genotypes. Retained markers in each ρ^- mutant are included in the thick arc of each circle, with the deleted segment being indicated by the thin arc. At the bottom of the diagram, two other possible orders of these markers in the mitochondrial genome are shown. In each case, these orders would generate one of the common classes of ρ^- mutants as the result of two deletions. After Gillham (1978).

not complement each other. Thus, two *mit*⁻ mutants affecting different gene products (e.g., a subunit of cytochrome oxidase and cytochrome *b*) behave as if they are unlinked and complement. Hence, the loci defined by recombination and complementation are congruent. In other cases noncomplementing *mit*⁻ mutants may behave as if they are unlinked, as in the case of mutations in the very long *cox1* gene.

The Chloroplast Genetics of *Chlamydomonas reinhardtii*

Harris (1989) in her authoratative monograph has chronicled virtually all aspects of the biology of *Chlamydomonas*. Organelle genetics and transformation in this alga are the subject of a comprehensive review by Boynton *et al.* (1992). Earlier work on chloroplast genetics in *Chlamydomonas* was detailed by Gillham (1978) and Sager (1972, 1977).

Life cycle. *C. reinhardtii* is a haploid, unicellular green alga. Each cell contains a single, large chloroplast within which are ca. 80 molecules of the 196-kb chloroplast genome.

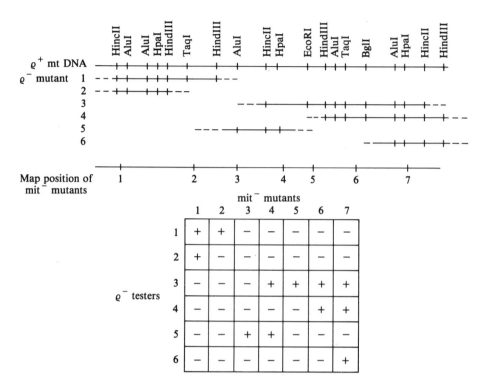

Fig. 7–4. Mapping *mit⁻* mutants using discriminating ρ⁻ mutants. By comparing the restriction patterns of the ρ⁻ mtDNA molecules with ρ⁺ mtDNA, the ρ⁺ mtDNA segments retained in each ρ⁻ mutant can be mapped with precision. The *mit⁻* mutants can then be mapped on the basis of their ability to recombine to give wild-type progeny with each ρ⁻ tester. If a *mit⁻* mutant recombines with a ρ⁻ mutant to yield wild-type progeny, the *mit⁻* mutant must map within the region of mtDNA retained by the ρ⁻ tester. The example given in the figure is hypothetical as are the restriction sites indicated on the ρ⁻ and ρ⁺ mtDNA molecules.

Cells of opposite mating type are similar in morphology and mating is controlled by a pair of alleles designated mating type plus (*mt⁺*) and mating type minus (*mt⁻*). Vegetative cells are converted into gametes by nitrogen starvation. On mixing, gametes of opposite mating type pair at the flagellar tips. The gametes then secrete an enzyme, autolysin, which causes them to shed their cell walls. Each *mt⁺* gamete elaborates a cylindrical appendage, the fertilization tubule, that attaches to a mating structure on the corresponding *mt⁻* gamete connecting the mating pair of *mt⁺* and *mt⁻* gametes by a cytoplasmic bridge. Mating pairs fuse to form diploid zygotes that elaborate new zygotic cell walls. Not only nuclei, but also chloroplasts fuse during the first 3 hr after mating. Zygotes are matured on nitrogen-free medium either in the light or on an alternating cycle of light (ca. 20 hr) and dark (ca. 5 days). Zygote germination is usually induced by transferring the zygotes to nitrogen-containing medium in the light. Meiosis and germination occur within 24 hr to yield four, haploid meiotic products. The nuclear mating type alleles segregate 2:2.

A small fraction of the zygotes formed in a cross (ca. 5%) fails to enter the sexual cycle and become vegetative diploids instead. *Chlamydomonas* diploids unlike yeast diploids have not been induced to undergo meiosis and, although heterozygous for mating type, behave phenotypically like mt^- gametes when starved for nitrogen.

C. reinhardtii can use either CO_2 or acetate as carbon sources. The cells can be grown in light with CO_2 as the sole carbon source *(phototrophic growth)*, in the light with both acetate and CO_2 as carbon sources *(mixotrophic growth)*, or in the dark with acetate as the sole carbon source *(heterotrophic growth)*. Since photosynthesis is dispensable as long as cells are grown on acetate, photosynthetic mutants can be readily isolated and maintained in this alga.

Inheritance of chloroplast and mitochondrial genes. Chloroplast and mitochondrial genes are transmitted in a uniparental manner by opposite mating types in *Chlamydomonas*, with the mt^+ parent normally transmitting the chloroplast genome and the mt^- parent the mitochondrial genome (Fig. 7–5).

The basic features of chloroplast gene transmission were known long before the discovery of cpDNA when Sager (1954) demonstrated the uniparental transmission of mutations resistant to high levels of streptomycin by the mt^+ parent. Over the years

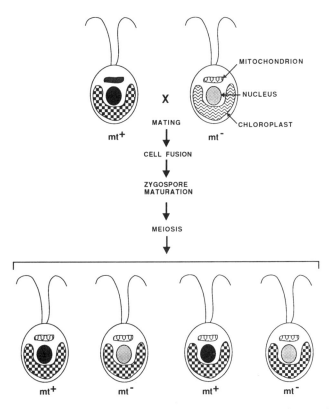

Fig. 7–5. The three basic modes of gene transmission in sexual zygotes of *C. reinhardtii*. Chloroplast genes are normally transmitted by the mt^+ parent (>90–95% of the zygotes) to all four meiotic products while the mt^- parent transmits the mitochondrial genome uniparentally (>99% of the zygotes). Nuclear genes segregate 2:2 reflecting the contributions of both parents.

additional mutations resistant to antibiotics were characterized showing the same pattern of inheritance. However, from the beginning it was apparent that occasional *exceptional zygotes* (ca. 1–10% of the total) occurred in which chloroplast genes from the mt^- parent were transmitted to the meiotic progeny. By analogy to chloroplast gene transmission in higher plants, zygotes transmitting chloroplast genes only from the mt^+ parent were called *maternal zygotes*. *Exceptional zygotes* segregating chloroplast genes from both parents were referred to as *biparental zygotes,* while those transmitting chloroplast genes only from the mt^- parent were named *paternal zygotes*. This terminology is imprecise for at least two reasons. First, the mt^+ parent elaborates the fertilization tubule for which the mt^- parent is the recipient. In this sense, the mt^+ parent is behaving like the male and the mt^- like the female. Second, mtDNA is transmitted uniparentally by the mt^- parent, which is therefore "maternal" for this species of DNA. A better terminology proposed by Mets (1980) and adopted by Harris (1989) is UP$^+$ and UP$^-$. These terms refer, respectively, to uniparental inheritance from the mt^+ and mt^- parents with BP alluding to biparental, but non-Mendelian inheritance.

The uniparental transmission of cpDNA by the mt^+ parent in *C. reinhardtii* was demonstrated unambiguously by Grant *et al.* (1980). They found that two small symmetric deletions in the inverted repeat were inherited only from the mt^+ parent. In the same year Mets (1980) and Lemieux *et al.* (1980) observed uniparental transmission of cpDNA in another pair of closely related species *C. eugametos* and *C. moewusii.*

Demonstration of a correlation in the pattern of inheritance of putative chloroplast markers and cpDNA does not constitute proof that the markers are in cpDNA molecule. This proof was provided by Dron *et al.* (1983). They showed that a uniparentally inherited mutation affecting RUBISCO activity was the result of a missense alteration in the chloroplast gene encoding the large subunit of the enzyme. Interspecific crosses between the interfertile species *C. reinhardtii* and *C. smithii* subsequently revealed that the entire chloroplast genome was uniparentally transmitted by the mt^+ parent (Boynton *et al.,* 1987). The chloroplast genomes of these two species are colinear and approximately the same size, but differ by many restriction fragment polymorphisms or RFLPs resulting from the presence of short dispersed repeats (Boynton *et al.,* 1992; Palmer *et al.,* 1985; Chapters 3 and 4). At the same time, Boynton *et al.* (1987) proved in these interspecific crosses that mtDNA from the mt^- parent was uniparentally inherited using two restriction site differences.

In the 1–5% of zygotes that differentiate into vegetative diploids instead of sexual zygotes (see Harris, 1989), the uniparental mechanism of organelle gene inheritance is attenuated. Thus, biparental diploids are frequent (50–90% of the total), although diploids transmitting the chloroplast genome from one parent or the other are also found. Diploids are classified as maternal, biparental, or paternal (or UP$^+$, BP, and UP$^-$) for chloroplast genes in the same way as meiotic zygotes. The frequency of diploids transmitting chloroplast genomes uniparentally from the mt^+ parent can be increased markedly by delaying the first mitotic division for 40 hr either by nitrogen starvation or incubation in the dark (Van Winkle-Swift, 1978).

Formally, the study of chloroplast gene segregation and recombination in biparental diploids is equivalent to examining these processes for yeast mitochondrial genes in vegetative diploids.

Chloroplast mutants. Like yeast mitochondrial mutants, chloroplast mutants in *Chlamydomonas* are broadly divisible into three classes (see Boynton *et al.,* 1992; Harris,

1989). The first includes mutations to antibiotic or herbicide resistance. The antibiotic-resistant mutations are localized in the rRNA genes or the gene encoding ribosomal protein S12. Mutations in the latter gene confer resistance to or dependence on streptomycin as do identical mutations in the homologous *E. coli* gene (Liu et al., 1989). The herbicide-resistant mutants map to the *psbA* gene encoding the D1 reaction center protein of PSII as do comparable mutations in cyanobacteria and higher plants.

Mutants belonging to the other two classes are acetate-requiring *(ac)* conditional carbon source-dependent mutants similar to the *mit* and *syn* mutants of yeast. These *ac* mutants cannot grow photosynthetically using CO_2 as a carbon source. Most of them belong to a class that affects components of the photosynthetic apparatus (e.g., RUBISCO large subunit, the β subunit of the chloroplast ATP synthase). Only two chloroplast *ac* mutants affecting chloroplast protein synthesis have been described (Shepherd et al., 1979).

As in yeast, methods exist for the specific induction of chloroplast mutants, for the selection of photosynthetic mutants and for complementation testing of chloroplast mutations. Shepherd *et al.* (1979) found that the thymidine analog 5-fluorodeoxyuridine behaves as a specific mutagen for chloroplast genes. Arsenate and metranidazole can be used to select nonphotosynthetic mutants. Such mutants can also be screened for using fluorescence assay techniques (see Harris, 1989; Bennoun and Delepelaire, 1982). Complementation of pairs of nonphotosynthetic chloroplast mutants have been assessed using fluorescence induction kinetics and delayed luminescence patterns (Bennoun *et al.*, 1980).

Mapping chloroplast genes in Chlamydomonas. Recombinational mapping of chloroplast genes in *Chlamydomonas* predates similar work with mitochondrial genes in yeast. However, physical resolution of the genetic map in yeast was largely completed by 1980, but has only taken shape for the chloroplast genome of *Chlamydomonas* within the past decade. In fact, with the exception of the inverted repeat region, there is no genetic map. Instead, mutations marking specific genes are scattered around the unique sequence regions of the chloroplast genome (Boynton et al., 1992). These are unlinked with the exception of mutations in the *psaB* and *rbcL* genes, which are tightly linked both genetically and physically (Girard-Bascou et al., 1987). Perhaps the major reason why progress was faster in correlating the genetic and physical maps of the yeast mitochondrial genome is the existence of a battery of useful petite deletion mutants encompassing all parts of the mitochondrial genome. Similar mutations do not exist in the *Chlamydomonas* chloroplast genome.

Detailed study of chloroplast gene segregation and recombination in *Chlamydomonas* became possible with the discovery that ultraviolet light irradiation of the mt^+ parent prior to mating permitted efficient transmission of chloroplast genes from the mt^- parent (Sager and Ramanis, 1967). Recombination and segregation of chloroplast genes could then be studied either by pedigree or zygote clone analysis among the progeny of biparental zygotes. These methodologies are similar to those used for yeast except that haploid meiotic progeny of sexual zygotes are usually analyzed in *Chlamydomonas*. In *Chlamydomonas* as in yeast both methods have advantages and limitations. Pedigree analysis, which can be carried out conveniently through the 16 cell stage following meiosis, permits the detailed enumeration of segregational and recombinational events. However, the method has the disadvantage that many cells are still heteroplasmic at the 16 cell stage, and relatively small numbers of progeny can be analyzed for recombination mapping. Zygote clone analysis obviates both of the latter difficulties, but does not permit determination of when or how segregational events took place.

The results and conclusions of genetic analysis of chloroplast genes in *Chlamydomonas* have been treated at length in earlier books (Gillham, 1978; Harris, 1989; Sager, 1972) and reviews (Adams *et al.*, 1976; Sager, 1977) so it would serve little purpose to go over the same ground again here. Instead, only the two major conclusions drawn from this work together with points of contention will be discussed. First, chloroplast genes in *Chlamydomonas*, like mitochondrial genes in yeast, segregate rapidly in pedigrees although segregation is not complete by the 16 cell stage. The major question here was whether chloroplast alleles from both parents segregated 1:1 among the progeny of biparental zygotes or whether there was a strong bias toward chloroplast alleles from the mt^+ among the progeny. Sager and Ramanis (1968), who observed a 1:1 segregation pattern, assumed that their results meant the chloroplast genome was probably diploid. Gillham *et al.* (1974), who saw a maternal bias, hypothesized that their results could be accounted for by a multicopy model in which there was an input bias favoring chloroplast genomes from the mt^+ parent in biparental zygotes.

Second, recombination mapping using pedigrees and zygote clones established that chloroplast gene markers were linked and that genetic maps of these markers could be constructed (Gillham, 1965; Sager and Ramanis, 1965, 1970). Sager and colleagues (Sager and Ramanis, 1976a,b; Singer *et al.*, 1976) concluded the chloroplast genetic map was circular. Their analysis made use of resistance mutations plus several other markers of differing phenotype and mapping methods developed on the basis of pedigree data and segregation rates of chloroplast markers in liquid medium. In contrast Harris *et al.* (1977), using zygote clone analysis, reported a linear map for chloroplast mutations to antibiotic resistance that could be made congruent with Sager's circle by cutting the circle at a specific point. Mets and Geist (1983) attempted to map five markers using a modified form of zygote clone analysis. They concluded that they could account for their data by displaying the markers on a circular map and assuming that all recombination occurred in the inverted repeat region. In fact three of the markers they used actually map physically in a linear array encompassing the *psbA* and 23S rRNA genes in the inverted repeat and the adjacent *rps12* gene in one of the unique sequence regions.

We now know that only genes in close physical proximity are likely to form linkage groups in the chloroplast genome. Thus, antibiotic resistance mutations in the rRNA genes in the inverted repeat form a linkage group where one map unit equals about 1 kb (Harris *et al.*, 1989). The *sr-u-sm2* mutation in the *rps12* gene, located in one of the unique sequence regions near the inverted repeat, may be linked to the rRNA genes, but this assignment is still tenuous. However, herbicide resistance mutations in the *psbA* gene, which lies adjacent to the 23S rRNA gene in the inverted repeat, are unlinked to 23S rRNA antibiotic resistance mutations. The reason is the existence of a recombination hotspot that has been localized to a 500-bp region in the vicinity of the 3'-end of the *psbA* gene (Newman *et al.*, 1992). The other linkage group identified in the *Chlamydomonas* chloroplast genome includes the *psaB* and *rbcL* genes (see above and Girard-Bascou *et al.*, 1987). These genes are separated by about 2 kb in one of the unique sequence regions. Mutations in them exhibit 1–4% recombination (0.5–2% recombination/kb). As is true for mitochondrial genes in yeast, the postulated upper limit of recombination for chloroplast genes (i.e., nonlinkage) is 25% (Chapter 9). Therefore, assuming an average of 1% recombination/kb and the absence of hotspots, genetic linkage should be detectable between markers separated by up to ca. 12–14 kb.

Three methods have been used to correlate the genetic and physical maps of the chloroplast genome in *Chlamydomonas*. The first involves direct sequencing of mutations known to affect specific genes. This has led to the mapping of nonphotosynthetic mutations such genes as *rbcL*, herbicide-resistance mutations in *psbA* gene, and antibiotic-resistance mutations in the rRNA genes. The second makes use of nonphotosynthetic deletion mutations in the chloroplast genome. These are usually isolated following growth of cells in 5-fluorodeoxyuridine. So far, all such mutants have proved to map either entirely within the inverted repeat (usually eliminating the *psbA* gene) or to extend out from the inverted repeat into the adjacent *atpB* gene. The latter deletions made it possible to assign a group of point and deletion mutations to the *atpB* gene (Woessner *et al.*, 1984).

The third method has been exploited in interspecific crosses of *C. eugametos* and *C. moewusii*, where extensive restriction fragment size and site differences exist (Lemieux *et al.*, 1981, 1984a,b; Lemieux and Lee, 1987). The mapping method relies on the demonstration that a genetic marker cosegregates with a specific restriction fragment. This method was also used by Robertson *et al.* (1990) in interspecific crosses of *C. reinhardtii* and *C. smithii* to localize two ATP synthase mutations to the chloroplast *atpE* gene.

Intermolecular and intramolecular recombination of chloroplast genomes. There are several pathways of chloroplast gene recombination (see Boynton *et al.*, 1992). Single exchange events between circular chloroplast genomes should yield molecular dimers and higher multimers. Such cointegrate molecules have been observed in flowering plants (Deng *et al.*, 1989). Resolution of cointegrate molecules will produce genetically detectable recombinants.

At least two intramolecular recombination pathways, both involving the inverted repeat, also exist. The first results in a 50:50 ratio of two isomeric forms of the chloroplast genome in which the single copy regions are reversed with respect to each other (Aldrich *et al.*, 1985; Palmer, 1983; Palmer *et al.*, 1985). Such intramolecular recombination is reminiscent of that seen for the 2-μm circular plasmid of yeast (Futcher, 1988). However, no single site in the chloroplast inverted repeat seems to correspond to the *FRT* region, the target of the FLP recombinase encoded by the 2-μm circle (Palmer *et al.*, 1985). The second recombination pathway results in efficient copy correction between sequences in the inverted repeat. When intermolecular recombination occurs for markers in the inverted repeat, molecules rapidly become homozygous either for the introduced allele or for the parental allele originally present in the molecule.

Mitochondrial genetics. *C. reinhardtii* is also becoming a useful model for the study of mitochondrial genetics. The organism has both cyanide-sensitive and -insensitive respiratory pathways, and the former pathway is dispensable when cells are grown in the light. Thus, Mendelian obligate photoautotropic mutants have been isolated with defects in cyanide sensitive respiration and cytochrome oxidase (Wiseman *et al.*, 1977). Growth of *C. reinhardtii* in either acriflavin or ethidium bromide converts the entire population to the minute colony phenotype when plated on medium lacking these dyes (Alexander *et al.*, 1974) because mtDNA is eliminated (Gillham *et al.*, 1987). Minute mutations die after 8–9 mitotic divisions. Matagne *et al.* (1989) isolated two obligate photoautotrophic survivors following acriflavin treatment with 1.5-kb terminal deletions in their mitochondrial genomes including most of the cytochrome *b* gene. Dorthu *et al.* (1992) have isolated additional deletions of this general type plus one which is internal. Randolph-Anderson *et al.* (1992) have further characterized one of the original terminal deletion mutations, *dum-1*. This mutant segregates lethal minute colonies at a frequency of ca. 4% per generation. Their data suggest that the deletion has extended at least into the adjacent *ndh4* gene in these minutes. Randolph-Anderson *et al.* (1992) hypothesize that this extension is lethal since loss of the *ndh4* gene product probably leads to loss of a functional NADH dehydrogenase and eliminates cyanide-insensitive as well as cyanide-sensitive respiration. Bennoun *et al.* (1991) succeeded in isolating point mutations to myxothiazol resistance mapping in the *cytb* gene. The mitochondrial genome of *C. reinhardtii* can also be transformed (see Organelle Gene Manipulation).

Organelle Genetics in Other Systems

Mitochondrial gene mutations have been isolated and studied in *Paramecium*, filamentous fungi, notably *Aspergillus, Neurospora,* and *Podospora*, in the fission yeast *Schizosaccharomyces* and in mammalian cells in tissue culture. Chloroplast gene mutations have also been studied in a number of flowering plants (e.g., *Oenothera, Nicotiana, Pelargonium*).

During the 1970s a series of interesting papers appeared on the mitochondrial genetics of *Paramecium tetraurelia* (reviewed by Gillham, 1978). Mitochondrial mutations to erythromycin and chloramphenicol resistance were isolated and used to study mitochondrial segregation and recombination following conjugation of mating pairs accompanied by cytoplasmic transfer. Two segregation patterns were observed. Either the two mitochondrial genotypes were relatively stable, and cells remained mixed for a great number of fissions, or one of the two mitochondrial genotypes was progressively eliminated yielding cells pure for the favored mitochondrial genotype after 20 to 40 fissions. However, experiments designed to detect mitochondrial gene recombination failed to yield any evidence for recombination.

Cytoplasmically inherited mutations affecting mitochondrial cytochromes have been known in *Neurospora crassa* for almost as long as petite mutations in yeast (Mitchell and Mitchell, 1952). Initially, it seemed that *N. crassa* would equal *S. cerevisiae* as a model in which to study mitochondrial genetics. This proved not to be the case, principally because mitochondrial mutants with distinctive and easily scorable phenotypes such as antibiotic resistance have not been isolated. On the other hand, investigations with *Neurospora* have provided fascinating and profound insights into topics such as the relationship between introns and plasmids (Chapter 10). In contrast, well-defined mitochondrial resistance mutations have been isolated in *Aspergillus nidulans* and used to study mitochondrial genome segregation (see Gillham, 1978; Turner *et al.*, 1982; Wolf and Del Giudice, 1988).

Much work has also been done on the mitochondrial genetics of the fission yeast *Schizosaccharomyces pombe* (Wolf, 1987; Wolf and Del Giudice, 1988). Mitochondrial mutations to antibiotic resistance have been isolated and characterized. In crosses these mutations do not segregate as rapidly as they do in *S. cerevisiae*. A cytoplasmic mutator has been identified in *S. pombe* that causes deletions and point mutations in mitochondrial genes leading to respiratory deficiency. The deletions fall in the *cox1, cox2, cox3,* and *cytb* genes. Since respiratory-deficient mutants equivalent to the *mit*$^-$ mutants of *S. cerevisiae* can be isolated in *S. pombe,* one would expect that vegetative petite mutants should also arise. Yet *S. pombe* is one of the so-called petite negative yeasts where ρ^- and ρ^0 mutations have proved difficult to obtain. Haffter and Fox (1992) finally succeeded in isolating two ρ^0 strains after very long incubation of cells in medium containing ethidium bromide. Each isolate contained a nuclear gene mutation whose sole phenotypic effect was to permit ρ^+ strains of *S. pombe* to undergo loss of mtDNA reproducibly. The nuclear mutations in the two ρ^0 strains affect different genes. Nuclear mutations have also been isolated in *Kluyveromyces lactis,* which convert this petite negative yeast to the petite positive phenotype (Chen and Clark-Walker, 1993).

Mitochondrial mutations have been isolated and studied in mouse, rat, and human tissue culture cells. Most of these mutants are resistant to antibiotics such as chloramphenicol and erythromycin. To demonstrate the cytoplasmic (presumably mitochondrial) nature of the of chloramphenicol resistance, Bunn *et al.* (1974) developed the technique of *cybrid* formation. In this method cells that are, for example, chloramphenicol resistant are enucleated with cytochalasin B . The resulting cytoplasts are fused with a chloramphenicol-sensitive cell line using sendai virus. Appropriate nuclear gene markers are employed to ensure that any unenucleated cells from the cytoplast population or nuclear hybrids between the two cell lines can be eliminated by the appropriate selection. If the recipient cells now become resistant to chloramphenicol the trait must be cytoplasmically transmitted since the donor cytoplasts were enucleated. Chloramphenicol-resistance mutations result from specific nucleotide substitutions occurring in the mitochondrial gene encoding the large rRNA molecule in positions identical or close to similar antibiotic-resistance mutations in yeast (Kearsay and Craig, 1982; Howell and Lee, 1989; Wallace *et al.,* 1982).

Chloramphenicol resistance has proved a particularly useful marker in studies of mitochondrial segregation in tissue culture cells. Wallace (1986) examined hybrid cell lines heteroplasmic for chloramphenicol-resistant and -sensitive mitochondria. He observed random (stochastic) partitioning of mitochondria to daughter cells. This phenomenon, referred to as replicative segregation, also occurs in muscle and fibroblast cell cultures from patients with Kearns–Sayre syndrome. These cell lines are heteroplasmic for normal and deleted mtDNA molecules (Shoffner and Wallace, 1990; Chapter 11). Wallace (1986) also reported that the expression threshold for chloramphenicol resistance in human hybrid cells was very sharp. Thus, hybrid cells with less that 10% mtDNA molecules carrying the chloramphenicol-resistance marker were killed by chloramphenicol during cloning whereas hybrids with 12% resistant mtDNAs cloned well in the presence of the drug. Similar results have been obtained for other mitochondrial traits (Howell *et al.,* 1984). These observations are important for they help to explain certain aspects of human mitochondrial diseases in that slight variations in ratios of mtDNA molecules in heteroplasmic cells can have large phenotypic effects (Chapter 11).

King and Attardi (1988) used a chloramphenicol-resistance marker to demonstrate that functional mitochondria could be injected into tissue culture cells partially depleted of mtDNA and then selected for. Colonies appeared on selective media containing chloramphenicol between 14 and 17 days after injection. A full complement of resistant mitochondria was present after a period of time corresponding to 20–25 cell generations.

The disadvantage of using chloramphenicol resistance is that the cells can have low respiratory capacity (King and Attardi, 1989). This happens because the activity of respiratory complexes, especially complex I, is reduced. This reduction is a consequence of less efficient mitochondrial translation, presumably resulting from defective assembly of mitochondrial ribosomal subunits. King and Attardi (1989) also discovered they could create ρ^0 cell lines lacking any mtDNA by prolonged incubation of cells in a medium containing ethidium bromide plus uridine and pyruvate. The latter two compounds had to be provided since the cells had lost a functioning respiratory chain. They injected chloramphenicol-sensitive or -resistant mitochondria into ρ^0 cells and selected for "transformants" in the absence of uridine. Colonies were first observed between 10 and 30 days. The efficiency of transformation was estimated to be ca. $1–3 \times 10^{-3}$ per injected cell. This technique has important applications for the study of mitochondrial diseases (Chapter 11).

In land plants plastome mutations affecting and, presumably, located within cpDNA have been known ever since the time of Baur and Correns. Their properties have been reviewed by Borner and Sears (1986). These mutations can be divided into two general classes. The first includes mutations defective in photosynthesis. One such mutation inactivates the *rbcL* gene in *Oenothera* by causing a deletion of 5 bp (Sears and Herrmann, 1985; Winter and Herrmann, 1988). A frame shift is thus introduced resulting in a stop codon seven DNA triplets later. Among the most intensely studied plastome mutations are those in the *psbA* gene, which confer herbicide resistance on the D-1 reaction center polypeptide of photosystem II. The second class of mutations affects components involved in chloroplast translation, notably ribosomes. These include chloroplast ribosome-deficient mutants and mutants resistant to antibiotics such as streptomycin (Chapter 13). Thus, antibiotic-resistant mutations can be isolated in tissue culture cells of *Nicotiana* following which resistant plants can be regenerated. These have proved indispensable as selectable markers for chloroplast transformation experiments and interspecies transfers of plastids (see below).

With the exception of *Oenothera* in which biparental transmission of plastids occurs, most plastid mutations have been isolated in flowering plant species in which maternal plastid transmission occurs (Chapter 8). This problem has been obviated in some species by protoplast fusion techniques, which make possible the production of cells with mixed populations of plastids and also mitochondria (see Medgyesy, 1990; Gillham *et al.*, 1991, for reviews). The most extensive work of this kind has been carried out with *Nicotiana*. Restriction fragment polymorphisms can be used to follow the fate of both cpDNA and mtDNA. In the former case genetic markers (e.g., antibiotic resistance) are available as well (see above). The most important conclusion to emerge is that while extensive recombination of mitochondrial genomes occurs in the absence of any selection, chloroplast gene recombination is extremely rare even with selection. The reason seems to be that chloroplast fusions hardly ever occur. So far, only two bona fide chloroplast recombinants have been reported from interspecific fusions despite several herculean attempts. One arose in a somatic hybrid between *Nicotiana tabacum* and *N. plumbaginifolia* (Medgyesy *et al.*, 1985) and the other in somatic hybrids between *N. tabacum* and *Solanum tuberosum* (Thanh and Medgyesy, 1989).

Although chloroplast recombinants are very infrequent in flowering plants, random segregation of chloroplasts occurs, permitting synthesis of new combinations of nuclear and chloroplast genomes. Techniques also exist for eliminating the donor nuclear genome, and cybrids can be selected using an appropriate chloroplast marker such as streptomycin resistance. Numerous interspecific and several intergeneric cybrids have

MODEL SYSTEMS AND GENE MANIPULATION 171

been created in this way (Chapter 8). While this same methodology results in mitochondrial transfer, the results are obfuscated by the frequent recombination of mitochondrial genomes following fusion and the lack of selectable mitochondrial markers. Nevertheless, cotransfer of selected chloroplast markers (e.g., antibiotic resistance) and unselected mitochondrial markers (i.e., cytoplasmic male sterility, Chapter 11) has resulted in 90–100% cotransfer of cytoplasmic male sterility in certain interspecific fusion combinations (Medgyesy, 1990). Consequently, this technique has important practical applications for cybrids, which are capable of regeneration in that the agriculturally valuable trait of mitochondrial male sterility can be transferred into new species (Chapter 11).

ORGANELLE GENE MANIPULATION

Two general methodologies have been used to manipulate organelle genes and their products. The first involves adding an organelle targeting sequence (presequence, Chapter 15) to a cloned organelle gene following which the chimeric construct is introduced into the nucleus by transformation. The protein encoded by the gene is then routed into the organelle via its presequence. The second method results in the direct replacement of a chloroplast or mitochondrial gene by its cloned homolog on a plasmid via organelle transformation.

Relocating Organelle Genes to the Nucleus

Several different instances in which engineered organelle genes have been relocated to the nucleus are reviewed by Nagley and Devenish (1989). The first involved subunit 8 of the yeast mitochondrial ATP synthase. The gene was first resynthesized completely to accommodate the coding differences between the mitochondrion and cytoplasm. The chemically synthesized gene contained changes in 30 out of 48 codons. The DNA sequence encoding the presequence of the *Neurospora* subunit 9 ATP synthetase gene was fused to the artificial subunit 8 gene so that a chimeric precursor protein could be formed. The host yeast cells used for introduction of the chimeric gene could not manufacture a functional subunit 8 because of mutations in the mitochondrial gene encoding this polypeptide. The chimeric gene was introduced into these host cells on a multicopy expression vector under the control of the yeast phosphoglycerokinasepromoter. Transformed host cells regained the ability to respire ethanol, an indication that ATP synthesis was proceeding normally. Immunochemical analysis of the ATP synthase of these transformed cells revealed that the imported version of subunit 8 was present. Similar experiments have been done with an intron-encoded maturase protein (bI4) of yeast. This intron-encoded protein is required for correct splicing of the fourth intron in the cytochrome *b* transcript and the fourth intron in the *cox1* transcript (Chapter 10). These experiments also employed the *Neurospora* subunit 9 presequence to target the gene of interest. The recipient cells had missense or nonsense mutations in the mitochondrial bI4 reading frame.

In other experiments the chloroplast *psbA* gene from *Amaranthus*, encoding the reaction center D1 protein of PSII and possessing an herbicide-resistance mutation, was fused the DNA sequence of a pea nuclear gene specifying the presequence of the small subunit of RUBISCO. This construct was introduced successfully into the nucleus of

Amaranthus by *Agrobacterium* Ti plasmid-mediated transformation. In this case the resident *psbA* gene was still present in the chloroplast but its D1 product was atrazine sensitive. Many of the transformants were resistant to atrazine levels toxic to sensitive control plants. However, the resistance phenotype was not retained for long periods of growth and exhibited variability probably because the numbers of chimeric genes introduced into different transformants fluctuated. The approach of transferring organelle genes to the nucleus was developed at about the same time that organelle transformation methods first became available. These gene relocation methodologies were designed largely to make possible engineering of organelle genes. Organelle transformation obviates the need for gene relocation for gene manipulation experiments. On the other hand, gene relocation experiments provide the unique opportunity to test whether certain organelle genes never found in the nucleus such *psbA* can be relocated there and function.

Organelle Transformation

The development of reliable and reproducible transformation technologies for both the chloroplast of *Chlamydomonas* (Boynton et al., 1988) and mitochondria of yeast (Johnston et al., 1988) late in the last decade made these organelles accessible to gene manipulation for the first time. Transformation technology has since been extended to the chloroplasts of *Nicotiana* (Svab et al., 1990) and the mitochondria of *Chlamydomonas* (Randolph-Anderson et al., 1992).

The technical achievement that made organelle transformation possible was the successful development of the biolistic device for shooting DNA into animal and plant cells by Sanford and colleagues (Klein et al., 1987; Sanford et al., 1987). The method employs a "gun" that fires DNA-coated microprojectiles at target cells or tissues maintained in a partial vacuum. The gun was originally powered by a gunpowder delivery system, but a gas discharge device has since been developed that employs a helium shock wave to accelerate the microprojectiles toward their targets. In transformation of *Chlamydomonas* and yeast, petri dishes covered with monolayers of cells are normally used as microprojectile recipients.

Chloroplast transformation in *Chlamydomonas* has been thoroughly characterized (for reviews see Boynton et al., 1990, 1992; Boynton and Gillham, 1992). Initially, the biolistic method was used to introduce a wild-type chloroplast DNA fragment that complemented *atpB* deletion and point mutations (Boynton et al., 1988). Subsequent experiments have yielded transformants for nonphotosynthetic deletion or point mutations in seven other chloroplast genes (Boynton and Gillham, 1992). Wild-type strains have also been transformed to antibiotic resistance using donor DNA carrying selectable resistance mutations in the rRNA genes (Newman et al., 1990). Similarly, donor fragments carrying herbicide-resistance mutations have been employed for *psbA* gene transformation experiments (Boynton et al., 1990; Przibilla et al., 1991).

Careful characterization of the transformants resulting from such experiments has revealed the following features of the chloroplast transformation process:

1. Transformation always occurs by homologous gene replacement without integration of vector sequences, unless the donor sequence fails to overlap both deletion endpoints in the recipient.

2. Exchange events lead to integration of donor DNA cluster near the vector:insert junction.
3. Genes introduced into the inverted repeat copy correct to yield molecules homozygous for the introduced gene.
4. Cells are initially heteroplasmic for transformed and untransformed chloroplast genomes, but segregate somatically to yield homoplasmic transformants.
5. Reduction of chloroplast genome copy number with 5-fluorodeoxyuridine (Wurtz et al., 1977) prior to bombardment is accompanied by an increase in certain classes of transformants (i.e., those involving deletions and antibiotic or herbicide resistance).
6. Transformation frequencies as high as 1×10^{-4} can be achieved.

Two systems for efficient targeted chloroplast gene disruption have also been developed in *Chlamydomonas*. The first makes use of a cotransformation method involving two plasmids (Newman et al., 1991). One plasmid carries a donor DNA insert with selectable markers for antibiotic resistance in the rRNA genes while the second has an insert with a disrupted photosynthesis gene. Approximately half of the resistant transformants prove to be initially heteroplasmic for chloroplast genomes in which the wild-type photosynthetic gene, has been replaced by the disrupted donor gene. Subsequent segregation yields homoplasmons for the disrupted photosynthetic gene, which can be recovered by single cell cloning. Cotransformation can also be used to introduce photosynthesis genes rendered nonfuctional by *in vitro* mutagenesis. A similar method has been used by Kindle et al. (1991) to introduce an altered, but functional photosynthetic gene.

The second method employs a selectable bacterial gene *(aadA)* coding for an enzyme that will detoxify antibiotics such as spectinomycin that are lethal to *C. reinhardtii* (Goldschmidt-Clermont, 1991). The bacterial gene is flanked by transcription and translation signals from the chloroplast genome of *Chlamydomonas,* which are themselves flanked by sequences from the gene to be disrupted. Transformants are selected for spectinomycin resistance and scored for the photosynthetic defect caused by insertion of the donor-resistance gene cassette into its target in the recipient. Successful introduction and expression of the *aadA* reporter gene under the control of flanking cpDNA sequences have paved the way for the introduction and expression of other reporter genes. Thus, the bacterial gene encoding the *E. coli* β-glucuronidase *(uidA)* or GUS has been similarly expressed in *Chlamydomonas* (Sakamoto et al., 1993). This is a highly popular reporter gene among plant molecular biologists whose product is easily assayed.

The biolistic system has also been used to effect plastid transformation in *Nicotiana* (Svab et al., 1990; Staub and Maliga, 1992; Svab and Maliga, 1993; see Maliga, 1993; Maliga et al., 1993 for reviews). Leaves are bombarded with plasmids carrying chloroplast antibiotic-resistance markers either in the 16 S rRNA genes (spectinomycin and streptomycin resistance) or in the *rps12* gene (streptomycin resistance) or with a chimeric plasmid containing the bacterial *aadA* gene and flanked by chloroplast sequences required for expression. Subsequently, the leaves are cut into sections and transferred to selective medium. Green calli forming on the antibiotic-bleached leaves are putative transformants and can be regenerated into whole plants. Using chloroplast-resistance markers transformation was initially an infrequent, but reproducible event with one transformed line being obtained for every 50 to 100 samples bombarded. However, a 100-fold increase was obtained when the chimeric *aadA* gene was used.

Two basic types of manipulations have thus far been carried out. In gene replacement experiments the donor plastid DNA integrates by homologous recombination as in *Chlamydomonas* followed by copy correction for those resistance markers in the inverted repeat. In these experiments two distinct transformation patterns have been observed. Two transformants were obtained that appeared to be homoplasmic for the whole or nearly the whole 6.2-kb donor fragment containing the 16 S rRNA mutation to spectinomycin resistance, an *rps12* mutation conferring resistance to streptomycin, plus many restriction site changes (Staub and Maliga, 1992). Other transformants, isolated using a plasmid with chloroplast 16 S rDNA mutations to spectinomycin and streptomycin resistance, were homoplasmic for the selected spectinomycin-resistance mutation, but behaved as segregating heteroplasmons with respect to streptomycin resistance (Svab *et al.*, 1990). The difference between these transformant classes is not understood, although differential copy correction has been proposed as an explanation (Maliga, 1993). The second manipulation, gene insertion, involves transformation with foreign genes such as *aadA* and *uidA* (Maliga, 1993; Maliga *et al.*, 1993; Staub and Maliga, 1993; Svab and Maliga, 1993). Again, integration is by homologous recombination since flanking cpDNA sequences are positioned 3' and 5' of the reporter gene to ensure proper expression.

O'Neill *et al.* (1993) have now reported obtaining stable chloroplast transformants following treatment of *N. plumbaginifolia* protoplasts with polyethylene glycol and selection on streptomycin plus spectinomycin medium using a cloned 16S rRNA gene from *Nicotiana tabacum* containing mutations conferring resistance to the two antibiotics. This technique has also been used successfully in *N. tabacum* (Goulds *et al.*, 1993).

Obviously, the potential application of chloroplast transformation systems to flowering plants is immense. Maliga (1993) points out that a desirable foreign gene making a useful protein could be inserted into the chloroplast genome where it will be amplified up to 10,000 copies per cell. Leaves could then be used as factories for production of the protein. However, before this dream is realized, a better understanding of the basic mechanisms controlling chloroplast gene expression will probably be required. These processes are highly regulated at the posttranscriptional, translational, and proteolytic levels (see Section 4).

Yeast mitochondrial transformation has so far proved less frequent and reliable than *Chlamydomonas* chloroplast transformation although the technique undoubtedly works (Butow and Fox, 1990; Fox *et al.*, 1988; Johnston *et al.*, 1988; Thorsness and Fox, 1990). Since the outset, investigators have used a cotransformation technique. A nuclear gene marker, usually the wild-type *URA3* gene, is employed to detect nuclear transformants, following which these transformants are screened to see which are also cotransformed for the mitochondrial marker. Thus, the recipient strain used by Johnston *et al.* (1988) had a nuclear *ura3* mutation plus a large deletion in the mitochondrial *cox1* gene. Transformation was effected by coprecipitating a 1:1 mixture of two plasmids onto the microprojectiles used for biolistic transformation. One plasmid contained the wild-type *URA3* gene and the second a functional, but novel form of the *cox1* gene lacking introns 1 and 2 (Chapter 10). From 6300 nuclear transformants six were also mitochondrial transformants. In the transformants analyzed, integration occurred by homologous recombination as in the case of chloroplast transformation. A special advantage of yeast is that plasmids containing mitochondrial genes of interest can be introduced into ρ^0 strains creating "designer" ρ^- strains (Fox et al., 1988). However, these mitochondrial transformants cannot be screened directly since they are not respiration competent. Instead, they must be crossed to mit^- strains with lesions in the gene of interest. If a transformant contains the desired mitochondrial plasmid, it can rescue the mit^- mutant by recombination in crosses. Use of this strategy and a plasmid containing wild-type URA3 and cox2 genes permitted Thorsness and Fox (1990) to introduce the URA3 gene into mitochondria and to measure the frequency of escape of this gene to the nucleus in ura3 mutant strains (Chapter 2, Organelle Gene Transfer).

Transformation of the mitochondrial genome of Chlamydomonas has also been effected using a mutant with a deletion extending into the cytb gene as a recipient and total wild-type mtDNA from C. reinhardtii or C. smithii as the donor (Randolph-Anderson et al., 1992). Transformants in this case were selected directly in the dark where the recipient strain is unable to grow on acetate medium because of the lesion in the respiratory chain caused by the cytb deletion. A notable feature of these experiments is that the transformants took a very long time (4 to 8 weeks of dark incubation) to appear. One wonders why similar direct selection experiments have not been reported in yeast.

Organelle transformation technology is still in its infancy, but the possible applications are tremendous. They are discussed in connection with specific experiments at various points in this book.

REFERENCES

Adams, G.M.W., Van Winkle-Swift, K.P., Gillham, N.W., and J.E. Boynton (1976). Plastid inheritance in Chlamydomonas reinhardtii. In The Genetics of Algae (R.A. Lewin, ed.), pp. 69–118. Blackwell, Oxford and University of California Press, Berkeley.

Aldrich, J., Cherney, B., Merlin, E., Williams, C., and L. Mets (1985). Recombination within the inverted repeat sequences of the Chlamydomonas reinhardtii chloroplast genome produces two orientation isomers. Curr. Genet. 9: 233–238.

Alexander, N., Gillham, N.W., and J.E. Boynton (1974). The mitochondrial genome of Chlamydomonas. Induction of minute colony mutations by acriflavin and their inheritance. Mol. Gen. Genet. 130: 275–290.

Baur, E. (1909). Das Wesen und die Erblichkeitsverhaltniss der "varietates albomarginatae hort" von Pelargonium zonale. Z. Verebungs. 1: 330–351.

Benoun, P., and P. Delepelaire (1982). Isolation of photosynthesis mutants in Chlamydomonas. In Methods in Chloroplast Molecular Biology (M. Edelman, N.-H. Chua, and R.B. Hallick, eds.), pp. 25–38. Elsevier/North Holland, Amsterdam.

Bennoun, P., Masson, A., and M. Delosme (1980). A method for complementation analysis of nuclear and chloroplast mutants of photosynthesis in Chlamydomonas. Genetics 95: 39–47.

Bennoun, P., Delosme, M., and U. Kuck (1991). Mitochondrial genetics of Chlamydomonas reinhardtii: Resistance mutations marking the cytochrome b gene. Genetics 127: 335–343.

Bertrand, H., and T.H. Pittenger (1972). Complementation among cytoplasmic mutants of Neurospora crassa. Mol. Gen. Genet. 117: 82–90.

Borner, T., and B.B. Sears (1986). Plastome mutants. Plant Mol. Biol. Reporter 4: 69–92.

Boynton, J.E., and N.W. Gillham (1992). Chloroplast transformation in Chlamydomonas. Methods Enzymol. 217: 510–536.

Boynton, J.E., Gillham, N.W., Harris, E.H., Tingle, C.L., Van Winkle-Swift, K., and G.M.W. Adams (1976). Transmission, segregation and recombination of chloroplast genes in Chlamydomonas. In Genetics and Biogenesis of Chloroplasts and Mitochondria (Th. Bucher, ed.), pp. 313–322. Elsevier/North Holland Press, Amsterdam.

Boynton, J.E., Harris, E.H., Burkhart, B.D., Lamerson, P.M., and N.W. Gillham (1987). Transmission of mitochondrial and chloroplast genomes in crosses of Chlamydomonas. Proc. Natl. Acad. Sci. U.S.A. 84: 2391–2395.

Boynton, J.E., Gillham, N.W., Harris, E.H., Hosler, J.P., Johnson, A.M., Jones, A.R., Randolph-Anderson, B.L., Robertson, D., Klein, T.M., Shark, K.B., and J.C. Sanford (1988). Chloroplast transformation in Chlamydomonas with high velocity microprojectiles. Science 240: 1534–1538.

Boynton, J.E., Gillham, N.W., Harris, E.H., Newman, S.M., Randolph-Anderson, B.L., Johnson, A.M., and A.R. Jones (1990). Manipulating the chloroplast genome of Chlamydomonas—Molecular genetics and transformation. In Current Research in Photosynthesis, Vol. III, (M. Baltscheffsky, ed.), pp. 509–516. Kluwer Academic Publishers, Amsterdam.

Boynton, J.E., Gillham, N.W., Newman, S.M., and E.H. Harris (1992). Organelle genetics and transformation in Chlamydomonas. In *Cell Organelles, Advances in Plant Gene Research* (R. Herrmann, ed.). pp. 3–64. Springer-Verlag, Wien, New York.

Bunn, C.L., Wallace, D.C., and J.M. Eisenstadt (1974). Cytoplasmic inheritance of chloramphenicol resistance in mouse tissue culture cells. *Proc. Natl. Acad. Sci. U.S.A.* 71: 1681–1685.

Butow, R.A., and T.D. Fox (1990). Organelle transformation: Shoot first, ask questions later. *Trends Biochem. Sci.* 15: 465–468.

Chen, X.-J., and G.D. Clark-Walker (1993). Mutations in *MG1* genes convert *Kluyveromyces lactis* into a petite positive yeast. *Genetics* 133: 517–525.

Clark-Walker, G.D. (1992). Evolution of mitochondrial genomes in fungi. *Int. Rev. Cytol.* 141: 89–127.

Clark-Walker, G.D., and A.W. Linnane (1966). In vivo differentiation of yeast cytoplasmic and mitochondrial protein synthesis with antibiotics. *Biochem. Biophys. Res. Commun.* 25: 8–13.

Coen, D., Deutsch, J., Netter, P., Petrochilo, E., and P.P. Slonimski (1970). Mitochondrial genetics. I. Methodology and phenomenology. *Symp. Soc. Exp. Biol.* 24: 449–496.

Correns, C. (1909). Vererbungsversuche mit blass (gelb) grunen und buntblattrigen sippen bei *Mirabilis jalapa, Urtica pilulifera* und *Lunaria annua*. *Z. Vererbungs.* 1: 291–329.

Deng, X.W., Wing, R.A., and W. Gruissem (1989). The chloroplast genome exists in multimeric forms. *Proc. Natl. Acad. Sci. U.S.A.* 86: 4156–4160.

Deutsch, J., Dujon, B., Netter, P., Petrochilo, E., Slonimski, P.P., Bolotin-Fukuhara, M., and D. Coen (1974). Mitochondrial genetics VI. The petite mutation in *Saccharomyces cerevisiae*: Interrelations between the loss of the ρ factor and the loss of the drug resistance mitochondrial markers. *Genetics* 76: 195–219.

De Zamaroczy, M., and G. Bernardi (1987). The AT spacers and the *var1* genes from the mitochondrial genomes of *Saccharomyces cerevisiae* and *Torulopsis glabrata*: Evolutionary origin and mechanism of formation. *Gene* 54: 1–22.

Dorthu, M.-P., Remy, S., Michel-Wolwertz, M.-R., Colleaux, L., Breyer, D. Beckers, M.-C., Englebert, S., Duyckaerts, C., Sluse, F.E., and R.F. Matagne (1992). Biochemical, genetic and molecular characterization of new respiratory-deficient mutants in *Chlamydomonas reinhardtii*. *Plant Mol. Biol.* 18: 759–772.

Dron, M., Rahire, M., Rochaix, J.-D., and L. Mets (1983). First DNA sequence of a chloroplast mutation: A missense alteration in the ribulosebisphosphate carboxylase large subunit gene. *Plasmid* 9: 321–324.

Dujon, B. (1981). Mitochondrial genetics and functions. In *Molecular Biology of the Yeast Saccharomyces: Life Cycle and Inheritance* (J.N. Strathern, E.W. Jones, and J.R. Broach, eds.), pp. 505–635. Cold Spring Harbor Laboratory, Cold Spring Harbor, NY.

Ephrussi, B. (1953). *Nucleo-cytoplasmic Relations in Micro-organisms*. Clarendon Press, Oxford.

Faye, G., Fukuhara, H., Grandchamp, C., Lazowska, J., Michel, F., Casey, J. Getz, G.S., Locker, J., Rabinowitz, M., Bolotin-Fukuhara, M., Coen, D. Deutsch, J., Dujon, B., Netter, P., and P.P. Slonimski (1973). Mitochondrial nucleic acids in the petite colonie mutants: Deletions and repetitions of genes. *Biochimie* 55: 779–792.

Flury, U., Mahler, H.R., and F. Feldman (1974). A novel respiration-deficient mutant of *Saccharomyces cerevisiae*. *J. Biol. Chem.* 249: 6130–6137.

Foury, F., and A. Tzagoloff (1978). Assembly of the mitochondrial membrane system. Genetic complementation of mit⁻ mutations in mitochondrial DNA of *Saccharomyces cerevisiae*. *J. Biol. Chem.* 253: 3792–3797.

Fox, T.D., Sanford, J.C., and T.W. McMullin (1988). Plasmids can stably transform yeast mitochondria lacking endogenous mtDNA. *Proc. Natl. Acad. Sci. U.S.A.* 85: 7288–7292.

Futcher, A.B. (1988). The 2 um circle plasmid of *Saccharomyces cerevisiae*. *Yeast*, 4: 27–40.

Gillham, N.W. (1965). Linkage and recombination between nonchromosomal mutations in *Chlamydomonas reinhardi*. *Proc. Natl. Acad. Sci. U.S.A* 54: 1560–1567.

Gillham, N.W. (1978). *Organelle Heredity*. Raven Press, New York.

Gillham, N.W., Boynton, J.E., and R.W. Lee (1974). Segregation and recombination of non-mendelian genes in *Chlamydomonas*. *Genetics* 78: 439–457.

Gillham, N.W., Boynton, J.E., and E.H. Harris (1987). Specific elimination of mitochondrial DNA from *Chlamydomonas* by intercalating dyes. *Curr. Genet.* 12: 41–47.

Gillham, N.W., Boynton, J.E., and E.H. Harris (1991). Transmission of plastid genes. In *The Molecular Biology of Plastids* (L. Bogorad and I.K. Vasil, eds.), pp. 55–92. Academic Press, San Diego.

Gingold, E.B. (1988). The replication and segregation of yeast mitochondrial DNA. In *Division and Segregation of Organelles* (S.A. Boffey and D. Lloyd, eds.), *Soc. Exp. Biol.* 35: 148–170.

Girard-Bascou, J., Choquet, Y., Schneider, M., Delosme, M., and M. Dron (1987). Characterization of a chloroplast mutation in the *psaA2* gene of *Chlamydomonas reinhardtii*. *Curr. Genet.* 12: 489–495.

Goldschmidt-Clermont, M. (1991). Transgenic expression of aminoglycoside adenine transferase in the chloroplast: A selectable marker for site-directed transformation of chlamydomonas. *Nucl. Acids. Res.* 19: 4083–4089.

Goulds, T., Maliga, P., and H.V. Koop (1993). Stable plastid transformation in PEG-treated protoplasts of *Nicotiana tabacum*. *Biotechnology* 11:95–97.

Grant, D.M., Gillham, N.W., and J.E. Boynton

(1980). Inheritance of chloroplast DNA in *Chlamydomonas reinhardtii*. *Proc. Natl. Acad. Sci. U.S.A.* 77: 6067–6071.

Haffter, P., and T.D. Fox (1992). Nuclear mutations in the petite-negative yeast *Schizosaccharomyces pombe* allow growth of cells lacking mitochondrial DNA. *Genetics* 131: 255–260.

Harris, E.H. (1989). *The Chlamydomonas Sourcebook*. Academic Press, San Diego.

Harris, E.H., Boynton, J.E., Tingle, C.L., and S.B. Fox (1977). Mapping of chloroplast genes involved in chloroplast ribosome biogenesis in *Chlamydomonas reinhardtii*. *Mol. Gen. Genet.* 155: 249–265.

Harris, E.H., Burkhart, B.D., Gillham, N.W., and J.E. Boynton (1989). Antibiotic resistance mutations in the chloroplast 16S and 23S rRNA genes of *Chlamydomonas reinhardtii*: Correlation of genetic and physical maps of the chloroplast genome. *Genetics* 123: 281–292.

Howell, N., and A. Lee (1989). Sequence analysis of mouse mitochondrial chloramphenicol-resistant mutants. *Somat. Cell Mol. Genet.* 15: 237–244.

Howell, N., Huang, P., and R.D. Kolodner (1984). Origin, transmission, and segregation of mitochondrial DNA dimers in mouse hybrid and cybrid cell lines. *Somat. Cell Mol. Genet.* 10: 259–274.

Jinks, J.L. (1964). *Extrachromosomal Inheritance.* Prentice-Hall, Englewood Cliffs, NJ.

Johnston, S.A., Anziano, P.Q., Shark, K., Sanford, J.C., and R.A. Butow (1988). Mitochondrial transformation in yeast by bombardment with microprojectiles. *Science* 240: 1538–1541.

Kearsay, S.E., and I.W. Craig (1982). Genetic basis of chloramphenicol resistance in mouse and human cell lines. In *Mitochondrial Genes* (P.P. Slonimski, P. Borst, and G. Attardi, eds.), pp. 117–120. Cold Spring Harbor Press, Cold Spring Harbor, NY.

Kindle, K.L., Richards, K.L., and D.B. Stern (1991). Engineering the chloroplast genome: Techniques and capabilities for chloroplast transformation in *Chlamydomonas reinhardtii*. *Proc. Natl. Acad. Sci. U.S.A.* 88: 1721–1725.

King, M.P., and G. Attardi (1988). Injection of mitochondria into human cells leads to a rapid replacement of the endogenous mitochondrial DNA. *Cell* 52: 811–819.

King, M.P., and G. Attardi (1989). Human cells lacking mtDNA: Repopulation with exogenous mitochondria by complementation. *Science* 246: 500–503.

Kirk, J.T.O., and R.A.E. Tilney-Bassett (1978). *The Plastids: Their Chemistry, Structure, Growth and Inheritance*, 2nd ed. Elsevier/North Holland, Amsterdam.

Klein, T.M., Wolf, E.D., Wu, R., and J.C. Sanford (1987). High-velocity microprojectiles for delivering nucleic acids into living cells. *Nature (London)* 327: 70–73.

Lazowska, J., Michel, F., Faye, G., Fukuhara, H., and P.P. Slonimski (1974). Physical and genetic organization of *petite* and *grande* yeast mitochondrial DNA. II. DNA-DNA hybridization studies and buoyant density determinations. *J. Mol. Biol.* 85: 393–410.

Lemieux, C., and R.W. Lee (1987). Nonreciprocal recombination between alleles of the chloroplast 23S rRNA in interspecific *Chlamydomonas* crosses. *Proc. Natl. Acad. Sci. U.S.A.* 84: 4166–4170.

Lemieux, C., Turmel, M., and R.W. Lee (1980). Characterization of chloroplast DNA in *Chlamydomonas eugametos* and *C. moewusii* and its inheritance in hybrid progeny. *Curr. Genet.* 2: 139–147.

Lemieux, C., Turmel, M., and R.W. Lee (1981). Physical evidence for recombination of chloroplast DNA in hybrid progeny of *Chlamydomonas eugametos* and *C. moewusii*. *Curr. Genet.* 3: 97–103.

Lemieux, C., Turmel, M., Seligy, V.L., and R.W. Lee (1984a). Chloroplast DNA recombination in interspecific hybrids of *Chlamydomonas*: Linkage between a nonmendelian locus for streptomycin resistance and restriction fragments coding for 16S rRNA. *Proc. Natl. Acad. Sci. U.S.A.* 81: 1164–1168.

Lemieux, C., Turmel, M., Seligy, V.L., and R.W. Lee (1984b). A genetical approach to the physical mapping of chloroplast genes in *Chlamydomonas*. *Can. J. Biochem. Cell Biol.* 62: 225–229.

Lewis, J. E., and C.W. Birky, Jr. (1984). Heteroplasmic yeast cells contain no selectable "hidden" mitochondrial alleles. *Curr. Genet.* 8: 81–84.

Linnane, A.W., Saunders, G.W., Gingold, E.B., and H.B. Lukins (1968). The biogenesis of mitochondria. V. Cytoplasmic inheritance of erythromycin resistance in *Saccharomyces cerevisiae*. *Proc. Natl. Acad. Sci. U.S.A.* 59: 903–910.

Liu, X.-Q., Gillham, N.W., and J.E. Boynton (1989). Chloroplast ribosomal protein gene *rps12* of *Chlamydomonas reinhardtii*: Wild-type sequence, mutation to streptomycin resistance and dependence, and function in *Escherichia coli*. *J. Biol. Chem.* 264: 16100–16108.

Locker, J., Rabinowitz, M., and G.S. Getz (1974). Tandem inverted repeats in mitochondrial DNA of petite mutants of *Saccharomyces cerevisiae*. *Proc. Natl. Acad. Sci. U.S.A.* 71: 1366–1370.

Maliga, P. (1993). Towards plastid transformation in flowering plants. *Trends Biotechnol.* 11: 101–107.

Maliga, P., Carrer, H., Kanevski, I., Staub, J., and Z. Svab (1993). Plastid engineering in land plants: a conservative genome is open to change. *Phil. Trans. R. Soc. Lond. B* 341: 449–454.

Marcou, D. (1954). Rajeunissement et arret de croissance chez *Podospora anserina*. *C. R. Acad. Sci. [D] (Paris)* 244: 661–663.

Marquardt, H. (1952). Neue Ergebnisse aus dem Gebiet der Plasmavererbung. *Umschaw. Wiss. Tech.* 52: 545–549.

Matagne, R.F., Michel-Wolwertz, M.R., Munaut, C., Duykaerts, C. and F. Sluse (1989). Induction and characterization of mitochondrial DNA mutants in *Chlamydomonas. J. Cell Biol.* 108: 1221–1226.

Medgyesy, P. (1990). Selection and analysis of cytoplasmic hybrids. In *Plant Cell Line Selection* (P.J. Dix, ed.), pp. 287–315. VCH, Weinheim.

Medgyesy, P., Fejes, E., and P. Maliga (1985). Interspecific chloroplast recombination in a *Nicotiana* somatic hybrid. *Proc. Natl. Acad. Sci. U.S.A.* 82: 6960–6964.

Mets, L. (1980). Uniparental inheritance of chloroplast DNA sequences in interspecific hybrids of *Chlamydomonas. Curr. Genet.* 2: 131–138.

Mets, L.J., and L.J. Geist (1983). Linkage of a known chloroplast gene mutation to the uniparental genome of *Chlamydomonas reinhardtii. Genetics* 105: 559–579.

Mitchell, M.B., and H.K. Mitchell (1952). A case of "maternal" inheritance in *Neurospora crassa. Proc. Natl. Acad. Sci. U.S.A.* 38: 442–449.

Nagley, P., and R.J. Devenish (1989). Leading organellar proteins along new pathways: The relocation of mitochondrial and chloroplast genes to the nucleus. *Trends Biochem. Sci.* 14: 31–35.

Newman, S. M., Boynton, J.E., Gillham, N.W., Randolph-Anderson, B.L., Johnson, A.M., and E.H. Harris (1990). Transformation of chloroplast ribosomal RNA genes in *Chlamydomonas:* Molecular and genetic characterization of integration events. *Genetics* 126: 875–888.

Newman, S.M., Gillham, N.W., Harris, E.H., Johnson, A.M., and J.E. Boynton (1991). Targeted disruption of chloroplast genes in *Chlamydomonas. Mol. Gen. Genet.* 230: 65–74.

Newman, S.M., Harris, E.H., Johnson, A.M., Boynton, J.E., and N.W. Gillham (1992). Nonrandom distribution of chloroplast recombination events in *Chlamydomonas reinhardtii:* Evidence for a hotspot and an adjacent cold region. *Genetics* 132: 413–429.

O'Neill, C., Horvath, G.V., Horvath, E., Dix, P.J., and P. Medgyesy (1993). Chloroplast transformation in plants: Polyethylene glycol (PEG) treatment of protoplasts is an alternative to biolistic delivery systems. *Plant J.* 3: 729–738.

Palmer, J.D. (1983). Chloroplast DNA exists in two orientations. *Nature (London)* 301: 92–93.

Palmer, J.D., Boynton, J.E., Gillham, N.W., and E.H. Harris (1985). Evolution and recombination of the large inverted repeat in *Chlamydomonas* chloroplast DNA. In *The Molecular Biology of the Photosynthetic Apparatus* (K.E. Steinback, S. Bonitz, C.J. Arntzen, and L. Bogorad, eds.), pp. 269–278. Cold Spring Harbor Press, Cold Spring Harbor, NY.

Przibilla, E., Heiss, S., Johanningmeier, U., and A. Trebst (1991). Site-specific mutagenesis of the D1 subunit of photosystem II in wild-type *Chlamydomonas. Plant Cell* 3: 169–174.

Putrament, A., Baranowska, H., and W. Prazmo (1973). Induction by manganese of mitochondrial antibiotic resistance mutations in yeast. *Mol. Gen. Genet.* 126: 357–366.

Randolph-Anderson, B.L., Boynton, J.E., Gillham, N.W., Harris, E.H., Johnson, A.M., Dorthu, M.-P., and R.F. Matagne (1992). Further characterization of the respiratory deficient *dum-1* mutation in *C. reinhardtii* and its use as a recipient for mitochondrial transformation. *Mol. Gen. Genet.* 236: 235–244.

Rizet, G. (1953). Sur l'impossibilité d'obtenir la multiplication végétative interrompue et illimitée de l'ascomycete *Podospora anserina. C.R. Acad. Sci. [D] (Paris)* 237: 838–840.

Robertson, D., Boynton, J.E., and N.W. Gillham (1990). Cotranscription of the wild-type chloroplast *atpE* gene encoding the CF_1/CF_0 epsilon subunit with the 3' half of the *rps7* gene in *Chlamydomonas reinhardtii* and characterization of frameshift mutations in *atpE. Mol. Gen. Genet.* 221: 155–163.

Rowlands, R.T., and G. Turner (1974). Recombination between extranuclear genes conferring oligomycin resistance and cold sensitivity in *Aspergillus nidulans. Mol. Gen. Genet.* 133: 151–161.

Sager, R. (1954). Mendelian and non-Mendelian inheritance of streptomycin resistance in *Chlamydomonas reinhardtii. Proc. Natl. Acad. Sci. U.S.A.* 40: 356–363.

Sager, R. (1972). *Cytoplasmic Genes and Organelles.* Academic Press, New York.

Sager, R. (1977). Genetic analysis of chloroplast DNA in *Chlamydomonas. Adv. Genet.* 19: 287–340.

Sager, R., and Z. Ramanis (1965). Recombination of nonchromosomal genes in *Chlamydomonas. Proc. Natl. Acad. Sci. U.S.A.* 53: 1053–1061.

Sager, R., and Z. Ramanis (1967). Biparental inheritance of nonchromosomal genes induced by ultraviolet irradiation. *Proc. Natl. Acad. Sci. U.S.A.* 58: 931–935.

Sager, R., and Z. Ramanis (1968). The pattern of segregation of cytoplasmic genes in *Chlamydomonas. Proc. Natl. Acad. Sci. U.S.A.* 61: 324–331.

Sager, R., and Z. Ramanis (1970). A genetic map of non-Mendelian genes in *Chlamydomonas. Proc. Natl. Acad. Sci. U.S.A.* 65: 593–600.

Sager, R., and Z. Ramanis (1976a). Chloroplast genetics of *Chlamydomonas.* I. Allelic segregation ratios. *Genetics* 83: 303–321.

Sager, R., and Z. Ramanis (1976b). Chloroplast genetics of *Chlamydomonas.* II. Mapping by cosegregation frequency analysis. *Genetics* 83: 323–340.

Sakamoto, W., Kindle, K.L., and D.B. Stern (1993). In vivo analysis of *Chlamydomonas petD* gene expression using stable transformation of β-glucuronidase translational fusions. *Proc. Natl. Acad. Sci. U.S.A.* 90: 497–501.

Sanford, J.C., Klein, T.M., Wolf, E.D., and N. Allen (1987). Delivery of substances into cells and tissues using a particle bombardment process. *Particulate Sci. Technol.* 5: 27–37.

Sears, B.B., and R.G. Herrmann (1985). Plastome mutation affecting the chloroplast ATP synthase

involves a post-transcriptional defect. *Curr. Genet.* 9: 521–528.
Shepherd, H.S., Boynton, J.E., and N.W. Gillham (1979). Mutations in nine chloroplast loci of *Chlamydomonas* affecting different photosynthetic functions. *Proc. Natl. Acad. Sci. U.S.A.* 76: 1353–1357.
Shoffner, J.M. IV, and D.C. Wallace (1990). Oxidative phosphorylation diseases. *Adv. Human Genet.* 19: 267–330.
Singer, B., Sager, R., and Z. Ramanis (1976). Chloroplast genetics of *Chlamydomonas* III. Closing the circle. *Genetics* 8: 341–354.
Slonimski, P.P., Pajot, P., Jacq, C., Foucher, M., Perrodin, G., Kochko, A., and A. Lamouroux (1978). Mosaic organization and expression of the mitochondrial DNA region controlling cytochrome *c* reductase and oxidase. I. Genetic, physical and complementation maps of the *box* region. In *Biochemistry and Genetics of Yeast: Pure and Applied Aspects* (M. Bacila, B.L. Horecker, and A.O.M. Stoppani eds.), pp. 339–368. Academic Press, New York.
Smith, S.E. (1988). Biparental inheritance of organelles and its implications in crop improvement. *Plant Breed. Rev.* 6: 361–393.
Staub, J.M., and P. Maliga (1992). Long regions of homologous DNA are incorporated into the tobacco plastid genome by transformation. *Plant Cell* 4: 39–45.
Staub, J., and P. Maliga (1993). Accumulation of D1 polypeptide in tobacco plastids is regulated *via* the untranslated region of *psbA* mRNA. *EMBO J.* 12: 601–606.
Svab, Z., and P. Maliga (1993). High-frequency plastid transformation in tobacco by selection for a chimeric *aadA* gene. *Proc. Natl. Acad. Sci. U.S.A.* 90: 913–917.
Svab, Z., Hajdukiewicz, P., and P. Maliga (1990). Stable transformation of plastids in higher plants. *Proc. Natl. Acad. Sci. U.S.A.* 87: 8526–8530.
Thanh, N.D., and P. Medgyesy (1989). Limited chloroplast gene transfer via recombination overcomes plastome-genome incompatibility between *Nicotiana tabacum* and *Solanum tuberosum*. *Plant Mol. Biol.* 12: 87–93.
Thomas, D.Y., and D. Wilkie (1968a). Inhibition of mitochondrial synthesis in yeast by erythromycin: Cytoplasmic and nuclear factors controlling resistance. *Genet. Res.* 11: 33–41.
Thomas, D.Y., and D. Wilkie (1968b). Recombination of mitochondrial drug resistance factors in *Saccharomyces cerevisiae*. *Biochem. Biophys. Res. Commun.* 30: 368–372.
Thorsness, P.E., and T.D. Fox (1990). Escape of DNA from mitochondria to the nucleus in *Saccharomyces cerevisiae*. *Nature (London)* 346: 376–379.
Turner, G., Earl, A.J., and D.R. Greaves (1982). Interspecies variation and recombination of mitochondrial DNA in *Aspergillus nidulans* species group and the selection of species-specific sequences by nuclear background. In *Mitochondrial Genes* (P.P. Slonimski, P. Borst, and G. Attardi, eds.), pp. 411–414. Cold Spring Harbor Press, Cold Spring Harbor, NY.
Tzagoloff, A. (1982). *Mitochondria*. Plenum Press, New York.
Tzagoloff, A., and C.L. Dieckmann (1990). *PET* genes of *Saccharomyces cerevisiae*. *Microbiol. Rev.* 54: 211–225.
Tzagoloff, A., Akai, A., and R.B. Needleman (1975a). Properties of cytoplasmic mutants of *Saccharomyces cerevisiae* with specific lesions in cytochrome oxidase. *Proc. Natl. Acad. Sci. U.S.A.* 72: 2054–2057.
Tzagoloff, A., Akai, A., and R.B. Needleman (1975b). Assembly of the mitochondrial membrane system: Isolation of nuclear and cytoplasmic mutants of *Saccharomyces cerevisiae* with specific defects in mitochondrial functions. *J. Bacteriol.* 122: 826–831.
VanWinkle-Swift, K.P. (1978). Uniparental inheritance is promoted by delayed division of the zygote in *Chlamydomonas*. *Nature (London)* 275: 749–751.
Wallace, D.C. (1986). Mitotic segregation of mitochondrial DNAs in human cell hybrids and expression of chloramphenicol resistance. *Somatic Cell Mol. Genet.* 12: 41–49.
Wallace, D.C., Oliver, N.A., Blanc, H., and C.W. Adams (1982). A system to study human mitochondrial genes: Applications to chloramphenicol resistance. In *Mitochondrial Genes* (P.P. Slonimski, P. Borst, and G. Attardi, eds.), pp. 105–116. Cold Spring Harbor Press, Cold Spring Harbor, NY.
Weiller, G., Schueller, C.M.E., and R.J. Schweyen (1989). Putative target sites for mobile G+C rich clusters in yeast mitochondrial DNA: Single elements in tandem arrays. *Mol. Gen. Genet.* 218: 272–283.
Wilkie, D., Saunders, G., and A.W. Linnane (1967). Inhibition of respiratory enzyme synthesis in yeast by chloramphenicol: Relationship between chloramphenicol tolerance and resistance to other antibacterial antibiotics. *Genet. Res. (Cambridge)* 10: 199–203.
Winter, P., and R.G. Herrmann (1988). A five-base-pair-deletion in the gene for the large subunit causes the lesion in the ribulose bisphosphate carboxylase/oxygenase-deficient plastome mutant sigma of *Oenothera hookeri*. *Bot. Acta* 101: 68–75.
Wiseman, A., Gillham, N.W., and J.E. Boynton (1977). Nuclear mutations affecting mitochondrial structure and function in *Chlamydomonas*. *J. Cell Biol.* 73: 56–77.
Woessner, J.P., Masson, A., Harris, E.H., Bennoun, P., Gillham, N.W., and J.E. Boynton (1984). Molecular and genetic analysis of the chloroplast ATPase of *Chlamydomonas*. *Plant Mol. Biol.* 3: 177–190.

Wolf, K. (1987). Mitochondrial genes of the fission yeast *Schizosaccharomyces pombe*. In *Gene Structure in Eukaryotic Microbes. SGM Special Publication 22* (J.R. Kinghorn, ed.), pp. 69–91. IRL Press, Oxford.

Wolf, K., and L. Del Giudice (1988). The variable mitochondrial genome of ascomycetes: Organization, mutational alterations and expression. *Adv. Genet.* 25: 185–308.

Wurtz, E.A., Boynton, J.E., and N.W. Gillham (1977). Perturbation of chloroplast DNA amounts and chloroplast gene transmission in *Chlamydomonas* by 5-fluorodeoxyuridine. *Proc. Natl. Acad. Sci. U.S.A.* 74: 4552–4556.

8

Transmission and Compatibility of Organelle Genomes

Maternal, paternal, or uniparental inheritance is a striking characteristic of chloroplast and mitochondrial genome transmission in most organisms. Evidently, uniparental transmission of organelles or organelle DNA evolved independently a number of times. Although uniparental transmission is the oldest observed characteristic of organelle genomes, the mechanisms that underly this transmission pattern are understood at a descriptive level in relatively few organisms and not at all in a molecular sense in any eukaryote. The same is true of the mechanisms determining compatibility between different nuclear and organellar genotypes

Transmission of chloroplast and mitochondrial genes in plants. The transmission patterns of chloroplast genes in crosses have been studied in many plants (see reviews by Gillham *et al.,* 1991; Kuroiwa, 1991; Sears, 1980; Smith, 1988; Tilney-Bassett, 1975; Whatley, 1982). In angiosperms, the pattern of plastid transmission has been documented for >50 genera of monocotyledons and dicotyledons. Although plastid transmission is often maternal, biparental transmission is not uncommon and has been observed in about 20 genera. In 11 of these, crosses of certain species regularly yield >5% of the progeny possessing paternal plastids (Smith, 1988). The potential for biparental transmission is even greater if the numerous additional species where plastid DNA can be found in generative or sperm cells of pollen are taken into account (Corriveau and Coleman, 1988).

Common to all angiosperms is the unequal mitotic division of the uninucleate pollen grain or microspore to yield a large vegetative cell and a much smaller generative cell (Sears, 1980). The generative cell undergoes a further division either before pollen is shed from the anthers or after pollination and germination. Two nonflagellated sperm cells result from mitotic division of the generative cell. They are completely surrounded by the vegetative cell cytoplasm of the pollen tube. Double fertilization results when one sperm nucleus fuses with the egg nucleus to form the zygote and the other fuses with the two female polar nuclei to yield the triploid endosperm.

On the basis of light and electron microscopic observations, Hagemann (1979, 1983, 1992) recognized four patterns of plastid transmission in angiosperms (Table 8–1). In the *Lycopersicon* type, exemplified by tomato (*L. esculentum),* all microspore

Table 8–1. Patterns of Plastid Transmission in Male Gametophytes, Gametes, and Crosses in Angiosperms

Transmission Pattern	Typical Species	Transmission of Male Plastids in Crosses
Lycopersicon type (male plastids excluded from generative cell)	*Mirabilis jalapa*	No
	Beta vulgaris	No
	Gossypium hirsutum	No
	Antirrhinum majus	Rare
	Petunia hybrida	Rare
	Lycopersicon esculentum	No
Solanum type (generative cell plastids degenerate)	*Solanum tuberosum*	Rare
Triticum type (sperm cell cytoplasm and plastids left outside egg)	*Triticum aestivum*	No
	Hordeum vulgare	No
Pelargonium type (male plastids transmitted to embryo)	*Pelargonium zonale*	Yes
	Oenothera spp.	Yes

From Gillham *et al.* (1991)

plastids are usually retained in the vegetative cell, and only chloroplasts from the maternal parent are transmitted to the progeny. However, paternal plastids are not always completely excluded in plants of the *Lycopersion* type. Thus, Cornu and Dulieu (1988) reported paternal transmission of chloroplasts in a single mutant line of *Petunia* (up to 2%). They also found that nuclear genes played a role in the process. The first partitioning of the cytoplasm during pollen mitosis is more equal in the *Solanum* (potato) type; thus the generative cell as well as the vegetative cell contains some plastids. Later on, the generative cell plastids are lost (or eliminated), so sperm cells normally do not contain plastids. In the *Triticum* (wheat) type, so far found only in grasses, plastids are found in the generative and sperm cells. When the sperm cell nucleus enters the egg cell, enucleated cell bodies containing mitochondria and chloroplasts are left outside (Mogensen, 1988). In the *Pelargonium* (geranium) type, also found in *Oenothera* (evening primrose), plastids are distributed to the generative and vegetative cells at the first pollen mitosis and biparental plastid transmission results.

This classification has been refined further depending on the presence or absence of nucleoids in generative and sperm cells (see Kuroiwa, 1991; Miyamura *et al.*, 1987). Based on organelle nucleoid behavior 18 plants studied were classified as belonging to one of three types. In the first, exemplified by *Nicotiana tabacum,* organelle nucleoids disappeared in the generative cells immediately after the first pollen mitosis. In the second, seen in *Triticum,* degeneration of organelle nucleoids took place between the second pollen mitosis and the initiation of sperm nucleus formation. In the third, typified by *Pelargonium,* organelle nucleoids were found in sperm cells and pollen tubes.

While these experiments cannot easily distinguish plastid and mitochondrial nucleoids, the loss of all nucleoids in the first two types of plants suggests that both species disappear and that actual elimination of organelle DNA may underlie the phenomenon of maternal inheritance. Experiments of this kind have been refined even further by coupling examination of nucleoids in generative cells to the use of specific probes for organelle DNA (Sodmergen *et al.*, 1992). In *Lillium longiflorum* organelle nucleoids are not visible in generative cells and they appear mostly degraded in vegetative cells as well. In contrast, there is no evidence for degradation of organelle nucleoids in *Pelargonium*. When total DNA from pollen and pollen tubes was probed with plastid- and mitochondrial-specific gene sequences, the results suggested these DNA species were largely degraded in *Lillium*, but not in *Pelargonium*.

The pattern of plastid transmission can also be influenced by nuclear genes. Thus, Tilney-Bassett (1988) proposed that the plastid transmission pattern in *Pelargonium* is under the control of a nuclear gene *(Pr)*. In Type I cultivars, which are homozygous for the *Pr1* allele, plastid transmission is predominantly maternal, but Type II cultivars, which are *Pr1Pr2* heterozygotes, yield nearly equal frequencies of maternal and paternal progeny with biparental offspring being rare. Tilney-Bassett *et al.* (1992) have since revised this model based on additional crosses. They suppose that there are actually two independently assorting nuclear genes each with a pair of alternative alleles (*Pr1/pr1* and *Pr2/pr2*). In this model all Type II segregants are single or double heterozygotes (e.g., *Pr1/pr1, Pr2/pr2; Pr1/pr1, Pr2/Pr2*) whereas Type I segregants are homozygotes (e.g., *Pr1/Pr1, pr2/pr2; pr/pr1, Pr2/Pr2*).

Transmission of mitochondria and mtDNA in angiosperms also appears to be generally maternal (Eckenrode and Levings, 1986; Kuroiwa, 1991; Sodmergen *et al.*, 1992). Furthermore, the maternal transmission of cytoplasmic male sterility, a mitochondrial trait in most plants (Chapter 11), has been documented in over 140 different plant species.

Conifers, in contrast to angiosperms, transmit their plastid DNA paternally. This transmission pattern has been demonstrated for several pines, Douglas fir, spruce, and redwood (see Gillham *et al.,* 1991; Hagemann, 1992; Neale and Sederoff, 1988). However, among six interspecific hybrids of two larch species, three showed strictly paternal inheritance of chloroplast RFLPs, while one was maternal and two were biparental for these physical markers (Szmidt *et al.,* 1987). The paternal pattern of cpDNA transmission seen in conifers is also reflected in light and electron microscopic investigations. Thus, Whatley (1982) reported that in conifers, the paternal plastids enter the egg and maternal plastids are often observed to degenerate. Maternal transmission of mtDNA is seen in interspecific crosses of *Pinus rigida* and *Pinus taeda*, but mtDNA is inherited paternally in redwood (Neale *et al.,* 1989). In *Chlamydomonas reinhardtii*, like *Pinus rigida*, the chloroplast and mitochondrial genomes are transmitted uniparentally by different parents (see *Transmission of organelle DNA in Chlamydomonas*).

Kuroiwa (1991) has reviewed extensive work done by his own group on transmission of nucleoids in *Chlamydomonas* and other green algae. In algae like *C. reinhardtii*, the coenocytic green alga *Dictyosphaeria cavernosa* and the giant unicellular green alga *Acetabularia calyculus* the gametes are isogamous and chloroplast nucleoids from one parent disappear following gametic and chloroplast fusion. During gametogenesis of the anisogamous alga *Bryopsis maxima* both the chloroplast and mitochondrial nucleoids from the male parent vanish. Likewise, in the oogamous alga *Chara corallina* chloroplast and mitochondrial nucleoids disappear during sperm formation. A similar scenario characterizes spermatogenesis in the fern *Pteris vittata*. In these latter three cases the data suggest that loss of male cpDNA occurs in two steps (Kuroiwa, 1991). The first step involves chloroplast division without cpDNA replication and the second leads to preferential degradation of cpDNA from the male. Finally, the motile spermatozoids of the fern *Marsilea* typically discard a cytoplasmic vesicle that includes the plastid(s) before the male gamete reaches the egg (see Sears, 1980 and Whatley, 1982, for reviews).

Transmission of mitochondrial DNA in animals and the question of paternal leakage. Mitochondrial DNA in most animals is thought generally to be maternally inherited (see Avise and Lansman, 1983, for a review). To test this hypothesis Lansman *et al.* (1983) performed a "critical experimental test." They analyzed 45- and 91-generation backcross progeny from interspecific matings of two moth species of the genus *Heliothis*. The notion underlying these experiments is that if the male parent contributes even small proportions of mtDNA molecules to the progeny, the repeated backcrossing scheme should enrich the total paternal mtDNA present in backcross hybrids. The sensitivity of the analysis was at the level of <1 molecule in 500. The experiment failed to detect any paternal mtDNA and set an upper limit of "paternal leakage" at about one molecule per 25,000 per generation in *Heliothis*. Similar studies in which two species of mice were backcrossed over 6–8 generations also indicated strict maternal inheritance with the level of resolution for paternal leakage being no more than 1/1000 maternal mtDNA molecules contributed (Gyllensten *et al.,* 1985). However, with the advent of the polymerase chain reaction, a much more sensitive system for the detection of rare molecules became available. Gyllensten *et al.* (1991) took advantage of this system to demonstrate that paternal leakage does occur after all in mice. The number of paternal mtDNA molecules that can enter the egg cytoplasm and persist is 1–4 per 100,000 maternal mtDNA molecules per generation. These extremely low levels of paternal leakage would not have been detected in the earlier experiments. The demonstration of paternal leakage by Gyllensten *et al.* (1991) also has potential evolutionary implications. The consequence in animal lineages where leakage occurs will be to increase the effective population size and to reduce estimates of the time when the most recent common ancestor existed relative to lineages assuming strict maternal inheritance.

In mussels *(Mytilus)* the inheritance of mtDNA appears to be biparental rather than strictly maternal (Hoeh *et al.,* 1991). Analyses of 150 individual *Mytilus* mtDNAs from 16 populations showed an unusually high frequency of heteroplasmy with 85 individuals being heteroplasmic (72 with two, 11 with three, and two with four different mtDNA types). Cytological studies of fertilization in *Mytilus* indicate that mitochondria are neither excluded from the egg nor degraded inside it, as they are in most other animals. Instead, they appear to mix and become morphologically indistinguishable from maternal mitochondria. The inference of biparental inheritance of mtDNA in bivalves seemingly extends to the freshwater bivalve *Anodonta* (Hoeh *et al.,* 1991) and perhaps other molluscan taxa as well. Zouros *et al.* (1992) have now demonstrated extensive contribution of paternal mtDNA in pair matings of *M. edulis* and *M. trossulus*. They estimate a paternal mtDNA transmission rate at least $1000\times$ greater than in mice. Limited biparental transmission of paternal mtDNA has also been observed in interspecific, but not intraspecific, crosses of *Drosophila* suggesting that biparental inheritance may be a phenomenon unique to interspecific hybrids in this genus (Kondo *et al.,* 1990). This explanation might also apply to *Mytilus* since the heteroplasmic individuals could be hybrids of *M. edulis* and *M. trossulus* (Hoeh *et al.,* 1991; Zouros *et al.,* 1992). In this regard, it is worth noting that biparental versus maternal plastids inheritance of *Pelargonium* is determined by a relatively simple two-gene, four-allele system.

Transmission of mtDNA in Paramecium, filamentous fungi, and slime molds. Gillham (1978) reviewed the mitochondrial genetics of *Paramecium*. In this ciliate, parental animals of different mating type conjugate. The two diploid micronuclei then undergo meiosis yielding eight meiotic products per parent, of which one survives in each parent. The polyploid macronucleus (somatic nucleus) subsequently disintegrates. Both of the surviving micronuclei divide once mitotically and each parent then donates a haploid micronucleus to its partner. These fuse to form diploid micronuclei that divide mitotically and also give rise to new macronuclei. Mitochondria are not normally transferred during this process although mitochondrial exchange can be effected by inducing formation of a cytoplasmic bridge as can be done with a dilute antiserum.

Fungal mitochondrial genomes usually exhibit uniparental–maternal transmission, but paternal inheritance has been demonstrated in the genus *Allomyces* (see Taylor, 1986). The definition of uniparental–maternal can depend more on morphology as opposed to actual sex or mating type. In *Neurospora crassa*, for example, there are two mating types, *A* and *a*. Mycelia of either mating type elaborate fruiting bodies (protoperithecia) on appropriate medium. After protoperithecia are produced, the sexual cycle is initiated by fertilization of the protoperithecia with conidia or mycelial fragments of opposite mating type. Manella *et al.* (1979) demonstrated strict uniparental inheritance of mtDNA from the protoperithecial parent. Thus, mtDNA transmission in *Neurospora* is independent of mating type, but is "maternal" in that the conidium behaves as a nuclear donor roughly equivalent to a sperm and the protoperithecium as the recipient analogous to an egg. Although mtDNA is uniparentally inherited in *Neurospora*, different combinations of mitochondrial and nuclear genomes are readily formed in heterokaryons where cytoplasmic mixing occurs, but nuclear fusion does not (see Chapter 11).

The situation in *Neurospora* differs from that in the basidiomycete *Coprinus cinereus* (May and Taylor, 1988). In this species, as in many agaric basidiomycetes, two haploid, monokaryotic colonies possessing compatible mating factors meet and after cell fusion exchange parental nuclei by a process of mitosis and nuclear migration. A dikaryon with two parental nuclei per hyphal segment results. However, mitochondria do not migrate so each parental mycelium retains its own resident mitochondrial genome. A mitochondrial mosaic results (i.e., the mycelium divides into sectors differing in mtDNA genotype).

Transmission of mtDNA has been examined in two slime molds. Mirfakhrai *et al.* (1990) used restriction fragment-length polymorphisms (RFLPs) to monitor the inheritance of mtDNA in the cellular slime mold *Polysphondylium pallidum*. In sexually competent species of slime molds such as *P. pallidum* two cells of opposite mating type in a cell aggregate fuse and form a diploid. The diploid then engulfs the peripheral cells in the aggregate, following which thick walls form around the resulting young macrocyst. The macrocyst, a true zygote, matures and undergoes meiosis, yielding haploid progeny that are released on breakage of the macrocyst walls during germination. When opposite mating types (*mat1* and *mat2*) of closely related strains of *P. pallidum* were crossed, the haploid progeny transmitted mtDNA from the *mat2* parent only regardless of their mating type. However, when more distantly related strains were crossed some progeny also transmitted mtDNA from the *mat1* parent. Thus, unlike *Neurospora*, but like *Chlamydomonas*, organelle genome transmission in *P. pallidum* is mating-type dependent.

A more complex situation obtains in the acellular slime mold *Physarum polycephalum* (Kawano and Kuroiwa, 1989). In *P. polycephalum* haploid isogametes fuse in pairs to form diploid zygotes. These develop into macroscopic, diploid plasmodia by successive mitotic cycles in the absence of cell division. Crossing is under the control of a mating type system, which comprises three unlinked loci: *matA, matB,* and *matC.* For efficient crossing, amoebae must carry different alleles at *matA* and *matB.* Heteroallelism at *matC* is also necessary when crosses are carried out under conditions of elevated pH or reduced ionic strength. The *matA* locus regulates the development of zygotes into plasmodia while *matB* and *matC* regulate zygote formation by influencing the probability that amoebae will fuse. Kawano and Kuroiwa (1989) found that transmission of mtDNA was uniparental in *P. polycephalum.* Their results also suggested that alleles at the *matA* locus regulated the transmission of mtDNA. The different *matA* alleles appear to operate in a hierarchy with *matA2* > *matA11* > *matA12* > *matA1* in determining mtDNA transmission. To complicate matters further, Kawano et al. (1991) report discovery of a 16-kb mitochondrial plasmid encoding at least one gene that promotes mitochondrial fusion in an mF^+ (for mitochondrial fusion) strain of *P. polycephalum.* In matings of the mF^+ and mF^- strains, transmission of mtDNA does not exhibit a simple uniparental inheritance pattern (Takano et al., 1992). Free plasmid is transmitted to the mitochondria regardless of the mtDNA transmission pattern and the plasmid DNA can also integrate by recombination into a specific region of sequence homology in the mF^- mitochondrial genome.

Transmission of organelle DNA in Chlamydomonas. Although the mechanisms underlying transmission of the chloroplast and mitochondrial genomes of *C. reinhardtii* have been the subject of much study and speculation, they are far from being understood (see reviews by Gillham et al., 1991; Harris, 1989; Kuroiwa, 1991; Sager and Grabowy, 1985).

Goodenough and colleagues (Galloway and Goodenough, 1985; Goodenough and Ferris, 1987) proposed a genetic model that explains the behavior of mating-type linked mutations affecting the mating process. The model also suggests how genes linked to mating type might control uniparental inheritance. This model was modified by Gillham et al. (1991) to emphasize the roles played by genes linked to mating type in the transmission of chloroplast and mitochondrial genomes (Fig. 8–1). The model proposes that gametogenesis triggered by nitrogen starvation results in the expression of specific genes whose products are required for protection of mt^+ cpDNA *(cpp)* and mt^- mtDNA *(mitp).* The *cpp* and *mitp* genes are linked, respectively, to the mt^+ and mt^- alleles. On fusion of cells of opposite mating type, products encoded by the *Z1* (linked to mt^+) and *Z2* (linked to mt^-) genes either form a complex or act sequentially to initiate the pathway leading to zygote formation. This triggers either activation or synthesis of proteins coded by the *cpd* and *mitd* genes required, respectively, for elimination of mt^- cpDNA and mt^+ mtDNA. The *cpd* gene is linked to mt^+ and *mitd* is linked to mt^-.

Although mutations affecting only one of the putative mating type linked genes *(cpd)* have so far been identified, the model is consistent with most observations. The predictions and the evidence are as follows:

1. The model predicts that mating type should actually be a chromosome region where recombination is restricted. Each mating type should have its own characteristic linked subset of genes. At least seven mating type-linked genes are known in *C.*

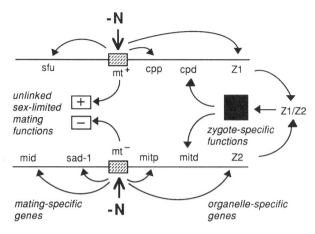

Fig. 8–1. Model for control of chloroplast and mitochondrial DNA transmission by the mating type locus in *Chlamydomonas reinhardtii*. Starvation for nitrogen (–N) triggers the differentiation of vegetative cells into gametes. The expression of the mating type alleles is postulated to result in the activation of several genes that are completely linked to mating type. They include mating-specific genes and genes specific for the uniparental transmission of the chloroplast and mitochondrial genomes by opposite mating types. The organelle-specific genes hypothesized to be activated in the gametes are *cpp*, linked to mt^+ and required for protection of cpDNA from this parent and *mitp*, linked to mt^- and necessary for protection of mtDNA from the mt^- parent. The zygotic part of the cycle is initiated following fusion of the gametic pair by the interaction of products of *Z1*, a gene linked to mt^+, and *Z2*, a gene linked to mt^-. This interaction, either directly or indirectly, results in activation of *cpd*, a gene whose product is required for elimination of mt^- cpDNA, and *mitd*, whose product is required for destruction of mt^+ mtDNA. Failure of the *Z1* and *Z2* products to interact results in vegetative diploid formation and biparental transmission of chloroplast and mitochondrial genes. From Gillham *et al*. (1991) as modified from Galloway and Goodenough (1985).

reinhardtii (Gillham *et al.*, 1991; Goodenough and Ferris, 1987; Harris, 1989). Some of these are marked by auxotrophic mutations; mutations at three affect mating functions carried out by gametes of one mating type or the other; and mutations at the *mat-3* locus allow cpDNA to be transmitted from the mt^+ parent, which would be expected for mutations in the *cpd* gene.

2. Failure of a mating pair to synthesize either the *Z1* or *Z2* products should interrupt progression into the meiotic zygote cycle and result in vegetative diploid formation. A corollary expectation is that the genes whose products are required for the destruction of organelle genomes from opposite parents should not be activated. Actual observations are consistent with both predictions. In vegetative diploids strict uniparental inheritance of organelle genomes no longer is obtained. Instead, vegetative diploids frequently exhibit biparental inheritance of organelle genomes and sometimes uniparental inheritance from the normally nontransmitting mt^- parent.

3. The cpDNA and mtDNA from the mt^+ and mt^- parents, respectively, should be destroyed in the zygote. This has been demonstrated for the respective DNAs (Boynton *et al.*, 1987) and for chloroplast nucleoids by Kuroiwa and colleagues (see Gill-

ham et al., 1991; Kuroiwa, 1991). Kuroiwa et al. found that chloroplast nucleoids from the mt^- parent disappear completely within 40 min after gamete fusion and apparently prior to fusion of both chloroplasts and nuclei. The remaining nucleoids fused to form a single large structure adjacent to the starch-containing pyrenoid. The disappearance of mt^- nucleoids does not result from preferential dispersion of these structures, making them difficult to visualize (Kuroiwa, 1991). Furthermore, in interspecific crosses of *C. reinhardtii* and *C. smithii*, in which cpDNA RFLPs can be scored, cpDNA from the mt^- largely disappears within 4 hr (Rosen et al., 1991).

Loss of mt^- nucleoids can be inhibited by UV irradiation of the mt^+ gametes prior to mating, a treatment that also increases the frequency of biparental zygotes. Incubation of young zygotes in the presence of inhibitors of RNA (actinomycin D) and protein (cycloheximide) synthesis also inhibits loss of mt^- nucleoids. These results are consistent with the hypothesis that a gene product required for destruction of cpDNA from the mt^+ parent is made in the young zygote. The expression of this gene or the proper functioning of its product would then be UV sensitive. In addition, Nakamura et al. (1988) found that the synthesis of six new polypeptides in young zygotes was inhibited if the mt^+, but not the mt^- parent was UV irradiated prior to mating. Unfortunately, these promising findings are countered by the observation that synthesis of these polypeptides is not blocked by growth and gametogenesis of mt^+ cells in 0.5 mM 5-fluorodeoxyuridine-containing medium (Nakamura and Kuroiwa, 1989). Under these conditions 70% of the mt^+ gametes lacked chloroplast nucleoids and destruction of mt^- nucleoids took place. However, This treatment has no substantial effect on viability, but results in the largely uniparental transmission of chloroplast genomes by the mt^- parent (Wurtz et al., 1977). Hence, synthesis of the six polypeptides in 5-fluorodeoxyuridine-grown cells would not appear necessarily to relate to elimination of mt^- chloroplast genomes.

Experiments with a homothallic species of *Chlamydomonas*, *C. monoica*, which is more closely related to the *C. eugametos/moewusii* species pair than to *C. reinhardtii*, provide further insight into the control of uniparental inheritance of chloroplast genes. VanWinkle-Swift and Aubert (1983) obtained genetic evidence compatible with a model that supposes mating type plays a direct role in controlling transmission of chloroplast genes. Their data were consistent with the notion that when mating type switches from mt^+ to mt^-, a cell loses its ability to transmit chloroplast genes and vice versa. VanWinkle-Swift and Hahn (1986) reported isolation of a mutant *(mt1-1)* that seems to be unable to protect its own cpDNA from destruction (VanWinkle-Swift and Salinger, 1988). In the mt^- mating type, this mutation has no effect and half of the chloroplast nucleoids disappear. When the switch to mt^+ occurs, *mt1-1* cannot protect its cpDNA from destruction, all nucleoids are seen to vanish from the zygote, and this condition is lethal.

There has been much speculation concerning the actual mechanisms that effect uniparental transmission of cpDNA in *Chlamydomonas reinhardtii*. This subject has been comprehensively reviewed by Gillham et al. (1991), Kuroiwa (1991), Sager et al. (1984), and Sager and Grabowy (1985), so only a brief recapitulation is required here. Sager and her colleagues proposed a restriction-modification model by analogy to bacteria. Chloroplast DNA from the mt^+ parent would be modified to protect against restriction while the unmodified cpDNA from the mt^- parent would be destroyed by a special restriction enzyme. Proof of the model requires identification of the posited activities.

Sager and colleagues reported that extensive methylation of cytosine residues occurred in cpDNA from mt^+, but not mt^- gametes and that a unique methyltransferase activity was present in gametes of the former mating type. However, other observations cast doubt on the significance of this extensive methylation. A nuclear mutant, *me-1*, was found that constitutively methylates >35% of the cytosine residues in cpDNA of either mating type, but does not alter the pattern of cpDNA transmission (Bolen *et al.*, 1982). Conversely, exposure of mt^+ cells undergoing gametogenesis to methylation inhibitors such as 5-azacytidine (azaC) greatly reduced methylation of cpDNA (Feng and Chiang, 1984). However, these hypomethylated mt^+ gametes transmitted their chloroplast genomes in the normal uniparental fashion. Sager and Grabowy (1985) subsequently argued that an additional round of methylation occurs in *me-1* gametes of the mt^+, but not the mt^- mating type. They also suggested that azaC may prevent the activation of the restriction enzyme required for cpDNA destruction so that unmodified cpDNA could be transmitted in the zygote. Despite these caveats the proposed correlation between extensive methylation and protection of mt^+ cpDNA from destruction seems more apparent than real, so the extensive methylation of mt^+ cpDNA very likely serves some other function related to the *Chlamydomonas* sexual cycle.

By observing the pattern of degradation of chloroplast nucleoids, Kuroiwa and colleagues (see Kuroiwa, 1991) came to the conclusion that DNases such as DNaseI rather than a restriction enzyme were involved in elimination of mt^- cpDNA in the zygote. They found that "digestion" of the nucleoids began synchronously within the chloroplast, that each nucleoid seemed to dissolve from its periphery toward the center, and that the whole process took about 20 min. The second observation was the key as a restriction endonuclease would not be expected to degrade the DNA, but simply make cuts in it. Therefore, chloroplast nucleoids might swell slightly, but would not be expected to disappear rapidly. Consistent with the digestion model Kuroiwa and colleagues identified a Ca^{2+} enzyme fraction (nuclease C) in gametes and young zygotes that degrades rather than restricts cpDNA. Six polypeptides have been identified in the nuclease C fraction. However, the simple hypothesis that one or more of these polypeptides is specific for mt^- cpDNA degradation is rendered unlikely by the observation of nuclease C in gametes of both mating types and vegetative cells (Kuroiwa, 1991).

A potential breakthrough in the identification of the nuclease thought to be responsible for elimination of mt^- cpDNA is the discovery of an acidic polypeptide in zygotes that binds to chloroplast nucleoids (Armbrust *et al.*, 1993). This polypeptide is encoded by the zygote-specific gene cluster *ezy-1*, which is linked to mating type and expressed almost immediately upon zygote formation. Expression of *ezy-1* is also selectively inhibited when mt^+, but not mt^- gametes are UV-irradiated. Thus, the *ezy-1* polypeptide has several of the characteristics expected for the long sought after nuclease and biochemical studies of this protein are to be awaited eagerly. Since *ezy-1* gene clusters are found linked to both mating type alleles, it seems unlikely that *ezy-1* and *mat-3* are identical. One possibility is that *mat-3* either directly or indirectly controls the expression of the *ezy-1* gene clusters.

WHY ARE ORGANELLE GENOMES TRANSMITTED UNIPARENTALLY?

Why are organelle genomes so often uniparentally transmitted? Why has this inheritance pattern apparently evolved several different times? Nobody really knows, but one hypothesis is that organelles compete against each other for transmission to future generations. Hence, uniparental transmission is the end result of "intragenomic conflict" between organelle genomes of maternal and paternal origin (see Cosmides and Tooby, 1981 and Werren et al., 1988 for discussions). The hypothesis of intragenomic conflict has been propelled to the forefront in tow important recent papers (Hurst, 1992; Hurst and Hamilton, 1992). Such a notion is bound to attract attention (Anderson, 1992). The basic idea is that after the evolution of cytoplasmic fusion of gametes, but prior to the emergence of mating types, a conflict arose between maternally and paternally derived organelle genes or genomes. The chloroplast or mitochondrial genes from one parent would attempt to annihilate those derived from the other. A nuclear control gene would then have evolved that left its associated organelle genome(s) susceptible to destruction. This gene would have spread through the population because it would prevent conflict between organelle genomes from different cells. Obviously, the gene could not spread forever or all organanelle genomes would have been destroyed. Therefore, genes evolved to establish binary mating types as in *Chlamydomonas* such that the organelle genome(s) of only one parent are destroyed.

COMPATIBILITY BETWEEN ORGANELLE AND NUCLEAR GENOMES

Compatibility of different combinations of organelle and nuclear genomes can be studied only in organisms where biparental transmission of organelles occurs or in cases where organelles from one species can be introduced artificially into the cytoplasm of another. The latter approach was used in three closely related species of the *Paramecium aurelia* complex that, for convenience, will be designated 1, 5, and 7 (see Gillham, 1978, for review). In these experiments donor mitochondria carrying an erythromycin-resistance marker were injected into erythromycin-sensitive cells following which the injected animals were selected for erythromycin resistance.

The major findings from these experiments were that mitochondria from species 1 and 5 could be readily transferred to any of the three species, but species 7 mitochondria could not be successfully transferred to species 1 or 5. After species 5 mitochondria had been established in species 7, they could not be reestablished in species 5 and only rarely in species 1. Similar results were obtained following transfer and selection of species 1 mitochondria in species 7. Therefore, when species 1 or 5 mitochondria colonized species 7, they acquired the compatibility characteristics of species 7. Similar but less striking results were observed in transfers from species 1 to species 5 and the reverse. Thus, species 1 mitochondria established in species 5 could be transferred to species 5 cells more easily than mitochondria from species 1. These experiments provide evidence for nuclear control of mitochondrial compatibility in *Paramecium*. Other results show that nuclear-mitochondrial incompatibility reactions can arise from an event as simple as a single gene mutation. Thus, in *P. tetraurelia* cells homozygous for the nuclear mutation *cl-1* are incompatible with wild-type mitochondria, but compatible with a different mitochondrial phenotype designated M^{cl}.

Formation of interspecific cybrids between plant cells in tissue culture followed by regeneration of whole plants provides another method for studying compatibility of organelle and nuclear genomes. Medgyesy (1990) discusses the question of whether there is a taxonomic limit to chloroplast transfer. In the family Solanaceae, intergeneric and intertribal cybrids have been produced in the *Nicotiana–Petunia, Nicotiana–Atropa,* and *Nicotiana–Salpiglossis* combinations. The plants derived from these cybrids show apparently normal greening and morphology although in the *Nicotiana–Atropa* cybrid, chloroplast and nucleus seem not to cooperate fully. This cybrid exhibits a marked reduction in amount of a thylakoid polypeptide, which is very abundant in the two parental species. Green cybrids could not be selected in the *Nicotiana–Solanum* combination. However, one of the two flowering plant cpDNA recombinants so far reported was isolated following fusion of irradiated protoplasts of *Solanum* having normal plastids with a light-sensitive plastome mutant of *N. tabacum* (Thanh and Medgyesy, 1989). Normal green *N. tabacum* plants resulted from this chloroplast recombinant and the "potacco" plastome proved stable during protoplast fusions and backcrosses. Thus, compatibility problems were overcome in this instance as the result of cpDNA recombination. These results are also of interest from the viewpoint of phylogeny since a recently published cpDNA phylogeny of the Solanaceae places *Atropa* and *Solanum* closer to *Nicotiana* than *Nicotiana* is to *Salpiglossis* (Olmstead and Palmer, 1992). Hence compatibility between chloroplast and nucleus was actually greatest for the phylogenetically least related cybrid *(Nicotiana–Salpiglossis).*

Biparental plastid transmission occurs in *Oenothera* and plastid compatibility and competition in crosses of these plants have been extensively reviewed elsewhere (Gillham, 1978; Gillham *et al.,* 1991; Hagemann, 1992; Kirk and Tilney-Bassett, 1978; Kutzelnigg and Stubbe, 1974). The seven pairs of chromosomes in *Oenothera* species are linked together in one or more chains or rings at the first meiotic division. This is because a series of subterminal reciprocal translocations prevent independent assortment and restrict recombination to the chromosome ends. Thus, nuclear genomes are, in essence, transmitted as units, making possible the study of defined plastome–genome interactions. From a study of plastome–genome interactions among 400 different combinations of the 10 North American and 14 European species in the subgenus *Euoenothera,* Stubbe (1964) concluded that there were five different plastid types in the subgenus and three basic haploid nuclear genomes that could be arranged in six diploid combinations. The compatibility responses between these plastid and nuclear genome combinations ranged from normal to lethal. Plastome IV was determined to be the most primitive since it was fully compatible with all nuclear combinations except one, with which it was partially compatible. The three haploid genomes defined on the basis of genome–plastome compatibility correspond to the three main species groups of *Euenothera* recognized by Cleland (1972) on taxonomic grounds. The five plastomes postulated by Stubbe are equivalent to five physically distinguishable chloroplast genomes based on restriction enzyme mapping (Gordon *et al.,* 1981, 1982). Plastomes I, III, and V exhibit the greatest differences. Plastome II is intermediate between plastomes IV and I as Stubbe proposed.

Since the physical differences that distinguish the five *Oenothera* plastome types are length mutations (>100 bp), they probably reside in intergenic regions. To test this hypothesis Wolfson *et al.* (1991) made sequence comparisons between the intergenic regions separating the ribosomal protein genes *rp116, rp114,* and *rps8* in four of the five plastome types. Sequence changes found include base pair substitutions, a 29-bp tandem duplication, and variation in the length of two poly(A) stretches. They concluded that plastomes I and II had a common progenitor as did plastomes III and IV. Wolfson *et al.* (1991) propose that a process of replication slippage may account for the origin of the 29-bp tandem duplication and other small deletions and insertions distinguishing different plastomes in *Oenothera.*

REFERENCES

Anderson, A. (1992). The evolution of sexes. *Science* 257: 324–326.

Armbrust, E.V., Ferris, P.J., and U.W. Goodenough (1993). A mating type-linked gene cluster expressed in Chlamydomonas zygotes participates in the uniparental inheritance of the chloroplast genome. *Cell* 74:801–811.

Avise, J.C., and R.A. Lansman (1983). Polymorphism of mitochondrial DNA in populations of higher animals. In *Evolution of Genes and Proteins* (M. Nei and R.K. Koehn, eds.), pp. 147–164. Sinauer Associates, Sunderland, MA.

Bolen, P.L., Grant, D.M., Swinton, D., Boynton, J.E., and N.W. Gillham (1982). Extensive methylation of chloroplast DNA by a nuclear gene mutation does not affect chloroplast gene transmission in *Chlamydomonas*. *Cell* 28: 335–343.

Boynton, J.E., Harris, E.H., Burkhart, B.D., Lamerson, P.M., and N.W. Gillham (1987). Transmission of mitochondrial and chloroplast genomes in crosses of *Chlamydomonas*. *Proc. Natl. Acad. Sci. USA* 84:2391–2395.

Cleland, R.E. (1972). *Oenothera: Cytogenetics and Evolution.* Academic Press, New York.

Cornu, A., and Dulieu (1988). Pollen transmission of plastid-DNA under genotype control in *Petunia hybrida* Hort. *J. Hered.* 79: 40–44.

Corriveau, J.S., and A.W. Coleman (1988). Rapid screening method to detect potential biparental inheritance of plastid DNA and results for over 200 angiosperm species. *Am. J. Bot.* 75: 1443–1458.

Cosmides, L.M., and J. Tooby (1981). Cytoplasmic inheritance and intragenomic conflict. *J. Theoret. Biol.* 89: 83–129.

Eckenrode, V.K., and C.S. Levings III (1986). Maize mitochondrial genes. *In vitro Cell. Dev. Biol.* 22: 169–176.

Feng, T.-Y., and K.-S. Chiang (1984). The persistence of maternal inheritance in *Chlamydomonas* despite hypomethylation of chloroplast DNA induced by inhibitors. *Proc. Natl. Acad. Sci. USA* 81: 3438–3442.

Galloway, R.E., and U.W. Goodenough (1985). Genetic analysis of mating locus linked mutations in *Chlamydomonas reinhardtii*. *Genetics* 111: 447–461.

Gillham, N.W. (1978). *Organelle Heredity.* Raven Press, New York.

Gillham, N.W., Boynton, J.E., and E.H. Harris (1991). Transmission of plastid genes. In *The Molecular Biology of Plastids* (L. Bogorad and I.K. Vasil, eds.), pp. 55–92. Academic Press, San Diego.

Goodenough, U.W., and P.J. Ferris (1987). Genetic regulation of development in *Chlamydomonas*. In *Genetic Regulation of Development* (W. Loomis, ed.), pp. 171–189. Alan R. Liss, New York.

Gordon, K.H.J., Crouse, E.J., Bohnert, H.J., and R.G. Herrmann (1981). Restriction endonuclease cleavage site map of chloroplast DNA from *Oenothera parviflora* (*Euoenothera* Plastome IV). *Theor. Appl. Genet.* 59: 281–296.

Gordon, K.H.J., Crouse, E.J., Bohnert, H.J., and R.G. Herrmann (1982). Physical mapping of differences in chloroplast DNA of the five wild-type plastomes in *Oenothera* subsection *Euoenothera*. *Theor. Appl. Genet.* 61: 373–384.

Gyllensten, U., Wharton, D., and A.C. Wilson (1985). Maternal inheritance of mitochondrial DNA during backcrossing of two species of mice. *J. Hered.* 76: 321–324.

Gyllensten, U., Wharton, D., Josefsson, A., and A.C. Wilson (1991). Paternal inheritance of mitochondrial DNA in mice. *Nature (London)* 352: 255–257.

Hagemann, R. (1979). Genetics and molecular biology of plastids of higher plants. *Stadler Symp.* 11: 91–115.

Hagemann, R. (1983). The formation of generative and sperm cells without plastids in angiosperms and the underlying mechanisms. In *Fertilization and Embryogenesis in Ovulated Plants* (O. Erdelska, ed.), pp. 97–99. Bratislava Slov. Acad. Sci., Bratislava.

Hagemann, R. (1992). Plastid genetics in higher plants. In *Plant Gene Research: Cell Organelles* (R. Herrmann, ed.), pp. 66–96. Springer-Verlag, Wien, New York.

Harris, E.H. (1989). *The Chlamydomonas Sourcebook.* Academic Press, San Diego.

Hoeh, W.R., Blakley, K.H., and Brown, W.M. (1991). Heteroplasmy suggests limited biparental inheritance of *Mytilus* mitochondrial DNA. *Science* 251: 1488–1490.

Hurst, L.D. (1992). Intragenomic conflict as an evolutionary force. *Proc. R. Soc. London* B, 248: 135–140.

Hurst, L.D., and W.D. Hamilton (1992). Cytoplasmic fusion and the nature of sexes. *Proc. R. Soc. London* B, 247: 189–194.

Kawano, S., and T. Kuroiwa (1989). Transmission patterns of mitochondrial DNA during plasmodium formation in *Physarum polycephalum*. *J. Gen. Microbiol.* 135: 1559–1566.

Kawano, S., Takano, H., Mori, K., and T. Kuroiwa (1991). A mitochondrial plasmid that promotes mitochondrial fusion in *Physarum polycephalum*. *Protoplasma* 160: 167–189.

Kirk, J.T.O., and R.A.E. Tilney-Bassett (1978). *The Plastids: Their Chemistry, Structure, Growth and Inheritance,* 2nd ed. Elsevier/North Holland, Amsterdam.

Kondo, R., Satta, Y., Matsuura, E.T., Ishiwa, H., Takahata, N., and S.I. Chigusa (1990). Incomplete maternal transmission of mitochondrial DNA in *Drosophila*. *Genetics* 126: 657–663.

Kuroiwa, T. (1991). The replication, differentiation, and inheritance of plastids with emphasis on the concept of organelle nuclei. *Int. Rev. Cytol.* 128: 1–62.

Kutzelnigg, H., and W. Stubbe (1974). Investigations on plastome mutants in Oenothera. *Sub-Cell. Biochem.* 3: 73–89.

Lansman, R.A., Avise, J.C., and M.D. Huettel (1983). Critical experimental test of the possibility of "paternal leakage" of mitochondrial DNA. *Proc. Natl. Acad. Sci. U.S.A.* 80: 1969–1971.

Manella, C.A., Pittenger, T. H., and A.M. Lambowitz (1979). Transmission of mitochondrial deoxyribonucleic acid in *Neurospora crassa* sexual crosses. *J. Bacteriol.* 137: 1449–1451.

May, G., and J.W. Taylor (1988). Patterns of mating and mitochondrial DNA inheritance in the agaric basidiomycete *Coprinus cinereus. Genetics* 118: 213–220.

Medgyesy, P. (1990). Selection and analysis of cytoplasmic hybrids. In *Plant Cell Line Selection* (P.J. Dix, ed.), pp. 287–315. VCH, Weinheim.

Mirfakhrai, M., Tanaka, Y., and K. Yanagisawa (1990). Evidence for mitochondrial DNA polymorphism and uniparental inheritance in the cellular slime mold *Polysphondylium pallidum:* Effect of intraspecies mating on mitochondrial DNA transmission. *Genetics* 124: 607–613.

Miyamura, S., Kuroiwa, T., and T. Nagata (1987). Disappearance of plastid and mitochondrial nucleoids during the formation of generative cells of higher plants revealed by fluorescence microscopy. *Protoplasma* 141: 149–159.

Mogensen, H.L. (1988). Exclusion of male mitochondria and plastids during syngamy in barley as a basis for maternal inheritance. *Proc. Natl. Acad. Sci. U.S.A.* 85: 2594–2597.

Nakamura, S., and T. Kuroiwa (1989). Selective elimination of chloroplast DNA by 5-fluorodeoxyuridine causing no effect on preferential digestion of male chloroplast nucleoids in *Chlamydomonas. Eur. J. Cell Biol.* 48: 165–173.

Nakamura, S., Sato, C., and T. Kuroiwa (1988). Polypeptides related to preferential digestion of male chloroplast nucleoids in *Chlamydomonas. Plant Sci.* 56: 129–136.

Neale, D.B., and R.R. Sederoff (1988). Inheritance and evolution of conifer organelle genomes. In *Genetic Manipulation of Woody Plants* (J.W. Hanover and E.E. Keathley, eds.), pp. 252–264. Plenum Press, New York.

Neale, D.B., Marshall, K.A., and R.R. Sederoff (1989). Chloroplast and mitochondrial DNA are paternally inherited in *Sequoia sempervirens* D. Don Endl. *Proc. Natl. Acad. Sci. U.S.A.* 86: 9347–9349.

Olmstead, R.G., and J.D. Palmer (1992). A chloroplast DNA phylogeny of the Solanaceae: Subfamilial relationships and character evolution. *Ann. Missouri Bot. Gard.* 79: 346–360.

Rosen, H., Newman, S.M., Boynton, J.E., and N.W. Gillham (1991). A nuclear mutant of *Chlamydomonas* that exhibits increased sensitivity to UV irradiation, reduced recombination of nuclear genes, and altered transmission of chloroplast genes. *Curr. Genet.* 19: 35–41.

Sager, R., and C. Grabowy (1985). Sex in *Chlamydomonas:* Sex and the single chloroplast. In *The Origin and Evolution of Sex* (H.O. Halvorson and A. Monroy, eds.), *MBL Lectures in Biology* 7: 113–121. Alan R. Liss, New York.

Sager, R., Sano, H., and C.T. Grabowy (1984). Control of maternal inheritance by DNA methylation in *Chlamydomonas. Curr. Top. Microbiol. Immunol.* 108: 157–172.

Sears, B.B. (1980). Elimination of plastids during spermatogenesis in the plant kingdom. *Plasmid,* 4: 233–255.

Smith, S.E. (1988). Biparental inheritance of organelles and its implications in crop improvement. *Plant Breed. Rev.* 6: 361–393.

Sodmergen, Suzuki, T., Kawano, S., Nakamura, S., Tano, S., and T. Kuroiwa (1992). Behavior of organelle nuclei (nucleoids) in generative and vegetative cells during maturation of pollen in *Lillium longiflorum* and *Pelargonium zonale. Protoplasma* 168: 73–82.

Stubbe, W. (1964). The role of the plastome in evolution of the genus *Oenothera. Genetica* 35: 28–33.

Szmidt, A.E., Alden, T., and J.-E. Hallgren (1987). Paternal inheritance of chloroplast DNA in *Larix. Plant Mol. Biol.* 9: 59–64.

Takano, H., Kawano, S., and T. Kuroiwa (1992). Constitutive homologous recombination between mitochondrial DNA and a linear mitochondrial plasmid in *Physarum polycephalum. Curr. Genet.* 22: 221–227.

Taylor, J.W. (1986). Fungal evolutionary biology and mitochondrial DNA. *Exp. Mycol.* 10: 259–269.

Thanh, N.D., and P. Medgyesy (1989). Limited chloroplast gene transfer via recombination overcomes plastome-genome incompatibility between *Nicotiana tabacum* and *Solanum tuberosum. Plant Mol. Biol.* 12: 87–93.

Tilney-Bassett, R.A.E. (1975). Genetics of variegated plants. In *Genetics and Biogenesis of Mitochondria and Chloroplasts* (C.W. Birky, Jr., P.S. Perlman, and T.J. Byers, eds.), pp. 268–308. Ohio State University Press, Columbus.

Tilney-Bassett, R.A.E. (1988). Inheritance of plastids in *Pelargonium.* In *The Division and Segregation of Organelles* (S.A. Boffey and D. Lloyd, eds.), pp. 115–129. Cambridge University Press, Cambridge.

Tilney-Bassett, R.A.E., Almouslem, A.B., and H.M. Amoatey (1992). Complementary genes control biparental inheritance in *Pelargonium. Theor. Appl. Genet.* 85: 317–324.

VanWinkle-Swift, K.P., and B. Aubert (1983). Uniparental inheritance in a homothallic alga. *Nature (London),* 303: 167–169.

VanWinkle-Swift, K.P., and J.-H. Hahn (1986). The search for mating-type-limited genes in the homothallic alga *Chlamydomonas monoica. Genetics* 113: 601–619.

VanWinkle-Swift, K.P., and A.P. Salinger (1988). Loss of mt^+-derived zygotic chloroplast DNA is associated with a lethal allele in *Chlamydomonas monoica. Curr. Genet.* 13: 331–337.

Werren, J.M., Nur, U., and C.-I. Wu (1988). Selfish genetic elements. *Trends Ecol. Evol.* 3: 297–302.

Whatley, J.M. (1982). Ultrastructure of plastid inheritance: Green algae to angiosperms. *Biol. Rev.* 57: 527–569.

Wolfson, R., Higgins, K.G., and B.B. Sears (1991). Evidence for replication slippage in the evolution of *Oenothera* chloroplast DNA. *Mol. Biol. Evol.* 8: 709–720.

Wurtz, E.A., Boynton, J.E., and N.W. Gillham (1977). Perturbation of chloroplast DNA amounts and chloroplast gene transmission in *Chlamydomonas* by 5-fluorodeoxyuridine. *Proc. Natl. Acad. Sci. U.S.A.* 74: 4552–4556.

Zouros, E., Freeman, K.R., Oberhauser-Ball, A., and G.H. Pogson (1992). Direct evidence for extensive paternal mitochondrial DNA inheritance in the marine mussel *Mytilis*. *Nature (London)* 359: 412–414.

9

Intracellular Genetics of Organelles

There is a major black box in the study of organelle genes and genomes. Simply put, this relates to the question of why organelle genes segregate rapidly and give interpretable patterns of recombination in those organisms where these processes can be studied. After all, there are several to many organelle genomes in each chloroplast and mitochondrion and these organelles themselves are often present in large numbers per cell.

The intracellular genetics of chloroplasts and mitochondria attracted most interest in the 1970s when genetic maps were being constructed for the mitochondrial and chloroplast genomes of yeast and *Chlamydomonas,* respectively. Since then most investigators have drifted off to study more tractable problems. A notable exception is C.W. Birky, Jr. We largely owe the development of an intracellular population genetic theory for chloroplasts and mitochondria to him and his collaborators.

THE ORIGIN OF ORGANELLE MUTATIONS

Antibiotic-resistance mutations lend themselves best to studies of the origin and expression of organelle mutations (see Birky, 1978; Dujon, 1981; Gillham, 1978 for reviews). They are easily selected and can be studied in organisms like yeast and *Chlamydomonas.* Two types of experiments derived from classical bacterial genetics, the Luria-Delbruck (1943) fluctuation test and the Newcombe (1949) respreading experiment, have been used to determine whether chloroplast and mitochondrial mutations arise spontaneously in the absence of antibiotic or whether these mutations are induced by exposure to the antibiotic. Both types of experiments indicate that chloroplast and mitochondrial mutations most likely arise in the absence of antibiotic, but that intracellular selection then occurs to ensure their expression. A striking feature of the fluctuation tests, as Dujon (1981) points out, is that interclonal variation for organelle gene mutations is not nearly as great as is observed in single copy genetic systems. In fact the interclonal variation for both chloroplast and mitochondrial mutations is only slightly greater than the mean (almost Poisson) or semiclonal (Gillham and Levine, 1962; Birky, 1973). One could interpret such results as suggesting that most mutations are antibiotic induced were it not for the fact that the same phenomenon has been observed both for chloroplasts and mitochondria and for antibiotics as different as streptomycin, erythromycin, and chloramphenicol. Furthermore, new mutations to antibiotic resistance in the chloroplast and mitochondrial genome are unlikely to be expressed immediately, but only after a segregation lag and intracellular selection.

Backer and Birky (1985) conducted a thorough reinvestigation of the origin question. They tested the roles of six hypothetical mechanisms for the origin of mitochondrial gene mutations to erythromycin resistance in yeast: (1) random partitioning of mitochondrial genomes at cell division, (2) intercellular selection for resistant cells, (3) intracellular random genetic drift of mitochondrial gene frequencies, (4) intracellular selection of antibiotic-resistant mitochondrial genomes on exposure to antibiotic, (5) induction of the mutations by the antibiotic, and (6) antibiotic-induced reduction of organelle genome ploidy permitting easier expression and fixation of new resistance mutations. Based on Luria–Delbruck fluctuation tests, Newcombe respreading experiments, and other experimental designs, Backer and Birky concluded intracellular selection plays the major role in producing erythromycin-resistant cells in the presence of the antibiotic. However, in the absence of erythromycin, the combined effects of random drift and random partitioning were most important in determining the fate of new mutations, most of which were lost rather than fixed.

In support of Backer and Birky's interpretation, intracellular selection for erythromycin-resistant mitochondria in the presence of antibiotic has also been clearly documented in *Paramecium* (reviewed by Gillham, 1978). Thus, when animals with resistant and sensitive mitochondria conjugate, the exconjugant derived from the sensitive conjugant will have a majority of sensitive mitochondria and resistant mitochondria will be in the minority. Exposure of such cells to antibiotic results in progressive transformation to the resistant phenotype over several days. These experiments demonstrate that intracellular selection can take place at the mitochondrial level, but they do not explain intraorganellar selection at the level of the mtDNA molecule. As Backer and Birky put it, "Intracellular selection is strongly favored by our experiments and others, but remains a phenomenon in search of a molecular mechanism."

Walsh (1992) published a theoretical treatment of the processes that may lead to the fixation of a new organelle mutation. He examines the expected substitution rate by obtaining the probability that a new mutant is fixed throughout the cell, allowing for arbitrary rates of genome turnover within the organelle and organelle turnover within the cell. The effects of gene conversion (possibly biased) and differences in genome and/or organelle replication rates are also considered. One significant finding is that if the rate of unbiased conversions is sufficiently strong, enough intracellular drift is created to overcome marked differences in the replication rates of wild-type and mutant genomes. Hence, organelle genomes with high conversion rates tend to be resistant to intracellular selection based on differences in genome replication or loss rates. However, in the absence of recombination, biased gene conversion and differences in genome replication rates do not affect the probability of fixation beyond the initial organelle. With organelle fusion and exchange, both processes influence the probability of fixation throughout the entire cell.

WHAT IS THE SEGREGATING UNIT?

Sorting out of organelle gene mutations during vegetative growth is one of their oldest observed characteristics, dating back to the discovery that sorting out of green and white plastids results in green and white leaf variegation. Since sorting-out (vegetative) segregation may occur at the level of the organelle DNA molecule, the nucleoid, or the organelle itself, it seems unlikely that a single mechanism can be involved. Rather, sorting out is better thought of as a failure to ensure a regular segregation, such as occurs in the case of nuclear chromosome separation at mitosis and meiosis. Stochastic models that posit random sorting out of organelles or organelle DNA molecules are appealing (Birky, 1983a). By the same token, subtle mechanisms must exist to limit the degree of randomness so as to ensure that following cell division, each daughter cell contains at least one or more organelles or organelle DNA molecules of a given kind.

What follows is a brief enumeration of the kinds of cellular events that might lead to sorting out of organelle genomes, nucleoids, and organelles based largely on views of Birky who has considered this topic in numerous reviews (1978, 1983a,b, 1991) and original articles. Birky (1978) refers to two general laws of organelle genetics: vegetative segregation and uniparental inheritance. He argues (1983a) that organelle genes are under relaxed cellular control. Thus, cells contain many copies of their organelle genomes, so the replication, recombination, and partitioning of these genomes do not have to be stringently controlled (i.e., each DNA molecule replicates once per cell cycle and each daughter cell receives one copy). This contrasts with nuclear genes where only two copies exist and there is no margin for error.

Relaxed control results in random changes in gene frequencies inside single cells or cell lineages. Birky (1983a) supposes that allele frequency variations can be used to describe not only vegetative segregation, but uniparental inheritance. Thus, both biparental and uniparental zygote clones can be obtained in yeast crosses. The variance of the distribution as well as the frequency of uniparental zygote clones can be enhanced experimentally. Birky (1983a, 1991) speculates that deterministic mechanisms have since been interposed to ensure uniparental inheritance on what may have been a primitive mechanism leading to partial uniparental inheritance based on relaxed cellular controls.

The mechanisms that may generate random drift in organelle gene frequencies under relaxed control have been explored by Birky (1978, 1983a, 1991). Random replication of organelle DNA could be involved in sorting out at the DNA level. If organelle DNA molecules are selected at random for replication, molecules of one genotype may be replicated more than once and the frequency of that genotype will increase. In cultured mouse cells, mtDNA molecules appear to be selected for replication, one or a few at a time, until the total number has been doubled (Bogenhagen and Clayton, 1977). Random selection of organelle DNA molecules for degradation would have the same effect. Repeated random pairing of DNA molecules for recombination will also result in random walks of gene frequencies because of gene conversion. Each of these events, separately or in combination, will lead to stochastic changes in chloroplast or mitochondrial allele frequencies, analogous to random drift in Mendelian population genetics. A consequence of fixation of an organelle allele within a given cell would be vegetative segregation. It also follows that random drift will be especially efficient in eliminating an organelle allele present in low frequency within a cell.

Organelle DNA molecules are bundled into nucleoids (Chapter 5). This means that the number of segregating units in a chloroplast or mitochondrion can be less than the number of DNA molecules per organelle (Table 5-1). Also, the number of nucleoids per organelle can vary (Chapter 5). Although this suggests that segregation of these structures may be somewhat random, Rose (1988) reviews experiments on cp and mtDNA segregation in several species that point toward a much more regular segregation of organelle DNA molecules to daughter chloroplasts and mitochondria.

Van Winkle-Swift (1980) proposed an attractive nucleoid segregation model for *Chlamydomonas*. The model supposes a nonrandom reductional distribution of chloroplast nucleoids to daughter cells, dictated solely by the spacial arrangement of parental nucleoids with respect to the plane of chloroplast division. Such a mechanism, if operative, could also serve to hasten the rate of segregation of chloroplast genomes in meiotic zygote progeny.

Vegetative segregation of mitochondrial markers among the progeny of yeast zygotes has also been extensively studied (see Birky, 1978; Birky *et al.*, 1982; Dujon, 1981; Gillham, 1978; Gingold, 1988 for reviews). Mitochondrial alleles for antibiotic resistance and sensitivity segregate rapidly following mating with the process being virtually complete within 15–20 cell divisions. Experiments reviewed by Dujon (1981) in which first zygotic buds were separated from zygotes by micromanipulation showed that frequencies of first buds homoplasmic for a given locus were not only relatively high, but a function of the input ratio of a particular cross (see Recombination of Organelle Genomes: The Phage Analogy Model). The latter could be predicted from random segregation since the greater the number of copies in a bud the greater the probability that the allele will be found in the bud.

Using an intentionally oversimplified hypothesis of random sampling, Dujon *et al.* (1974) calculated the number of segregating units entering each bud as three. Birky *et al.* (1978) and Waxman and Birky (1982) arrived at estimates of 2–20 segregating units in the mother cell and 1–4 per bud. Since Sena *et al.* (1976) reported that there are about 100 mtDNA molecules per zygote and more than 40 mtDNA molecules per bud, the mtDNA molecules must be "packaged" in some way. In the yeast mitochondrion as in the chloroplast of *Chlamydomonas*, nucleoids may help to explain this numerical paradox. During G_1, after mtDNA synthesis has occurred, 30 or so small mitochondria appear to fuse to form one, large mitochondrion (Fig. 5–1 and Kuroiwa, 1982). This fusion is accompanied by nucleoid fusion. When the latter mitochondrion divides, each of the daughters contains an elongate nucleoid. The daughter mitochondria subsequently fragment to yield smaller mitochondria, each of which possesses at least one spherical nucleoid. These observations help in two ways if they also apply to yeast zygotes. First, they suggest that the number of nucleoids per mitochondrion is probably small. Second, they provide physical evidence for nucleoid fusion that would open the opportunity for mitochondrial gene recombination. On the other hand, one is left with the question of how mtDNA molecules partition with the nucleoids in such a way as to permit recombination while preventing extensive mixing. After all, homoplasmy can be achieved rapidly during bud segregation from zygotes.

There is also the possibility that a few "master molecules" exist. These would be the true genetic copies. While this may seem farfetched, there are two suggestive observations. Bendich and Smith (1990) have made moving pictures of chloroplast and mitochondrial DNA molecules being fractionated by pulsed-field electrophoresis. For watermelon chloroplasts, they report identifying three species of cpDNA: linear molecules representing monomeric to tetrameric lengths of the 155-kb unit genome, much longer linears of at least 1200 kb, and the expected circular molecules. Linear molecules of 50–100 kb were also observed in mtDNA preparations from yeast. Similar large linear molecules of two to seven genome units were also reported for the 19-kb mtDNA molecules of *Candida glabrata* (Maleszka *et al.*, 1991; also see Chapter 5). Both Bendich and Smith and Maleszka *et al.* argue that the existence of such high-molecular-weight species may help to explain the ploidy paradox whereby the number of genetic units is so much smaller than the measured quantity of organelle DNA. These "master" molecules, then, would be the true segregating species whose behavior is being studied in crosses. Such a possibility was first raised many years ago by Sager and Ramanis (1968) on the basis of genetic data from crosses involving *Chlamydomonas* chloroplast genes (reviewed by Gillham, 1978; Harris, 1989; Sager, 1977).

Finally, most cells have several to many copies of their chloroplasts and mitochondria that must partition at cell division (see review by Birky, 1983b). Hennis and Birky (1984) examined partitioning of chloroplasts of the marine alga *Olisthodiscus,* a favorable subject for such an investigation since the plastids are easily counted and the cells can be grown synchronously. Cells of this alga contain 8–34 plastids with a mean of around 20. Analysis of the data supported one model in which daughter cells receiving large numbers of chloroplasts go directly to the next division without replicating their chloroplasts, while cells with very small numbers of chloroplasts go through two rounds of chloroplast replication before dividing. The result was that parental chloroplasts were partitioned equally in about 76% of the divisions, while in the remaining 24% the deviations from equality were very small. Equal partitioning of chloroplasts also seems to occur in flowering plants (Possingham *et al.*, 1988).

Quantitative analysis of mitochondrial partitioning has also been done cytologically during spermatogenesis in several scorpion species (reviewed by Birky, 1983b). In *Centrurus exilicauda* the mitochondria of the primary spermatocytes aggregate to form a ring that lies on the surface of the spindle at the equator. The ring elongates as the chromosomes separate at anaphase. The ends of the ring then break and the two long filaments are cut in half by the cleavage furrow. Each secondary spermatocyte now has two sausage-shaped mitochondria that again lie parallel to the second meiotic spindle and are cut in half by the second cleavage furrow so that each spermatid contains two mitochondria of similar volume. Partitioning of mitochondria during spermatogenesis in two other scorpions, *Opisthacanthus elatus* and *Hadrurus hirsutus,* proceeds somewhat differently. The primary spermatocytes contain 24 roughly spherical mitochondria that do not divide during spermatogenesis, but are partitioned to the four spermatids. Partitioning is not uniform and equal, so some spermatids contain more or less than six mitochondria, the predicted number per spermatid if partitioning were absolutely precise as in *Centrurus.* However, mitochondrial partitioning in *Opisthocanthus* and *Hadrurus* is not strictly random either as the numbers of mitochondria per spermatid cluster closely around a mean of six.

So where are we? The facts are that physical copy numbers are large and genetic copy numbers are small. In between there are many intriguing observations and a lot of elegant theory. But in the end the absolutely satisfying correlation seen between the behavior of nuclear genes and chromosomes remains elusive for chloroplasts and mitochondria.

PERSISTENT HETEROPLASMY AND ITS CAUSES

Heteroplasmy in its strictest sense refers to the presence of at least two different organelle genomes within a chloroplast or mitochondrion. These may differ by as little as a single pair of alleles. Heteroplasmons normally segregate rapidly to yield homoplasmons fulfilling the general law of vegetative segregation, but there are exceptions.

Persistent heteroplasmy in the *Chlamydomonas* chloroplast fits the strict definition as there is but a single chloroplast. The most careful studies of persistent heteroplasmy in this alga indicate that it is an "unnatural" state. Of particular note is the work of Spreitzer and his colleagues. Spreitzer *et al.* (1984) isolated revertants of a UGA nonsense mutation in the chloroplast *rbcL* gene encoding the RUBISCO large subunit. On minimal medium these revertants were wild type in phenotype, but the levels of RUBISCO holoenzyme and carboxylase activity were only about 45% of wild-type. When these revertants were transferred to acetate medium in the dark, permissive for growth of RUBISCO mutants, they segregated wild-type and acetate-requiring cells. The acetate-requiring segregants were genetically stable, but the wild-type segregants were not.

To explain these data, Spreitzer *et al.* (1984) proposed a model according to which all ca. 80 copies of the chloroplast genome in each revertant would contain the UGA nonsense mutation in the *rbcL* gene. However, a certain percentage of these chloroplast genome copies would also possess a second mutation in a suppressor *(sup)* gene. Under permissive growth conditions for the *rbcL* nonsense mutation, the *sup* gene would not be required and the sup^+ allele would be selected in preference to its mutant allele in subsequent mitotic divisons. Hence, progeny homoplasmic for chloroplast genomes carrying the sup^+ allele would segregate. These progeny would be acetate requiring since all chloroplast genomes possess the *rbcL* nonsense mutation. This phenomenon of "heteroplasmic suppression" was then reported for a UAG nonsense mutation in the *rbcL* gene (Spreitzer and Chastain, 1987).

Yu and Spreitzer (1992) determined the molecular basis for stable heteroplasmicity in the UAG nonsense mutation. As predicted, the original nonsense mutation was still present in the revertant, but the strain proved to be heteroplasmic for two forms of the chloroplast-encoded $tRNA^{Trp}$ gene. One had the normal CCA anticodon while the other was mutated to CUA, which can recognize the UAG nonsense codon. On minimal medium only cells heteroplasmic for chloroplast genomes carrying both of these tRNA genes will survive. The suppressor tRNA is required to read the UAG nonsense codon in the mRNA encoded by the *rbcL* mutant while the normal tRNA is necessary for reading tryptophan codons in all mRNAs. Under nonselective conditions homoplasmons having only chloroplast genomes with the mutant *rbcL* gene and the normal $tRNA^{Trp}$ will survive as RUBISCO function is dispensable. However, cells homoplasmic for chloroplast genomes containing the mutated $tRNA^{Trp}$ will die since they cannot read tryptophan codons in chloroplast mRNAs. Boynton *et al.* (1992) review other instances of persistent heteroplasmons in *C. reinhardtii*.

Most instances of heteroplasmy are identified at the cellular level since most cells possess multiple copies of their chloroplasts and mitochondria. Several cases of persistent heteroplasmy have been reported on genetic grounds for yeast mitochondrial genes. Thus, Treat-Clemmons and Birky (1983) verified the continued presence of heteroplasmons using genetic markers distributed over 17% of the yeast mitochondrial genome. Even after three to five subclonings (ca. 60–100 generations) some cells remained heteroplasmic. However, these heteroplasmons segregated less than 10% heteroplasmic progeny after one or two subclonings so they are not persistent in the sense that they can give rise only to heteroplasmons. Treat-Clemmons and Birky (1983) believe their observations are compatible with the hypothesis that vegetative segregation is partly due to random partitioning of mtDNA molecules, or groups of molecules, between daughter cells at cytokinesis. This hypothesis predicts that a very small proportion of heteroplasmic cells will be present even after many cell divisions, but that their proportion will decrease with time.

Because mitochondrial genomes in most animals are maternally inherited, the existence of heteroplasmy in both vertebrates and insects has attracted considerable attention. Solignac et al. (1987 and references therein) categorize the different heteroplasmies seen into five groups: (1) nucleotide site heteroplasmy [humans (see also Chapter 11), cow, fish (Bentzen et al., 1988), Drosophila], (2) existence of a variable number of nucleotides in a homopolymer nucleotide stretch (rat and cow), (3) intraindividual continuous length variation with a wide range (up to 1 kb) in a given region of a molecule (frog, rabbit), (4) discrete length variation which usually results from variability of a tandemly repeated sequence in the control region [Drosophila, crickets (Rand and Harrison, 1989), weevils (Boyce et al., 1989), frogs, fishes, newts, lizards], and (5) large deletions (mouse, human beings, see Chapter 11).

Avise (1991) reviewed the experimental studies of Solignac et al. (1984, 1987) and Rand and Harrison (1986) on the segregation rates of heteroplasmic lines in Drosophila mauritiana and crickets. Avise concludes that an average of 10–100 animal generations may be required for a single maternal lineage to achieve homoplasmy while several hundred animal generations may be required for a population of heteroplasmic individuals to achieve this state. Avise (1991) points out that under neutrality theory, and, in the absence of overriding mutational pressure, an increase in the number of germ-cell divisions per organism (g), or a decrease in the number of segregating units per germ cell (N_{mt}) will shorten the transition time during which any initial heteroplasmy is converted to homoplasmicity by genetic drift in subsequent animal generations. In insects g is ca. 10, but in mammals g may be as high as 50.

The work of Hauswirth and Laipis (Ashley et al., 1989; Hauswirth and Laipis, 1982a,b; Laipis et al., 1988) is indicative of much a faster segregation rate in cattle than predicted. Thus, Ashley et al. (1989) followed transmission of mtDNA through four generations of Holstein cows finding substantial shifts in heteroplasmy levels between single generations. Their results indicated that a return to homoplasmy can occur within two to three generations. Ashley et al. (1989) calculate that the number of segregating units is between 20 and 100 depending on whether g equals 10 or 50. Hauswirth and Laipis (e.g., Laipis et al., 1988) suggested a mechanistic explanation for the rapid segregation of mitochondrial genotypes they see in Holstein lineages. They point out that mitochondria and mtDNA undergo significant amplification during follicular development of the bovine oocyte, leading to a reduction of the genome/organelle ratio to near one. Hauswirth and Laipis hypothesize that partitioning of these mitochondria of low genome copy number into dividing cells of the early embryo could result in the chance segregation of variant subpopulations of mtDNA species. Consequently, cells forming the embryonic inner cell mass, in contrast to extraembryonic tissues, could receive very different ratios of heteroplasmic mtDNA molecules. Hauswirth and Laipis believe that different rates of mtDNA segregation seen in insects and vertebrates may reflect distinctions in the cytoplasmic partitioning process in early development.

This hereditary bottleneck is also evident in studies of Koehler et al. (1991). These investigators followed mtDNA segregation in leukocytes in a different lineage of Holstein cattle. They found that one bovine sequence variant can be replaced by another within a single generation. However, Howell *et al.* (1992) query whether the hereditary bottleneck need always be narrow. They followed segregation of a silent polymorphism in the human *ndh6* gene in a human maternal lineage comprising eight individuals and spanning three generations. Heteroplasmy persisted in all eight people and there was little variation between individuals and generations. As Howell *et al.* conclude, "the results presented here underscore the emerging picture of mitochondrial gene segregation as a complex and dynamic process requiring further analysis, both theoretical and experimental."

How do heteroplasmies arise in the first place? One of the most commonly detected heteroplasmies involves variations in the number of tandemly repeated sequences within the control region. In discussing their work on weevils of the genus *Pissodes,* where the variation in the magnitude of size class differences, the number of size classes possible within single individuals, and the abundance of heteroplasmy in all individuals is unprecedented, Boyce *et al.* (1989) consider the possible role of slipped mispairing. They suggest that slipped mispairing within or between mtDNA molecules during replication could play a role in the generation of size variation. Slipped mispairing involves the generation of insertions and deletions through the mispairing of complementary sequences of tandem repeats.

Two very important classes of heteroplasmies have not entered into this discussion. They are the heteroplasmies that accompany mitochondrial diseases in filamentous fungi and human beings. These mitochondrial diseases are considered in Chapter 11.

RECOMBINATION OF ORGANELLE GENOMES, THE PHAGE ANALOGY MODEL

Intermolecular recombination of chloroplast genomes in *Chlamydomonas* and mitochondrial genomes in yeast is formally nonreciprocal (Gillham, 1965; Van Winkle-Swift and Birky, 1978). The phage analogy model to explain the recombination of mitochondrial genomes in yeast was developed by Dujon *et al.* (1974) in analogy to the Visconti–Delbruck (1953) theory that describes bacteriophage recombination. Since the phage analogy model and its consequences have been thoroughly reviewed elsewhere (Birky, 1978; Dujon, 1981; Gillham, 1978), an abbreviated account will be provided here.

In a phage cross the host bacteria are infected with viruses of different genotype at known average multiplicities of infection. The titers of the parent viruses are readily established by plating at appropriate dilution on sensitive indicator bacteria. Thus, the average *input* fraction can be measured accurately and varied experimentally with ease. No such technique is available for titrating the relative input fraction of organelle genomes in a cross; therefore the input must be deduced indirectly. One of the important assumptions of the phage analogy model is that in a mitochondrial cross the relative input fraction of mitochondrial genomes from each parent will be proportional to the output as scored by cellular phenotype. Essential to this assumption is the absence of selective advantage for any mitochondrial genome at the intracellular level or of specific mitochondrial genotypes at the intercellular level. If these conditions are realized, the relative input of mitochondrial genomes can be deduced from the ratio of any pair of mitochondrial alleles with this ratio being similar for all pairs of alleles in a given cross.

There is abundant evidence that the predicted *coordinate output* of alleles occurs. A number of experimental treatments of parental cells prior to crossing modify the output over a broad range. All of these treatments always affect the output allele ratios in a coordinated fashion and are most simply interpreted as variations of the input ratio.

A second assumption of the model is that random pairing and recombination occur between the different mtDNA molecules in the cell (a panmictic pool). If this is true, the upper limit of recombination for markers that are unlinked genetically should be 20–25% rather than 50% because half of all pairings occur between molecules of like genotype. This is the maximum recombination frequency actually observed. Also, random pairing predicts that the frequency of recombinants between a pair of markers should vary with the input ratio since the probability of pairing between two molecules of given genotypes varies with the frequency of each genotype in the intracellular population. This prediction is met by the experimental results.

The third assumption is that there are multiple rounds of pairing and recombination. In yeast mitochondrial crosses, markers distant from each other by only short physical distances are genetically unlinked. In fact, a series of crosses indicated that genetic linkage is apparent only between two mitochondrial markers separated from each other by less than 1 kb. These results suggest a very high efficiency of mitochondrial recombination and are consistent with the notion of multiple rounds of pairing and recombination. Estimates of the number of mating rounds vary between one and several.

A final assumption is random segregation of mitochondrial genomes.

The phage analogy model not only predicts many of the observed results of mitochondrial crosses in yeast, but has proved applicable to crosses involving chloroplast genes in *Chlamydomonas* as well (Boynton *et al.,* 1976). For example, output frequencies reflect input ratios as they do in yeast and 20–25% recombination between markers appears to be characteristic of unlinked markers as is also true in yeast. On the other hand, the phage analogy model is admittedly an oversimplification, and a more complete model would need to take into account modes of replication and segregation of organelle DNA molecules. Perhaps the most remarkable aspect of organelle genome segregation and recombination is that these processes exhibit characteristics compatible both with large and small numbers of copies. A unified theory underlying segregation and recombination of organelle genes must deal with this paradox.

POPULATION GENETIC THEORY OF ORGANELLES

The mathematical tools of theoretical population genetics have been applied increasingly to organelles. Successively more realistic and complete treatments have been devised for chloroplasts and mitochondria using classic parameters from the Fisher–Wright model including mutation, selection, drift, and migration. Much of this work has been reviewed by Birky (1991). He points out that these theoretical treatments have generally dealt with multicellular organisms having separate sexes. The organism is considered to be diploid when organelle and nuclear genes are compared. An infinite alleles model is normally used so that the unit of study, generally referred to as a gene, should be larger than a single base pair.

Birky (1991) feels the following conclusions are now well established from such studies:

1. The average time required for fixation or loss of alleles as the result of intracellular random drift is short. Therefore, most cells are homoplasmic most of the time unless there is a high mutation rate or a substantial paternal contribution of organelles to the zygote.
2. Much of population genetic theory for nuclear genes can be adapted to organelles by assuming different effective numbers of genes. Thus, because organelle genes are usually transmitted uniparentally and because organelle genomes are generally identical, individuals are counted as having only one copy of each gene instead of two as for the nucleus.
3. The rate and path of approach to equilibrium are different for diversity within and between individuals and populations.
4. The effective population size and migration rate are usually lower than for nuclear genes in the same population for the reasons given under two above.

Recombination between genes residing in different chloroplasts and mitochondria is rare in most organisms because of uniparental inheritance and vegetative segregation without organelle fusion in some (e.g., land plant chloroplasts). This contrasts markedly with nuclear genes where unlinked genes undergo independent assortment and linked genes recombine. This difference has significant theoretical ramifications (Birky, 1991). For example, an important consequence of strong linkage is the phenomenon of hitchhiking. An advantageous mutation (the driver) is fixed by selection and carries with it one or more linked neutral mutations that might not have been fixed (hitchhikers). In its most extreme form (periodic selection) the driver and its linked hitchhikers may become fixed, eliminating alternative alleles in the population and reducing genetic diversity to zero. Diversity is then built up again by mutation until the next periodic selection event.

An important population genetic question, dealt with in an experimental sense in Chapter 8, is the interaction between organelle and nuclear genes. Because of their close cooperation in building vital assemblages in the respiratory and photosynthetic electron transport chains, these interactions must be very finely tuned at both the structural and regulatory levels. Birky (1991) reviews the work of Asmussen and colleagues on this subject. Because organelle and nuclear genes are, by definition, always unlinked, any departure from random associations between organelle and nuclear genotypes (linkage disequilibrium) must either result from historical factors (i.e., the population has not had time to go to equilibrium or assortative mating occurs) or selection is at work. The important conclusion is that all disequilibrium between nuclear and cytoplasmic genes goes to zero with random mating, or with positive assortative mating (as can occur in hybrid zones where females mate preferentially with males of the same species). This can be determined by nuclear genotype alone or jointly by a cytoplasmic locus and multiple nuclear genes that might be responsible for interspecies differences. Different models can be distinguished by the rates of decay of various disequilibria.

In his summary Birky (1991) points out that the theory so far developed has been successful in explaining why organelle gene diversity can be high in a population of organisms, but is usually zero within organisms. Theory is also being used to help explain differences in evolutionary rates of organelle genes in plants and animals and among genomes within taxa. A theoretical framework is essential to the use of organelle genomes in studying biogeography and phylogenetics. As Birky says, "for many of these applications, theory and experiment are moving hand in hand as they should."

REFERENCES

Ashley, M.V., Laipis, P.J., and W.W. Hauswirth (1989). Rapid segregation of heteroplasmic bovine mitochondria. *Nucl. Acids Res.* 17: 7325–7331.

Avise, J.C. (1991). Ten unorthodox perspectives on evolution prompted by comparative population genetic findings on mitochondrial DNA. *Annu. Rev. Genet.* 25: 45–69.

Backer, J.S., and C.W. Birky, Jr. (1985). The origin of mutant cells: mechanisms by which *Saccharomyces cerevisiae* produces cells homoplasmic for new mitochondrial mutations. *Curr. Genet.* 9: 627–640.

Bendich, A.J., and S.B. Smith (1990). Moving pictures and pulsed-field gel electrophoresis show linear DNA molecules from chloroplasts and mitochondria. *Curr. Genet.* 17: 421–425.

Bentzen, P., Leggett, W.C., and G.G. Brown (1988). Length and restriction site heteroplasmy in the mitochondrial DNA of American Shad *(Alosa sapidissima). Genetics* 118: 509–518.

Birky, C.W., Jr. (1973). On the origin of mitochondrial mutants: evidence for intracellular selection of mitochondria in the origin of antibiotic-resistant cells in yeast. *Genetics* 74:421–432.

Birky, C.W., Jr. (1978). Transmission genetics of mitochondria and chloroplasts. *Annu. Rev. Genet.* 12: 471–512.

Birky, C.W., Jr. (1983a). Relaxed cellular controls and organelle heredity. *Science* 222: 468–475.

Birky, C.W., Jr. (1983b). The partitioning of cytoplasmic organelles at cell division. *Int. Rev. Cytol. Suppl.* 15: 49–89.

Birky, C.W., Jr. (1991). Evolution and population genetics of organelle-genes: Mechanisms and models. In *Evolution at the Molecular Level* (R.K. Selander, A.G. Clark and T.S. Whittam, eds.), pp. 112–134. Sinauer, Sunderland, MA.

Birky, C.W., Jr., Strausberg, R.L., Perlman, P.S., and J.L. Forster (1978). Vegetative segregation of mitochondria in yeast: Estimating parameters using a random model. *Mol. Gen. Genet.* 158: 251–261.

Birky, C.W., Jr., Acton, A.R., Dietrich, R., and M. Carver (1982). Mitochondrial transmission genetics: Replication, recombination and segregation of mitochondrial DNA and its inheritance in crosses. In *Mitochondrial Genes* (P.P. Slonimski, P. Borst, and G. Attardi, eds.), pp. 333–348. Cold Spring Harbor Laboratory, Cold Spring Harbor, NY.

Bogenhagen, D., and D.A. Clayton (1977). Mouse L cell mitochondrial DNA molecules are selected randomly for replication through the cell cycle. *Cell* 11: 719–727.

Boyce, T.M., Zwick, M.E., and C.E. Aquadro (1989). Mitochondrial DNA in the bark weevils: Size, structure and heteroplasmicity. *Genetics* 123: 825–836.

Boynton, J.E., Gilham, N.W., Harris, E.H., Tingle, C.L., VanWinkle-Swift, K., and G.M.W. Adams (1976). Transmission, segregation, and recombination of chloroplast genes in *Chlamydomonas.* In *Genetics and Biogenesis of Chloroplasts and Mitochondria* (Th. Bucher, W. Neupert, W. Sebald, and S. Werner eds.), pp. 313–322. Elsevier/North-Holland, Amsterdam.

Boynton, J.E., Gillham, N.W., Newman, S.M., and E.H. Harris (1992). Organelle genetics and transformation in *Chlamydomonas*. In *Cell Organelles, Advances in Plant Gene Research* (R. Herrmann, ed.). pp. 3–64 Springer-Verlag, Wien, New York.

Dujon, B. (1981). Mitochondrial genetics and functions. In *Molecular Biology of the Yeast Saccharomyces: Life Cycle and Inheritance* (J.N. Strathern, E.W. Jones and J.R. Broach, eds.), pp. 505–635. Cold Spring Harbor Laboratory, Cold Spring Harbor, NY.

Dujon, B., Slonimski, P.P., and L. Weill (1974). Mitochondrial genetics. IX. A model for recombination and segregation of mitochondrial genomes in yeast. *Genetics* 78: 415–437.

Gillham, N.W. (1965). Linkage and recombination between nonchromosomal mutations in *Chlamydomonas reinhardi. Proc. Natl. Acad. Sci. U.S.A.* 54: 1560–1567.

Gillham, N.W. (1978). *Organelle Heredity.* Raven Press, New York.

Gillham, N.W., and R.P. Levine (1962). Studies on the origin of streptomycin resistant mutants in *Chlamydomonas reinhardi. Genetics* 47: 1463–1474.

Gingold, E.B. (1988). The replication and segregation of yeast mitochondrial DNA. In *Division and Segregation of Organelles* (S.A. Boffey and D. Lloyd, eds.), *Soc. Exp. Biol.* 35: 148–170.

Harris, E.H. (1989). *The Chlamydomonas Sourcebook.* Academic Press, San Diego.

Hauswirth, W.W., and P.J. Laipis (1982a). Rapid variation in mammalian mitochondrial genotypes: Implications for the mechanism of maternal inheritance. In *Mitochondrial Genes* (P.P. Slonimski, P. Borst, and G. Attardi, eds.), pp. 137–141. Cold Spring Harbor Laboratory, Cold Spring Harbor, NY.

Hauswirth, W.W., and P.J. Laipis (1982b). Mitochondrial DNA polymorphism in a maternal lineage of Holstein cows. *Proc. Natl. Acad. Sci. U.S.A.* 79: 4686–4690.

Hennis, A.S., and C.W. Birky, Jr. (1984). Stochastic partitioning of chloroplasts at cell division in the alga *Olisthodiscus* and compensating control of chloroplast replication. *J. Cell Sci.* 70: 1–15.

Howell, N., Halvorson, S., Kubacka, I., McCullough, D.A., Bindoff, L.A., and D.M. Turnbull (1992). Mitochondrial gene segregation in mammals: Is the bottleneck always narrow? *Hum. Genet.* 90: 117–120.

Koehler, C.M., Lindberg, G.L., Brown, D.R., Beitz, D.C., Freeman, A.E. Mayfield, J.E., and A.M. Myers (1991). Replacement of bovine mitochondrial DNA by a sequence variant within one generation. *Genetics* 129: 247–255.

Kuroiwa, T. (1982). Mitochondrial nuclei. *Int. Rev. Cytol.* 75: 1–59.

Laipis, P.J., Van de Walle, M.J., and W.W. Hauswirth (1988). Unequal partitioning of bovine mitochondrial genotypes among siblings. *Proc. Natl. Acad. Sci. U.S.A.* 85: 8107–8110.

Luria, S.E., and M. Delbruck (1943). Mutations of bacteria from virus sensitivity to virus resistance. *Genetics* 28: 491–511.

Maleszka, R., Skelly, P.J., and G.D. Clark-Walker (1991). Rolling circle replication of DNA in yeast mitochondria. *EMBO J.* 10: 3923–3929.

Newcombe, H.B. (1949). Origin of bacterial variants. *Nature (London)* 164: 150–151.

Possingham, J.V., Hashimoto, H., and J. Oross (1988). Factors that influence plastid division in higher plants. In *Division and Segregation of Organelles* (S.A. Boffey and D. Lloyd, eds.), pp. 1–20. Cambridge University Press, Cambridge.

Rand, D.M., and R.G. Harrison (1986). Mitochondrial DNA transmission genetics in crickets. *Genetics* 955–970.

Rand, D.M., and R.G. Harrison (1989). Molecular population genetics of mtDNA size variation in crickets. *Genetics* 121: 551–569.

Rose, R.J. (1988). The role of membranes in the segregation of plastid DNA. In *Division and Segregation of Organelles* (S.A. Boffey and D. Lloyd, eds.), pp. 171–195. Cambridge University Press, Cambridge.

Sager, R. (1977). Genetic analysis of chloroplast DNA in *Chlamydomonas*. *Adv. Genet.* 19: 287–340.

Sager, R., and Z. Ramanis (1968). The pattern of segregation of cytoplasmic genes in *Chlamydomonas*. *Proc. Natl. Acad. Sci. U.S.A.* 61: 324–331.

Sena, E.P., Welch, J., and S. Fogel (1976). Nuclear and mitochondrial DNA replication during synchronous mating in *Saccharomyces cerevisiae*. *Science* 194: 433–435.

Solignac, M., Genermont, J., Monnerot, M., and J.-C. Mounolou (1984). Genetics of mitochondria in *Drosophila*: mtDNA inheritance in heteroplasmic strains of *D. mauritiana*. *Mol. Gen. Genet.* 197: 183–188.

Solignac, M., Genermont, J., Monnerot, M., and J.-C. Mounolou (1987). *Drosophila* mitochondrial genetics: Evolution of heteroplasmy through germ line cell divisions. *Genetics* 117: 687–696.

Spreitzer, R.J., and C.J. Chastain (1987). Heteroplasmic suppression of an amber mutation in the *Chlamydomonas* chloroplast gene that encodes the large subunit of ribulosebisphosphate carboxylase/oxygenase. *Curr. Genet.* 11: 611–616.

Spreitzer, R.J., Chastain, C.J., and W.L. Ogren (1984). Chloroplast gene suppression of defective ribulosebisphosphate carboxylase/oxygenase in *Chlamydomonas reinhardtii*. *Curr. Genet.* 9: 83–89.

Treat-Clemmons, L.G., and C.W. Birky, Jr. (1983). Persistent heteroplasmic cells for mitochondrial genes in *Saccharomyces cerevisiae*. *Curr. Genet.* 7: 489–492.

Van Winkle-Swift, K.P. (1980). A model for the rapid vegetative segregation of multiple chloroplast genomes in *Chlamydomonas*: Assumptions and predictions of the model. *Curr. Genet.* 1: 113–125.

Van Winkle-Swift, K.P., and C.W. Birky, Jr. (1978). The nonreciprocality of organelle gene recombination in *Chlamydomonas reinhardtii* and *Saccharomyces cerevisiae*. *Mol. Gen. Genet.* 166: 193–209.

Visconti, N., and M. Delbruck (1953). The mechanism of genetic recombination in phage. *Genetics* 38: 5–33.

Walsh, J.B. (1992). Intracellular selection, conversion bias, and the expected substitution rate of organelle genes. *Genetics* 130: 939–946.

Waxman, M.F., and C.W. Birky, Jr. (1982). Partial pedigree analysis of yeast mitochondrial genes during vegetative reproduction. *Curr. Genet.* 5: 171–180.

Yu, W., and R.J. Spreitzer (1992). Chloroplast heteroplasmicity is stabilized by an amber-suppressor tryptophan $tRNA_{CUA}$. *Proc. Natl. Acad. Sci. U.S.A.* 89: 3904–3907.

10

Introns, Mobile Elements, and Plasmids

Phylogenetic arguments for early or late origin of group I and group II introns (Chapter 4) are blurred to some extent by accumulating evidence that many of these elements are or were mobile in chloroplasts and mitochondria. Thus, mobile group I introns direct their own "homing" to homologous genes lacking them through the aegis of intron-encoded endonucleases (Clark-Walker, 1992; Dujon, 1989; Lambowitz and Belfort, 1993; Pel and Grivell, 1993) while group II introns move via an intermediate generated by a reverse transcriptase (Lambowitz, 1989; Lambowitz and Belfort, 1993; Lambowitz and Perlman, 1990). Autonomous mitochondrial plasmids in *Neurospora* and *Podospora* appear to be DNA copies of excised group I and group II introns (Lambowitz and Belfort, 1993). These elements either encode reverse transcriptase-like proteins or replicate by reverse transcription.

Introns are not the only mobile elements found in organelles. In yeast there is evidence for the mobility of G+C-rich elements (Clark-Walker, 1992). Different sized alleles of the "expandable" *var1* mitochondrial gene of yeast make proteins that vary correspondingly in length (Butow *et al.*, 1985). This size variation can be traced to the presence of certain elements that can be transferred from "long" to "short" alleles by preferential gene conversion.

INTRONS

This section begins by considering the comparative structures of different classes of introns and then turns to the subject of self-splicing and protein-assisted splicing of the group I and group II introns found in mitochondria and chloroplasts. Later in the chapter introns return once again in the context of mobile elements and fungal mitochondrial plasmids.

Intron Structure

Four different classes of introns are currently recognized (Cech, 1990; Perlman *et al.*, 1990) of which only two, group I and II introns, found in organelle DNA, are discussed in detail. *Nuclear pre-mRNA* introns are abundant in the protein-coding nuclear genes of some groups of eukaryotes, but they are presently unknown in several eukaryotic groups including the earliest protistan lineages (Palmer and Logsdon, 1991). Little of the internal sequence is required beyond maintenance of a certain minimum length. Splicing of nuclear pre-mRNA introns is elaborate and requires assembly of the spliceosome, a complex of small RNA-protein particles (snRNPs) called U1–U6, which enter into the splicing pathway at different points. Within the spliceosome structure, two ordered reactions rearrange the pre-mRNA substrate, resulting in release of the spliced exons and the excised intron (Fig. 10–1). *Nuclear tRNA intron* splicing involves sequential enzymatic reactions (Fig. 10–1). These introns have been identified in nuclear tRNA genes from many eukaryotes. They always interrupt the mature tRNA sequence at the same site one nucleotide away from the 3' side of the anticodon, next to a conserved purine. Introns are also found in some chloroplast tRNA genes, but these are group I or group II introns and are spliced by mechanisms distinct from those involved in splicing nuclear pre-tRNAs.

The secondary structure of *group I* introns was determined by Davies *et al.* (1982) and Michel *et al.* (1982) based on comparative sequence analysis and is discussed in several reviews (Burke, 1988; Cech, 1988, 1990; Lambowitz and Belfort, 1993; Perlman *et al.*, 1990; Saldanha *et al.*, 1993; Waring and Davies, 1984). The conserved core of group I introns is made up of nine helices or paired regions (P1–P9 in Fig.10–2a). Conserved sequences are designated *P, Q, R,* and *S* (Davies *et al.*, 1982). *P* pairs with *Q* to form P4 and a portion of *R* pairs with part of *S* to form P7. The rest of the sequences

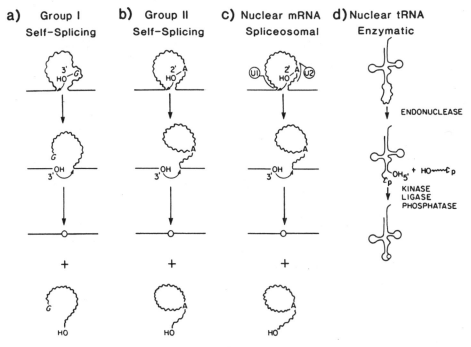

Fig. 10–1. Splicing mechanisms of the four major groups (a–d) of precursor RNAs. Wavy lines indicate introns, smooth lines represent flanking exons. For nuclear mRNA splicing, many components assemble with the pre-mRNA to form the spliceosome; only two, the U1 and U2 small nuclear ribonucleoproteins, are shown. From Cech (1990) with permission.

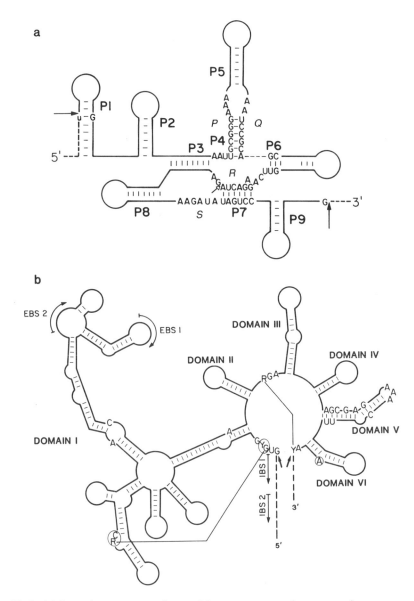

Fig. 10–2. (a) Secondary structure of group I introns as proven by comparative sequence analysis. A typical structure indicating conserved base-paired elements P1–P9. *P, Q, R,* and *S* represent conserved sequence elements and are represented by their most common nucleotide sequences. Dashed line between P4 and P6 is added to make the diagram less crowded, and does not indicate any omission of nucleotides. Filled arrow, 5′ and 3′ splice sites. Open arrow, site of insertion of extra stem–loop(s) P7.1 and P7.2 in group IA introns. From Cech (1990) with permission. (b) The core secondary structure of group II introns (see Michel and Dujon, 1983). Sections of the structure are identified as domains bounded by inverted repeat sequences. Pairing interactions have been identified involving intron-binding sites (IBS1 and IBS2) within the 5′ exon (E1) and exon-binding sites (EBS1 and EBS2) within D1 of the intron (Jacquier and Michel, 1987). A second pairing involves CR and YG where R and Y equal a purine and a pyrimidine respectively. Filled arrows, 5′ and 3′ splice sites. Dashed lines at the 3′ and 5′ ends in A and B represent exon sequences. From Cech, unpublished, with permission.

required to form the P helices are not conserved. Nevertheless, the same secondary structure has been confirmed for over 50 group I introns. Davies *et al.* (1982) proposed that an intron sequence capable of pairing with the 3'-end of the upstream exon and the 5'-end of the downstream exon is always present. This "internal guide sequence" (IGS) is usually near the 5' boundary and may align the two exon boundaries for splicing. Pairing of the IGS and the 5' exon yields the P1 structure, which is important for splicing. Since P1 involves exon sequences that are not conserved except for the U residue at the exon–intron boundary, the IGS for each group I intron differs. Pairing with the 3' exon is not as yet supported by experimental evidence. Group I introns also frequently contain extra sequences that can include long ORFs encoding proteins required for intron splicing or mobility. These extra sequences form stem–loop structures that are peripheral to the core and much more variable between introns. Michel and Westhof (1990) have aligned 87 sequences of self-splicing group I introns. With the aid of stereochemical modeling, they have constructed a three-dimensional model of the catalytic core of the group I intron. Based on their sequence comparisons, Michel and Westhof have classified group I introns into two major subgroups (IA and IB).

Group I intron splicing proceeds via two consecutive trans-esterification reactions (Fig. 10–1). The first is initiated by a guanosine residue or a 5' phosphorylated form such as guanosine monophosphate as the attacking nucleophile. This reaction releases as intermediates the 5' exon with a 3'-hydroxyl and a linear intervening sequence (IVS)-3' exon with the G added to the intron 5'-end. The 5' exon, now terminating in a free 3'-hydroxyl group, then attacks the phosphorus atom at the 3' splice site. Ligation of the exons and excision of the intron then result.

RNA-mediated self-splicing via this pathway was first worked out for the nuclear pre-rRNA of *Tetrahymena* by Cech and colleagues (Kruger *et al.*, 1982). Later this pathway was shown to be applicable to self-splicing of precursor RNAs containing group I introns from a variety of sources including a *Neurospora crassa* mitochondrial pre-mRNA and several yeast mitochondrial pre-mRNAs (see Cech, 1990, for references). Self-splicing of group I introns *in vitro* has also been reported for the *Chlamydomonas reinhardtii* chloroplast genes encoding the large rRNA and the large subunit of RUBISCO (Herrin *et al.*, 1990; Thompson and Herrin, 1991). The group I intron in the former gene contains an open reading frame encoding an endonuclease required for its own transposition (Durrenberger and Rochaix, 1991 and see *Mobile group I introns*). Splicing of this intron *in vivo* depends on at least one nuclear gene product (Herrin *et al.*, 1990). Self-splicing group I introns have also been identified in pre-mRNAs of bacteriophages T4 and SP01, and in nuclear pre-rRNA of the slime mold *Physarum polycephalum* and the fungus *Pneumocystis carinii*.

Michel *et al.* (1982) pointed out that fungal mitochondrial genes also contain a second, rarer intron type subsequently identified in certain chloroplast genes as well (see Sugiura, 1992, for review). The secondary structures of these *group II introns* have been derived using a comparative approach (see Michel *et al.*, 1989; Perlman *et al.*, 1990; Saldanha *et al.*, 1993; Woolford and Peebles, 1992 for review and discussion). A set of 70 different group II introns was compared and conserved base pairings were sought in spite of primary sequence divergence.

The core secondary structure of a group II intron is like a wheel with six spokes (helices) (Fig. 10–2b). This structure has nothing in common with group I intron secondary structure. Sections of the structure are identified as domains bounded by inverted repeat sequences. Nucleotides between domains are designated. The function of some of the domains has been established (Saldanha et al., 1993). Pairing interactions have been identified that involve intron binding sites (IBS1 and IBS2) within the 5' exon (E1) and exon-binding sites (EBS1 and EBS2) within domain I of the intron. The most critical interaction is base pairing of EBS1 and IBS1. This pairing is not conserved in sequence, but the 5' splice site is always after a specific base pair in the structure. Two additional base pairing interactions, EBS2–IBS2 and CR–YG (Fig. 10–2b), also contribute to positioning of the 5' splice site. Saldanha et al. (1993) point out that assembly of the 3' splice site involves at least four interactions: (1) docking of domain VI to the core, (2) base pairing of the terminal nucleotide of the intron and a nucleotide between domains II and III (Y–R, Fig. 10–2b), (3) base pairing of the first nucleotide of the 3' exon and the nucleotide preceding EBS1 in a "guide" interaction, and (4) another undefined interaction, also involving the first nucleotide in the 3' exon.

Group II introns have a different set of short conserved boundary sequences than group I introns. The 5' conserved sequence has the consensus GUGYG, but, unlike group I introns, there is no conserved nucleotide in the 5' exon. The 3' conserved sequence is the dinucleotide AU or AC. Unlike group I introns, most group II introns do not encode proteins. However, where ORFs are present, they are in-frame with the preceding exon. These ORFs specify large proteins of 450 to greater than 750 amino acids with sequence similarity to retroviral reverse transcriptases (Lambowitz and Belfort, 1993; Michel and Lang, 1985; Xiong and Eickbush, 1988).

Like group I introns, group II introns undergo a pair of trans-esterification reactions (Fig. 10–1). The initiating nucleophile for group II introns is the 2'-hydroxyl of an adenosine residue in domain 6. The first step is attack by this hydroxyl on the 5' splice junction. This releases the 5' exon with a 3'-hydroxyl end and a lariat IVS-3' exon intermediate. The free 5' exon attacks the 3' splice junction to produce the spliced exons plus the excised IVS lariat. The phosphate at the splice junction is derived from the first nucleotide of the second exon, while the 2'-5'-phosphodiester comes from the first IVS nucleotide. Self-splicing of several yeast mitochondrial group II introns has been shown to occur (e.g., Peebles et al., 1986; Hebbar et al., 1992), but only at nonphysiological temperatures (e.g., 40–45°C) and salt concentrations (100 mM magnesium plus potassium or ammmonium at 50 mM or higher). Hebbar et al. (1992) conclude that the requirement for high salt concentrations for efficient in vitro splicing of group II introns may result because the high negative charge of the RNA may destabilize secondary or tertiary structures necessary for the self-splicing reaction. High salt concentrations may mimic functions provided by maturases and other proteins involved in splicing in vivo.

Similarities exist between the splicing of group II and nuclear pre-mRNA introns (see Jacquier, 1990; Lambowitz and Belfort, 1993; Sharp, 1991). For both group II and nuclear pre-mRNA introns the attacking group is the 2'-hydroxyl of an adenosine within the intron and the second transesterification step results in the release of a lariat IVS. On the other hand, the sequences surrounding the 5' junction of the nuclear pre-mRNA intron are only known to pair with a single sequence in U1 snRNA. In contrast, in group II introns the exon upstream of the 5' junction contains the IBS1 and IBS2 located near the 3'-end of the exon that bind with the EBS1 and EBS2 sequences, respectively, in domain 1 of the intron. Moreover, the 5' consensus sequences of group II introns are not surrounded by structural motifs showing obvious similarities to those seen in nuclear pre-mRNA IVSs and required for base pairing to U1 snRNA. Jacquier (1990) notes that the branch site is in a similar environment for both nuclear pre-mRNA and group II introns (i.e., bulged out from a helix). This observation strengthens the hypothesis that similar splicing mechanisms result from comparable RNA structures driven in *trans* for the pre-mRNAs by the snRNAs and in *cis* by RNA structure in the group II introns. However, studies of the interactions of sequences surrounding the 5' splice site have not revealed similarities between the structures involved in 5' junction recognition between group II and pre-mRNA introns.

Sharp (1991) in his short article entitled "Five Easy Pieces" brings to bear an additional perspective. As discussed later in this chapter, the *psaA* chloroplast gene in *Chlamydomonas* exists as three widely scattered exons that possess flanking sequences reminiscent of group II intron sequences. For exons one and two, duplex pairing of these intron sequences plus a small RNA *(tscA)*, also encoded by the chloroplast genome and essential for *psaA* assembly, will yield a prototypical group II intron structure. Sharp suggests that "division of a group II intron into 'five easy pieces' could have generated the precursors for the five snRNAs that form the spliceosome in the nuclei of eukaryotic cells." This suggestion is supported the the observation that domain V and a domain I substructure, C1, function in trans (Lambowitz and Belfort, 1993). This would be consistent with the evolution of group II intron domains into snRNAs.

Only when the structures of the catalytic sites of these two groups of introns are known will it be possible to determine whether the two splicing mechanisms are equivalent and, in particular, whether pre-mRNA splicing is essentially a self-splicing reaction. Progress toward this end should come from experiments of the kind reported by Koch *et al.* (1992). They have systematically deleted different domains of the self-splicing group II intron aI5c of the yeast *cox1* gene. They conclude from their experiments that the binding site for domain 5 is within domain 1 and that the complex of 5' exon, domain 1 and domain 5 plus connecting sequences represent the essential catalytic core of this intron.

A novel class of introns (group III) has been described for *Euglena* chloroplast genes (Christopher and Hallick, 1989). *Euglena* group III introns are relatively uniform in size (95–109 bp) and share common features with each other that differentiate them from group I and II introns. Hallick and colleagues (Copertino and Hallick, 1991; Copertino *et al.,* 1991, 1992; Drager and Hallick, 1993) have also described "twintrons" (introns within introns) in the chloroplast genome of *Euglena.* They have described four classes of twintrons.

1. A twintron in the *psbF* gene consists of a group II intron inserted into domain V of a second group II intron. The internal 618 nucleotide intron of the *psbF* pre-mRNA is first excised from the 1042 nucleotide intron of the *psbF* pre-mRNA, following which the remaining 424 nucleotide group II intron is spliced out to yield mature *psbF* mRNA.
2. A second twintron in the *rps3* gene consists of a 311 nucleotide group II intron inserted at the base of a domain VI-like structure within a 98 nucleotide group III intron.
3. Three introns in the *rpoC1* gene and one within *rpl16* are composed of group III introns within other group III introns. Removal of the internal group III intron proceeds excision of the external intron.

4. There is a complex four-intron twintron in the *rps18* gene consisting of four group III introns that are excised in four separate splicing reactions.

Copertino *et al.*, (1992) present an illuminating discussion of twintron evolution. They suppose that introns must have been inserted into *Euglena* chloroplast genes in the course of evolution (i.e., introns late) and that twintrons evolved by further insertion of introns into preexising introns (i.e., introns even later). They speculate that "one possible mechanism for twintron formation is a three-step process of reverse splicing, reverse transcription and homologous recombination." There is strong evidence for the existence of such a mechanism for insertion for fungal mitochondrial introns as discussed later in this chapter. Copertino *et al.*, (1992) also point out that a size of ca. 100 nucleotides characterizes all group III introns and note that "group III introns appear as abbreviated group II introns that contain domain VI minimally attached to the 5' splice site, lacking most of the remaining structural elements." They suggest that many of the cis-elements required for group II intron splicing may be supplied in trans for group III introns noting Sharpe's (1991) speculation discussed above on the potential evolutionary transition from group II to nuclear pre-mRNAs. Copertino *et al.*, (1992) speculate that the *Euglena* group III introns may represent the "closest relative of organelle introns to the nuclear pre-mRNA introns."

The Role of Proteins in Facilitating Splicing

Mitochondrial and nuclear gene mutations affecting intron splicing were discovered for the introns of the *cytb* and *coxI* mosaic genes of yeast before there was physical evidence for these introns. As Perlman *et al.* (1990) point out, by 1977 researchers studying the structure and function of the *cytb* gene had made use of genetic data to conclude that this gene must have interrupted coding sequences. Judicious characterization of specific mutants has continued to play an important role in the identification of cis-acting sequences and trans-acting factors ever since.

Splicing the cytb and coxI transcripts in yeast. *Cytb* and *coxI* are two of the three intron-containing genes in yeast (Table 10–1). Splicing of their transcripts has been scrutinized intensively (for reviews Burke, 1988; Cech, 1985; Dujon, 1981; Grivell, 1989; Lambowitz and Belfort, 1993; Lambowitz and Perlman, 1990; Pel and Grivell, 1993; Perlman *et al.*, 1990; Saldanha *et al.*, 1993; Tabak and Grivell, 1986; Tzagoloff and Myers, 1986; Wolf and Del Giudice, 1988). Both genes show remarkable strain-dependent variation in structure. In strain K14-4A *coxI* is >10 kb and has 8 exons separated by 7 introns. Two optional introns, which split exon 5 into three parts in strain KL14-4A, are missing in strain D273-10B while another two, between exons 1 and 2 and 4 and 5 in D273-10B, are absent in *S. carlsbergensis*, a close relative of *S. cerevisiae*. Long and short forms of the *cytb* gene are also found in different yeast strains. In its long form, *cytb* consists of six exons separated by five introns spanning a segment of ca. 7.5 kb, but in its short form exons 1 to 4 are fused and only the last two introns are present. The long forms of both genes contain type I and type II introns. Some introns contain ORFs in frame with the upstream exon that encode *maturase* proteins required for their own splicing.

Mitochondrial mutants played a key role in the discovery of maturases (Fig. 10–3) and in the identification of intron sequences important in the splicing process. Muta-

Table 10–1. Classification of Introns in Yeast Mitochondrial Genes

Gene	Intron	Type	ORF	Mitochondrial ORF function
cyt b	bI1[a]	II	−	
	bI2	I	+	Maturase
	bI3	I	+	Maturase
	bI4	I	+	Maturase
	bI5[a]	I	−	
cox1	aI1[a]	II	+	Maturase
	aI2[a]	II	+	Maturase
	aI3[a]	I	+	Endonuclease
	aI4	I	+	Endonuclease/maturase
	aI5a	I	+	Endonuclease
	aI5b	I	+[b]	Unknown
	aI5c[a]	II	−	
LSU	LSU.1	I	+	Endonuclease

From Pel and Grivell (1993).

[a] In vitro self-splicing demonstrated (modified after Lambowitz and Perlman, 1990).

[b] In contrast to the other ORFs in the *cytb* and *cox1* genes the ORF in intron aI5b is not in frame with the upstream exon.

tions in the *cytb (COB)* gene either affect cytochrome *b* alone *(cob)*, confer resistance to antibiotics (e.g., antimycin, *ana*), or cause deficiencies in cytochrome oxidase as well *(box)* (Fig. 10–3). The latter defect arises because *box* mutants are unable to synthesize normal amounts of cytochrome oxidase subunit I. There are 10 *box* loci whose numbering reflects order of discovery and not position in *cytb* (Fig. 10–3). All *box* mutants are defective in splicing and map in both exons and introns. The exon mutants *(box4,5, box1, box6)* behave as if they are in a single cistron, consistent with the fact that they encode different regions of the same polypeptide. However, the intron mutants are of two kinds. Mutants in certain loci *(box3, box7, box10)* complement both the exon mutants and each other. This suggests that they are recessive and affect different genes. Mutations at the other loci *(box2, box8, box9)* are cis dominant and complementation tests cannot be done. The existence of these two classes of intron mutants is explained as follows. The trans-complementing mutants are located in maturase-coding regions. Hence, a *box10* mutant will complement a *box3* mutant because the former mutant makes a functional bI2 maturase while the latter synthesizes a wild-type bI3 maturase (Fig. 10–3; Table 10–1). The *box7* mutants affect the maturase encoded by bI4. The cis-dominant *box* mutants cannot be complemented because they interfere with intron folding and prevent proper splicing. Most of the exon mutants at the *box4,5, box1*, and *box6* loci synthesize polypeptides that are shorter than apocytochrome *b*. These proteins may be fragments of the mature CYTB protein, resulting from premature chain termination.

The pleiotropic deficiency in cytochrome oxidase subunit I observed in *box* mutants brings us to the group I intron bI4 of the *cytb* gene (Table 10–1). The maturase encoded by this intron is required not only for splicing bI4, but also to splice aI4, another group I intron in the *cox1* gene (Table 10–1). Thus, *box* mutations, all of which map upstream of or within bI4, with the exception of *box6* (Fig. 10–3), eliminate synthesis of cytochrome oxidase subunit I. The likely reason is that intron and exon

INTRONS, MOBILE ELEMENTS, AND PLASMIDS

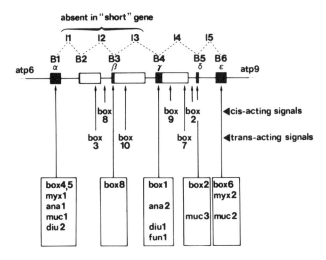

Fig. 10-3. The two forms of the *cytb* gene of *S. cerevisiae*. B1–B6 represent exons and bI1 through bI5 equal introns. The *box* loci are numbered 1–10 in historical order. The *box* mutants cause deficiencies in cytochrome b and cytochrome oxidase subunit I. Several of the *box* loci are found in introns. Mutants in these loci may either affect intron folding (cis-acting signals) or maturase function (trans-acting signals). The exon loci are surrounded by open boxes. They include the remaining *box* loci plus mutational sites conferring resistance to myxothaizol *(myx1, myx2)*, antimycin *(ana1, ana2)*, mucidin *(muc1, muc2, muc3)*, diuron *(diu1, diu2)*, and funiculosin *(fun1)*. From Wolf and Del Giudice (1988) with permission.

mutants upstream fail to splice the *cytb* pre-mRNA properly since they cannot make specific maturases. This interferes with the processing pathway of *cytb* mRNA so that the bI4 maturase is never activated.

Three different groups of *box* mutants map in intron bI4 (Fig. 10–3). The *box7* and *box9* loci are contained within the maturase reading frame while the *box2* locus is near the 3′-end of the intron. The *box7* mutants are trans-acting recessives while the *box2* and *box9* mutants are cis-acting dominants. Thus, the *box7* mutants affect maturase function whereas *box2* and *9* mutants prevent proper base pairing of the *R* and *S* sequences and unconserved E and E′ sequences in helix P3, respectively (Fig. 10–2). Hence, *box2* and *9* mutants are unable to fold bI4 properly so the maturase encoded by this intron cannot be released to splice intron aI4 resulting in the *box* phenotype.

Although the bI4 maturase is required to splice intron aI4 of *cox1*, the aI4 intron also contains an in-frame ORF with the upstream exon encoding a protein closely related to the bI4 maturase (Table 10–1). The aI4 protein also functions as a site-specific nuclease that promotes its own transfer to *cox1* genes lacking it (see *Mobile group I introns*). The maturase function of this protein can be activated by a base pair substitution mutation, *mim-2* (Dujardin et al., 1982). The maturase activated by *mim2* also suppresses *box7* mutations blocking the function of the bI4 maturase. Thus, the *mim2* maturase helps to catalyze splicing of both the aI4 and bI4 introns.

Conversion of the aI4 endonuclease to a maturase by a base pair substitution illustrates an important point. Comparative analysis reveals that three group I maturases share a dodecapeptide motif designated LAGLI-DADG. This motif is associated with a larger family of site-specific endonucleases that mediate group I intron mobility (Lambowitz and Belfort, 1993; Pel and Grivell, 1993).

The group II introns aI1 and aI2 in the *cox1* gene encode maturases that are related to reverse transcriptases (see Kennell *et al.*, 1993; Lambowitz and Belfort, 1993; Lambowitz and Perlman, 1990). These reverse transcriptase activities are found in ribonucleoprotein particles from yeast mitochondria, are highly specific for the introns and their flanking regions, and will use *cox1* pre-mRNA or excised intron RNAs as templates. Thus, group II introns aI1 and aI2 appear to be retroelements and the reverse transcriptases encoded by them seem closely related to the enzyme specified by the Mauriceville mitochondrial plasmid of *Neurospora* discussed later in this chapter. Most other group II intron ORFs also resemble reverse transcriptases including those in the mitochondrial genome of *Marchantia*.

Nuclear-encoded proteins also facilitate mitochondrial intron splicing in yeast (Table 10–2 also see reviews by Burke, 1988 and Grivell, 1989). The *CBP2* gene encodes a protein required to splice bI2 (Table 10–2) (Gampel *et al.*, 1989). The protein may stabilize bI2 in the catalytically active form. Mutants in the *MSS18* gene, encoding a 268 amino acid protein, prevent processing of the aI5b intron in the *cox1* gene (Seraphin *et al.*, 1988). Disruption of this *MSS18* results in a leaky respiratory-deficient phenotype so some splicing can occur *in vivo* in the absence of the *MSS18* gene product. *MSS18* mutants retain this leaky phenotype even when their mitochondrial genomes lack introns, suggesting that the *MSS18* gene product may function in mRNA translation as well as splicing (Grivell, 1989). This is one of several nuclear gene-encoded proteins that have functions other than splicing.

Table 10–2. Nuclear-encoded Proteins Required for Splicing of Mitochondrial Introns

Gene	Introns Affected	Other Functions
Category 1		
NAM2	aI4,bI4	mt leucyl tRNA synthetase
CBP2	bI2	None
MRS1/PET157	aI5b,bI3	None
PET54	aI5b	*cox3* translation
MRS2	Group II	Unknown mt function
CYT18	Group I in *N. crassa*	mt tyrosyl tRNA synthetase
YTS1	Group I in *P. anserina*	mt tyrosyl tRNA synthetase
SUV3	aI5b,bI3	RNA metabolism/translation?
Category 2		
MSS51	aI1,aI2,aI4,aI5c	*cox1* translation
MSS18	aI5b	Unknown function
MSS116	several *cox1, cytb*	RNA unwindase?
mRF-1	aI1,aI2	Release factor
Category 3		
MRS3	bI1	mt carrier
MRS4	bI1	mt carrier
NAM1/MTF2	bI2,aI1	Transcription/translation factor
NAM7/UPF1	bI2,aI1	RNA helicase?
NAM8	bI2,aI1	Unknown function

From Pel and Grivell (1993).

The *MSS51* gene product is required for splicing of *cox1* pre-mRNA (Faye and Simon, 1983; Simon and Faye, 1984). The *cox1* pre-mRNA, which accumulates in *pet* mutants in this gene, contains the aI1, aI2, aI4, and aI5c introns (Table 10–2) with aI3 being self-splicing. Since processing of aI4 and aI5c is dependent on translation of upstream exons, the *MSS51* gene product probably acts on the group II aI1 and/or aI2 introns and is only indirectly required for splicing the two group I introns, aI4 and aI5c. Like the *MSS18* gene product, the *MSS116* gene product seems to have a dual role in splicing and translation (Seraphin et al., 1989). The protein encoded by *MSS116* is required for splicing of several introns in the *cytb* and *cox1* genes (Table 10–2). This protein belongs to a family that includes initiation factor eIF4A and several other proteins thought to have helicase activity. As expected for a multifunctional protein, the respiratory defect caused by a null *mss116* mutation is not suppressed in an intronless strain. The *MRS1* gene product is required for excision of two group I introns, aI5B and bI3 (Bousquet et al., 1990) while the *MRS2* protein is essential for excision of all mitochondrial group II introns in yeast (Wiesenberger et al., 1992).

Like the *MSS18* and *MSS116* gene products, the *MRS2* protein must have another function since strains with a disrupted *MRS2* gene, but devoid of mitochondrial introns, display the *pet*⁻ phenotype. Other dual function genes are *PET54* and *NAM1*. The product of the former gene affects translation of the mitochondrially encoded *cox3* gene (Chapter 17) and also splicing of aI5b with the functions being separable (Valencik et al., 1989; Valencik and McEwen, 1991). *NAM1* encodes a mitochondrial transcription factor and also seems to be involved in splicing (Ben-Asher et al., 1989; Lisowsky, 1990). A dominant suppressor mutation *(SUP-101)* mapping in the *MRS3* gene suppresses a splicing-deficient mutant in the group II intron bI1, but disruption of this gene has no effect on wild type (Schmidt et al., 1987). This is also true of the *MRS4* gene (Wiesenberger et al., 1991).

One of the most interesting nuclear genes is *NAM2*. Dominant mutations in this gene suppress *box7* maturase mutations in the bI4 intron (Dujardin et al., 1980). The product of these suppressor mutations activates the latent aI4 maturase described earlier so that splicing of both the aI4 and bI4 introns occurs normally (Dujardin et al., 1983). Cloning and sequencing of the *NAM2* gene revealed that the gene encodes the mitochondrial leucyl-tRNA synthetase (Herbert et al., 1988). This protein seems to have the dual functions of leucyl-tRNA charging and facilitating RNA splicing as does the mitochondrial tyrosyl-tRNA synthetase encoded by the *cyt-18* gene of *Neurospora*. (see below). However, the role of the *NAM2* gene product in splicing is not straightforward. The function is activated by a dominant mutation, which in turn activates a normally latent maturase (although the cloned wild-type gene on a multicopy plasmid has weak suppressor activity, Labouesse et al., 1987).

In order to rationalize the many different trans-acting factors affecting mitochondrial mRNA splicing Pel and Grivell (1993) have divided the nuclear genes encoding them into three categories (Table 10–2). The first includes genes encoding proteins likely to be directly involved in splicing, the second genes specifying proteins that may be involved in RNA maturase synthesis, and the third includes genes which were isolated as multicopy suppressors of mitochondrial splicing defects.

Nuclear genes required for mitochondrial intron splicing in Neurospora. The large mitochondrial rDNA gene of *Neurospora crassa* contains a single group I intron. Within this intron is a reading frame encoding the small subunit mitochondrial r-protein S-5 (see Lambowitz et al., 1985). This protein is required for assembly of small subunits. The primary transcript of the large rRNA gene is a 35S (5.6-kb) molecule from which the 2.3-kb intron sequence is removed to yield the mature 25S (3.3-kb) rRNA molecule. This intron will self-splice *in vitro*, but not *in vivo* (Garriga and Lambowitz, 1984). Its splicing depends on the products of at least three nuclear genes (*cyt-4*, *cyt-18*, and *cyt-19*), mutants of which accumulate the 35S precursor. The *cyt-4* mutants are cold-sensitive and the 35S RNA that accumulates in them at 25°C has a 110-nucleotide 3' extension. At 37°C, unspliced 35S RNA no longer accumulates, but the mutants remain defective in 3'-end synthesis (Garriga et al., 1984). The *cyt-4* mutants are also pleiotropic, having defects in a number of other mitochondrial RNA-processing pathways (Dobinson et al., 1989). These defects include inhibition of some 5'- and 3'-end processing reactions and defective splicing of the two group I *cytb* introns. The protein coded by *cyt-4* (Turcq et al., 1992) bears strong similarity to proteins in *S. cerevisiae* and *Schizosaccharomyces pombe*, which may have protein phosphatase functions. The CYT-4 protein is found in mitochondria and is truncated or deficient in *cyt-4* mutants. The results suggest that RNA splicing and processing reactions are regulated by protein phosphorylation.

The *cyt-18* and *cyt-19* genes encode trans-acting factors required for splicing the same set of mitochondrial group I introns (Lambowitz et al., 1985). Temperature-sensitive mutants in these genes accumulate 35S RNA and unspliced mRNA precursors of the *cytb* and *atp6* genes. While the CYT-19 protein has not been identified, *cyt-18* encodes the mitochondrial tyrosyl-tRNA synthetase (tyrRS; Akins and Lambowitz, 1987; Lambowitz and Perlman, 1990; Majumder et al., 1989). The *cyt-18-1* and *cyt-18-2* mutants, defective in both splicing and aminoacylation, have the same missense mutation ($Gly_{127} \rightarrow Glu$). The affected residue is in the nucleotide binding fold. This region is conserved in the tyrRS enzymes from *E. coli* and yeast and *N. crassa* mitochondria.

Studies of partial revertants and *in vitro* mutagenesis of the *cyt-18* gene followed by synthesis of the mutant tyrRS using chimeric plasmids in *E. coli* have permitted further dissection of the enzyme into functional domains (Cherniack et al., 1990; Kittle et al., 1991; Lambowitz and Perlman, 1990). Secondary mutations that restore splicing activity are clustered towards the N-terminus of the CYT-18 protein in a span of ca. 65 amino acids. This region is absent from the yeast mitochondrial and *E. coli* tyrRS enzymes. Experiments with the CYT-18 protein made in *E. coli* show that other regions of the protein, including the putative C-terminal tRNA-binding domain, are also required for splicing. Several mutations made *in vitro* near the ATP-binding site result in loss of tyrRS and tyrosyl adenylation activities of the CYT-18 protein while substantial splicing activity remains. Thus, normal catalytic activity of the protein is not required for splicing.

The finding that the CYT-18 protein is needed to splice a number of unrelated group I introns raises the possibility that the protein interacts with conserved sequences or structures in those introns (Lambowitz and Perlman, 1990). A reasonable hypothesis would be that the protein recognizes regions that resemble the normal substrate $tRNA^{Tyr}$. Consistent with this hypothesis is the finding that regions likely to participate in binding include parts of the nucleotide binding fold and the C-terminal domain. Both of these regions are also involved in binding $tRNA^{Tyr}$ in the comparable bacterial tyrRSs (Kittle et al, 1991; Kamper et al., 1992). The observation that regions required for splicing are distributed throughout the CYT-18 protein suggests that the protein makes multiple contacts with the intron RNA, as it very likely does in tRNA binding (Bedouelle, 1990). CYT-18 also possesses the nonconserved N-terminal domain in which partial revertants having splicing activity cluster. This domain is missing from the *E. coli* and yeast mitochondrial enzymes, both of which lack splicing activity (Lambowitz and Perlman, 1990). Since this domain is also found in the bifunctional protein from *Podospora anserina* (Kamper *et al.*, 1992), it may be a relatively recent addition to the tyrRS protein following the divergence of the filamentous ascomycetes and yeast (Cherniack *et al.*, 1990; Kamper *et al.*, 1992).

Two other studies have pretty much clinched the argument that the intron and tRNA binding domains of the CYT18 protein are related (Guo and Lambowitz, 1992; Mohr *et al.* 1992). The CYT18 protein appears to promote splicing by binding to the intron core and stabilizing the core in a conformation necessary for catalysis. Furthermore, the catalytic core bears structural resemblance to tRNA molecules. Of particular note is the finding that the splicing activity of tyrRS requires the amino-terminal domain missing from the yeast and bacterial enzymes. This is consistent with the hypothesis set forth above that acquisition of splicing function by this enzyme may be of relatively recent evolutionary origin.

Not only does the CYT-18 protein promote splicing of a variety of group I mitochondrial introns in *N. crassa*, but this protein will also promote reverse splicing *in vitro* under physiologically relevant conditions (Mohr and Lambowitz, 1991). Thus, CYT-18 may contribute to intron mobility since a reverse transcriptase has also been identified in *Neurospora* mitochondria (see Mobile Elements). Other experiments show that the wild-type CYT-18 protein will suppress mutations in the distantly related bacteriophage T4 *td* intron (Mohr *et al.*, 1992). These results support the hypothesis that CYT-18 is able to recognize certain common features between the fungal and bacteriophage introns and can stabilize the catalytically active core of the intron in the above T4 *td* mutants. Thus, CYT-18 differs markedly from the maturases described earlier for the yeast *cytb* and *cox1* genes. They are generally specific for one or two introns while CYT-18 acts generally to facilitate group I intron splicing.

Trans-splicing of group II introns. The *psaA* gene of *Chlamydomonas reinhardtii* consists of three widely separated exons on the chloroplast genome (see Rochaix, 1992a,b; Rochaix *et al.* 1991 for reviews). The first exon is 50 kb away from the second exon, which is removed by 90 kb from exon 3. Transcription of these exons is discontinuous and they are flanked by typical group II intron consensus sequences (Kuck *et al.*, 1987). Both *psaA* introns are split into 5' and 3' regions. Nuclear mutants affecting maturation of *psaA* mRNA belong to at least 14 complementation groups and fall in one of three functional classes (Choquet *et al.*, 1988; Goldschmidt-Clermont *et al.*, 1990). Class A mutants belong to five different complementation groups. They cannot splice the intron sequences between exons 2 and 3 transcripts, but can splice those separating exons 1 and 2. Therefore, one of the two resulting RNAs contains the spliced exons 1 and 2 while the second contains the unspliced exon 3 (Fig. 10–4A). Class B mutants fall in two loci and are unable to perform either splicing reaction so they accumulate unspliced transcripts of all three exons (Fig. 10–4A). The exon 2 transcript also contains the *psbD* mRNA, which cannot be removed because of the splicing defect. Mutations in seven genes result in the class C phenotype. These mutants cannot splice exon 1 and 2 transcripts, but they are able to splice exons 2 and 3. These mutants accumulate two tran-

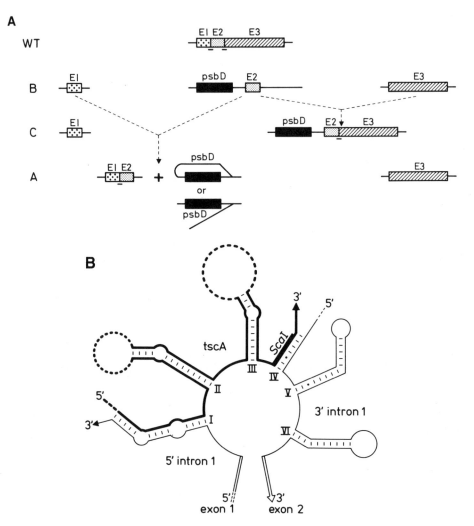

Fig. 10–4. (A) Three classes of nuclear gene splicing mutants for the chloroplast *psaA* gene of *Chlamydomonas reinhardtii*. E1, E2, and E3 represent the transcripts of the three *psaA* exons. Each exon is transcribed into a pre-mRNA. The *psbD* gene is cotranscribed with exon 2. Because of the dispersion of the exons along the chloroplast genome both introns i1 and i2 are disrupted. The 5'- and 3'-ends of the introns are indicated. Mutants from class B (2 loci) are deficient in both trans-splicing reactions. Class C mutants (7 loci) are unable to splice exons 1 and 2, but can splice exons 2 and 3. Class A mutants (5 loci) are unable to splice exons 2 and 3, but can splice exons 1 and 2. Class A mutants and wild-type cells also accumulate intron 1 containing the *psbD* transcript. From Rochaix et al. (1990) with permission. (B) Secondary structure model for the split intron 1 of the chloroplast *psaA* gene of *Chlamydomonas reinhardtii*. In this model, the conserved core structure of the group II intron is composed of three separate transcripts: the precursor of exon 1 (with the 5' part of intron 1), the precursor of exon 2 (with the 3' part of intron 1), and the tscA RNA (thick line). The sequence corresponding to the *Sca*I restriction site in the *tscA* gene (black bar) base pairs with the 3' part of intron 1 to form helix IV. The large loops in domains II and III (dotted) are not drawn to scale. From Goldschmidt-Clermont (1991) with permission.

scripts. One contains the exon 1 message and the other the spliced exon 2–3 mRNA plus the *psbD* message, which cannot be removed by splicing.

A chloroplast gene product *(tscA)* is also required for splicing exons 1 and 2 (Goldschmidt-Clermont *et al.,* 1991; Goldschmidt-Clermont, 1991). The *tscA* gene product is a small chloroplast RNA that acts in the first trans-splicing reaction. The *tscA* RNA probably base pairs with the two separate exon precursors and forms part of the catalytic core of the group II intron, which must be trans-spliced to bring exons 1 and 2 together (Fig. 10–4B also see Sharp, 1991).

The *psaA* gene of land plants is continuous, but the *rps12* gene is divided into two parts that must be trans-spliced (see Sugiura, 1991, 1992 for reviews). In tobacco there is one copy of the 5'-*rps12* exon in the unique sequence region. This exon is separated by 28 kb from two copies of the 3'-*rps12* coding sequence in the inverted repeat. The 3' segment is itself broken into two exons by a 536-bp cis-spliced intron. The two transcripts are spliced in trans and the intron between the two 3' exons is excised to yield the mature transcript. Thus this transcript requires both cis- and trans-splicing steps to mature.

Trans-splicing is not confined to chloroplasts and examples of trans-splicing are to be found in three different *ndh* genes of flowering plant mitochondria (see Bonen, 1993, for a review). RNA editing (Chapter 14) seems to be involved in improving intron folding in trans-spliced transcripts of certain higher plant mitochondrial genes. The *ndh2* gene of the *Oenothera* mitochondrial genome is split into five exons (Binder *et al.,* 1992). The first two and the last three are coded in distant locations in the genome and transcribed separately. Formation of the mature mRNA requires three cis- and one trans-splicing event. The cis-spliced exons are separated by group II introns. The sequences adjacent to the trans-spliced exons b and c also have the characteristics of group II introns. Three RNA editing events are observed in this split group II intron. Two of these events allow additional base pairings in the secondary structure. They may promote the trans-splicing reaction. The *ndh1* genes of *Oenothera, Petunia,* and wheat are broken into five separate exons (see Bonen, 1993; Gray *et al.,* 1992; Hanson and Folkerts, 1992, for reviews). Bonen (1993) points out that in dicotyledons such as *Oenothera* and broad bean, the terminal two exons are separated by a conventional group II intron (which may encode a RNA maturase with sequence similarity to retroviral reverse transcriptases). However, in wheat and petunia these exons are widely separated in the mitochondrial genome. The discontinuity is upstream of the maturase ORF in wheat, but is downstream in petunia. Thus, rearrangement mutations seem to have occurred during the course of angiosperm mtDNA evolution that resulted in a shift from a cis- to a trans-splicing motif. In *Oenothera* editing of two nucleotides in the 3'-part of *ndh1* trans-splicing group II sequences improves folding of domain VI and may be required for trans-splicing (Wissinger *et al.,* 1991).

MOBILE ELEMENTS

As Dujon (1989) points out in a review, mobility of group I introns was described genetically long before the introns themselves were discovered. The best known of these introns is the ω group I intron in the mitochondrial 21S rRNA gene of yeast. This intron is transmitted to the homologous site in intronless ω⁻ strains by "intron homing." The homing process is accompanied by gene conversion of closely linked antibiotic resistance markers in the ω^- strain to the phenotype(s) present in the ω^+ strain.

Gene conversion associated with the expandable *var-1* gene yeast provides yet another mechanism for moving mitochondrial DNA sequences from donor to recipient strains (Butow et al., 1985). Again, genetic analysis coupled with characterization of the "long," "short," and various intermediate forms of the VAR protein emanating from mitochondrial crosses defined the phenomenon before its molecular basis was understood. Since that time other mobile elements, notably G+C clusters in noncoding regions of the yeast mitochondrial genome, have been identified (see Clark-Walker, 1992; Weiller et al., 1989). Not to be outdone, group II introns also appear to have developed mechanisms designed to ensure their mobility (see Lambowitz, 1989; Lambowitz and Belfort, 1993). These introns appear to insert into mtDNA by a process of reverse transcription. Studies of mobile elements, particularly in mitochondria, have led the way in defining the mechanisms by which introns can be transferred from one gene to another.

Mobile Group I Introns

Dujon (1989) distinguishes three mechanisms of intron mobility. *Intron homing* is the process by which an intron in one gene is introduced into the same site in a homologous gene lacking the intron. This process occurs with very high efficiency. *Intron loss* is the precise deletion of an intron from an intron-containing gene. Intron loss occurs with low efficiency, but can be selected for in mutants where the intron cannot be spliced. *Intron transposition* is the addition of an intron to a specific site in a nonhomologous gene lacking that intron or to a different site in the homologous gene. This process almost certainly occurs as witnessed by the distribution of closely related introns in different genes, but remains to be experimentally demonstrated.

The mobility of the mitochondrial rRNA large subunit group I intron (LSU.1) of yeast has been studied for many years and serves as the model for later studies (see Gillham, 1978; Dujon, 1981; Dujon and Jacquier, 1983; Dujon et al., 1986; Dujon, 1989; Lambowitz and Belfort, 1993; Pel and Grivell, 1993; Perlman and Butow, 1989 for reviews). Mutations exhibiting polar behavior and biased transmission in crosses map in the gene encoding the large rRNA of the yeast mitochondrion. They are resistant to antibiotics such as chloramphenicol (cap^r) and erythromycin (ery^r). The responsible locus was named ω and all yeast strains can be classified as ω^+ or ω^- based on the frequencies of reciprocal recombinants for polar markers. For example, in a cross of the type $cap^r\ ery^r \times cap^s\ ery^s$, reciprocal recombinants will be in approximately equal frequency (i.e., $cap^r\ ery^s = cap^s\ ery^r$) if both strains are either ω^+ or ω^-. In contrast, if the $cap^r\ ery^r$ parent is ω^+ and the $cap^s\ ery^s$ parent is ω^-, $cap^r\ ery^s \gg cap^s\ ery^r$ recombinants. If the ω alleles are reversed with respect to the antibiotic resistance markers, $cap^r\ ery^s \ll cap^s\ ery^r$ recombinants. The three polar loci defined by antibiotic resistance markers were designated *rib-1, rib-2,* and *rib-3* with *rib-1* being the most polar and *rib-3* the least polar. In $\omega^+ \times \omega^-$ crosses the most asymmetric outputs of reciprocal recombinants are seen for *rib-1* and *rib-2* and least for *rib-2* and *rib-3*. In $\omega^+ \times \omega^-$ crosses most recombinants are ω^+ and the ω^- allele is virtually eliminated from the population.

The explanation of the polarity of transmission of resistance markers in $\omega^+ \times \omega^-$ crosses and the disappearance of ω^- alleles is the existence in ω^+ strains of the 1.1-kb

LSU.1 intron. This intron contains a 235 amino acid open reading frame (ORF) that encodes the I-SceI endonuclease responsible for homing of the intron into ω^- strains (Table 10–3). This ORF is not in frame with the preceding exon and is inserted in stem P8 of the intron core structure (Fig. 10–2a). The I-SceI endonuclease causes a double strand break in the intronless LSU gene at a location corresponding to the intron insertion site (Dujon et al., 1985; Zinn and Butow, 1985). The enzyme, which has been synthesized in *E. coli* after the universal code equivalent of the ORF had been constructed (Colleaux et al., 1986), makes a staggered 4-bp cut in a recognition sequence that is at least 18 bp long (Colleaux et al., 1988).

The *rib1* locus, defined by mutations to chloramphenicol resistance, can be divided into two tightly linked clusters of cap^r mutations designated *rib-1-A* and *rib-1-B* (Fig. 10–5). Mutations at the *rib-1-B* locus map within the insertion site and do not exhibit polarity in crosses to ω^+ strains. These ω^n mutations delimit essential nucleotides within the insertion site, replacement of which prevents cleavage of the recognition sequence. Many mutations have also been constructed *in vitro* to define the recognition site further. In ω^- strains the *rib3* locus, defined by mutations to resistant to erythromycin and related antibiotics, maps about 670 bp upstream of the *rib-1-B* locus; the *rib-1-A* locus is 50 to 60 bp downstream of *rib-1-B;* and the *rib-2* locus, also defined by erythromycin resistance mutations, is ca. 110 bp downstream of *rib-1-A* (Fig. 10–5). During intron insertion, gene conversion of adjacent sequences to ω^+ can occur in both directions with the probability of conversion declining as a function of distance. This explains the phenomenon of polarity. The probability of conversion at the *rib-1-A* locus is highest with rates of conversion of *rib-2* mutations being intermediate and *rib-3* mutations low. Thus, polarity is really bidirectional and not unidirectional as originally thought.

The mobility of the Sc LSU.1 intron remained a curiosity for a long time for, as Dujon (1989) explains, demonstration of mobility depends on three criteria, none of which is sufficient by itself:

1. Existence of intron-containing and intron-less forms of the gene coupled with the possibility of genetic exchange.
2. Existence within the intron of an internal ORF encoding a functional double stranded endonuclease.
3. Existence within the intronless gene of the appropriate recognition sequence for that endonuclease.

Table 10–3. Summary of Mobile Group I Introns in Chloroplasts and Mitochondria

Organism	Genome	Gene	Intron	Endonuclease
Saccharomyces cerevisiae	mt*	L-rDNA	LSU.1	I-SceI
	mt	cox-1	cox1a-I3	I-SceIII
	mt	cox-1	cox1a-I4	I-SceII
	mt	cox-1	cox1a-I5	I-SceIV
Chlamydomonas eugametos	cp	L-rDNA	LSU-5	I-CeuI
Chlamydomonas reinhardtii	cp	L-rDNA	LSU	I-CreI
Chlamydomonas smithii	mt	cytb	cobI1	I-CsmI

Modified from Lambowitz and Belfort (1993).

*mt = mitochondrial; cp = chloroplast

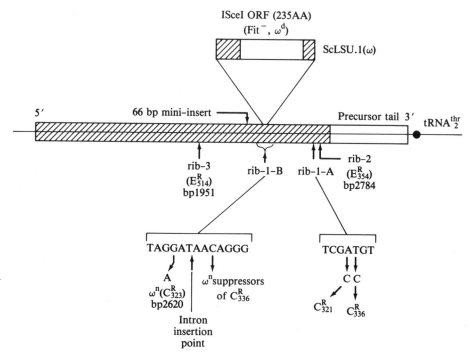

Fig. 10–5. Physical and genetic map of the large rRNA gene of *S. cerevisiae* showing the position where the mobile group I intron Sc LSU.1 (ω) inserts. The intron is indicated above the gene and the ORF that encodes the double-stranded endonuclease I-*Sce*I is unshaded. Mutations called Fit⁻ or ω^d in the intron block endonuclease function. The positions of the antibiotic resistance loci *rib-1-A*, *rib-1-B*, *rib-2*, and *rib-3* are also shown together with antibiotic resistance mutations (E^R, erythromycin resistance, C^R, chloramphenicol resistance) that identify them. The base sequences including the *rib-1-A* and *rib-1-B* loci are also shown. Of particular interest is the *rib-1-B* locus containing the insertion point for Sc LSU.1. The ω^N mutations, which cause either chloramphenicol resistance or can suppress this phenotype, also prevent insertion of Sc LSU.1 into the intronless large subunit gene of ω^- strains.

Most of these criteria, especially the first one, are not met simultaneously. This is illustrated by the case of a mobile intron within the 23S rRNA gene of the *Chlamydomonas* chloroplast (Durrenberger and Rochaix, 1991) which can self-splice *in vitro* (Herrin *et al.*, 1990). This intron contains an internal open reading frame and all known strains of *C. reinhardtii* have the intron. This means that an intronless gene has to be created to demonstrate mobility. Durrenberger and Rochaix achieved this by making a cDNA from the mature 23S rRNA. Expression of the ORF in *E. coli* yielded an endonuclease that cleaved the cDNA at or close to the exon junction sequence. However, to demonstrate that the intron could home *in vivo*, Durrenberger and Rochaix (1991) had to take a novel approach. The cDNA with the homing site was inserted into a vector containing the chloroplast *atpB* gene plus the downstream sequence from *psbD* and used to transform an *atpB* deletion mutant. Characterization of the chloroplast DNA of the transformants revealed that the 23S rRNA intron had inserted into the intronless homing site present in the cDNA downstream of the *atpB* gene. The I-*Cre*I endonuclease encoded by this intron has been isolated and its recognition sequence is 24 bp in length (Thompson et al., 1992). Screening of natural variants of this recognition sequence reveals that the I-*Cre*I enzyme will tolerate single and even multiple base pair changes within the recognition site (Durrenberger and Rochaix, 1993).

Several other mobile group I introns have also been identified in chloroplasts and mitochondria (Table 10-3). The chloroplast gene encoding the 23S rRNA of *Chlamydomonas eugametos* contains six group I introns and is located in the inverted repeat as in other chloroplast genomes (Turmel et al., 1991). Intron 5 exhibits mobility in interspecific crosses of *C. eugametos* and *C. moewusii* and is transmitted to both copies of the intronless recipient gene (Lemieux and Lee, 1987). Intron 5 encodes a double-stranded endonuclease (Gauthier et al., 1991). This enzyme recognizes a sequence of 19 bp centered around the insertion site and produces a staggered cut 5 bp downstream from the insertion site (Marshall and Lemieux, 1991, 1992).

Chlamydomonas reinhardtii is interfertile with *C. smithii*. Boynton et al. (1987) observed that in vegetative diploids, where mtDNA from both parents is transmitted (Chapter 7), the group I intron in the *C. smithii cytb* gene becomes integrated into the intronless *C. reinhardtii* gene with very high efficiency. The *C. smithii* intron contains an ORF with 36% amino acid identity with I-*Sce*1 (Colleaux et al., 1990). The aI4 intron of the *cox1* gene is also mobile and encodes an endonuclease I-*Sce*2 with properties similar to I-*Sce*1 (Delahodde et al., 1989; Wenzlau et al., 1989). This enzyme also has latent maturase activity (see *Splicing the cytb and cox1 transcripts in yeast*). Obviously, this would be advantageous if the intron failed to self-splice or the organism to which it was transferred lacked the machinery to splice it. As Dujon (1989) puts it, "the balance acts in favor of the presence of introns so long as this presence is phenotypically silent."

Mobile group I introns are also known in phage T4 and the nuclear rDNA gene of the slime mold *Polysphondylium polycephalum*, and are likely in at least two *Neurospora* mitochondrial genes (Dujon, 1989; Lambowitz and Belfort, 1993).

Classification of the endonucleases encoded by group I introns reveals that the majority are characterized by the two dodecapeptide repeat LAGLI-DADG mentioned earlier and shared with several maturases (Lambowitz and Belfort, 1993). However, a second class of group I introns have the motif GIY-10/11aa-YIG together with longer repeated sequences.

Splicing-deficient mutants are also respiration deficient, but respiratory-competent revertants are readily obtained (Dujon, 1989). These revertants frequently delete the entire defective intron plus upstream and downstream introns. The revertable mutants include ones that impair splicing by affecting ORFs in the *cytb* bI2 and bI3 introns (Gargouri et al., 1983), exon mutants interfering with formation of the P1 stem of bI4 (Perea and Jacq, 1985), and nuclear gene mutations affecting splicing of the *cox1* aI5b intron (Hill et al., 1985; Seraphin et al., 1988). Because group II introns encode proteins with homologies with retroviral reverse transcriptases (Michel and Lang, 1985) an attractive hypothesis to explain group I intron loss is that these reverse transcriptases make cDNA copies of the spliced mRNA. This is followed by homologous recombination between exon sequences whereby the cDNA replaces the resident gene. Levra-Juillet et al. (1989) provided evidence for this model in the case of the aI5b intron. This intron cannot be spliced in nuclear *mss18* mutants (Table 10-2), but can be deleted by reversion in these strains. Deletion will not occur unless aI1 or aI2 are present in the gene. These are the only two group II introns in the yeast mitochondrial genome. By the same token group I intron insertion could occur by reverse splicing into an intronless mRNA followed by synthesis of a cDNA copy by reverse transcriptase and homologous recombination with the intronless resident gene (Lambowitz and Belfort, 1993). Reverse splicing of group I and group II introns has been demonstrated *in vitro* (see Lambowitz and Belfort, 1993).

Mobile Elements Associated with the Yeast Mitochondrial var-1 Gene

The properties of the *var1* gene have been reviewed by Butow *et al.* (1985) and Wolf and Del Giudice (1988). The VAR1 protein is a protein of the small subunit of the mitochondrial ribosome and is required for its assembly (Terpstra *et al.*, 1979). The protein ranges in size from 40 to 44 kDa and the gene encoding the protein is extremely A+T rich (ca. 90% A+T). Variation in size of the *var1* protein can be explained by the association of two types of genetic elements with the *var1* gene. They account for all the observed forms of the VAR1 protein (Fig. 10-6). These elements are designated a and b with variant forms of b termed b_p (for b partial). This genetic model for the structure of the *var1* gene was formulated from results obtained from crosses. For example, when a strain having the largest detectable VAR1 protein (44 kDa), is crossed

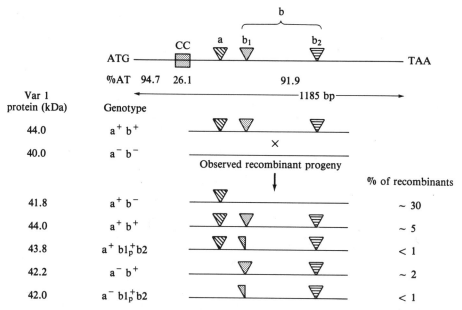

Fig. 10-6. The structure of the long form of the *var1* gene of *S. cerevisiae* showing DNA insertions and the different recombinant progeny genotypes obtained following a cross between strains carrying the long and short forms of the gene. The box labeled *cc* indicates the common G+C cluster found in all *S. cerevisiae* strains, but not in *S. capensis* (Wenzlau and Perlman, 1990). The inverted triangles indicate the a (▽), b_1 (▽), b_{1p} (▼), and b_2 (▽) insertion sequences (see text for details). The b_1 and b_2 elements together constitute the b element and do not recombine. The percent of each recombinant class is also shown. Adapted from Butow *et al.* (1985).

to a strain containing the smallest protein (40 kDa), six classes of progeny are obtained corresponding to the six variant forms of the protein actually observed. The predominant nonparental class (20–30% of the progeny) contains the *a* element alone (Fig. 10–6). The second important nonparental class contains only the *b* element and is much less frequent (ca. 2% of the progeny). Other nonparental classes are accounted for by the presence of b_p. The *b* element itself is physically subdivisible into two parts (Fig. 10–6), but these normally recombine as a unit.

The molecular events underlying this genetic model are well understood. The *var1* gene lacking both the *a* and *b* elements and encoding the 40-kDa protein is ca. 90% A+T so about 83% of the amino acid codons are A+T rich with one-third of the amino acids in the protein being asparagines (AAT). Nearly 40% of the G+C residues are concentrated 186 bp downstream of the initiator ATG in the common cluster *(cc)* (Fig. 10–6). This cluster is present in all laboratory strains of *S. cerevisiae* surveyed (Hudspeth *et al.*, 1984), but not in *S. capensis* (Wenzlau and Perlman, 1990), a point that will be returned to shortly. The physical equivalents of the *a*, *b*, and b_p elements are in-frame DNA insertions coding for additional amino acids that increase the size of the VAR1 protein. The *a* insert correlates with the presence of a new G+C cluster 146 bp downstream from *cc*. The *a* cluster is identical in sequence to the common cluster, but in opposite orientation. The *b* elements correlate with in-frame insertions of AAT (asparagine) codons. The b_1 element is four AAT codons and inserts in a region containing eight AATs. The b_p element is derived from b_1 and contains two AAT codons. The b_2 element includes six AAT codons and inserts into a region with five AATs. Although b_1 and b_2 are 350 bp apart, *b* element insertion involves both sets of AAT codons.

The structure of the *a* element and its insertion are reminiscent of transposition. At the ends of this optional G+C rich cluster are 9-bp terminal repeats. The cluster is flanked by a 22-bp pure AT sequence, but the "recipient" sequence is only 20 bp, suggesting that a 2-bp duplication occurs upon insertion of the *a* element. A mitochondrial gene product is not required for transmission of the *a* element as insertion occurs in crosses of vegetative petite mutants that are unable to carry out mitochondrial protein synthesis (Zinn *et al.*, 1988).

Wenzlau and Perlman (1990) showed that the *cc* cluster is also a mobile and optional G+C-rich cluster. This cluster is not found in *Saccharomyces capensis* as mentioned above, but is transmitted to the *var1* gene of *S. capensis* in crosses to *S. cerevisiae*. *S. capensis* also lacks the *a* element, which appears to convert independently of the *cc* element in crosses to *S. cerevisiae*. Since double-strand breaks are known to enhance the frequency of homologous recombination in yeast, Zinn *et al.* (1988) examined the *var1* gene for double strand breaks and detected them at the boundaries of the *cc* and *a* elements. However, cuts were not observed in cells lacking the *a* element. Since these experiments were done with vegetative petite cells the enzyme(s) making the cuts must be coded in the nucleus.

G+C clusters constitute the most abundant repetitive elements in yeast mtDNA and many of them are optional, thus contributing to the polymorphism of the mitochondrial genome of *S. cerevisiae* (de Zamoroczy and Bernardi, 1986). Weiller *et al.* (1989) made a survey of such sites in various yeast strains and discovered that most of the 26 dimorphic or polymorphic sites differ by the presence or absence of a majority class element, which they call M. The M class terminates with an AGGAG motif. Weiller *et al.* compared sequences with and without the GC clusters and found that elements of subclasses M1 and M2 are inserted 3' to a TAG motif flanked by an A+T-rich sequence. M3 elements, in contrast, occur in tandem arrays of two to four G+C clusters and are inserted 3' to the AGGAG terminal sequence of a preexisting cluster. Weiller *et al.* therefore regard the TAG element as being part of the target site for M1 and M2 elements. The AGGAG motif of a preexisting cluster would then constitute the target site for M3 cluster insertion. The dinucleotide AG is common to both target sites so Weiller *et al.* suggest that it is duplicated during G+C cluster insertion. Weiller *et al.* also recognize two minor classes of G+C clusters. The G class includes five highly homologous clusters that occur in *ori* sequences of yeast mtDNA (see de Zamoroczy *et al.*, 1984). One of these elements is a true optional cluster, but the others may also be optional as they occur only in half of the very similar *ori* sequences. The V class elements consist of a group of four G+C-rich elements including the *a* insert in the *var1* gene.

As Butow *et al.* (1985) note in their review of the *var1* gene, "jumping GC clusters may be a normal feature of the yeast mitochondrial genome."

Mobility of Group II Introns

Clark-Walker (1992) reviews the evidence for group II intron mobility. In yeast Meunier *et al.* (1990) reported evidence for biased transmission of the aI1 and aI2 group II introns. Skelly *et al.* (1991) observed that the group II aI1 intron in *Kluyveromyces lactis* also is transmitted with high frequency. These genetic observations, like similar findings with group I introns, are consistent with mobility, but unlike the latter introns an endonuclease is not involved. The mechanism of group II intron transmission differs and probably involves reverse splicing followed by reverse transcription of the unspliced precursor mRNA. The intron-containing cDNA would then replace the resident intronless gene by homologous recombination. This scenario, while promising, is still rather indirect, but there are three suggestive pieces of supporting evidence. First, is the finding of open reading frames in some group II introns resembling retroviral reverse transcriptases (Michel and Lang, 1985; Xiong and Eickbush, 1988). Second, the observations by Levra-Juillet *et al.* (1989) cited earlier (see *Mobile group I introns*) show that excision of a group I intron, aI5b, in the *cox1* gene of yeast depends on the presence of at least one of the two group II introns in this gene. These results provide indirect evidence that the group II intron encoded reverse transcriptase may synthesize a cDNA from mature mRNA. Third, the bI1 group II intron of yeast can reverse splice *in vitro* into the fused exons of the mature mRNA (Augustin *et al.*, 1990; Morl and Schmelzer, 1990).

Finally, the senDNA α, associated with senescence in the filamentous fungus *Podospora anserina* (Chapter 11), is a precisely excised group II intron that encodes a reverse transcriptase-like protein (Osiewacz and Esser, 1984). However, it is not known whether the replication of the senDNA α proceeds via reverse transcription or DNA replication.

MITOCHONDRIAL PLASMIDS

Circular and linear plasmids have been isolated from the mitochondria of filamentous fungi and flowering plants (reviewed by Esser *et al.*, 1986; Hanson and Folkerts, 1992;

Lambowitz and Belfort, 1993; Lambowitz et al., 1986; Lonsdale, 1989; Lonsdale and Grienenberger, 1992; Meinhardt et al., 1990; Nargang, 1985; Samac and Leong, 1989). Studies of these plasmids in *Neurospora* and maize have been particularly rewarding. Investigation of the circular Mauriceville and Varkud mitochondrial plasmids in *Neurospora* revealed that these plasmids retain conserved sequence elements characteristic of group I introns and encode a reverse transcriptase believed to function in their replication. These observations, discussed shortly, blur the distinction between introns and plasmids. In maize, certain linear mitochondrial plasmids seem to possess properties akin to episomes. Thus, they exist free in certain cytoplasms, apparently integrate into others, and linearize the mitochondrial genome in yet another cytoplasm. The remainder of this chapter is devoted to discussing these plasmids. The properties of the equally interesting linear *kalilo* and *maranhar* mitochondrial plasmids of *Neurospora* are reviewed in the next chapter because these plasmids produce severe mitochondrial syndromes.

The Circular Mitochondrial Plasmids of Neurospora: An Evolutionary "Missing Link"?

The circular mitochondrial plasmids of *N. crassa* are small DNA molecules ranging from 3.6 to 5.2 kb (see Lambowitz et al., 1986 for a review). These plasmids belong to three homology groups called Mauriceville, Fiji, and LaBelle. The 3.6-kb Mauriceville plasmid was first isolated from an *N. crassa* strain called Mauriceville-1c. Subsequently, a 3.7-kb plasmid belonging to the same homology group was identified in a strain of *N. intermedia* from Varkud, India. The major transcripts of both plasmids are full length linear RNAs. The Varkud plasmid also synthesizes a major transcript of 4.9 kb containing the full-length 3.7-kb transcript plus a 1.2-kb terminal extension. Two findings from DNA sequencing relate these plasmids to mitochondrial introns (Akins et al., 1988; Lambowitz et al., 1986; Nargang, 1986; Nargang et al., 1984). First, both plasmids contain cognates of conserved nucleotide sequence elements [P3 (5'), P, Q, R, P3 (3'), and S] that characterize group I introns (Fig. 10–2a). Second, both plasmids possess a highly conserved ORF that could encode a large polypeptide of 710 amino acids. This ORF uses the distinctive codons typifying intron ORFs in contrast to mitochondrial genes and, like group II intron ORFs, contains the seven short blocks of conserved amino acids that characterize reverse transcriptases. Hence, these plasmids have structural features reminiscent of group I introns, but encode a reverse transcriptase-like protein similar to group II intron ORFs.

The finding that the ORFs resemble reverse transcriptases coupled with the fact that the plasmid transcripts were full length suggested these elements might replicate by reverse transcription. Direct evidence for this proposition has been obtained. Mitochondria from *Neurospora* strains containing the Mauriceville and Varkud plasmids were found to possess a reverse transcriptase activity, but this activity was missing from a normal wild-type strain lacking the plasmid (Kuiper and Lambowitz, 1988). The reverse transcriptase proved to be highly specific for endogenous plasmid RNA in ribonucleoprotein preparations. The enzyme synthesizes a full-length minus-strand DNA beginning at the 3'-end of the plasmid transcript. The 3'-end of the transcript has tRNA-like characteristics, including a 3'-terminal CCA and potential secondary structure similar to the tRNA-like structures at the 3'-ends of plant RNA viruses. This structure was shown in brome mosaic virus to be essential for initiation of minus-strand RNA synthesis by the viral RNA polymerase.

Based on these observations, Kuiper and Lambowitz (1988) proposed a replication mechanism for the Mauriceville and Varkud plasmids. The first step in their scheme is analogous to synthesis of a full length minus strand by a plant RNA virus except that the full length strand in this case is DNA whose synthesis is catalyzed by the plasmid-encoded reverse transcriptase. Following minus strand synthesis, an RNAse H activity, which does not seem to be associated with the plasmid-encoded protein (Kuiper et al., 1990), would digest the RNA template. DNA polymerase would then catalyze synthesis of the plus strand. Finally, a DNA ligase would circularize double-stranded plasmid DNA by blunt end ligation. Kuiper and Lambowitz (1988) propose that this replication mechanism might be ancestral to that of retroviruses.

Kuiper et al. (1990) further characterized the reverse transcriptase. Unlike retroviral reverse transcriptases, which are synthesized as fusion proteins (Varmus and Brown, 1989), extensive proteolytic processing of the 81-kDa protein from *Neurospora* mitochondria does not seem to be required for reverse transcriptase activity. The ORF sequences upstream and downstream of the putative reverse transcriptase domain also lack gene sequences characteristic of other retroid elements. The region downstream of the reverse transcriptase corresponding to the carboxy terminal half of the protein is highly conserved in Mauriceville and Varkud plasmids. This region may have additional activities associated with plasmid replication such as ligation.

In accord with the earlier findings, Wang et al. (1992) have released the reverse transcriptase of the Mauriceville plasmid from mitochondrial ribonucleoprotein particles and partially purified the activity. They show directly that the enzyme synthesizes full-length cDNA copies of *in vitro* transcripts beginning at the 3' end and with a preference for the 3' tRNA-like structure. Wang et al. imagine that the the Mauriceville reverse transcriptase could have evolved into a retroviral-type reverse transcriptase by adapting its ability to recognize the 3' tRNA-like structure to bind cellular tRNAs used as retroviral primers and acquiring associated RNase H, protease, and integrase coding regions. Consistent with this hypothesis, Wang and Lambowitz (1993) report that the purified Mauriceville reverse transcriptase does indeed lack RNase H activity, but that *Neurospora* mitochondria contain an endogenous RNase H activity that is present in ribonucleoprotein particles prior to their purification. The reverse transcriptase encoded by the Mauriceville and Varkud plasmids in phylogenetically related to the enzymes specified by group II introns (Lambowitz and Belfort, 1993).

To determine whether the Mauriceville and Varkud plasmids could transpose into the mitochondrial genome, Akins et al. (1986) isolated growth variants of the strains carrying these plasmids. The variants were obtained by subjecting these strains to continuous vegetative growth, which selects for the accumulation of suppressive mtDNAs in *Neurospora* mitochondria (Chapter 11). The rationale behind the selection scheme was 2-fold. First, essentially all mtDNA mutations arising in *Neurospora* have been found to predominate over wild-type mtDNA in heteroplasmons whether they involve very small alterations or gross deletions. Second, integration of the *kalilo* plasmid gives rise to defective mitochondrial genomes, which become suppressive (Chapter 11). Consistent with this rationale, the 12 growth variants isolated all contained altered mitochondrial plasmids and 10 of them possessed defective mtDNAs that had undergone large alterations including deletions and insertions. At least three of the mutants had

INTRONS, MOBILE ELEMENTS, AND PLASMIDS

mitochondrial plasmid sequences integrated into them. Akins *et al.* (1986) concluded that these plasmid DNA alterations present in the 12 growth variants had caused the plasmids to become suppressive to mtDNA in the same way that amplified segments of the mitochondrial genome itself can become suppressive to the intact mitochondrial genome in *Neurospora* and other filamentous ascomycetes (Chapter 7). Impairment of the rate of mycelial growth results. The suppressive plasmids generally contain short insertions at or near the major 5' transcription start site that included a mitochondrial tRNA sequence. The suppressive behavior of plasmids appears to result from the ability of the inserted tRNA sequences to cause overproduction of plasmid transcripts (Akins *et al.*, 1988).

Akins *et al.* (1986) argue that their results provide evidence that the suppressive plasmids arise by a mechanism that includes a reverse transcription step, that the plasmids integrate into the mtDNA in a manner consistent with integration via an RNA intermediate, and that a site-specific DNA endonuclease may have catalyzed some of the integration events. They also note that the presence of mitochondrial tRNAs in the suppressive plasmids is reminiscent of incorporation of "host" tRNAs by RNA viruses. As Nargang *et al.* (1984) noted, the Mauriceville and Varkud plasmids, with their group I intron motifs and their ability to reverse transcribe, may belong to a class of genetic elements that was or is the progenitor of mtDNA introns. Other possibilities are that the plasmids were derived from excised introns or from autonomous elements that recombined with group I introns (Lambowitz and Belfort, 1993).

There is a final twist to this story arising from studies of the LaBelle plasmid isolated from *N. intermedia*. This 4.1-kb circular plasmid posseses a single long ORF that could encode a protein of 132 kDa (Pande *et al.*, 1989). This ORF contains seven short stretches of amino acids that match conserved sequence blocks characteristic of reverse transcriptases. Schulte and Lambowitz (1991) report that a unique DNA polymerase activity, apparently encoded by this ORF, probably functions in the replication of the LaBelle plasmid and appears to be derived from a reverse transcriptase. However, Li and Nargang (1993) examined the sequence encoding the DNA polymerase of the Fiji mitochondrial plasmid of *Neurospora*. Comparison to the LaBelle polymerase reading frame reveals the presence of sequences characteristic of the B DNA polymerase family with the reverse transcriptase motifs being poorly conserved. Hence, Li and Nargang believe the reverse transcriptase motifs are not significant. Nargang *et al.* (1992) also report a "footprint" of LaBelle in normal mitochondrial genomes of different *Neurospora* species. LaBelle is currently found in only one *N. intermedia* strain, so Nargang *et al.* suppose that this 1.6-kb region of *Neurospora* mtDNA was derived by insertion of part of the LaBelle plasmid prior to the divergence of *Neurospora* species. Subsequently, the plasmid itself is presumed to have been lost from all species except *N. intermedia*. Interestingly, this plasmid-derived sequence currently contains the promoter used for transcription of the mitochondrial small rRNA gene.

Maize Mitochondrial Plasmids and Episomes

The properties of plant mitochondrial plasmids and, in particular, the linear and circular plasmids of the maize mitochondrion have been considered in several reviews (Hanson and Folkerts, 1992; Lonsdale, 1989; Lonsdale and Grienenberger, 1992; Newton, 1988). The cytoplasmic male sterile (CMS) varieties of maize (see Chapter 11) have been

Table 10–4. Mitochondrial Plasmids of Maize and Teosinte

Species	Conformation	Cytoplasm	Replicon	Size (kb)
Maize	Circular	N, cms-C,T,S	—	1.9
(Zea mays)			—	1.4
		cms-C	—	1.5
			—	1.4
	Linear	All	2.3L	2.3
			2.1L	2.15
		cms-S	S1	6.4
			S2	5.4
		RU	R1	7.4
			R2	5.4
Teosinte	Linear	—	D1	7.4
(Zea diploperennis)			D2	5.4

Modified from Lonsdale (1989).

designated *cms-T* (Texas), *cms-C,* and *cms-S*. The plasmids of these strains have been characterized in addition to those of normal male fertile maize (N), 12 Latin American races with male fertile cytoplasms designated (RU), and the related teosinte species *Z. diploperennis* (Timothy et al., 1983; Weissinger et al., 1982). The circular plasmids of maize are unrelated to the mitochondrial genome though sequences homologous to the widely distributed 1.9-kb plasmid (Table 10–4) have been reported in the nuclear genome (Abbott et al., 1985). This plasmid also hybridizes to a specific RNA species. Several of these small, circular plasmids have been sequenced and they bear no homology between maize and other species (e.g., broadbean and sugar beet). Lonsdale (1989) concludes that these circular plasmids may represent selfish replicons.

Of greater interest are the S, R, and D episomes of maize (Table 10–4). Because these linear molecules have related sequences in the mitochondrial genome and have been shown to integrate, they have been called episomes (Schardl et al., 1984). S2, R2, and D2 are identical to each other and R1 is identical to D1. The free form of each episome has a polypeptide linked to the 5′-phosphate of a short terminal repeat, but the gene encoding the protein remains to be identified (Lonsdale and Grienenberger, 1992). S1 seems to have arisen from an intermolecular recombination event between episomes of the R1 and R2 types. This means that *cms*-S maize, which possesses the S1 and S2 plasmids (Table 10–4), must originally have contained R1 and R2. The terminal part of the R1 sequence (R1*), lost during the creation of S1, is found as an integrated sequence at two positions in the *cms*-S mitochondrial genome. This sequence is identical to the first 187 bp of the 208-bp terminal inverted repeats of S1 and S2 so regions of homology are created at both sites in the mitochondrial genome. Recombination between the main mitochondrial genome and the S1 or S2 replicons at the R1* sequence is thought to result in linearization of the mitochondrial genome such that S1 or S2 is attached to one terminus and the R1* sequence at the other (Schardl et al., 1984, 1985). Several ORFs have been detected in the S episomes and antisera raised to the putative proteins coded by two of these ORFs reveal the presence of polypeptides. In the S2 plasmid one ORF *(urf1)* encodes a protein with homology to a viral RNA polymerase

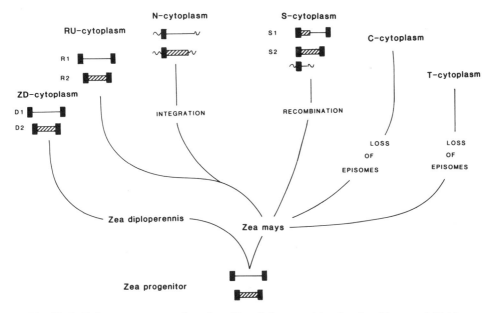

Fig. 10–7. Episome sequences of teosinte *(Zea diploperennis)* and maize *(Zea mays).* Evidence indicates that acquisition of the mitochondrial self-replicating DNAs related to the R and D episomes preceded the evolutionary divergence of *Z. diploperennis* and *Z. mays.* Cytoplasmic evolution led to five recognized cytoplasmic types in *Z. mays.* Only in the RU and S-cytoplasms are free-replicating episomes maintained. In the N, *cms*-C, and *cms*-T cytoplasms free-replicating episomes have been lost. Only in the N-cytoplasm are sequences related to R1 and R2 present in the mitochondrial genome, though the *cms*-C and *cms*-T cytoplasms may contain short sequences related to terminal inverted repeats. From Lonsdale (1989) with permission.

while a second ORF *(urf3)* specifies a polypeptide with homology to DNA polymerase (Lonsdale and Grienenberger, 1992).

The maize N-cytoplasm possesses integrated sequences in the mitochondrial genome that seem to derive from R1 and R2 episomes of the RU strain (Table 10–4). These replicons share no sequence homology apart from their terminal inverted repeats, one of which is missing for each integrated replicon. Plants with the *cms*-C and *cms*-T cytoplasms lack these plasmids altogether (Table 10–4). Figure 10–7 depicts an evolutionary scheme that rationalizes these different events.

Finally, Bedinger *et al.* (1986) found another linear plasmid present in all races of maize. This plasmid exists in 2.3 and 2.15 kb forms with the latter resulting from a deletion which maps to one end of the plasmid. This plasmid is absent from teosintes, which instead have homologous sequences in the mitochondrial genome. This homology is restricted to a chloroplast DNA fragment containing the genes for tRNA^{Trp} and tRNA^{Pro}, including the intergenic region between them. This plasmid appears to be essential since it contains the only known functional copy of a tRNA^{Trp} gene in the maize mitochondrial genome (Leon *et al.,* 1989). The protein encoded by the plasmid ORF1 gene is a candidate for the 5′-terminal protein bound to the linear DNA of this plasmid (Leon *et al.,* 1992).

REFERENCES

Abbott, A.G., O'Dell, M., and R.B. Flavell (1985). Quantitative variation in components of the maize mitochondrial genome between plants with different male-sterile cytoplasms. *Plant Mol. Biol.* 4: 233–240.

Akins, R.A., and A.M. Lambowitz (1987). A protein required for splicing group I introns in Neurospora mitochondria is mitochondrial tyrosyl-tRNA synthetase or a derivative thereof. *Cell* 50: 331–345.

Akins, R.A., Kelley, R. L., and A.M. Lambowitz (1989). Characterization of mutant mitochondrial plasmids of *Neurospora* spp. that have incorporated tRNAs by reverse transcription. *Mol. Cell. Biol.* 9: 678–691.

Akins, R.A., Kelley, R.L., and A.M. Lambowitz (1986). Mitochondrial plasmids of *Neurospora*: Integration into mitochondrial DNA and evidence for reverse transcription in mitochondria. *Cell* 47: 505–516.

Akins, R.A., Grant, D.M., Stohl, L.L., Bottorff, D.A., Nargang, F.E., and A.M. Lambowitz (1988). Nucleotide sequence of the Varkud mitochondrial plasmid of *Neurospora* and synthesis of a hybrid transcript with a 5' leader derived from mitochondrial RNA. *J. Mol. Biol.* 204: 1–25.

Augustin, S., Muller, M.W., and R.J. Schweyen (1990). Reverse self-splicing of group II intron RNAs *in vitro*. *Nature (London)* 343: 383–386.

Bedinger, P., de Hostos, E.L., Leon, P., and V. Walbot (1986). Cloning and characterization of a linear 2.3 kb mitochondrial plasmid of maize. *Mol. Gen. Genet.* 205: 206–212.

Bedouelle, H. (1990). Recognition of tRNA^{tyr} by tyrosyl-tRNA synthetase. *Biochimie* 72: 589–598.

Ben Asher, E., Groudinsky, O., Dujardin, G., Altamura, N., Kermorgant, M., and P.P. Slonimski (1989). Novel class of nuclear genes involves in both mRNA splicing and protein synthesis in *Saccharomyces cerevisiae* mitochondria. *Mol. Gen. Genet.* 215: 517–528.

Binder, S., Marchfelder, A., Brennicke, A., and B. Wissinger (1992). RNA editing in trans-splicing intron sequences of nad2 mRNAs in *Oenothera* mitochondria. *J. Biol. Chem.* 267: 7615–7623.

Bonen, L. (1993). *Trans*-splicing of pre-mRNA in plants, animals and protists. *FASEB J.* 7: 40–46.

Bousquet, I., Dujardin, G., Poyton, R.O., and P.P. Slonimski (1990). Two group I mitochondrial introns in the *cob-box* and *coxI* genes require the same *MRS1/PET157* nuclear gene product for splicing. *Curr. Genet.* 18: 117–124.

Boynton, J.E., Harris, E.H., Burkhart, B.D., Lamerson, P.M., and N.W. Gillham (1987). Transmission of mitochondrial and chloroplast genomes in crosses of *Chlamydomonas*. *Proc. Natl. Acad. Sci. USA* 84: 2391–2395.

Burke, J.M. (1988). Molecular genetics of group I introns: RNA structures and protein factors required for splicing—a review. *Gene* 73: 273–294.

Butow, R.A., Perlman, P.S., and L.I. Grossman (1985). The unusual *var1* gene of yeast mitochondrial DNA. *Science* 228: 1496–1501.

Cech, T.R. (1985). Self-splicing RNAs: Implications for evolution. *Int. Rev. Cytol.* 93: 3–22.

Cech, T.R. (1988). Conserved sequences and structures of group I introns: Building an active site for RNA catalysis—a review. *Gene* 73: 259–271.

Cech, T.R. (1990) Self-splicing of group I introns. *Annu. Rev. Biochem.* 59: 543–568.

Cherniack, A.D., Garriga, G., Kittle, J.D., Jr., Akins, R.A., and A.M. Lambowitz (1990). Function of Neurospora mitochondrial tyrosyl-tRNA synthetase in RNA splicing requires an idiosyncratic domain not found in other synthetases. *Cell* 62: 745–755.

Choquet, Y., Goldschmidt-Clermont, M., Girard-Bascou, J., Kuck, U., Bennoun, P., and J.-D. Rochaix (1988). Mutant phenotypes support a *trans*-splicing mechanism for the expression of the tripartite *psaA* gene in the *C. reinhardtii* chloroplast. *Cell* 52: 903–913.

Christopher, D.A., and R.B. Hallick (1989). *Euglena gracilis* chloroplast ribosomal protein operon: A new chloroplast gene for ribosomal protein L5 and description of a novel organelle intron category designated group III. *Nucl. Acids Res.* 17: 7591–7608.

Clark-Walker, G.D. (1992). Evolution of mitochondrial genomes in fungi. *Int. Rev. Cytol.* 141: 89–127.

Colleaux, L., d'Auriol, L., Betermier, M., Cottarel, G., Jacquier, A., Galibert, F., and B. Dujon (1986). Universal code equivalent of a yeast mitochondrial reading frame is expressed into *E. coli* as a specific double strand endonuclease. *Cell* 44: 521–533.

Colleaux, L., d'Auriol, L., Galibert, F., and B. Dujon (1988). Recognition and cleavage site of the intron encoded omega transposase. *Proc. Natl. Acad. Sci. U.S.A.* 85: 6022–6026.

Colleaux, L., Michel-Wolwertz, M.-R., Matagne, R.F., and B. Dujon (1990). The apocytochrome *b* gene of *Chlamydomonas smithii* contains a mobile intron related to both *Saccharomyces* and *Neurospora* introns. *Mol. Gen. Genet.* 223: 288–296.

Copertino, D.W., and R.B. Hallick (1991). Group II twintron: An intron within an intron in a chloroplast cytochrome b-559 gene. *EMBO J.* 10: 433–442.

Copertino, D.W., Christopher, D.A., and R.B. Hallick (1991). Group II twintron: An intron within an intron in a chloroplast cytochrome b-559 gene. *EMBO J.* 10: 433–442.

Copertino, D.W., Shigeoka, S., and R.B. Hallick (1992). Chloroplast group III twintron excision utilizing multiple 5'- and 3'-splice sites. *EMBO J.* 11: 5041–5050.

Davies, R.W., Waring, R.B., Ray, J.A., Brown, T.A., and D. Scazzocchio (1982). Making ends meet: A model for RNA splicing in fungal mitochondria. *Nature (London)* 300: 719–724.

Delahodde, A., Goguel, V., Becam, A.M., Creusot, F., Perea, J., Banroques, J., and C. Jacq (1989). Site-specific DNA endonuclease and RNA maturase activities of two homologous intron-encoded proteins of yeast mitochondria. *Cell* 56: 431–441.

de Zamoroczy, M., and G. Bernardi (1986). The GC clusters of the mitochondrial genome of yeast and their evolutionary origin. *Gene* 41: 1–22.

de Zamoroczy, M., Faugeron-Fonty, G., Baldacci, G., Goursot, R., and G. Bernardi (1984). The ori sequences of the mitochondrial genome of a wild-type yeast strain: number; location; orientation and structure. *Gene* 32: 439–457.

Dobinson, K.F., Henderson, M., Kelley, R.L., Collins, R.A., and A.M. Lambowitz (1989). Mutations in nuclear gene *cyt-4* of *Neurospora crassa* result in pleiotropic defects in processing and splicing of mitochondrial RNAs. *Genetics* 123: 97–108.

Drager, R.G., and R.B. Hallick (1993). A complex twintron is excised as four individual introns. *Nucl. Acids Res.* 21: 2389–2394.

Dujardin, G., Pajot, P., Groudinsky, O., and P.P. Slonimski (1980). Long range control circuits within mitochondria and between nucleus and mitochondria. I. Methodology and phenomenology of suppressors. *Mol. Gen. Genet.* 179: 469–482.

Dujardin, G., Jacq, C., and P.P. Slonimski (1982). Single base substitutions in an intron of oxidase gene compensates splicing defects of the cytochrome *b* gene. *Nature (London)* 298: 628–632.

Dujardin, G., Labouesse, M., Netter, P., and P.P. Slonimski (1983). Genetic and biochemical studies of a nuclear suppressor NAM2: Extraneous activation of a latent pleiotropic maturase. In *Mitochondria 1983: Nucleo-Mitochondrial Interactions* (R.J. Schweyen, K. Wolf, and F. Kaudewitz, eds.), pp. 233–250, de Gruyter, Berlin.

Dujon, B. (1981). Mitochondrial genetics and functions. In *Molecular Biology of the Yeast Saccharomyces: Life Cycle and Inheritance* (J.N. Strathern, E.W. Jones, and J.R. Broach, eds.), pp. 505–635. Cold Spring Harbor Laboratory, Cold Spring Harbor, NY.

Dujon, B. (1989). Group I introns as mobile genetic elements: Facts and mechanistic speculations—a review. *Gene* 82: 91–114.

Dujon, B., and A. Jacquier (1983). Organization of the mitochondrial 21S rRNA gene in *Saccharomyces cerevisiae:* Mutants of the peptidyl transferase center and nature of the *omega* locus. In *Mitochondria 1983* (R.J. Schweyen, K. Wolf, and F. Kaudewitz, eds.), pp. 389–403. de Gruyter, Berlin.

Dujon, B., Cottarel, G., Colleaux, L., Betermier, M., Jacquier, A., d'Auriol, L., and F. Galibert (1985). Mechanism of integration of an intron within a mitochondrial gene: A double strand break and the transposase function of an intron encoded protein as revealed by in vivo and in vitro assays. In *Achievements and Perspectives of Mitochondrial Research,* Vol. II (E. Quagliariello, E.C. Slater, F. Palmieri, C. Saccone, and A.M Kroon, eds.), pp. 215–225. Elsevier, Amsterdam.

Dujon, B., Colleaux, L., Jacquier, A., Michel, F., and C. Monteilhet (1986). Mitochondrial introns as mobile genetic elements: The role of intron encoded proteins. In *Extrachromosomal Elements in Lower Eukaryotes* (R.B. Wickner, A. Hinnebush, A.M. Lambowitz, I.C. Gunsalus, and A. Hollaender, eds.), pp. 5–27. Plenum Press, New York.

Durrenberger, F., and J.-D. Rochaix (1991). Chloroplast ribosomal intron of *Chlamydomonas reinhardtii:* In vitro self-splicing, DNA endonuclease activity and in vivo mobility. *EMBO J.* 10: 3495–3501.

Durrenberger, F., and J.-D. Rochaix (1993). Characterization of the cleavage site of the I-*Cre*I DNA endonuclease encoded by the chloroplast ribosomal intron of *Chlamydomonas reinhardtii. Mol. Gen. Genet.* 236: 409–414.

Esser, K., Kuck, U., Lang-Hinrichs, C., Lemke, P., Osiewacz, H.D., Stahl, U., and P. Tudzynski (1986). *Plasmids of Eukaryotes. Fundamentals and Applications.* Springer Verlag, Berlin.

Faye, G., and M. Simon (1983). Analysis of a yeast nuclear gene involved in the maturation of mitochondrial pre-messenger RNA of the cytochrome oxidase subunit I. *Cell* 32: 77–87.

Gampel, A., Nishikimi, M., and A. Tzagoloff (1989). CBP2 protein promotes in vitro excision of a yeast mitochondrial group I intron. *Mol. Cell. Biol.* 9: 5424–5433.

Gargouri, A., Lazowska, J., and P.P. Slonimski (1983). DNA-splicing of introns in the gene: A general way of reverting intron mutations. In *Mitochondria 1983* (R.J. Schweyen, K. Wolf, and F. Kaudewitz, eds.), pp. 259–268. de Gruyter, Berlin.

Garriga, G., and A.M. Lambowitz (1984). RNA splicing in Neurospora mitochondria: Self-splicing of a mitochondrial intron in vitro. *Cell* 39: 631–641.

Garriga, G., Bertrand, H., and A.M. Lambowitz (1984). RNA splicing in Neurospora mitochondria: Nuclear mutants defective in both splicing and 3' end synthesis of the large rRNA. *Cell* 36: 623–634.

Gauthier, A., Turmel, M., and C. Lemieux (1991). A group I intron in the chloroplast large subunit rRNA gene of *Chlamydomonas eugametos* encodes a double-strand endonuclease that cleaves the homing site of this intron. *Curr. Genet.* 19: 43–47.

Gillham, N.W. (1978). *Organelle Heredity.* Raven Press, New York.

Goldschmidt-Clermont, M. (1991). Transgenic expression of aminoglycoside adenine transferase in the chloroplast: A selectable marker for site-directed transformation of chlamydomonas. *Nucl. Acids Res.* 19: 4083–4089.

Goldschmidt-Clermont, M., Girard-Bascou, J., Choquet, Y., and J.-D. Rochaix (1990). Trans-splicing mutants of *Chlamydomonas reinhardtii. Mol. Gen. Genet.* 223: 417–425.

Goldschmidt-Clermont, M., Choquet, Y., Girard-Bascou, J., Michel, F., Schirmer-Rahire, M., and J.-D. Rochaix (1991). A small chloroplast RNA may be required for *trans*-splicing in Chlamydomonas reinhardtii. *Cell* 65: 135–143.

Gray, M.W., Hanic-Joyce, P.J., and P.S. Covello (1992). Transcription, processing and editing in plant mitochondria. *Annu. Rev. Plant Physiol. Plant Mol. Biol.* 43: 145–175.

Grivell, L.A. (1989). Nucleo-mitochondrial interactions in yeast mitochondrial biogenesis. *Eur. J. Biochem.* 182: 477–493.

Guo, Q., and A.M. Lambowitz (1992). A tyrosyl-tRNA synthetase binds specifically to the group I intron catalytic core. *Genes and Development* 6: 1357–1372.

Hanson, M.R., and O.F. Folkerts (1992). Structure and function of the higher plant mitochondrial genome. *Int. Rev. Cytol.* 141: 129–172.

Hebbar, S.K., Belcher, S.M., and P.S. Perlman (1992). A maturase-encoding group IIA intron of yeast mitochondria self-splices *in vitro*. *Nucl. Acids Res.* 20: 1747–1754.

Herbert, C.J., Labouesse, M., Dujardin, G., and P.P. Slonimski (1988). The NAM2 proteins from *S. cerevisiae* and *S. douglasii* are mitochondrial leucyl-tRNA synthetases, and are involved in mRNA splicing. *EMBO J.* 7: 473–483.

Herrin, D.L., Chen, Y.-F., and G.W. Schmidt (1990). RNA splicing in *Chlamydomonas* chloroplasts. *J. Biol. Chem.* 265: 21134–21140.

Hill, J., McGraw, P., and A. Tzagoloff (1985). A mutation in yeast mitochondrial DNA results in a precise excision of the terminal intron of the cytochrome *b* gene. *J. Biol. Chem.* 260: 3235–3238.

Hudspeth, M.E.S., Vincent, R.D., Perlman, P.S., Shumard, D.S., Treisman, L.O., and L.I. Grossman (1984). Expandable *var1* gene of yeast mitochondrial DNA: In-frame insertions can explain the strain-specific protein size polymorphism. *Proc. Natl. Acad. Sci. U.S.A.* 81: 3148–3152.

Jacquier, A. (1990). Self-splcing group II and nuclear pre-mRNA introns: How similar are they? *Trends Biochem. Sci.* 15: 351–354.

Jacquier, A., and F. Michel (1987). Multiple exon-binding sites in class II self-splicing introns. *Cell* 50: 17–29.

Kamper, U., Kuck, U., Cherniack, A.D., and A.M. Lambowitz (1992). The mitochondrial tyrosyl-tRNA synthetase of *Podospora anserina* is a bifunctional enzyme active in protein synthesis and RNA splicing. *Mol. Cell Biol.* 12: 499–511.

Kennell, J.C., Moran, J.V., Perlman, P.S., Butow, R.A., and A.M. Lambowitz (1993). Reverse transcriptase activity associated with maturase-encoding group II introns in yeast mitochondria. *Cell* 73: 133–146.

Kittle, J.D., Jr., Mohr, G., Gianelos, J.A., Wang, H., and A.M. Lambowitz (1991). The *Neurospora* mitochondrial tyrosyl-tRNA synthetase is sufficient for group I intron splicing *in vitro* and uses the carboxy-terminal tRNA-binding domain along with other regions. *Genes Dev.* 5: 1009–1021.

Koch, J.L., Boulanger, S.C., Dib-Haji, S.D., Hebbar, S.K., and P.S. Perlman (1992). Group II introns deleted for multiple substructures retain self-splicing activity. *Mol. Cell. Biol.* 12: 1950–1958.

Kruger, K., Grabowski, P.J., Zaug, A.J., Sands, J., Gottschling, D.E., and T.R. Cech (1982). Self-splicing RNA: Autoexcision and autocyclization of the ribosomal RNA intervening sequence of *Tetrahymena*. *Cell* 31: 147–157.

Kuck, U., Choquet, Y., Schneider, M., Dron, M., and P. Bennoun (1987). Structural and transcription analysis of two homologous genes for the P_{700} chlorophyll a-apoproteins in *Chlamydomonas reinhardtii*: Evidence for in vivo *trans* splicing. *EMBO J.* 6: 2185–2195.

Kuiper, M.T.R., and A.M. Lambowitz (1988). A novel reverse transcriptase activity associated with mitochondrial plasmids of Neurospora. *Cell* 55: 693–704.

Kuiper, M.T.R., Sabourin, J.R., and A.M. Lambowitz (1990). Identification of the reverse transcriptase encoded by the Mauriceville and Varkud mitochondrial plasmids of *Neurospora*. *J. Biol. Chem.* 265: 6936–6943.

Labouesse, M., Herbert, C.J., Dujardin, G., and P.P. Slonimski (1987). Three suppressor mutations which cure a mitochondrial RNA maturase deficiency occur at the same codon in the open reading frame of the nuclear *NAM2* gene. *EMBO J.* 6: 713–721.

Lambowitz, A.M. (1989). Infectious introns. *Cell* 56: 323–326.

Lambowitz, A.M., and M. Belfort (1993). Introns as mobile elements. *Annu. Rev. Biochem.* 62: 587–622.

Lambowitz, A.M., and P.S. Perlman (1990). Involvement of aminoacyl-tRNA synthetases and other proteins in group I and group II intron splicing. *Trends in Biochem. Sci.* 15: 440–444.

Lambowitz, A.M., Akins, R.A., Garriga, G., Henderson, M., Kubelik, A.R., and K.A. Maloney (1985). Mitochondrial introns and mitochondrial plasmids of *Neurospora*. In *Achievements and Perspectives of Mitochondrial Research* (E. Quagliariello, E.C. Slater, F. Palmieri, C. Saccone, and A.M. Kroon eds.), vol. II, pp. 237–247. Elsevier, Amsterdam.

Lambowitz, A.M., Akins, R.A., Kelley, R.L., Pande, S., and F.E. Nargang (1986). Mitochondrial plasmids of *Neurospora* and other filamentous fungi. In *Extrachromosomal Elements in Lower Eukaryotes* (R.B. Wickner, A. Hinnebusch, A.M. Lambowitz, and A. Hollaender, eds.), pp. 83–92. Plenum Press, New York.

Lemieux, C., and R.W. Lee (1987). Nonreciprocal recombination between alleles of the chloroplast 23S rRNA gene in interspecific *Chlamydomonas* crosses. *Proc. Natl. Acad. Sci. U.S.A.*, 84: 4166–4170.

Leon, P., Walbot, V., and P. Bedinger (1989). Molecular analysis of the linear 2.3 kb plasmid of maize mitochondria: Apparent capture of tRNA genes. *Nucl. Acids Res.* 17: 4089–4099.

Leon, P., O'Brien-Vedder, C., and V. Walbot (1992). Expression of ORF1 of the linear 2.3 kb plasmid of maize mitochondria: product localization and similarities to the 130 kDa protein encoded by the S2 episome. *Curr. Genet.* 22: 61–67.

Levra-Juillet, E., Boulet, A., Seraphin, B., Simon, M., and G. Faye (1989). Mitochondrial introns aI1 and/or aI2 are needed for the in vivo deletion of intervening sequences. *Mol. Gen. Genet.* 217: 168–171.

Li, Q. and F.E. Nargang (1993). Two *Neurospora* mitochondrial plasmids encode DNA polymerases containing motifs characteristic of family B DNA polymerases but lack the sequence Asp-Thr-Asp. *Proc. Natl. Acad. Sci. USA* 90:4299–4303.

Lisowsky, T. (1990). Molecular analysis of the mitochondrial transcription factor mtf2 of *Saccharomyces cerevisiae*. *Mol. Gen. Genet.* 220: 186–190.

Lonsdale, D.M. (1989). The plant mitochondrial genome. In *The Biochemistry of Plants*, Vol. 15, (A. Marcus, ed.), pp. 229–295. Academic Press, San Diego.

Lonsdale, D.M., and J.M. Grienenberger (1992). The mitochondrial genome of plants. In *Plant Gene Research: Cell Organelles* (R. Herrmann, ed.), pp. 183–218. Springer-Verlag, Wien.

Majumder, A.L., Akins, R.A., Wilkinson, J.G., Kelley, R.L., Snook, A.J., and A.M. Lambowitz (1989). Involvement of tyrosyl-tRNA synthetase in splicing of group I introns in *Neurospora crassa* mitochondria: Biochemical and immunochemical analyses of splicing activity. *Mol. Cell. Biol.* 9: 2089–2104.

Marshall, P., and C. Lemieux (1991). Cleavage pattern of the homing endonuclease encoded by the fifth intron in the chloroplast large subunit rRNA-encoding gene of *Chlamydomonas eugametos*. *Gene* 104: 241–245.

Marshall, P., and C. Lemieux (1992). The I*Ceu*I endonuclease recognizes a sequence of 19 base pairs and preferentially cleaves the coding strand of the *Chlamydomonas moewusii* chloroplast large subunit rRNA gene. *Nucl. Acids Res.* 20: 6401–6407.

Meinhardt, F., Kempken, F., Kamper, J., and K. Esser (1990). Linear plasmids among eukaryotes: Fundamentals and application. *Curr. Genet.* 17: 89–95.

Meunier, B., Tian, G.L., Macadre, C., Slonimski, P.P., and J. Lazowska (1990). Group II introns transpose in yeast mitochondria. In *Structure, Function, and Biogenesis of Energy Transfer Systems* (E. Quagliariello, S. Papa, F. Palmieri, and C. Saccone, eds.), pp. 169–174. Elsevier Science, Amsterdam.

Michel, F., and B. Dujon (1983). Conservation of RNA secondary structures in two intron families including mitochondrial-, chloroplast-, and nuclear-encoded members. *EMBO J.* 2: 33–38.

Michel, F., and B.F. Lang (1985). Mitochondrial class II introns encode proteins related to the reverse transcriptases of retroviruses. *Nature (London)* 316: 641–643.

Michel, F., and E. Westhof (1990). Modelling of the three-dimensional architecture of group I catalytic introns based on comparative sequence analysis. *J. Mol. Biol.* 216: 585–610.

Michel, F., Jacquier, A., and B. Dujon (1982). Comparison of fungal mitochondrial introns reveals extensive homologies in RNA secondary structure. *Biochimie* 64: 867–881.

Michel, F., Umesono, K., and H. Ozeki (1989). Comparative and functional anatomy of group II catalytic introns. *Gene* 82:5–30.

Mohr, G., and A.M. Lambowitz (1991). Integration of a group I intron into a ribosomal RNA sequence promoted by a tyrosyl-tRNA synthetase. *Nature (London)* 354: 164–167.

Mohr, G., Zhang, A., Gianelos, J.A., Belfort, M., and A.M. Lambowitz (1992). The Neurospora CYT-18 protein suppresses defects in the phage T4 *td* intron by stabilizaing the catalytically active structure of the intron core. *Cell* 69: 483–494.

Morl, M., and C. Schmelzer (1990). Integration of group II intron bI1 into foreign RNA by reversal of the self-splicing reaction in vitro. *Cell* 60: 629–636.

Nargang, F.E. (1985). Fungal mitochondrial plasmids. *Exp. Mycol.* 9: 285–293.

Nargang, F.E. (1986). Conservation of a long open reading frame in two *Neurospora* mitochondrial plasmids. *Mol. Biol. Evol.* 3: 19–28.

Nargang, F.E., Bell, J.B., Stohl, L.L., and A.M. Lambowitz (1984). The DNA sequence and genetic organization of a Neurospora mitochondrial plasmid suggest a relationship to introns and mobile elements. *Cell* 38: 441–453.

Nargang, F.E., Pande, S., Kennell, J.C., Akins, R.A., and A.M. Lambowitz (1992). Evidence that a 1.6 kilobase region of *Neurospora* mtDNA was derived by insertion of part of the LaBelle mitochondrial plasmid. *Nucl. Acids Res.* 20: 1101–1108.

Newton, K.J. (1988). Plant mitochondrial genomes: organization, expression and variation. *Annu. Rev. Plant Physiol. Plant Mol. Biol.* 39: 503–532.

Osiewacz, H.D., and K. Esser (1984). The mitochondrial plasmid of *Podospora anserina*: A mobile intron of a mitochondrial gene. *Curr. Genet.* 8: 299–305.

Palmer, J.D., and J.M. Logsdon, Jr. (1991). The recent origins of introns. *Curr. Opinion Genet. Dev.* 1: 470–477.

Pande, S., Lemire, E.G., and F.E. Nargang (1989). The mitochondrial plasmid from *Neurospora intermedia* strain LaBelle-1b contains a long open reading frame with blocks of amino acids characteristic of reverse transcriptases and related proteins. *Nucl. Acids Res.,* 17: 2023–2042.

Peebles, C.L., Perlman, P.S., Mecklenburg, K.L., Petrillo, M.L., Tabor, J.H., Jarrell, K.A., and H.-L. Cheng (1986). A self-splicing RNA excises an intron lariat. *Cell* 44: 213–223.

Pel, H.J., and L.A. Grivell (1993). The biology of yeast mitochondrial introns. *Mol. Biol. Reports* 18: 1–13.

Perea, J., and C. Jacq (1985). Role of the 5' hairpin structure in the splicing accuracy of the fourth intron of the yeast *cob-box* gene. *EMBO J.* 4: 3281–3288.

Perea, J., Desdouets, C., Schapira, M., and C. Jacq (1993). I-*Sce*III: a novel group I intron-encoded endonuclease from the yeast mitochondria. *Nucl. Acids Res.* 21: 358.

Perlman, P.S., and R.A. Butow (1989). Mobile introns and intron-encoded proteins. *Science* 246: 1106–1109.

Perlman, P.S., Peebles, C.L., and C. Daniels (1990). Different types of introns and splicing mechanisms. In *Intervening Sequences in Evolution and Development* (E.M. Stone and R.J. Schwarz, ed.), pp. 112–161. Oxford University Press, New York.

Rochaix, J.-D. (1992a). Control of plastid gene expression in *Chlamydomonas reinhardtii*. In *Plant Gene Research: Cell Organelles* (R.G. Herrmann, ed.), pp. 249–274, Springer-Verlag, Wien.

Rochaix, J.-D. (1992b). Post-transcriptional steps in the expression of chloroplast genes. *Annu. Rev. Cell Biol.* 8: 1–28.

Rochaix, J.-D., Goldschmidt-Clermont, M., Choquet, Y., Kuchka, M., and J. Girard-Bascou (1991). Nuclear and chloroplast genes involved in the expression of specific chloroplast genes of *Chlamydomonas reinhardtii*. In *Plant Molecular Biology* 2 (R.G. Herrmann and B. Larkins, eds.), pp. 401–409. Plenum Press, New York.

Saldanaha, R., Mohr, G., Belfort, M., and A.M. Lambowitz (1993). Group I and group II introns. *FASEB J.* 7: 15–24.

Samac, D.A., and S.A. Leong (1989). Mitochondrial plasmids of filamentous fungi: Characteristics and use in transformation vectors. *Mol. Plant-Microbe Interact.* 2: 155–159.

Schardl, C.L., Lonsdale, D.M., Pring, D.R., and K.R. Rose (1984). Linearization of maize mitochondrial chromosomes by recombination with linear episomes. *Nature (London)* 310: 292–296.

Schardl, C.L., Pring D.R., and D.M. Lonsdale (1985). Mitochondrial DNA rearrangements associated with fertile revertants of S-type male-sterile maize. *Cell* 43: 361–368.

Schmidt, C., Sollner, T., and R.J. Schweyen (1987). Nuclear suppression of a mitochondrial RNA splice defect: Nucleotide sequence and disruption of the *MRS3* gene. *Mol. Gen. Genet.* 210: 145–152.

Schulte, U., and A.M. Lambowitz (1991). The LaBelle mitochondrial plasmid of *Neurospora intermedia* encodes a novel DNA polymerase that may be derived from a reverse transcriptase. *Mol. Cell. Biol.* 11: 1696–1706.

Seraphin, B., Simon, M., and G. Faye (1988). MSS18, a yeast nuclear gene involved in splicing of intron aI5b of the mitochondrial *cox1* transcript. *EMBO J.* 7: 1455–1464.

Seraphin, B., Simon, M., Boulet, A., and G. Faye (1989). Mitochondrial splicing requires a protein from a novel helicase family. *Nature (London)* 337: 84–87.

Sharp, P.A. (1991) "Five easy pieces." *Science* 254: 663.

Simon, M., and G. Faye (1984). Steps in processing of the mitochondrial cytochrome oxidase subunit I pre-mRNA affected by a nuclear mutation in yeast. *Proc. Natl. Acad. Sci. U.S.A.* 81: 8–12.

Skelly, P.J., Hardy, C.M., and G.D. Clark-Walker (1991). A mobile group II intron of a naturally occurrring rearranged, mitochondrial genome in *Kluyveromyces lactis. Curr. Genet.* 20: 99–114.

Sugiura, M. (1991). Transcript processing in plastids: Trimming, cutting, splicing. In *The Molecular Biology of Plastids* (L. Bogorad and I.K. Vasil, eds.), pp. 125–137. Academic Press, San Diego.

Sugiura, M. (1992). The chloroplast genome. *Plant Mol. Biol.* 19: 149–168.

Tabak, H.F., and L.A. Grivell (1986). RNA catalysis in the excision of yeast mitochondrial introns. *Trends Genet.* 2: 51–54.

Terpstra, P., Zanders, E., and R.A. Butow (1979). The association of *var1* with the 38S mitochondrial ribosomal subunit of yeast. *J. Biol. Chem.* 254: 12653–12661.

Thompson, A.J., and D.L. Herrin (1991). *In vitro* self-splicing reactions of the chloroplast group I intron Cr.LSU from *Chlamydomonas reinhardtii* and *in vivo* manipulation via gene-replacement. *Nucl. Acids Res.* 19: 6611–6618.

Thompson, A.J., Yuan, X., Kudlicki, W., and D.L. Herrin (1992). Cleavage and recognition pattern of a double strand specific endonuclease (I-*Cre*I) encoded by the chloroplast 23S rRNA intron of *Chlamydomonas reinhardtii. Gene* 119: 247–251.

Timothy, D.H., Levings, C.S. III, Hu, W.W.L., and M.M. Goodman (1983). Plasmid-like mitochondrial DNAs in diploperennial teosinte. *Maydica* 28: 139–149.

Turmel, M., Boulanger, J., Schnare, M.N., Gray, M.W., and C. Lemieux (1991). Six group I introns and three internal transcribed spacers in the chloroplast large subunit ribosomal RNA gene of the green alga *Chlamydomonas eugametos. J. Mol. Biol.* 218: 293–311.

Turcq, B., Dobinson, K.F., Serizawa, N., and A.M. Lambowitz (1992). A protein required for RNA processing and splicing in *Neurospora* mitochondria is related to gene products involved in cell cycle protein phosphatase functions. *Proc. Natl. Acad. Sci. U.S.A.* 89: 1676–1680.

Tzagoloff, A., and A.M. Myers (1986). Genetics of

mitochondrial biogenesis. *Annu. Rev. Biochem.* 55: 249–285.

Valencik, M., and J.E. McEwen (1991). Genetic evidence that different functional domains of the *PET54* gene product facilitate expression of the mitochondrial genes *COX1* and *COX3* in *Saccharomyces cerevisiae. Mol. Cell. Biol.* 11:2399–2405.

Valencik, M., Kloeckener-Gruissem, B., Poyton, R.O., and J.E. McEwen (1989). Disruption of the yeast nuclear *PET54* gene blocks excision of mitochondrial intron aI5b from pre-mRNA for cytochrome *c* oxidase subunit I. *EMBO J.* 8: 3899–3904.

Varmus, H.E., and P. Brown (1989). Retroviruses. In *Mobile DNA* (D.E. Berg and M.M. Howe, eds.), pp. 53–108. American Society for Microbiology, Washington, D.C.

Wang, H., and A.M. Lambowitz (1993). Reverse transcription of the Mauriceville plasmid of *Neurospora:* lack of ribonuclease H activity associated with reverse transcriptase and possible use of mitochondrial ribonuclease H. *J. Biol. Chem.* 268: 18951–18959.

Wang, H., Kennell, J.C., Kuiper, M.T.R., Sabourin, J., Saldanha, R., and A.M. Lambowitz (1992). The Mauriceville plasmid of *Neurospora crassa:* characterization of a novel reverse transcriptase that begins cDNA synthesis at the 3' end of template RNA. *Mol. Cell. Biol.* 12: 5131–5144.

Waring, R.B., and R.W. Davies (1984). Assessment of a model for intron RNA secondary structure relevant to RNA self-splicing—a review. *Gene* 28: 277–291.

Weiller, G., Schueller, C.M.E., and R.J. Schweyen (1989). Putative target sites for mobile G+C rich clusters in yeast mitochondrial DNA: Single elements and tandem arrays. *Mol. Gen. Genet.* 218: 272–283.

Weissenger, A.D., Timothy, D.H., Levings, C.S. III, Hu, W.W.L., and M.M. Goodman (1982). Unique plasmid-like DNAs from indigenous maize races of Latin America. *Proc. Natl. Acad. Sci. U.S.A.* 79: 1–5.

Wenzlau, J.M., and P.S Perlman (1990). Mobility of two optional G+C-rich clusters of the *var1* gene of yeast mitochondrial DNA. *Genetics* 126: 53–62.

Wenzlau, J.M., Saldanha, R.J., Butow, R.A., and P.S. Perlman (1989). A latent maturase is also an endonuclease needed for intron mobility. *Cell* 56: 421–430.

Wiesenberger, G., Link, T.A., von Ahsen, U., Waldherr, M., and R.J. Schweyen (1991). MRS3 and MRS4, two suppressors of mtRNA splicing defects in yeast, are new members of the mitochondrial carrier family. *J. Mol. Biol.* 217: 23–37.

Wiesenberger, G., Waldherr, M., and R.J. Schweyen (1992). The nuclear gene MRS2 is essential for the excision of group II introns from yeast mitochondrial transcripts *in vivo. J. Biol. Chem.* 267: 6963–6969.

Wissinger, B., Schuster, W., and A. Brennicke (1991). *Trans* splicing in Oenothera mitochondria *nad1* mRNAs are edited in exon and *trans*-splicing group II intron sequences. *Cell* 65: 473–482.

Wolf, K., and L. Del Giudice (1988). The variable mitochondrial genome of Ascomycetes: Organization, mutational alterations, and expression. *Adv. Genet.* 25: 185–308.

Woolford, J.L. Jr., and C.L. Peebles (1992). RNA splicing in lower eukaryotes. *Curr. Opinion Genet. Dev.* 2: 712–719.

Xiong, Y., and T.H. Eickbush (1988). Similarity of reverse transcriptase-like sequences of viruses, transposable elements, and mitochondrial introns. *Mol. Biol. Evol.* 5: 675–690.

Zinn, A.R., and R.A. Butow (1985). Nonreciprocal exchange between alleles of the yeast mitochondrial 21S rRNA gene: Kinetics and the involvement of a double-strand break. *Cell* 40: 887–895.

Zinn, A.R., Pohlman, J.K., Perlman, P.S., and R.A. Butow (1988). *In vivo* double-strand breaks occur at recombinogenic G+C-rich sequences in the yeast mitochondrial genome. *Proc. Natl. Acad. Sci. U.S.A.* 85: 2686–2690.

11

Mitochondrial Diseases

This chapter focuses on mitochondrial diseases, notably in filamentous fungi, human beings, and flowering plants. Studies of mitochondrial dysfunction in yeast and filamentous fungi, demonstration of the uniparental–maternal pattern of inheritance of mitochondrial genomes in most organisms (Chapter 8), and experiments on mitochondrial genome segregation in heteroplasmic animals such as cows (Chapter 9) have provided the fundamental background information that make human mitochondrial syndromes readily interpretable. In contrast, mitochondrial malfunctions in flowering plants stand apart. Most of them result in the phenomenon called cytoplasmic male sterility (CMS). CMS in the best characterized cases results from the synthesis of novel proteins in the mitochondria of afflicted plants. These proteins are the products of unique chimeric genes. One might rightly query why this chapter does not include diseases of chloroplasts. The answer is that although syndromes similar to certain of those described here probably exist in chloroplasts, they have received nowhere near such detailed scrutiny as have mitochondrial diseases.

MITOCHONDRIAL SYNDROMES OF FILAMENTOUS FUNGI

The experimental study of mitochondrial disease in filamentous fungi such as *Neurospora* is relevant to human mitochondrial disease because these organisms, like ourselves, are obligate aerobes. This means that while these fungi can tolerate a certain fraction of defective mitochondrial genomes, a total lack of mitochondrial genome function is probably lethal. Several classes of disease-causing mutations can be distinguished: point mutations or small deletions in the mitochondrial genome, preferential increases in subgenomic molecules containing only a portion of the entire mitochondrial genome (as in the stopper mutants of *Neurospora*), amplification of specific segments of the mitochondrial genome (as in the case of senescence in *Podospora*), and insertion of plasmids into the mitochondrial genome (for reviews see Dujon and Belcour, 1989; Griffiths, 1992; Kuck, 1989; Wolf and Del Giudice, 1988).

Point Mutations and Small Deletions in *Neurospora*

The most famous of *Neurospora* mitochondrial mutants *[poky]* (*[mi-1]*) was first described by Mitchell and Mitchell in 1952 as a slow-growing respiration-deficient strain. Identification of the primary defect in this mutant has taken almost 40 years, largely because the mutant is very pleiotropic in phenotype (see Gillham, 1978 for a review). Rifkin and Luck (1971) were the first to show that the many pleiotropic effects in *[poky]* were attributable to a defect in mitochondrial protein synthesis. They reported that the number of mitochondrial ribosomes was greatly reduced in *[poky]* as compared to wild type due to a deficiency in small subunits. Later, Akins and Lambowitz (1984) demonstrated that *[poky]* had a four base deletion at the 5'-end of the mitochondrial 19S rRNA. Aberrant 5'-ends were also detected in *[poky]* 19S rRNA with these ends being heterogeneous and mapping 38–45 nucleotides downstream of the normal 5'-end. At the time Akins and Lambowitz thought that the deficiency in 19S rRNA in *[poky]* cultures might result from impaired processing of this molecule rRNA or to its instability. This was a reasonable hypothesis since the sequence at the 5'-end of wild-type 19S rRNA can be folded into a small hairpin that would be destabilized by the *[poky]* deletion, but the story does not end here. Kennell and Lambowitz (1989) developed an *in vitro* transcription system for *Neurospora* mtDNA to enable them to identify transcription start sites. They defined a 15-nucleotide consensus sequence at the transcription initiation sites for three different genes including the 19S rRNA gene. The four nucleotides deleted in the *[poky]* mutation are part of this consensus sequence. Kennell and Lambowitz found that this deletion drastically reduced promoter activity *in vitro* so they argued that the residual 19S rRNA transcripts seen *in vivo* probably initiate from an upstream promoter or promoters. Consequently, longer than normal rRNA precursors are made that are subsequently processed or degraded to yield rRNAs shorter than the wild-type 19S rRNA. Reduced synthesis of 19S rRNA transcripts would account for the deficiency in mitochondrial small ribosomal subunits and the pleiotropic *[poky]* phenotype.

Complementation between mitochondrial mutants in *Neurospora* can be measured (see Gillham, 1978). The technique employs nuclear markers to force heterokaryon formation between mycelia homoplasmic for different mitochondrial mutations. Bertrand and Pittenger (1972) used this method to classify the then-known slow-growing mitochondrial mutants into three complementation groups. The *[poky]* mutant belongs to complementation group I together with the mutants *[exn-1]* (for extranuclear) through *[exn-4], [SG-1]* (for slow growth) and *[SG-3]*, and *[stp-B1]* (for stopper). Akins and Lambowitz (1984) observed the same 4-bp deletion in the 19S rRNA promoter in all mutants analyzed. Whether this means that the same mutation has occurred repeatedly or whether *[poky]* mitochondrial genomes have contaminated other mycelia yielding the remaining supposedly distinct group I mutants has not been resolved.

Like other mitochondrial mutants in filamentous fungi, the different group I mutants are suppressive to wild type. This means that in heteroplasmons mutant mitochondrial genomes gradually replace wild-type molecules. Data combined from two separate studies revealed that at least 38 of 42 heterokaryons constructed between *[poky]* and wild type acquired the mutant phenotype (Manella and Lambowitz, 1978, 1979). Jinks (1966) pointed out long ago that suppressivity was to be expected in mitochondrial mutants of filamentous fungi because a new mitochondrial mutation will affect only a single mtDNA molecule. Hence, the mutation will pass unnoticed unless dominant to wild type or because the mutant mtDNA molecules can outcompete wild type. For *[poky]* Akins and Lambowitz (1984) suggest that more rapid replication of mtDNA might occur because the inhibition of mitochondrial protein synthesis in *[poky]* mitochondria may prevent synthesis of a mitochondrially encoded protein that negatively regulates mtDNA replication. Alternatively, rapid unidirectional gene conversion might spread the *poky* phenotype.

The phenotypes of group I mutants are often complicated by secondary effects on the mitochondrial genome. For instance, Manella *et al.* (1979) reported two variant types of mtDNA in *poky* subcultures (IIa and HI-10). Type IIa subcultures possess a mixture of mtDNA molecules, many of which have a tandemly repeated 2.1-kb fragment. The number of repeats per molecule varies from zero to eight. A derivative of this strain amplifies a large (27-kb) sequence from the mitochondrial genome that includes the rRNA genes. Similarly, a derivative of a second group I mutant, *[SG-3]-551*, was found to amplify a somewhat shorter mtDNA sequence (ca. 20 kb) from the same region of the mitochondrial genome (Bertrand *et al.*, 1980). This strain was also deficient in intact mitochondrial genomes. In both cases the amplified molecules lacked genes encoding important components of the respiratory chain including cytochrome *b* and cytochrome oxidase subunits 1 and 2.

Six nonallelic nuclear suppressors partially alleviate the deficiency in mitochondrial ribosomal small subunits in *[poky]* and other group I mutants (Collins and Bertrand, 1978). Two suppressors studied in combination with *[poky]* were found to lack normal 19S rRNA 5'-ends, but had an increased concentration of mitochondrial small ribosomal subunits nevertheless (Akins and Lambowitz, 1984).

Complementation group II of Bertrand and Pittenger contains but a single mutant *[mi-3]*. This mutant does not complement group I mutants, but mutations in groups I and II complement group III. Lemire and Nargang (1986) demonstrated that *[mi-3]* resulted from a missense mutation in the *cox1* gene, but, in contrast to *[poky]*, *[mi-3]* is probably not suppressive (Hawse *et al.*, 1990). Resolution of the *[mi-3]* or wild-type phenotype in heterokaryons may depend solely on random factors (Hawse *et al.*, 1990).

Preferential Increases of Subgenomic Molecules in *Neurospora*

Group I mutants having amplified pieces of mtDNA are reminiscent of the group III stopper mutants. The stoppers are so named because they have an irregular stop–start growth phenotype in which mycelial growth commences, then stops, then starts again (McDougall and Pittenger, 1969; Bertrand and Pittenger, 1969). All stoppers contain populations of defective mtDNA molecules that vary in concentration depending on whether the mutant is growing or not. During the stopped phase of growth a circular DNA molecule of 21–24 kb containing the rRNA and most of the tRNA genes predominates (Dujon and Belcour, 1989). Remarkably, this sequence is also retained in the group I "nonstoppers" *[poky]* H1-10 and *[SG-3]*-551 described above (Bertrand *et al.*, 1980). Other stopper mutants (e.g., *[stp A18t]*) are characterized by smaller deletions or insertions elsewhere in the mtDNA.

The stopper mutant [E35] possesses two populations of aberrant mitochondrial genomes (DeVries *et al.* 1985, 1986a,b). The smaller molecule has a 40-kb deletion. The larger molecule possesses a 5-kb deletion that is overlapped by the 40-kb deletion and includes the *ndh2* and *ndh3* genes. These results suggest that *stp*[E35] ought to be lacking in NADH dehydrogenase activity and that seems to be the case. The mutant is also deficient in cytochrome *b* and cytochrome oxidase, implying that complex I may be required for the assembly of the other two complexes on the inner membrane. The deletion breakpoints of the two subgenomic mtDNA molecules of stopper [E35] are located in direct repeats consisting of a nearly pure stretch of nonalternating G+C base pairs. This may mean that intramolecular recombination within these repeated sequences led to the formation of the two subgenomic species.

Experiments with yet another stopper mutant have provided further evidence for the role of intramolecular recombination in producing subgenomic mtDNA molecules (Gross *et al.*, 1984). In this stopper two molecular species are found of 21 and 41 kb. The former species predominates during the stopped phase of growth and the latter species appears on resumption of growth. The two subgenomic circles arise by reciprocal recombination at or near two directly repeated tRNA^{Met} sequences. Their frequency apparently depends on relative replication rates from origins unique to each molecule. Gross *et al.* (1989a,b) isolated a phenotypic revertant of their stopper mutant (*stp-ruv*). The mitochondrial genome of the revertant is divided into two nonoverlapping subgenomic circles of 21 and 36 kb. This new and more stable combination of subgenomic circles is associated with deletions of 1 and 4 kb in the 36-kb circle accounting for its smaller size relative to the 41-kb circle in the original stopper mutant. Gross *et al.* (1989a) believe that the deletions may have arisen from the single recombinational event generating the 36-kb circle. The deletions remove both copies of tRNA^{Met} from the 36-kb circle making the revertant dependent on the single tRNA^{Met} present in the 21-kb circle for mitochondrial protein synthesis. This is reminiscent of the apparent dependence of maize mitochondrial protein synthesis on the chloroplast-like tRNA^{Trp} carried by a 2.3-kb plasmid (Chapter 10). In addition the 4-kb deletion removes the *ndh2* and *ndh3* genes, which are also absent from the 21-kb circle and from the [E35] stopper mutant as discussed above. The 21-kb mtDNA molecule of *stp-ruv* appears to undergo regional amplification yielding plasmid-like supercoiled circles varying in size from subunit length to very high multimers (Gross *et al.*, 1989b).

Both cyanide-sensitive and -insensitive respiratory pathways require complex I function (see Chapter 1), so survival of stopper [E35] and *stp-ruv* following deletion of the *ndh2* and *ndh3* genes is puzzling. However, *stp-ruv* is reported to be capable of cyanide-insensitive respiration despite the absence of the two NADH-dehydrogenase subunits encoded by the two missing *ndh* genes (Gross *et al.* 1989a). These two observations are in apparent conflict and it would seem important to measure complex I function and cyanide-insensitive respiration in one or both of the mutants.

Senescence and Other Diseases Arising from Amplification of mtDNA Sequences

Indefinite vegetative propagation is not possible in a number of fungi because the mycelium eventually sickens and dies (see Esser and Tudzynski, 1980). Senescence is under the control of cytoplasmic, presumably mitochondrial, genes and can be avoided only by introducing cycles of sexual propagation.

Senescence has been studied for many years in *Podospora anserina*. The literature from the "premolecular" phase is reviewed by Gillham (1978). The basic observations are these. When a mycelium of any geographic race of *P. anserina* is kept in continuous culture, it will grow at a rate of 7 mm/day for a variable distance (5–100 cm), depending on the geographic race. The growth rate then suddenly decreases and, 1 to 2 days later, the mycelium ceases to proliferate. The terminal hyphae become small, vacuolated, and incapable of further extension. Marcou (1961) and later Smith and Rubenstein (1973a,b) showed that the vegetative period of growth prior to senescence could be divided into three stages. The first, or nonsenescent state, is variable in length and ends with a random event. Following this event the mycelium enters into a fixed growth period (the incubation period). This is terminated by morphologically visible senescence and death. These parameters vary in different *P. anserina* strains. Senescence is contagious as shown by grafting senescent to normal mycelia and, in the absence of nuclear transfer, converts wild-type mycelia to the senescent phenotype (Marcou, 1961). In senescing cultures, the cytochrome aa_3 content is gradually reduced, indicating that mitochondrial function is being affected (Belcour and Begel, 1978). When a senescent strain is used as the paternal parent, the progeny are normal, but in the reciprocal cross the progeny spores in every ascus, each of which originates from a separate meiotic event, are either all normal or all senescent. The latter result is thought to be obtained because the senescent maternal parent has a mixture of normal and senescent mitochondria.

The molecular biology of senescence is a fascinating story (see Dujon and Belcour, 1989; Griffiths, 1992; Wolf and Del Giudice, 1988, for reviews). The mitochondrial genome of *P. anserina* is large and ranges from 87 to 100 kb in different races (Clark-Walker, 1992). In comparison its relatives *Neurospora crassa* and *Aspergillus nidulans* have mitochondrial genomes of 60–73 and 33–40 kb, respectively. Mitochondrial genes in *Podospora* are also rich in introns, with *cox1* alone having 16 (Cummings *et al.*, 1989). Whereas mtDNA isolated from young mycelia always contains randomly broken linear molecules of various sizes and a few ca. 90-kb circular molecules corresponding to intact mitochondrial genomes, DNA isolated from senescent mycelia contains numerous small circular molecules (Dujon and Belcour, 1989; Griffiths, 1992). The latter molecules comprise a multimeric series in which a reiterated sequence is arranged in head-to-tail fashion . The size of the monomeric circles is in good agreement with the repeat size as determined by restriction enzyme analysis.

These so-called senDNAs are derived from the mitochondrial genome (Fig. 11–1). senDNA α, also called plDNA, is 2539 bp long and corresponds to the first intron of the *cox1* gene (Osiewacz and Esser, 1984). In the short-lived *smt*⁻ and *A* wild-type strains this group II intron is excised, amplified, and circularized in most senescent cultures (Dujon and Belcour, 1989). At both excision sites inverted and direct repeats have been identified (Cummings and Wright, 1983; Osiewacz and Esser, 1984). senDNA α contains a long open reading frame adjacent to and in frame with the upstream exon. The protein encoded by this ORF has significant if scattered homologies with retroviral and retrotransposon reverse transcriptases, as is true of other class II fungal mitochondrial introns (Michel and Lang, 1985). While a transcript of 3.2 kb hybridizes with all exons of the *cox1* gene, two different transcripts of 2.5 and 2.7 kb (Kuck *et al.*, 1985) or 2.4 and 2.5 kb (Wright and Cummings, 1983a) hybridize to senDNA α. These two transcripts seem to correspond to the linear and lariat forms of group II intron RNA (Schmidt *et al.*, 1987). An RNA-dependent DNA polymerase activity has also been reported from middle-aged, but not young cultures of *P. anserina* (Steinhilber and Cummings, 1986).

Several other regions of the *Podospora* mitochondrial genome also yield senDNAs and these vary in length from 4 to 20 kb (Fig. 11–1). Each of the sen DNAs from the β region possesses a common 1100-bp-long segment (Koll *et al.*, 1985). This DNA species originates from an intergenic region downstream of the *cox1* gene in which

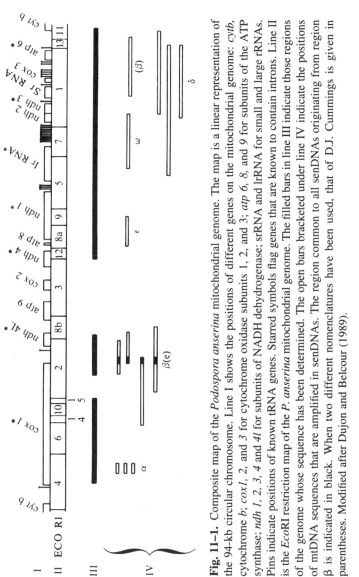

Fig. 11-1. Composite map of the *Podospora anserina* mitochondrial genome. The map is a linear representation of the 94-kb circular chromosome. Line I shows the positions of different genes on the mitochondrial genome: *cytb*, cytochrome *b*; *cox1*, *2*, and *3* for cytochrome oxidase subunits 1, 2, and 3; *atp 6*, *8*, and *9* for subunits of the ATP synthase; *ndh 1*, *2*, *3*, *4* and *4l* for subunits of NADH dehydrogenase; srRNA and lrRNA for small and large rRNAs. Pins indicate positions of known tRNA genes. Starred symbols flag genes that are known to contain introns. Line II is the *Eco*RI restriction map of the *P. anserina* mitochondrial genome. The filled bars in line III indicate those regions of the genome whose sequence has been determined. The open bars bracketed under line IV indicate the positions of mtDNA sequences that are amplified in senDNAs. The region common to all senDNAs originating from region β is indicated in black. When two different nomenclatures have been used, that of D.J. Cummings is given in parentheses. Modified after Dujon and Belcour (1989).

intramolecular recombination occurs frequently (Dujon and Belcour, 1989). The only property common to all senDNAs is the presence of at least one intron open reading frame (Belcour et al., 1986), but the significance of this observation is not known especially in view of the large number of introns present in the mitochondrial genome of *P. anserina*.

There is direct evidence of two kinds that the senescent phenotype can be suppressed by interference with the formation and propagation of senDNA. First, transfer of a senescent culture to medium containing ethidium bromide "rejuvenates" the culture (Koll et al., 1984). This treatment not only permits growth of the senescent mycelium, but reverses the process so that on transfer to medium lacking ethidium bromide the life span of the rejuvenated culture is similar to that of a young culture. At the molecular level rejuvenation is paralleled by the loss of senDNA. Presumably, ethidium bromide intercalates preferentially into the senDNA molecules and prevents their replication or causes their destruction in much the same way that this compound "cures" yeast of its mitochondrial genome (Chapter 7).

The second kind of evidence is genetic. Since senescence is a terminal state, longevity mutants can be selected from senescent cultures. In three long life mutants *(mex1, mex5, mex7)* sequencing experiments showed that most of the first intron of the *cox1* gene had been deleted plus a few base pairs from the upstream exon (Belcour and Vierny, 1986). These deletions prevent elaboration of senDNA α, but the mutants also lacked any cytochrome aa_3 peak on spectroscopic analysis. In the second class of mutants *(ex1, ex2)* almost the entire *cox1* gene is deleted including intron 1 (Schulte et al., 1988, 1989). The mitochondrial genome of *ex1* is organized in heteromeric, subgenomic circles lacking *cox1*. In the *mex3* and *mex6* mutants *cox1* and intron 1 are not deleted (Koll et al., 1987). Instead, a region called β near the 3'-end of *cox1* is fused either to a region called ϵ *(mex3)* or a region called γ *(mex6)*. Nuclear double mutants that interfere with excision of senDNA α *(gr viv)* or its amplification *(i viv)* are also long lived (Tudzynski et al., 1982). Lastly, an 8.4-kb linear mitochondrial plasmid, pAL 2-1, has been discovered in the long lived AL2 strain of *P. anserina* (Hermans and Osiewacz, 1992). The presence of this plasmid is correlated with mtDNA rearrangements and delayed amplification of senDNA α (Osiewacz et al., 1989). In summary, genetic intervention with senDNA amplification by any of several routes results in longevity. These observations strongly support the hypothesis that senDNA α is a major culprit in *Podospora* senescence.

Other recently obtained indirect *in vivo* evidence supports the hypothesis that a reverse transcriptase encoded by the intron α-encoded protein may be involved in the senescence response (Sainsard-Chanet et al., 1993). Using the polymerase chain reaction (PCR), *cox1* DNA molecules lacking introns could be detected in young and senescent cultures. These intron-deleted molecules were far more abundant in senescent than in young cultures. Comparisons between senescent wild-type strains and mutants such as *mex1* and *mex5* lacking the reverse transcriptase reading frame did not reveal intron-deleted molecules. Since the *mex* mutants are long-lived, the obvious conclusion is that reverse transcription, presumably related to senDNA α amplification, is an important part of the senescence process.

The rejuvenation, genetic, and PCR experiments support the hypothesis that amplification of senDNAs derived from the mitochondrial genome is responsible for the senescence phenotype, but the mechanism by which the senDNAs cause senescence is still a mystery. Since most of the *mex* and *ex* mutants lack cytochrome oxidase activity, an important unanswered question is how they respire. In general, the thinking is that the mutants survive via cyanide-insensitive respiration that bypasses cytochrome oxidase (e.g., Schulte *et al.,* 1988), but this remains to be proved.

There are two footnotes to the senescence story in *Podospora*. Wright and Cummings (1983b) claimed that during senescence both senDNA α and senDNA β were transposed to the nucleus and integrated into nuclear DNA. This observation was incorporated into a model of senescence that assumed these senDNAs or "mobile introns" could become integrated into either the mitochondrial or nuclear genomes and thereby block essential functions (Kuck *et al.,* 1985). In fact, Koll (1986) repeated the experiments of Wright and Cummings (1983b) and could find no evidence for integration of senDNA α into the nucleus. In another paper, transformation of *P. anserina* protoplasts with pBR322 recombinant plasmids containing senDNA α was reported (Stahl *et al.,* 1982), but these experiments could not be repeated (Sainsard-Chanet and Begel, 1986).

The phenotype of the ragged mutants of *Aspergillus amstelodami* can also be traced to amplification of mtDNA sequences (see Dujon and Belcour, 1989; Griffiths, 1992; Kuntzel *et al.,* 1982). These mutants exhibit slow and erratic mycelial growth and a deficiency in cytochromes *b* and aa_3. Preparations of mtDNA from the ragged mutants possess amplified sequences in the form of multiple head-to-tail repeats of 0.9 to 2.7 kb in addition to intact wild-type mitochondrial genomes. The amplified sequences in most ragged mutants come from two segments of the mitochondrial genome (regions 1 and 2), with six of seven mutants tested containing region 2 amplifications. Whether these amplified sequences are related in any way to introns is not known.

Lastly, the nuclear natural death mutant *(nd)* of *Neurospora crassa* appears to destabilize the mitochondrial genome of this fungus (Seidel-Rogol *et al.,* 1989). This recessive, pleiotropic mutant has phenotypic and molecular defects similar to those seen in senescence, including deficiencies in cytochromes aa_3 and *b* and accumulation of gross rearrangements, and large deletions in the mitochondrial genomes of vegetatively propagated mycelia. As Seidel-Rogol *et al.* (1989) put it, "the *nd* mutation is unique in that it inactivates a nuclear gene that encodes a mitochondrial component that normally protects mitochondrial DNA from rearrangement."

The Killer Plasmids of *Neurospora intermedia*

Names like *kalilo* and *maranhar* evoke images of soft, warm winds, palm trees, spices, and the South Seas. Nothing could be farther from the truth. *Kalilo* is a Hawaiian word meaning dying or hovering between life and death. *Kalilo* and *maranhar* are linear plasmids that are poorly adapted parasites and whose integration into the mitochondrial genome leads ultimately to death of the organism. *Kalilo* is the most thoroughly studied of the two plasmids (Bertrand *et al.,* 1985, 1986; Griffiths, 1992; Griffiths and Bertrand, 1984; Myers *et al.,* 1989; Chan *et al.,* 1991; Vierula *et al.,* 1990). This plasmid is found in about 30% of field-isolated strains of *N. intermedia* from the Hawaiian island of Kauai. These isolates, in contrast to wild type, develop a series of symptoms characteristic of fungal senescence including a progressive decline in mitochondrial function, loss of female fertility, an exponential decline in hyphal growth rate, and death of the mycelium at its frontier. The *kalilo* plasmid responsible for this syndrome is an 8.6-kb linear, double-stranded DNA molecule *(kalDNA)*. This element has long (1366-bp) terminal inverted repeats (TIRs). The 5'-terminal nucleotides at the ends of the TIRs are each covalently linked through a phosphodiester bond to a 120-kDa protein. This organization is reminiscent of the linear maize mitochondrial plasmids described in Chapter 10. The TIRs may be origins of plasmid replication as well as sites for the integration of the element into mtDNA.

When the *kalDNA* nucleotide sequence is translated using the *Neurospora* mitochondrial genetic code, two long, nonoverlapping ORFs are detected that are oriented in opposite directions. ORF1 encodes a protein with marked sequence similarity to bacteriophage T7 RNA polymerase while ORF2 specifies a DNA polymerase. This signature is characteristic of invertrons (Chan *et al.*, 1991). This is a term coined by Sakaguchi (1990) to describe those prokaryotic and eukaryotic linear DNA plasmids, viruses, and transposons that have the same organization as *kalilo*. The adenoviruses are perhaps the best characterized invertrons.

Integration of *kalDNA* into the mitochondrial genome is thought to trigger the process of senescence (Bertrand *et al.*, 1985, 1986; Myers *et al.*, 1989). The plasmid inserts at various points in the mitochondrial genome including within the large rRNA gene. Integration of the entire plasmid occurs with the exception of no more than 20 bp, which are lost from each end (Chan *et al.*, 1991). This process generates very large inverted repeats of mtDNA flanking the insert (Dasgupta *et al.*, 1988). During subsequent growth, mitochondrial genomes containing integrated kalDNA accumulate at the expense of their wild-type counterparts so that these lethal mitochondrial genomes eventually predominate in the mycelium. Direct confirmation of the suppressive nature of *kalilo* comes from the demonstration that senescence is readily transmitted in heterokaryons with normal strains of *N. intermedia* and *N. crassa* (Griffiths *et al.*, 1990).

One confusing part of the *kalilo* story now seems to have been resolved. Bertrand *et. al.* (1986) reported that the *kalilo* plasmid could not be detected free in mitochondria, but only in nuclear fractions. This observation was at variance with observed maternal inheritance of *kalilo* and the fact that the two ORFs in the plasmid are translatable using only the *Neurospora* mitochondrial code and not the universal genetic code. More recently free plasmid has been isolated from mitochondria treated with DNase following protease treatment (summarized by Chan *et al.*, 1991). Disruption of mitochondria with the nonionic detergent Triton X-100, which is added to the medium used to extract nuclei from mycelia, results in the artifactual association of plasmid with chromatin and probably accounts for the earlier results.

The *maranhar* plasmid was found in several field-collected isolates of *N. crassa* from Aarey, India (Court *et al.*, 1991). Strains carrying this 7-kb linear DNA plasmid also undergo senescence. Like *kalilo* the *maranhar* plasmid has inverted terminal repeats with a protein covalently bound to each 5'-terminus. As in the case of *kalilo*, mitochondrial genomes with integrated *maranhar* sequences accumulate during subculture. Sequencing of *maranhar* reveals that it, like *kalilo*, possesses two ORFs (Court and Bertrand, 1992). As in *kalilo*, ORF1 specifies a single subunit RNA polymerase and ORF2 a DNA polymerase. Two major transcripts corresponding to these ORFs as well as their protein products have been detected (Court and Bertrand, 1993). In spite of these apparent similarities, *kalilo* and *maranhar* are not closely related and the proteins encoded by the two ORFs do not have a great deal of sequence similarity.

HUMAN MITOCHONDRIAL SYNDROMES: BIOCHEMICAL PLEIOTROPY, AGING, AND DEGENERATIVE DISEASE

It is testimony to our overarching preoccupation with ourselves that the human mitochondrial genome is rapidly becoming the best defined mitochondrial genome genetically next to yeast (see reviews by Poulton, 1992; Schapira, 1993; Shoffner and Wallace, 1990; Wallace, 1989, 1992a). While the study of diseases traceable to lesions in human mtDNA has intrinsic significance of its own, a more global question also emerges. Thus, Wallace (1992a,b), one of the leading investigators in the field, argued that mutations in human mtDNA may be associated with a variety of degenerative diseases and ultimately with the aging process itself. We explore this question following a discussion of the specific kinds of syndromes attributable to alterations in human mtDNA.

Point Mutations in mtDNA and Human Disease

The discovery of the mtDNA lesions responsible for specific diseases has proceeded by leaps and bounds over the past few years. Missense mutations can result in defects in specific mitochondrial structural genes or in mitochondrial protein synthesis because of alterations in different mitochondrial tRNA genes (see Wallace, 1992a).

As expected, missense mutations causing mitochondrial diseases typically show maternal inheritance. However, this inheritance pattern is not always completely straightforward. Progeny may be heteroplasmic for wild-type and mutant mtDNA molecules and have a wild-type phenotype. Expression may vary for other reasons such as nuclear genetic background, sex, and environment. Also, key members of a pedigree may have died.

Missense mutations in structural genes genome typically fall into two phenotypic classes: Leber's hereditary optic neuropathy (LHON) and neurogenic muscular weakness, ataxia, and retinitis pigmentosa (NARP) (Shoffner and Wallace, 1990; Wallace, 1992a). LHON is a late-onset form of acute or subacute blindness resulting from optic nerve degeneration in which central vision is rapidly lost in both eyes, but peripheral vision is retained. The commonest mtDNA mutation resulting in LHON (at nucleotide pair 11,778) changes the 340th amino acid of the ND4 protein from arginine to histidine (Table 11–1). This alteration occurs in about 50% of LHON cases. Among the progeny of mothers heteroplasmic for LHON, blindness will occur if segregation of the mtDNA molecules at meiosis increases the proportion of mutant mitochondrial genomes in any of the offspring toward 100%.

Other missense mutations can also cause the LHON phenotype (Table 11–1). Thus, mutations in *ndh1* (bps 3460 and 4160) also produce this phenotype as does a mutation in *cytb* (bp 15257). Five additional mutations appear to contribute to LHON by interacting with other mutations in the mitochondrial genome (Johns and Berman, 1991; Brown et al., 1992). These occur in the *ndh1* (bp 4216), *ndh2* (bps 4917, 5244), *ndh5* (bp 13708), and *cytb* (bp 15812) genes (Table 7–1). The contribution of each mutation to the phenotype can be difficult to sort out. For example, in 24 LHON patients with the *ndh1* mutation at bp 4216, 10 also had the bp 4917 mutation in *ndh2* (42%), nine the bp 13708 mutation in *ndh5* (37%), four the classical bp 11778 mutation in *ndh4* plus the bp 13708 mutation (16%) and one the bp 11778 mutation (4%). Except in the case of bp 11778 the question is whether each mutation in these double mutants contributes to the LHON phenotype or whether certain mutations are "hitchhikers" (Chapter 9). Suggestive evidence that each mutation may contribute to the LHON phenotype comes from a lineage studied by Brown et al. (1992) in which four mutations accumulated sequentially in the *ndh 5*, *cytb* (two mutations), and *ndh2* genes increased the probability of blindness.

A second site mutation may also act to suppress the effect of a mitochondrial mutation leading to the LHON phenotype. Thus, the LHON-causing *ndh1* mutation at bp 4160 is more severe than other LHON-producing mutations. However, a branch of the pedigree in which this mutation was studied has a milder form of the disease and also possesses a second *ndh1* missense mutation at bp 4136 (Howell et al., 1991).

A missense mutation at bp 8993 in the *atp6* gene results in the NARP syndrome (Holt et al., 1990). As mentioned earlier NARP, like LHON, is highly pleiotropic and among its many phenotypic manifestations causes a form of retinitis pigmentosa, ataxia, seizures, dementia, and developmental delay.

Mutations in tRNA genes (Table 11–1) have even more severe effects than those affecting protein-coding genes because they reduce synthesis of all mitochondrially encoded proteins. The syndromes caused by these diseases have the acronyms MERRF (ragged-red fiber disease), MELAS (mitochondrial encephalomyopathy, lactic acidosis, and strokelike symptoms), and MMC (maternally inherited myopathy and cardiomyopathy). Other diseases that also may involve alterations in mitochondrial tRNA genes are several ocular myopathies and lethal infantile mitochondrial myopathy (LIMM).

MERRF (ragged red fiber disease) is so named because the muscle fibers have a ragged red appearance when stained with Gomori-modified trichrome dye. The mitochondria in these fibers are abnormal and show paracrystalline arrays and degeneration of the cristae. Severely affected patients exhibit uncontrollable periodic jerking, dementia, and sometimes respiratory failure (Wallace, 1992a). The mutation responsible for the disease results from an $A+T \rightarrow G+C$ change at bp 8344 in the $T \psi C$ loop of $tRNA^{Lys}$ (Table 11–1). This, in turn, causes a reduction in mitochondrial protein synthesis that primarily affects complexes I and IV, possibly because these complexes have the greatest number of mitochondrially encoded subunits. The severity of the defects in respiration and oxidative phosphorylation caused by the mutation varies in different individuals and is directly related to the severity of clinical manifestations. This variation can, in part, be correlated with age. Maternal relatives of MERRF patients are usually heteroplasmic for the mutation as are their progeny. Individuals under 20 generally do not express the MERRF phenotype unless >95% of the mitochondrial genomes possess the mutation whereas severe effects of the disease can be seen in 60- to 70-year-old people with 85% mutant mtDNA molecules.

About 80% of MELAS cases arise because of an $A+T \rightarrow G+C$ mutation at bp 3243 in the dihydrouridine loop of $tRNA^{Leu}_{UUR}$ (Table 11–1). As is true of MERRF patients, individuals with the MELAS syndrome are heteroplasmic for mtDNA molecules carrying the mutation. This mutation also inactivates the binding site for the mitochondrial transcription termination factor (mTERF) that lies within the $tRNA^{Leu}_{UUR}$ gene (see Chapter 12 and Kruse et al., 1989). MELAS patients suffer from reversible, periodic stroke-like episodes. Interestingly a different syndrome, MMC, is attributed to a second mutation in the $tRNA^{Leu}_{UUR}$ gene at bp 3260 (Table 11–1). This mutation is in the stem of the anticodon loop (Zeviani et al., 1991) outside the transcription terminator-binding sequence. The bp 3260 mutation results in Complex I and IV defects that are proportional to the percentage of mutant mtDNAs in the affected individual. A third mutation has been reported at position 3302 near the 3' end to tRNA$^{Leu}_{UUR}$ that is associated with abnormal mitochondrial RNA processing (Bindoff et al., 1993). In comparing skeletal muscle and skin fibroblast tissue important differences in mRNA processing were observed that lead to a pronounced effect of the mutation on mitochondrial function in skeletal tissue. Other syndromes that may be associated with specific mitochondrial tRNA alterations are certain cases of LIMM. LIMM includes a heterogeneous group of diseases resulting in severe neonatal lactic acidosis and other defects that result in death within a few months of birth from respiratory failure, and ocular myopathy (Table 11–1).

Extreme care has to be exercised in relating alterations in specific tRNA genes to mitochondrial myopathies because of the possibility of hitchhiking effects. Thus Lauber et al. (1991) report a case of ocular myopathy associated with chronic intestinal pseudoobstruction in which a combination of mutations in three tRNA genes was found (Table

Table 11–1. Point Mutations of the Human Mitochondrial Genome

Gene	Base Pair Change	Amino Acid Change	Exists as Homo- or Heteroplasmon	Myopathy[a]
ndh1	3460 (GC → AT)	Ala → Thr	Homoplasmon	LHON
	4136 (AT → GC)	Tyr → Cys	Homoplasmon	LHON
	4160 (TA → CG)	Leu → Pro	Homoplasmon	LHON
	4216 (TA → CG)	Tyr → His	Homoplasmon	LHON-associated
ndh2	4917 (AT → GC)	Asp → Asn	Homoplasmon	LHON-associated
	5244 (GC → AT)	Gly → Ser	Heteroplasmon	LHON-associated
ndh4	11778 (GC → AT)	Arg → His	Homo- or heteroplasmon	LHON
ndh5	13708 (GC → AT)	Ala → Thr	Homoplasmon	LHON-associated
cytb	15257 (GC → AT)	Asp → Asn	Homoplasmon	LHON
	15812 (GC → AT)	Val → Met	Homoplasmon	LHON-associated
atp6	8993 (TA → GC)	Leu → Arg	Heteroplasmon	NARP
tRNA$^{Leu}_{UUR}$	3243 (AT → GC)	NA[b]	Heteroplasmon	MELAS
	3260 (AT → GC)	NA	Heteroplasmon	MMC
tRNA^{Ile}	4317 (AT → GC)	NA	?	FICP
tRNA^{Lys}	8344 (AT → GC)	NA	Heteroplasmon	MERRF
tRNA^{Gly}	10006 (AT → GC)	NA	Homoplasmon	CIPO
tRNA$^{Ser}_{AGY}$	12246 (CG → GC)	NA	Homoplasmon	CIPO
tRNA$^{Leu}_{CUN}$	12308 (AT → GC)	NA	Homoplasmon	CIPO
			Heteroplasmon	CPEO
tRNA^{Thr}	15923 (AT → GC)	NA	?	LIMM
	15924 (AT → GC)	NA	?	LIMM

Data from Wallace (1992a).

[a]Myopathy abbreviations: LHON (Leber hereditary optic neuropathy); NARP (neurogenic muscle weakness, ataxia, and retinitis pigmentosa); MELAS (mitochondrial encephalomyopathy, lactic acidosis, and strokelike episodes); MMC (maternal myopathy and cardiomyopathy); MERRF (myoclonic epilepsy and ragged red muscle fibers); CIPO (chronic intestinal pseudoobstruction with myopathy and ophthalmoplegia); CPEO (progressive chronic external ophthalmoplegia); FICP (fatal infantile cardiomyopathy plus, a MELAS-associated cardiomyopathy); LIMM (lethal infantile mitochondrial myopathy).

[b]NA, not applicable.

11–1). One of these was a mutation at np 12308 in the tRNA$^{Leu}_{CUN}$ gene. This mutation is a polymorphism that occurs with high frequency (16%) in the general population van den Ouweland et al. (1992) reported subsequently.

Experiments correlating the inheritance of mitochondrial myopathies and mtDNA inheritance provide strong circumstantial evidence for causality, but they do not prove it. Tissue culture cells and mitochondrial injection (Chapter 7; King and Attardi, 1989) have been used to provide direct evidence that specific mitochondrial mutations result in respiration deficiency. These experiments also make possible the characterization of the effects of mitochondrial lesions in a specific genetic background under standardized physiological conditions. Such experiments have been informative for both the MERRF and MELAS mutations. In the case of MERRF the mitochondrial injection experiments showed that the genetic defect responsible for the syndrome was transferred with mitochondria from an afflicted individual into different human cell lines lacking mtDNA (Chomyn et al., 1991). The transformants obtained with different myoblasts from the donor contained either mutated or wild-type mtDNA (i.e., the donor was heteroplasmic, probably at the cellular level). Lastly, the occurrence of abnormalities in the pattern of synthesis of mitochondrially encoded proteins by these mitochondria could be correlated with respiration-deficient phenotype of the mutant mitochondria.

Similar results were also obtained when mitochondria from MELAS patients were injected into tissue culture cells (Chomyn et al., 1992). These experiments also beautifully illustrate the "threshold effect" (e.g., Shoffner and Wallace, 1990) whereby cell phenotype does not change significantly until mutant mtDNA molecules reach a high proportion within the cell. In the case of MELAS, the threshold at which the mutant phenotype converts to wild type is around 6% wild-type mtDNA molecules with full protection against the mutant phenotype being reached when a cell has 10% wild-type mitochondrial genomes (Fig. 11–2). These results provide striking evidence that a cell can tolerate a relatively high number of defective mitochondrial genomes without an obvious change in phenotype. This may be quite important since MELAS mtDNA molecules appear to enjoy a replicative advantage to wild type mtDNA. Thus, when ρ^0 human cells were injected with mitochondria from a cell line heteroplasmic for the MELAS mutation, they underwent a rapid shift toward the pure mutant phenotype (Yoneda et al., 1992). However, no significant change in cell growth rate was noted as the population shifted from wild type to mutant. This replicative advantage of the mutant mtDNA molecules may extrapolate to real life. Kobayashi et al. (1992) report that MELAS patients and their siblings tend to have higher proportions of mutant mtDNAs than their mothers. Thus, MELAS mtDNA molecules appear to amplify selectively in pedigrees heteroplasmic for the mutation.

Deletions

Most mtDNA deletion mutations are spontaneous and have no family history suggesting that they arise *de novo* during development (Wallace, 1992a,b). They result in Kearns–Sayre syndrome (KSS), chronic external ophthalmoplegia plus (CEOP), and Pearson marrow/pancreas syndrome. All are serious diseases with Pearson syndrome generally being fatal in childhood. Over 100 human mtDNA deletions have been examined and their general characteristics are summarized in Fig. 11–3. The vast majority (95%)

MITOCHONDRIAL DISEASES

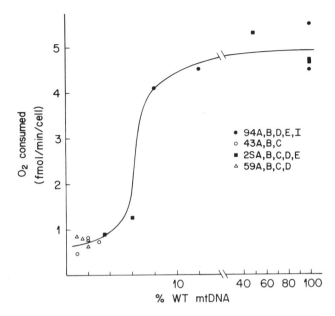

Fig. 11–2. Demonstration that a small minority of wild-type mitochondrial genomes is sufficient to protect ρ^0 tissue culture cells containing a mixture of wild-type and MELAS mitochondrial genomes following injection of ρ^0 cells with mitochondria from a MELAS patient. The numbers 94 and 43 represent clones derived from one patient, 2S clones are from the second patient, and 59 clones are from the third patient. From Chomyn et al. (1992).

remove sequences between the origins of heavy (O_H) and light (O_L) strand replication (Fig. 11–3, right side), but a few delete sequences between O_L and O_H (Fig. 11–3, left side). The deletions extend to, but do not encompass O_L including bps 5721–5781 and O_H including the L-strand promoter (P_L, bps 392–445) through the termination-associated sequences (TAS) for 7 S RNA (Chapter 3) at bps 16157–16172. Four deletions remove the H-strand promoter (P_H), but not P_L, showing that H-strand transcription is not required for maintenance of mtDNA, but that P_L is important for mtDNA replication.

Apparently identical deletions can give rise to different syndromes. Moraes et al. (1989) characterized deletions in 32 patients and found seemingly identical deletions in two pairs of patients of 7.5 and 7.6 kb, respectively. In each case one member of the pair had typical Kearns-Sayre syndrome and the other had ocular myopathy. Presumably, this results from different proportions of deleted to wild-type mtDNA molecules in the individuals of each pair, but other genetic and environmental distinctions could be involved as well.

The most common deletion (Fig. 11–3), found in 30–50% of all patients, is flanked by 13 nucleotide pair direct repeats (5'-ACCTCCCTCACCA) at nucleotide pairs 8468 and 13446 and removes 4997 base pairs of sequence (Schon et al., 1989; Shoffner et al., 1989). A second, less common deletion (7.4 kb) occurs between the 12-bp direct repeat (5'-CATCAACAACCG) at positions 8648 and 16084 (Fig. 11–3). Mita et al. (1990) classified these and other deletions as belonging to class I, signaling the fact

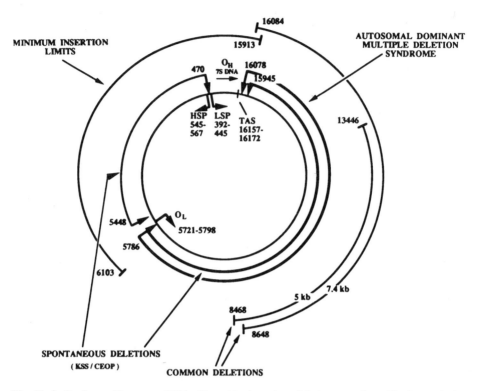

Fig. 11–3. Regions of human mtDNA affected by insertion–deletion mutations. The inner circle shows the position of the heavy strand promoter (HSP), the light strand promoter (LSP), the origin of heavy strand replication (O_H), the origin of light strand replication (O_L), and the termination sequence for 7S RNA synthesis (TAS). The next arcs outward, reading clockwise from O_H to O_L and O_L to O_H, show the maximum extent of deletions mapped in CEOP, KSS, and Pearson syndrome patients and in autosomal dominant KSS pedigrees. The outer O_H to O_L arcs show the positions of the common 4997-bp (5-kb) and 7436-bp (7.4-kb) deletions. The outer O_L to O_H arc indicates the minimum area encompassed by insertion mutations. From Wallace (1992a).

that they are precisely flanked by direct repeats. Class II deletions, in contrast, are not. The class II deletions are either flanked imprecisely by perfect direct repeats, or else obvious repeat elements are absent immediately adjacent to the deletion breakpoints. The mechanisms by which the human mtDNA deletion mutations arise are unknown, but recombination between flanking direct repeats has been suggested for the class I deletions (Mita et al., 1990). Such a model would certainly be compatible with the mechanisms by which subgenomic circles are thought to arise in plant mitochondria (Chapter 3) and certain stopper mutants of *Neurospora*. Another mechanism that could produce these deletions is slip-replication (Shoffner et al., 1989). According to this mechanism mispairing of direct repeats during mtDNA replication would lead to formation of deleted and normal daughter molecules. Topoisomerase cleavage is yet another mechanism proposed for the origin of deletion mutations (Nelson et al., 1989).

Copy Number Mutations

Wallace (1992a) mentions two instances in which mtDNA depletion may be responsible for mitochondrial myopathy. The first case is apparently genetic involving two siblings and a cousin who died of lethal mitochondrial myopathy (Moraes *et al.,* 1991). The muscle of one sibling had only 2% of the normal mtDNA level while the liver of the cousin had 12% of the wild-type number of mitochondrial genomes. In the other case a phenocopy of the mtDNA depletion syndrome was observed in AIDS patients treated with Zidovudine, which blocks both viral and mtDNA replication (Simpson *et al.,* 1989). Patients treated with the drug suffer a reduction in mtDNA amounts of 22–78% and develop the ragged-red fiber syndrome (Arnaudo *et al.,* 1991). However, some caution needs to be exercised in interpreting such experiments. Lansman and Clayton (1975) reported that photodamage of mtDNA substituted with 5-bromodeoxyuridine caused extensive inhibition of mitochondrial RNA synthesis, but the rate of mitochondrial protein synthesis remained unchanged for at least 48 hr after RNA synthesis had been suppressed 85%. These experiments and others discussed later suggest that general mechanisms of translational control may exist in plastids and mitochondria that buffer against parallel declines in organelle genome and mRNA levels. Such mechanisms may serve to protect organelles and cells against variations in genome number due to the stochastic processes.

Nuclear Gene Mutations

Nuclear gene mutations probably are responsible for many diseases resulting in respiratory deficiency, but are more difficult to recognize because they are not maternally inherited like mitochondrial genes (Shoffner and Wallace, 1990). This means that indirect criteria such as pedigree analysis demonstrating Mendelian inheritance are required for their identification. Also, primary defects are likely to be more difficult to assess for such diseases. One need only recall that over 215 different PET genes have been identified in yeast. Despite this there are indications that nuclear gene mutations are responsible for a number of potential mitochondrial myopathies (Shoffner and Wallace, 1990).

One particularly interesting class of nuclear gene mutations includes autosomal dominants that seem to cause multiple mtDNA deletions in afflicted individuals (Wallace, 1992a). These effects are reminiscent of those caused by the *Neurospora* nuclear death mutant described earlier.

Mitochondrial Mutation, Degenerative Disease, and Aging

There are striking parallels between certain mitochondrial myopathies in filamentous fungi and human beings. Accumulation of defective mtDNAs or imbalance in subgenomic molecules leads to the *Neurospora* stopper phenotype. Amplification of defective mtDNA molecules causes the ragged phenotype in *Aspergillus* and amplification of small segments of mtDNA in *Podospora,* notably senDNA α, causes senescence. In each of these cases the accumulation of defective mtDNA molecules or pieces of these molecules results in the cessation of growth and/or death. In humans accumulation of defective mtDNA molecules results in severe mitochondrial myopathies such as KSS.

There is, however, an important difference between fungi and humans. In filamentous fungi mitochondria seem to mix more or less randomly in the mycelium whereas in human beings tissue segregation of mitochondria and/or their genomes seems to occur, as evidenced in a disease like MERRF where only some fibers have the ragged red phenotype and others appear normal.

Wallace (1992a,b) reviewed the evidence that aging and degenerative disease may be related to declines in oxidative phosphorylation and damage to the human mitochondrial genome. Symptoms in most human mtDNA diseases do not appear until late in life, after which they progress in severity. Oxidative phosphorylation normally declines with age. This decline appears correlated with increases in the number of cytochrome oxidase-negative skeletal muscle and heart fibers, as well as reductions in both the respiration rate and activities associated with complexes I and IV. These metabolic alterations could be manifestations of changes in the mitochondrial genome with increasing age. In one study Corrall-Debrinski et al. (1991) presented evidence that aging human hearts progressively accumulated both the common (4997 bp) and less common (7436 bp) mtDNA deletions after age 35. This led Wallace to argue that mtDNA damage accumulates with age as the result of oxidation by superoxide anions and hydrogen peroxide (oxygen radicals), which are byproducts of oxidative phosphorylation. The notion is that as oxidative damage increases, mtDNA replication becomes inhibited. This prolongs the time that replicating mtDNA is triple stranded and increases the probability of mtDNA deletions via a slipped mispairing mechanism. Molecules containing these deletions would subsequently become enriched due to their replicative advantage.

Wallace notes that according to this hypothesis, tissues experiencing chronic inhibition of respiration and oxidative phosphorylation would be expected to exhibit increased damage to mtDNA. To test this hypothesis Corrall-Debrinski et al. (1991) compared mtDNA damage in normal and ischemic hearts. In ischemic heart disease, coronary artery occlusion starves the heart of nutrients and oxygen essential for mitochondrial function. Corrall-Debrinski et al. (1991) found that mtDNA molecules with the common 4997-bp deletion in control heart tissue were virtually undetectable up to age 35, but the proportion increased with age, reaching 0.0035% in an 85 year old. Ischemic heart tissue, in contrast, possessed 0.02 to 0.85% deleted molecules.

While these data tend to support Wallace's arguments, the fraction of deleted mtDNA molecules is still extremely low. Given the threshold effect described earlier (e.g., Fig. 11–2), it is hard to see how such small numbers of abnormal mtDNA molecules can have much of a phenotypic effect even allowing for the fact that only a single mitochondrial abnormality, the 4997-bp deletion, is being assayed. One possibility is that the deleted molecules become represented in extraordinarily high proportions in certain cells or tissues as the result of somatic segregation.

Human mitochondrial diseases are often highly pleiotropic. They most obviously involve those tissues that rely to the greatest extent on mitochondrial energy production such as the central nervous system, heart, skeletal muscle, pancreatic islands, kidney, and liver. Wallace (1992a) points out that one or more of these characteristics is found in a variety of common degenerative diseases. These include epilepsy, cardiac disease, Parkinson's disease, Alzheimer's disease, Huntington's chorea, and the form of diabetes mellitus that becomes expressed with maturity. He mentions, for example, that Parkinson's disease exhibits many of the characteristics expected for a mitochondrial myopathy including tremor, impaired movements, and occasional dementia. Furthermore, defects in mitochondrial complexes have been noted in Parkinson's patients, and mtDNA deletions are known to accumulate in human brains with age. Wallace sums up by pointing out that "Parkinsonism might result from a variety of factors that inhibit OXPHOS: deleterious mutations in mitochondrial or nuclear OXPHOS genes, environmental toxins, or a combination of factors." In the final analysis, the aging process and different degenerative diseases are phenotypically complex. Wallace (1992a,b) has drawn attention to one potentially important component, the mitochondrion, in which loss or reduction in function may have significant phenotypic consequences and Cooper *et al.* (1992) target complex I as being particularly sensitive to mtDNA defects.

MITOCHONDRIAL SYNDROMES OF FLOWERING PLANTS

The nonchromosomal stripe mutants of maize have the most in common with mitochondrial diseases described earlier in this chapter. These mutants result because of deletions affecting specific mitochondrial genes. However, they have been studied in much less detail than the mitochondrial disease called cytoplasmic male sterility (CMS), in which plants are normal in appearance, but fail to shed functional pollen. This syndrome has been described in over 140 different species of plants (Laser and Lersten, 1972) and is characteristically maternally inherited. However, only in the cases of the T cytoplasm of maize *(cms-T)* and *Petunia* has a direct connection been established between an alteration in the mitochondrial genome and CMS (Hanson, 1991). In these cases novel proteins encoded by chimeric mitochondrial genes appear to be responsible for the CMS phenotype. Such chimeric genes are quite common in flowering plant mitochondrial genomes and arise as the result of recombination within the mitochondrial genome (Hanson and Folkerts, 1992). Chimeric genes have also been identified in other male sterile cytoplasms, but whether their products are responsible for the male sterile phenotype remains to be established. To date subgenomic molecules or amplified segments of mtDNA have not been implicated in mitochondrial disease in flowering plants.

The Nonchromosomal Striped Mutants of Maize

Nonchromosomal stripe (NCS) mutants of maize are characterized by poor growth, decreased yields, and variable leaf striping (see Newton, 1988 and Hanson and Folkerts, 1992 for reviews). The NCS mutants possess alterations in essential mitochondrial genes and survive only as heteroplasmons. Thus, NCS2 has an alteration in the *ndh4* gene, NCS5 and NCS6 have different deletions in the *cox2* gene, and NCS3 has an alteration in the cotranscribed r-protein genes encoding S3 and L16. In each case the deletion events can be traced to recombination between short repeated sequences. They do not appear to involve large segments of the mitochondrial genome, as is true of many mitochondrial deletion mutants of fungi and humans.

NCS mutations can easily be mistaken for plastid mutations because the yellow and white striping of the leaves is accompanied by plastid structural abnormalities and reduced carbon fixation. These effects on plastid phenotype, together with maternal inheritance of plastid and mitochondrial genomes in many flowering plants, cause one to wonder whether certain presumed plastid mutations are actually mitochondrial mutations in disguise. Studies of the NCS mutants are likely to provide valuable information on the segregation of mitochondria and mitochondrial genomes in flowering plants.

Cytoplasmic Male Sterility

CMS has been studied the most in maize, *Petunia*, *Brassica* (cabbage, cauliflower, turnip etc.), and to a lesser extent in a variety of other agriculturally or horticulturally important plants (reviews by Bonen and Brown, 1993; Braun *et al.*, 1992; Hanson, 1991; Hanson and Conde, 1985; Hanson and Folkerts, 1992; Hanson *et al.*, 1991; Laughnan and Gabay-Laughnan, 1983; Levings, 1990; Newton, 1988; Pring and Lonsdale, 1985). CMS can arise either in intraspecific crosses or as the result of interspecific hybridization (alloplasmic CMS). CMS is agriculturally important in the production of hybrid seed because slower, uneconomic methods such as hand emasculation before pollination are avoided. For example, commercial production of hybrid sorghum seed depends on use of a CMS system. The flowers are self-fertile, but so small that they are difficult to hand emasculate. Hence, a CMS plant is used as the female parent in a cross, and any seed formed by the female parent must be the result of cross-pollination. Because CMS is maternally inherited, the F_1 generation plants from such a cross will be unable to set seed since they are male sterile. Fortunately, in most cases, dominant nuclear restorer *(Rf)* genes have been discovered, which result in male fertility even though the mitochondria are male sterile. When these genes are introduced from the pollen parent, the male sterile phenotype is suppressed and the F_1 hybrid progeny are fertile even though they contain the "male sterile" cytoplasm.

Cytoplasmic male sterility in maize. CMS was first reported in maize by Rhoades (1931). Since then other CMS isolates have been obtained (Beckett, 1971) and they have been assigned to three groups: *cms-T* (Texas), *cms-C* (Charrua), and *cms-S* (USDA), based on the genetics of fertility restoration. Plants with the *cms-T* cytoplasm are restored by dominant alleles of two genes *Rf1* and *Rf2*. Plants heterozygous for these loci produce normal pollen and are said to be *sporophytically* restored. This means that the genetic constitution of the diploid, sporophytic tissue of the anther rather than the haploid, gametophytic pollen grain determines pollen development. Plants with the *cms-C* cytoplasm are also restored sporophytically, but only a single dominant allele at the *Rf4* locus is required (Braun *et al.*, 1992; Laughnan and Gabay-Laughnan, 1983). In the case of *cms-S* a single dominant allele at the *Rf3* locus is sufficient to produce a fertile plant. Restoration is *gametophytic* in this case since in an *Rf3rf3* heterozygote the 50% of the pollen carrying the *rf3* allele will be aborted.

The molecular basis of CMS in maize is best understood in the case of *cms-T*. This cytoplasm was so extensively used for the production of hybrid maize that by 1970 about 85% of the hybrid corn crop carried this cytoplasm (see Braun *et al.*, 1992; Korth *et al.*, 1991; Levings and Siedow, 1992). Unfortunately, in that same year the widespread use of *cms-T* was also directly responsible for the disastrous epidemic of Southern corn leaf blight to which this cytoplasm is uniquely susceptible. Over 15% of the corn crop in the southern and corn belt regions of the United States was lost at an estimated cost of over one billion dollars. Southern corn leaf blight is caused by a fungus called *Bipolaris maydis* race T, formerly known as *Helminthosporium maydis*, race T. A second fungal pathogen, *Phyllosticta maydis* (yellow corn leaf blight), also uniquely infests *cmsT* maize, but its effects are less serious as the fungus is restricted to the cooler northern regions of the United States.

Plants with the *cmsT* cytoplasm synthesize a unique 13-kDa protein (Forde et al., 1978). This polypeptide is encoded by one of two ORFs, T-*urf13*, in a 3547-bp mtDNA fragment from *cmsT* that was selected for study because of its unique and differentially abundant transcripts (Dewey et al., 1986). T-*urf13* has a chimeric structure and consists of 5' flanking sequences with significant similarity to the 5' flanking region of the *atp6* gene (see Braun et al., 1992; Levings and Brown, 1989, for a review and Fig. 11–4A). The coding region of T-*urf13* consists of 88 codons with homology to an untranscribed 3' flanking region of the 26S rRNA gene *(rrn26)* and 18 codons with homology to the coding region of *rrn26*. T-*urf13* and *atp6* are believed to have similar promoters and complete copies of *atp6* and *rrn26* are found elsewhere in the *cms-T* mitochondrial genome. Adjacent to and downstream of T-*urf13*, *urf25*, is an open reading frame present in normal maize mitochondrial genomes as well as in *Nicotiana* and *Petunia*, but whose function is unknown (Braun et al., 1992; Hanson, 1991). This gene is also chimeric in both normal and *cms-T* lines as its 3'-terminus including the UAG stop codon are comprised of sequences derived from a chloroplast tRNA^{Arg} gene. The recombinational events that may have led to the 3547-bp sequence of the *cms-T* mitochondrial genome containing *T-urf13* and *urf25* are depicted in Fig. 11–4A.

Direct evidence that *T-urf13* encodes the 13-kDa polypeptide is the demonstration that antibodies elicited against chemically synthesized oligopeptides, corresponding to portions of the *T-urf13* predicted amino acid sequence, immunoprecipitated the 13-kDa polypeptide (URF13) from *cms-T,* but not normal mitochondria (Dewey et al., 1987; Wise et al., 1987). Experiments with reversions to male fertility show that synthesis of URF13 in *cms-T* plants is directly responsible both for CMS and susceptibility to the toxin *(Bmt)* produced by *Bipolaris maydis* (see Braun et al., 1992; Levings and Brown, 1989; Newton, 1988 for reviews). In plants regenerated from callus culture reversion from male sterility to fertility often occurs and is invariably accompanied by resistance to *Bmt*. The revertants generally have lost a 6.6-kb *Xho*I fragment that contains the *T-urf13* reading frame. In the *cms-T* revertants *urf-25* remains unaltered. In one exceptional case a 5-bp insertion within the *T-urf13* reading frame was responsible for the revertant phenotype. This insertion caused a frameshift, resulting in a premature stop codon and a truncated 13-kDa protein (Wise et al., 1987).

Although practically nothing is known concerning the mechanism by which URF13 causes male sterility, a lot is known about the mechanism by which the protein promotes susceptibility to *Bmt* and also the pathotoxin from *Phyllosticta maydis (Pm)* (see Korth et al., 1991; Levings and Siedow, 1992 for reviews). URF13 is localized in the inner mitochondrial membrane. When expressed in *E. coli*, the protein localizes to the plasma membrane of this bacterium. Like *cms-T* plants, *E. coli* cells synthesizing URF13 become susceptible to toxin, so *E. coli* serves as a useful model for understanding how URF13 promotes toxin sensitivity. Labeled *Pm* toxin has been found to bind specifically to URF13 in *E. coli* and in *cms-T* mitochondria. These binding experiments also show that *Bmt* competes for binding at the same site as methomyl, a structurally unrelated compound that is the active ingredient in the DuPont insecticide Lannate. Methomyl also causes swelling of *cms-T* mitochondria. Braun et al. (1989) demonstrated that the interaction between *Bmt* and URF13 in *E. coli* permeabilizes the plasma membrane because URF13 then becomes a channel-forming protein. Massive leakage

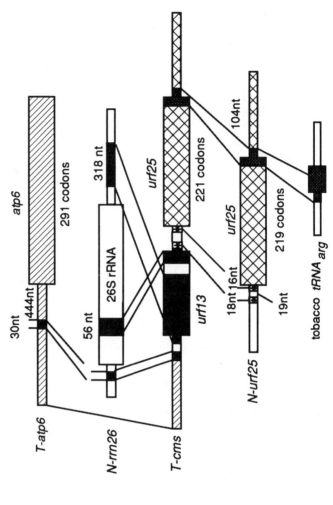

Fig. 11–4. Structures of the chimeric genes responsible for the CMS phenotype in *cms-T* maize and *Petunia*. (A) Structure of maize T-*urf13*. Tall rectangles indicate open reading frames. Portions of T-*urf13* are derived from a normal *atp6* and 26S rRNA genes. The intact genes are also found elsewhere in the *cms-T* mitochondrial genome, but the *urf25* gene downstream of *urf13* is the only copy present in *cms-T*. Chloroplast DNA sequences derived from an arginine tRNA gene are found at the 3'-terminus of the *urf25* gene from both a normal fertile maize line and *cms-T*.

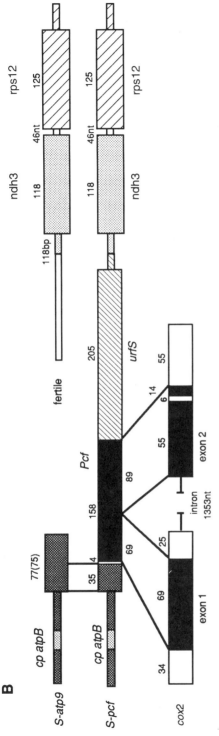

(B) Structure of the *Petunia pcf* mitochondrial gene responsible for male sterility. Numbers above tall rectangles indicate number of codons. The DNA sequence of S-*atp9* predicts an open reading frame of 77 codons, but RNA editing shortens it to 75 codons. The first 35 codons of *pcf* are identical to *atp9*, followed by a four-codon junction region probably created by recombination with *cox2*. Regions of both *cox2* exons are present without any intron sequence; the junction between the two exons in *pcf* matches a putative abnormal splicing site. The *cox2* region of *pcf* is followed by an unidentified reading frame called *urfS*. Normal *atp9* and *cox2* genes are present in the *Petunia* CMS line, but the only copies of *ndh3* and *rps12* are downstream of *pcf* and they appear normal. From Hanson (1991).

of small ions through the plasma membrane occurs as a result. A similar response to *Bmt* was reported in *cms-T* mitochondria (Holden and Sze, 1984).

Levings and Siedow (1992) review experimental evidence and theoretical considerations that have led to a model of URF13 organization within the membrane. The model supposes that there are three membrane-spanning helical domains, with the amino terminus of the protein extending into the mitochondrial intermembrane space and the carboxyl terminus protruding into the matrix. To explain leakage of compounds such as NAD out of the cell in response to toxin, a hydrophilic channel as large as 1.5 nm in diameter must be postulated. A pore of 0.8 to 1.2 nm in diameter would require association of URF13 monomers in at least trimeric or tetrameric structures. Toxin would be required to open the pore.

The URF13 amino acids responsible for toxin interaction are being defined by experiments in which *in vitro* mutagenized *T-urf13* is expressed in *E. coli* (Braun et al., 1989; Korth et al., 1991; Levings and Siedow, 1992). Yeast can be made toxin and methomyl sensitive by constructing a chimeric gene composed of T-*urf13* and the mitochondrial presequence of the *Neurospora atp9* gene (Glab et al., 1990, 1993; Huang et al., 1990). Cells transformed with this construct process the hybrid protein, and URF13 can be shown to localize to the mitochondrial membrane fraction. Unfortunately, the lack of a maize mitochondrial transformation system makes it impossible to determine whether the same amino acid residues that are responsible for toxin binding are those that play a role in cytoplasmic male sterility.

The dominant nuclear alleles *Rf1* and *Rf2* are required to restore male fertility to *cms-T* plants. *Rf1* specifically affects expression of *T-urf13* by altering the transcriptional products of this gene and reducing abundance of the URF13 protein by 80% (Braun et al., 1992; Dewey et al., 1987; Kennell et al., 1987). As expected, there is no effect of the *rf1* allele on *T-urf13* expression. At present, there is no evidence that *Rf2* affects *T-urf13* expression so it is not known why this allele is also required to restore fertility. Interestingly, *cms-T* plants are susceptible to *Bmt* irrespective of the presence of *Rf1* or *Rf2* (Lim and Hooker, 1972), suggesting that the threshold levels of URF13 required for toxin sensitivity may be lower than those needed to confer the CMS phenotype.

The basis for male sterility in *cms-C* and *cms-S* strains remains to be established. In *cms-C* plants the *atp6* gene has an N-terminal fusion with the *atp9* 5' flanking region plus 13 codons of *atp9* followed by 147 codons derived from cpDNA. In a second fusion, the 5' flank plus 184 codons of *atp6* are fused to 275 codons of the *cox2* gene (Braun et al., 1992; Levings and Dewey, 1988; Fragoso et al., 1989). Neither of these chimeric genes seems to be directly related to the *cms-C* phenotype. Plants having the *cms-S* cytoplasm also possess the linear S1 and S2 episomes (Chapter 10). Initially, it seemed that the presence of these plasmids might be related to the CMS phenotype. Cytoplasmic revertants in one nuclear background (M825) appeared to have lost S1 and S2 with S2 apparently integrating into the mitochondrial genome (see Levings and Brown, 1989; Newton, 1988). However, the presence of these plasmids is probably unrelated to the CMS phenotype since cytoplasmic revertants in the WF9 nuclear background retain the S episomes. Also, the fertile RU strains of maize possess the closely related R1 and R2 plasmids.

Cytoplasmic male sterility in Petunia. As in *cms-T* maize, a novel protein produced by a chimeric mitochondrial gene has been shown to be the culprit responsible for CMS in *Petunia* (see Hanson, 1991; Hanson and Folkerts, 1992; Hanson et al., 1991 for reviews). The abnormal *Petunia* gene, called *pcf* (Petunia CMS-associated fused), contains the 5' flanking region of *atp9* and part of its coding region, a portion of each exon of *cox2*, and an unidentified reading frame called *urf-S* (Fig. 11–4B). Within the noncoding-transcribed region of *atp9* is a short sequence of high similarity to the chloroplast *atpB* gene. The chimeric *pcf* gene is inserted next to the *ndh3* and *rps12* genes. Transcripts of the *pcf* gene initiate at one of three points in the *atp9* 5' flanking region. One of these transcripts is specifically reduced in fertile strains carrying the dominant nuclear *Rf* allele. Synthetic peptide antibodies have been made to detect the *urfS* portion of the protein encoded by the *pcf* gene. Use of these antibodies revealed the presence of a 25-kDa protein in CMS lines that was absent in fertile lines and much reduced in fertility restored lines. Unlike URF13 this protein is present in soluble as well as membrane fractions from mitochondria. While these results strongly suggest that the 25-kDa protein is responsible for male sterility in *Petunia*, the clinching argument would be the isolation of fertile revertants lacking the gene.

In isonuclear CMS and fertile lines of *Petunia* there is a consistent reduction in alternative oxidase activity (Chapter 1) in isolated mitochondria, cells, suspension cultures, and immature anthers of CMS lines. Suspension cultures from restored lines are also restored in alternative oxidase activity. Hence, a correlation can be established between the restoration of pollen development, an increase in alternative oxidase, and a decrease in abundance of the *pcf* gene product. Perhaps alternative oxidase activity is important for proper functioning of the tapetal layer and ultimately for pollen development.

Other cases of cytoplasmic male sterility in flowering plants. Though most information on CMS has so far been obtained for *Petunia* and maize, specific regions of the mitochondrial genome have now been implicated in CMS in several other species, including *Brassica* Polima, sunflower, and bean (Singh and Brown, 1991; Horn et al., 1991; Laver et al., 1991; MacKenzie et al., 1988). In *Brassica* Polima, the presence of either one or two restorer genes results in monocistronic transcripts of *atp6*, instead of the dicistronic *orf224/atp6* transcripts found in CMS lines (Singh and Brown, 1991).

Hernould et al. (1993) have succeeded in mimicking CMS by introducing edited and unedited versions of the wheat *atp9* gene fused to the coding sequence for the yeast *cox4* transit peptide into the tobacco nucleus. CMS occurred only with the unedited version of the chimeric gene that exhibited the expected Mendelian inheritance. Hernould et al. hypothesize that unedited ATP9 protein, or protein fragments, compete with the normal version present in tobacco mitochondria to create a mixture of functional and nonfunctional ATP synthase complexes. They suppose that CMS results because energy requirements are maximal in the tapetal cells during microsporogenesis.

REFERENCES

Akins, R.A., and A.M. Lambowitz (1984). The [*poky*] mutant of *Neurospora* contains a 4-base-pair deletion at the 5' end of the mitochondrial small rRNA. *Proc. Natl. Acad. Sci. U.S.A.* 81: 3791–3795.

Arnaudo, E., Dalakis, M., Shanske, S., Moraes, C.T., DiMauro, S., and E. Schon (1991). Depletion of muscle mitochondrial DNA in AIDS patients with zidovudine-induced myopathy. *Lancet* 337: 508–510.

Beckett, J.B. (1971). Classification of male-sterile cytoplasms in maize (*Zea mays* L.). *Crop Sci.* 11: 724–727.

Belcour, L., and O. Begel (1978). Lethal mitochondrial genotypes in *Podospora anserina:* A model for senescence. *Mol. Gen. Genet.* 163: 113–123.

Belcour, L., and C. Vierny (1986). Variable DNA splicing sites of a mitochondrial intron: Relationship to the senescence process in *Podospora. EMBO J.* 5: 609–614.

Belcour, L., Koll, F., Vierny, C., Sainsard-Chanet, A., and O. Begel (1986). Are proteins encoded in mitochondrial introns involved in the process of senescence in the fungus *Podospora anserina?* In *Modern Trends in Aging Research* (Y. Courtois, B. Faucheux, B. Forette, D.L. Knook, and J.A. Treton, eds.), pp. 63–71. John Libbey Eurotext, London.

Bertrand, H., and T.H. Pittenger (1969). Cytoplasmic mutants selected from continuously growing cultures of *Neurospora crassa. Genetics,* 61: 643–659.

Bertrand, H., and T.H. Pittenger (1972). Isolation and classification of extranuclear mutants of *Neurospora crassa. Genetics* 71: 521–533.

Bertrand, H., Chan, B. S.-S., and A.J.F. Griffiths (1985). Insertion of a foreign nucleotide sequence into mitochondrial DNA causes senescence in Neurospora intermedia. *Cell* 41: 877–884.

Bertrand, H., Griffiths, A.J.F., Court, D.A., and C.K. Cheng (1986). An extrachromosomal plasmid is the etiological precursor of kalDNA insertion sequences in the mitochondrial chromosome of senescent Neurospora. *Cell* 47: 829–837.

Bertrand, H., Collins, R.A., Stohl, L.L., Goewert, R.R., and A.M. Lambowitz (1980). Deletion mutants of *Neurospora crassa* mitochondrial DNA and their relationship to the "stop-start" growth phenotype. *Proc. Natl. Acad. Sci. U.S.A.* 77: 6032–6036.

Bindoff, L.A., Howell, N., Poulton, J., McCullough, D.A., Morten, K.J., Lightowlers, R.N., Turnbull, D.M., and K. Weber (1993). Abnormal RNA processing associated with a novel tRNA mutations in mitochondrial DNA. *J. Biol. Chem.* 268: 19559–19564.

Bonen, L., and G.G. Brown (1993). Genetic plasticity and its consequences: perspectives on gene organization and expression in plant mitochondria. *Can. J. Bot.* 71: 645–660.

Braun, C.J., Siedow, J.N., Williams, M.E., and C.S. Levings III (1989) Mutations in the maize mitochondrial T-*urf13* gene eliminate sensitivity to a fungal pathotoxin. *Proc. Natl. Acad. Sci. U.S.A.* 86: 4435–4439.

Braun, C.J., Brown, G.G., and C.S. Levings III (1992). Cytoplasmic male sterility. In *Plant Gene Research: Cell Organelles* (R.G. Herrmann, ed.), pp. 219–245. Springer-Verlag, Wien.

Brown, M.D., Voljavec, A.S., Lott, M.T., Torroni, A., Yang, C.-C., and D.C. Wallace (1992). Mitochondrial DNA complex I and III mutations associated with Leber's hereditary optic neuropathy. *Genetics* 130: 163–173.

Chan, B. S.-S., Court, D.A., Vierula, P.J., and H. Bertrand (1991). The *kalilo* linear senescence-inducing plasmid of *Neurospora* is an invertron and encodes DNA and RNA polymerases. *Curr. Genet.* 20: 225–237.

Chomyn, A., Meola, G., Bresolin, N., Lai, S.T., Scarlato, G., and G. Attardi (1991). In vitro genetic transfer of protein synthesis and respiration defects to mitochondrial DNA-less cells with myopathy-patient mitochondria. *Mol. Cell. Biol.* 11: 2236–2244.

Chomyn, A., Martinuzzi, A., Yoneda, M., Daga, A., Hurko, O., Johns, D., Lai, S.T., Nonaka, I., Angelini, C., and G. Attardi (1992). MELAS mutation in mtDNA binding site for transcription termination factor causes defects in protein synthesis and in respiration but no change in levels of upstream and downstream mature transcripts. *Proc. Natl. Acad. Sci. U.S.A.* 89: 4221–4225.

Clark-Walker, G.D. (1992). Evolution of mitochondrial genomes in fungi. *Int. Rev. Cytol.* 141: 89–127.

Collins, R.A., and H. Bertrand (1978). Nuclear suppressors of the *[poky]* cytoplasmic mutant in *Neurospora crassa.* III. Effects on other cytoplasmic mutants and on mitochondrial ribosome assembly in *[poky] Mol. Gen. Genet.* 161: 267–273.

Cooper, J.M., Mann, V.M., Krige, D., and A.H.V. Schapira (1992). Human mitochondrial complex I dysfunction. *Biochim. Biophys. Acta* 1101: 198–203.

Corral-Debrinski, M., Stepien, G., Shoffner, J.M., Lott, M.T., Kanter, K., and D.C. Wallace (1991). Hypoxemia is associated with mitochondrial DNA damage and gene induction. *J. Am. Med. Assoc.* 266: 1812–1816.

Court, D.A., Griffiths, A.J.F., Kraus, S.R., Russell, P.J., and H. Bertrand. (1991). A new senescence-inducing mitochondrial linear plasmid in field-isolated *Neurospora crassa* strains from India. *Curr. Genet.* 19: 129–137.

Court, D.A., and H. Bertrand (1992). Genetic organization and structural features of *maranhar,* a senescence-inducing linear mitochondrial plasmid of *Neurospora crassa. Curr. Genet.* 22: 385–397.

Court, D.A., and H. Bertrand (1993). Expression of the open reading frames of a senescence-inducing, linear mitochondrial plasmid of *Neurospora crassa. Plasmid* 30: 51–66.

Cummings, D.J., and R.M. Wright (1983). DNA sequence of the excision sites of a mitochondrial plasmid from senescent *Podospora anserina. Nucl. Acids Res.* 11: 2111–2119.

Cummings, D.J., Michel, F., and K.L. McNally (1989). DNA sequence analysis of the 24.5 kilobase pair cytochrome oxidase subunit I mitochondrial gene from *Podospora anserina:* A gene with sixteen introns. *Curr. Genet.* 16: 381–406.

Dasgupta, J., Chan, B.S.-S., and H. Bertrand (1988). *Kalilo* insertion sequences from the senescent strains of *Neurospora intermedia* are flanked by long inverted repeats of mitochondrial DNA. *Genome* (Suppl. 1): 318.

De Vries, H., De Jonge, J.C., and C. Schrage (1985).

The *Neurospora* mitochondrial stopper mutant, [E35], lacks two protein genes indispensable for the formation of complexes I, III and IV. In *Achievements and Perspectives of Mitochondrial Research*, Vol. II (E. Quagliariello, E.C. Slater, F. Palmieri, C. Saccone, and A.M Kroon, eds.), pp. 285–292. Elsevier, Amsterdam.

De Vries, H., Schrage, C., and J.C. De Jonge (1986a). The mitochondrial DNA of *Neurospora crassa:* Deletion by intramolecular recombination and the expression of mitochondrial genes. In *Extrachromosomal Elements in Lower Eukaryotes* (R.B. Wickner, A. Hinnebusch, A.M. Lambowitz, I.C. Gunsalus, and A. Hollaender, eds.), pp. 57–65. Plenum Press, New York.

De Vries, H., Alzner-DeWeerd, B., Breitenberger, C.A., Chang, D.D., de Jonge, J., and U.L. RajBhandary (1986b). The E35 stopper mutant of *Neurospora crassa:* Precise localization of deletion endpoints in mitochondrial DNA and evidence that the deleted DNA codes for a subunit of NADH dehydrogenase. *EMBO J.* 5: 779–785.

Dewey, R.E., Levings III, C.S., and D.M Timothy (1986). Novel recombinations in the male mitochondrial genome produce a unique transcriptional unit in the Texas male-sterile cytoplasm. *Cell* 44: 439–449.

Dewey, R.E., Timothy, D.H., and C.S. Levings III (1987). A mitochondrial protein associated with cytoplasmic male sterility in the T cytoplasm of maize. *Proc. Natl. Acad. Sci. U.S.A.* 84: 5374–5378.

Dujon, B., and L. Belcour (1989). Mitochondrial DNA instabilities and rearrangements in yeasts and fungi. In *Mobile DNA* (D.E. Berg and M.M. Howe eds.), pp. 861–878. American Society for Microbiology, Washington, DC.

Esser, K., and P. Tudzynski (1980). Senescence in Fungi. In *Senescence in Plants* Vol. 2 (K.V. Thimann, ed.), pp. 67–83. CRC Press, Boca Raton, FL.

Forde, B.G., Oliver, R.J.C., and C.J. Leaver (1978). Variation in mitochondrial translation products associated with male-sterile cytoplasms in maize. *Proc. Natl. Acad. Sci. U.S.A.* 75: 3841–3845.

Fragoso, L.L., Nichols, S.E., and C.S. Levings III (1989). Rearrangements in maize mitochondrial genes. *Genome* 31: 160–168.

Gillham, N.W. (1978). *Organelle Heredity.* Raven Press, New York.

Glab, N., Wise, R.P., Pring, D.R., Jacq, C., and P. Slonimski (1990). Expression in *Saccharomyces cerevisiae* of a gene associated with cytoplasmic male sterility from maize: Respiratory dysfunction and uncoupling of yeast mitochondria. *Mol. Gen. Genet.* 223: 24–32.

Glab, N., Petit, P.X., and P.P. Slonimski (1993). Mitochondrial dysfunction in yeast expressing the cytoplasmic male sterility T-urf13 gene from maize: Analysis at the population and individual cell level. *Mol. Gen. Genet.* 236: 299–308.

Griffiths, A.J.F., (1992). Fungal senescence. *Annu. Rev. Genet.* 26: 351–372.

Griffiths, A.J.F., and H. Bertrand (1984). Unstable cytoplasms in Hawaiian strains of *Neurospora intermedia. Curr. Genet.* 8: 387–398.

Griffiths, A.J.F., Kraus, S.R., Barton, R., Court, D.A., Myers, C.J., and H. Bertrand (1990). Heterokaryotic transmission of senescence plasmid DNA in *Neurospora. Curr. Genet.* 17: 139–145.

Gross, S.R., Hsieh, T.-s., and P.H. Levine (1984). Intramolecular recombination as a source of mitochondrial chromosome heteromorphism in Neurospora. *Cell* 38: 233–239.

Gross, S.R., Mary, A., and P.H. Levine (1989a). Change in chromosome number associated with a double deletion in the *Neurospora crassa* mitochondrial chromosome. *Genetics* 121: 685–691.

Gross, S.R., Levine, P.H., Metzger, S., and G. Glaser (1989b). Recombination and replication of plasmid-like derivatives of a short section of the mitochondrial chromosome of *Neurospora crassa. Genetics* 121: 693–701.

Hanson, M.R. (1991). Plant mitochondrial mutations and male sterility. *Annu. Rev. Genet.* 25: 461–486.

Hanson, M.R., and M.F. Conde (1985). Functioning and variation of cytoplasmic genomes: Lessons from cytoplasmic-nuclear interactions affecting male fertility in plants. *Int. Rev. Cytol.* 94: 213–267.

Hanson, M.R., and O.F. Folkerts (1992). Structure and function of the higher plant mitochondrial genome. *Int. Rev. Cytol.* 141: 129–172.

Hanson, M.R., Connett, M.B., Folkerts, O., Izhar, S., McEvoy, S.M., Nivison, H.T., and K.D. Pruitt (1991). Cytoplasmic male sterility in Petunia. In *Plant Molecular Biology,* Vol. 2 (R.G. Herrmann and B. Larkins, eds.), pp. 383–399. Plenum Press, New York.

Hawse, A., Collins, R.A., and F.E. Nargang (1990). Behavior of *[mi3]* mutation and conversion of polymorphic mtDNA markers in heterokaryons of *Neurospora crassa. Genetics* 126: 63–72.

Hermans, J., and H.D. Osiewacz (1992). The linear mitochondrial plasmid pAL2-1 of a long-lived *Podospora anserina* mutant is an invertron encoding a DNA and RNA polymerase. *Curr. Genet.* 22: 491–500.

Hernould, M., Suharsono, S., Litvak, M., Araya, A., and A. Mouras (1993). Male-sterility induction in transgenic tobacco plants with an unedited *atp9* mitochondrial gene from wheat. *Proc. Natl. Acad. Sci. U.S.A.* 90: 2370–2374.

Holden, M.J., and H. Sze (1984). Helminthosporium maydis T toxin increased membrane permeability to Ca^{2+} in susceptible corn mitochondria. *Plant Physiol.* 75: 235–237.

Holt, I.J., Harding, A.E., Petty, R.K., and J.A. Morgan-Hughes (1990). A new mitochondrial disease associated with mitochondrial DNA heteroplasmy. *Am. J. Hum. Genet.* 46: 428–433.

Horn, R., Kohler, R.H., and K. Zetsche (1991). A mitochondrial 16 kDa protein is associated with cytoplasmic male sterility in sunflower. *Plant Mol. Biol.* 17: 29–36.

Howell, N., Kubacka, I., Xu, M., and D.A. McCullough (1991). Leber's hereditary optic neuropathy: Involvement of the mitochondrial ND1 gene and evidence for an intragenic suppressor mutation. *Am. J. Hum. Genet.* 48: 935–942.

Huang, J., Lee, S.-H., Medici, R., Hack, E., and A.M. Myers (1990). Expression in yeast of the T-URF13 protein from Texas-male sterile maize mitochondria confers sensitivity to methomyl and to Texas-cytoplasm-specific fungal toxins. *EMBO J.* 9: 339–347.

Jinks, J.L. (1966). Mechanisms of inheritance. 4. Extranuclear inheritance. In *The Fungi*, Vol. 2 (G.C. Ainsworth and A.S. Sussman, eds.), pp. 619–660. Academic Press, New York.

Johns, D.R., and J. Berman (1991). Alternative, simultaneous complex I mitochondrial DNA mutations in Leber's hereditary optic neuropathy. *Biochem. Biophys. Res. Commun.* 174: 1324–1330.

Kennell, J.C., and A.M. Lambowitz (1989). Development of an in vitro transcription system for *Neurospora crassa* mitochondrial DNA and identification of transcription initiation sites. *Mol. Cell. Biol.* 9: 3603–3613.

Kennell, J.C., Wise, R.P., and D.R. Pring (1987). Influence of nuclear background on transcription of a maize mitochondrial region associated with Texas male sterile cytoplasm. *Mol. Gen. Genet.* 210: 399–406.

King, M.P., and G. Attardi (1989). Human cells lacking mtDNA: Repopulation with exogenous mitochondria by complementation. *Science* 246: 500–503.

Kobayashi, Y., Ichihashi, K., Ohta, S., Nihei, K., Kagawa, Y., Yanagisawa, M., and M.Y. Momai (1992). The mutant mitochondrial genes in mitochondrial myopathy, encephalomyopathy, lactic acidosis and stroke-like episodes (MELAS) were selectively amplified through generations. *J. Inher. Metab. Dis.* 15: 803–808.

Koll, F. (1986). Does nuclear integration of mitochondrial sequences occur during senescence in *Podospora? Nature (London)* 324: 597–599.

Koll, F., Begel, O., Keller, A.-M., Vierny, C., and L. Belcour (1984). Ethidium bromide rejuvenation of senescent cultures of *Podospora anserina*: Loss of senescence-specific DNA and recovery of normal mitochondrial DNA. *Curr. Genet.* 8: 127–134.

Koll, F., Belcour, L., and C. Vierny (1985). A 1100bp sequence of mitochondrial DNA is involved in senescence process in *Podospora*: Study of senescent and mutant cultures. *Plasmid* 14: 106–117.

Koll, F., Begel, O., and L. Belcour (1987). Insertion of short poly $d(A)d(T)$ sequences at recombination junctions in mitochondrial DNA of *Podospora*. *Mol. Gen. Genet.* 209: 630–632.

Korth, K.L., Struck, F., Kaspi, C.I., Siedow, J.N., and C.S. Levings III (1991). Topological orientation of the membrane protein URF13. In *Plant Molecular Biology 2* (R.G. Herrmann and B. Larkins, eds.), pp. 375–381. Plenum Press, New York.

Kruse, B., Narasimhan, N., and G. Attardi (1989). Termination of transcription in human mitochondria: Identification and purification of a DNA binding protein factor that promotes termination. *Cell* 58: 391–397.

Kuck, U. (1989). Mitochondrial DNA rearrangements in *Podospora anserina*. *Exp. Mycol.* 13: 111–120.

Kuck, U., Osiewacz, H.D., Schmidt, U., Kappelhoff, B., Schulte, E., Stahl, U., and K. Esser (1985). The onset of senescence is affected by DNA rearrangements of a discontinuous mitochondrial gene in *Podospora anserina*. *Curr. Genet.* 9: 373–382.

Kuntzel, H., Kochel, H.G., Lazarus, C.M. and H. Lunsdorf (1982). Mitochondrial genes in *Aspergillus*. In *Mitochondrial Genes* (P.P. Slonimski, P. Borst, and G. Attardi, eds.), pp. 391–403. Cold Spring Harbor Laboratory, Cold Spring Harbor, NY.

Lansman, R.A., and D.A. Clayton (1975). Mitochondrial protein synthesis in mouse L-cells: Effect of selective nicking of mitochondrial DNA. *J. Mol. Biol.* 99: 777–793.

Laser, K.D., and N.R. Lersten (1972). Anatomy and cytology of microsporogenesis in cytoplasmic male sterile angiosperms. *Bot. Rev.* 38: 425–454.

Lauber, J., Marsac, C., Kadenbach, B., and P. Seibel (1991). Mutations in mitochondrial tRNA genes: A frequent cause of neuromuscular disease. *Nucl. Acids Res.* 19: 1393–1397.

Laughnan, J.R., and S. Gabay-Laughnan (1983). Cytoplasmic male sterility in maize. *Annu. Rev. Genet.* 17: 27–48.

Laver, H.K., Reynolds, S.J., Monegar, F., and C.J. Leaver (1991). Mitochondrial genome organization and expression associated with cytoplasmic male sterility in sunflower *(Helianthus annuus)*. *Plant J.* 1: 185–194.

Lemire, E.G., and F.E. Nargang (1986). A missense mutation in the *oxi3* gene of the *[mi-3]* extranuclear mutant of *Neurospora crassa*. *J. Biol. Chem.* 261: 5610–5615.

Levings, C.S. III (1990). The Texas cytoplasm of maize: Cytoplasmic male sterility and disease susceptibility. *Science* 250: 942–947.

Levings, C.S. III, and G.G. Brown (1989). Molecular biology of plant mitochondria. *Cell* 56: 171–179.

Levings, C.S. III, and R.E. Dewey (1988). Molecular studies of cytoplasmic male sterility in maize. *Phil Trans. R. Soc., London B* 319: 93–102.

Levings, C.S. III, and J.N. Siedow (1992). Molecular basis of disease susceptibility in the Texas cytoplasm of maize. *Plant Mol. Biol.* 19: 135–147.

Lim, S.M., and A.L. Hooker (1972). Disease determinant of Helminthosporium maydis Race T. *Phytopathology* 62: 968–971.

Mackenzie, S.A., Pring, D.R., Bassett, M.J., and

C.D. Chase (1988). Mitochondrial DNA rearrangement associated with fertility restoration and cytoplasmic male sterility in *Phaseolus vulgaris*. *Proc. Natl. Acad. Sci. U.S.A.* 85: 2714–2717.

Manella, C.A., and A.M. Lambowitz (1978). Interaction of wild-type and *[poky]* mitochondrial DNA in heterokaryons of *Neurospora*. *Biochem. Biophys. Res. Commun.* 80: 673–679.

Manella, C.A., and A.M. Lambowitz (1979). Unidirectional gene conversion associated with two insertions in *Neurospora crassa* mitochondrial DNA. *Genetics* 93: 645–654.

Manella, C.A., Goewert, R.R., and A.M. Lambowitz (1979). Characterization of variant Neurospora crassa mitochondrial DNAs which contain tandem reiterations. *Cell* 18: 1197–1207.

Marcou, D. (1961). Notion de longevite et nature cytoplasmique de determinent de la senescence. *Ann. Sci. Nat. Bot.* 2: 653–764.

McDougall, K.J., and T.H. Pittenger (1969). A cytoplasmic variant of *Neurospora crassa*. *Genetics* 61: 551–565.

Michel, F., and B.F. Lang (1985). Mitochondrial class II introns encode proteins related to the reverse transcriptases of retroviruses. *Nature (London)* 316: 641–643.

Mita, S., Tizzuto, R., Moraes, C.T., Shanske, S., Arnaudo, E., Fabrizi, G.M. Koga, Y., DeMauro, S., and E.A. Schon (1990). Recombination via flanking direct repeats is a major cause of large-scale deletions of human mitochondrial DNA. *Nucl. Acids Res.* 18: 561–567.

Mitchell, M.B., and H.K. Mitchell (1952). A case of "maternal" inheritance in *Neurospora crassa*. *Proc. Natl. Acad. Sci. U.S.A.* 38: 442–449.

Moraes, C.T., DiMauro, S., Zeviani, M., Lombes, A., Shanske, S., Miranda, A.F., Nakase, H., Bonilla, E., Werneck, L.C., Servidei, S., Nonaka, I., Koga, Y., Spiro, A.J., Brownell, A.K.W., Schmidt, B., Schotland, D.L., Zupanc, M., DeVivo, D.C., Schon, E.A., and L.P. Rowland (1989). Mitochondrial DNA deletions in progressive external ophthalmoplegia and Kearns-Sayre syndrome. *N. Engl. J. Med.* 320: 1293–1299.

Moraes, C.T., Shanske, S., Trilschler, H.J., Aprille, J.R., Andreetta, F., Bonilla, E., Schon, E.A., and B. DiMauro (1991). mtDNA depletion with variable tissue expression: A novel genetic abnormality in mitochondrial diseases. *Am. J. Hum. Genet.* 48: 492–501.

Myers, C.J., Griffiths, A.J.F., and H. Bertrand (1989). Linear *kalilo* DNA is a *Neurospora* mitochondrial plasmid that integrates into the mitochondrial DNA. *Mol. Gen. Genet.* 220: 113–120.

Nelson, I., d'Auriol, L., Galibert, F., Ponsot, G., and P. Lestienne (1989). Identification nucleotidique et modele cinetique d'une deletion heteroplasmique de 4666 paires de bases de l'ADN mitochondrial dans le syndrome de Kearns-Sayre. *C. R. Acad. Sci.* 309: 403–407.

Newton, K.J. (1988). Plant mitochondrial genomes: Organization, expression and variation. *Annu. Rev. Plant Physiol. Plant Mol. Biol.* 39: 503–532.

Osiewacz, H.D., and K. Esser (1984). The mitochondrial plasmid of *Podospora anserina:* A mobile intron of a mitochondrial gene. *Curr. Genet.* 8: 299–305.

Osiewacz, H.D., Hermanns, J., Marcou, D., Triffi, M., and K. Esser (1989). Mitochondrial DNA rearrangements are correlated with a delayed amplification of the mobile intron (plDNA) in a long-lived mutant of *Podospora anserina*. *Mut. Res.* 219: 9–15.

Poulton, J. (1992). Mitochondrial DNA and genetic disease. *BioEssays* 14: 763–768.

Pring, D.R., and D.M. Lonsdale (1985). Molecular biology of higher plant mitochondrial DNA. *Int. Rev. Cytol.* 97: 1–46.

Rifkin, M.R., and D.J.L. Luck (1971). Defective production of mitochondrial ribosomes in the poky mutant of *Neurospora crassa*. *Proc. Natl. Acad. Sci. U.S.A.* 68: 287–290.

Rhoades, M.M. (1931). Cytoplasmic inheritance of male sterility in maize. *Science* 73: 340–341.

Sainsard-Chanet, A., and O. Begel (1986). Transformation of yeast and *Podospora:* Innocuity of senescence-specific DNAs. *Mol. Gen. Genet.* 204: 443–451.

Sainsard-Chanet, A., Begel, O., and L. Belcour (1993). DNA deletion of mitochondrial introns is correlated with the process of senescence in *Podospora anserina*. *J. Mol. Biol.* 234: 1–7.

Sakaguchi, K. (1990). Invertrons, a class of structurally and functionally related genetic elements that includes linear DNA plasmids, transposable elements and genomes of adeno-type viruses. *Microbiol. Rev.* 54: 66–74.

Schmidt, U., Kosack, M., and U. Stahl (1987). Lariat RNA of a group II intron in a filamentous fungus. *Curr. Genet.* 12: 291–295.

Schapira, A.H.V. (1993). Mitochondrial disorders. *Curr. Opinion in Genetics and Development* 3: 457–465.

Schon, E.A., Rizzuto, R., Moraes, C.T., Nakase, H., Zeviani, M., and S. DiMauro (1989). A direct repeat is a hotspot for large-scale deletion of human mitochondrial DNA. *Science* 244: 346–349.

Schulte, E., Kuck, U., and K. Esser (1988). Extrachromosomal mutants from *Podospora anserina:* Permanent vegetative growth in spite of multiple recombination events in the mitochondrial genome. *Mol. Gen. Genet.* 211: 342–349.

Schulte, Kuck, U., and K. Esser (1989). Multipartite structure of mitochondrial DNA in a fungal longlife mutant. *Plasmid* 21: 79–84.

Seidel-Rogol, B.L., King, J., and H. Bertrand (1989). Unstable mitochondrial DNA in natural-death nuclear mutants of *Neurospora crassa*. *Mol. Cell. Biol.* 9: 4259–4264.

Shoffner, J.M. IV, and D.C. Wallace (1990). Oxidative phosphorylation diseases. *Adv. Hum. Gene.* 19: 267–330.

Shoffner, J.M., Lott, M.T., Voljavec, A.S., Soueidan,

S.A., Costigan, D.A., and D.C. Wallace (1989). Spontaneous Kearns-Sayre/chronic external ophthalmoplegia plus syndrome associated with a mitochondrial DNA deletion: A slip-replication model and metabolic therapy. *Proc. Natl. Acad. Sci. U.S.A.* 86: 7952–7956.

Simpson, M.V., Chin, C.D., Keilbaugh, S.A., Lin, T.-S., and W.H. Prusoff (1989). Studies on the inhibition of mitochondrial DNA replication by 3'-azido-3'-deoxythymidine and other nucleoside analogs which inhibit HIV-1 replication. *Biochem. Pharmacol.* 38: 1033–1036.

Singh, M., and G.G. Brown (1991). Suppression of cytoplasmic male sterility by nuclear genes alters expression of a novel mitochondrial gene region. *Plant Cell* 3: 1349–1362.

Smith, J.R., and I. Rubenstein (1973a). The development of senescence in *Podospora anserina*. *J. Gen. Microbiol.* 76: 283–296.

Smith, J.R., and I. Rubenstein (1973b). Cytoplasmic inheritnce of the timing of "senescence" in *Podospora anserina*. *J. Gen. Microbiol.* 76: 297–304.

Stahl, U., Tudzynski, P., Kuck, U., and K. Esser (1982). Replication and expression of a bacterial-mitochondrial hybrid plasmid in the fungus *Podospora anserina*. *Proc. Natl. Acad. Sci. U.S.A.* 79: 3641–3645.

Steinhilber, W., and D.J. Cummings (1986). A DNA polymerase activity with characteristics of a reverse transcriptase in *Podospora anserina*. *Curr. Genet.* 10: 389–392.

Tudzynski, P., Stahl, U., and K. Esser (1982). Development of a eukaryotic cloning system in *Podospora anserina*. I. Long-lived mutants as potential recipients. *Curr. Genet.* 6: 219–222.

van den Ouweland, J.M.W., Bruining, G.J., Lindhout, D., Wit, J.-M., Veldhuyzen, B.F.E., and J.A. Maassen (1992). Mutations in mitochondrial tRNA genes: Non-linkage with syndromes of Wolfram and chronic progressive external ophthalmoplegia. *Nucl. Acids Res.* 20: 679–682.

Vierula, P.J., Cheng, C. K., Court, D.A., Humphrey, R.W., Thomas, D.Y., and H. Bertrand (1990). The *kalilo* senescence plasmid of *Neurospora intermedia* has covalently-linked 5' terminal proteins. *Curr. Genet.* 17: 195–201.

Wallace, D.C. (1989). Mitochondrial DNA mutations and neuromuscular disease. *Trends Genet.* 5: 9–13.

Wallace, D.C. (1992a). Diseases of the mitochondrial DNA. *Annu. Rev. Biochem.* 61: 1175–1212.

Wallace, D.C. (1992b). Mitochondrial genetics: A paradigm for aging and degenerative diseases? *Science* 256: 628–632.

Wise, R.P., Pring, D.R., and B.G. Gengenbach (1987). Mutation to fertility and toxin insensitivity in Texas (T) cytoplasm maize is associated with a frameshift in a mitochondrial open reading frame. *Proc. Natl. Acad. Sci. U.S.A.* 84: 2858–2862.

Wolf, K., and L. Del Giudice (1988). The variable mitochondrial genome of Ascomycetes: Organization, mutational alterations, and expression. *Adv. Genet.* 25: 185–308.

Wright, R.M., and D.J. Cummings (1983a). Transcription of a mitochondrial plasmid during senescence in *Podospora anserina*. *Curr. Genet.* 7: 457–464.

Wright, R.M., and D.J. Cummings (1983b). Integration of mitochondrial gene sequences within the nuclear genome during senescence in a fungus. *Nature (London)* 302: 86–88.

Yoneda, M., Chomyn, A., Martinuzzi, A., Hurko, O., and G. Attardi (1992). Marked replicative advantage of human mtDNA carrying a point mutation that causes the MELAS encephalomyopathy. *Proc. Natl. Acad. Sci. U.S.A.* 89: 11164–11168.

Zeviani, M., Gellera, C., Antozzi, C., Rimoldi, M., Morandi, L. et al. (1991). *Lancet* 338: 143–147.

PART IV

EXPRESSION AND BIOGENESIS

12

Transcription and mRNA Processing in Organelles

Amino acid incorporation by isolated chloroplasts and mitochondria was first reported in the mid-1950s. However, interpretation of these early studies soon became clouded because of the possible presence of contaminating bacteria, and, in the case of mitochondria, because incorporation by isolated microsomes also had to be taken into account. By the early 1970s the ability of isolated organelles to synthesize proteins was well established. These kinds of experiments together with an ever-increasing knowledge of the gene contents of chloroplast and mitochondrial genomes (Chapter 3) made it apparent that chloroplasts and mitochondria synthesize discrete subsets of their own proteins. Furthermore, these proteins, without exception, proved to be organelle gene products. Historically, this was an important finding because in the 1970s there was a rather protracted argument in the literature concerning whether mRNA molecules, proteins, or both could be imported and exported from chloroplasts and mitochondria. This question has largely resolved itself since. Currently, there is no reason to believe that mRNAs are ever imported into organelles although a great variety of proteins is (Chapter 15). However, as discussed earlier (Chapter 3) and later in Chapter 13, tRNA transport mechanisms seem to have evolved independently for mitochondria in several groups of organisms and, apparently, in plastids as well.

The apparatus involved in the transcription and translation of organelle DNAs has been characterized in ever increasing detail. In 1962 Lyttleton reported the discovery of ribosomes in spinach chloroplasts. Mitochondrial ribosomes were reported in 1967 in *Neurospora* by Kuntzel and Noll and in rat liver by O'Brien and Kalf. In 1964 Kirk found that broad bean chloroplasts could form polynucleotides and that their synthesis required the presence of all four ribonucleoside triphosphates. Incorporation could be abolished either by deoxyribonuclease or actinomycin D, the classic inhibitor of RNA polymerase. In the same year Luck and Reich published essentially similar results for *Neurospora* mitochondria. Since then, organelle DNA transcripts have received detailed scrutiny as has the machinery required for their expression.

An important early finding was that organelle protein synthesis is sensitive to antibiotics such as chloramphenicol, which block protein synthesis on bacterial (70S-type) ribosomes, but resistant to inhibitors like cycloheximide that act on eukaryotic

(80S-type) ribosomes. Inhibitor experiments became the vogue in the late 1960s. The notion was that inhibitors of translation on 70S and 80S ribosomes could be used to distinguish between those proteins made in chloroplasts and mitochondria and those destined for import into these organelles. Although many useful experiments were reported, the methodology was abused. Short-term direct effects were not always distinguished from long-term secondary effects whose interpretation was ambiguous. Consequently, there was a certain amount of disenchantment with the use of inhibitors amongs students of organelle gene function. Today, the results of inhibitor experiments are generally accepted as valid as long as these experiments are carefully controlled and designed to avoid secondary effects. In fact, experiments with inhibitors and with isolated organelles lead to similar conclusions concerning the sites of synthesis of chloroplast and mitochondrial proteins. Much of this early work was reviewed by Gillham (1978).

We now know a great deal about organelle gene expression. This chapter on transcription and mRNA processing in organelles is the first of three on the mechanisms involved in organelle gene expression.

RNA POLYMERASES AND ASSOCIATED FACTORS

Mitochondrial and chloroplast RNA polymerases are strikingly different in structure with the enzyme from the former organelle bearing resemblance to certain bacteriophage polymerases while the chloroplast enzyme or enzymes share similarity with comparable enzymes from *E. coli* and cyanobacteria The mitochondrial RNA polymerases from human, *Xenopus laevis,* and yeast mitochondria and their promoter requirements have been characterized (see reviews by Clayton 1991a,b; Clayton, 1992; Jaehning, 1993; Schinkel and Tabak, 1989; Shadel and Clayton, 1993). In each case a core polymerase and specificity factor are involved. The nuclear *RPO41* gene encoding the 145-150 kDa yeast core enzyme has been isolated and sequenced (Jaehning, 1993; Masters *et al.,* 1987; Shadel and Clayton, 1993). The derived amino acid sequence bears striking similarity to the single polypeptide enzymes of phages T3 and T7. This is especially true of the regions having the greatest similarity between the two phage enzymes. There are no regions of homology to any of the *E. coli* RNA polymerase subunits. The yeast specificity factor is encoded by a second nuclear gene *(MTF1)* and controls correct promoter recognition (Jaehning, 1993). So far, the yeast mitochondrial RNA polymerase is the only such enzyme to have been purified to homogeneity (Jaehning, 1993; Clayton, 1992).

The human mitochondrial H- and L-strand promoters are bipartite, consisting of a core region at the transcription initiation sites plus a second upstream region to which the mitochondrial transcription factor (mtTF1) binds (Fig. 5–4). mtTF1 activates transcription as a result of DNA binding conceivably by inducing physical changes at the promoter such as bending, possibly in concert with other proteins required for the transcription process. The nuclear gene encoding mtTF1 has been isolated and expressed in *E. coli.* The 204 amino acid protein contains two domains characteristic of high mobility group (HMG) proteins, termed HMG boxes 1 and 2, and a 42 amino acid mitochondrial targeting presequence. mtTF1 is most closely related to human hUBF, a protein known to be important in RNA polymerase I transcription in the nucleus. This raises the possibility that mitochondrial and RNA polymerase I transcription systems might have a shared identity. Mouse mtTF1 is similar in size and specificity to the human factor and can be swapped for human mtTF1 *in vitro.* Correct transcript initiation ensues although at a reduced level.

In *Xenopus,* the two promoters function bidirectionally rather unidirectionally as in human mtDNA. Furthermore, the promoters are smaller and closer together while in chicken there is but a single bidirectional promoter (Shadel and Clayton, 1993). In *Xenopus* there also appears to be a dissociable transcription factor acting in concert with the core polymerase. The absence of an apparent requirement for an upstream sequence in *Xenopus* promoters may indicate that the transcription factor and core polymerase interact differently than is true of these two proteins in mammals.

A second factor in yeast, ABF2, is an HMG-box protein that has some similarity to the human mtTF1 (Jaehning, 1993; Xu and Clayton, 1992). However, ABF2 is not required for selective initiation of transcription in yeast, does not stimulate transcription *in vitro,* and disruption of the *ABF2* gene does not result in loss of RNA polymerase activity or loss of mitochondrial function as long as cells are grown on a nonfermentable carbon source (Jaehning, 1993). On the other hand, mtDNA is lost in yeast strains having a disrupted *ABF2* gene when these cells are grown on a fermentable carbon source. When the human gene encoding mtTF1 is introduced into such a strain, the gene is expressed and mtDNA no longer disappears when cells are grown on a fermentable carbon source (Parisi *et al.,* 1993). These results suggest that human mtTF1 is probably the functional equivalent of yeast ABF2. As is often the case in naming genes and proteins, nomenclatorial confusion is beginning to arise (i.e., the human and yeast mTF1s seem to have different roles). To nip this problem in the bud, Clayton and colleagues (Shadel and Clayton, 1993; Xu and Clayton, 1992) proposed naming the yeast proteins sc-mtRNA polymerase (core enzyme), sc-mTFA (ABF2), sc-mtTFB (MTF1), etc. The homologous human proteins have the prefix h-.

Chloroplast RNA polymerases are much more complex than the yeast mitochondrial enzyme, sharing many characteristics of *E. coli* RNA polymerase (see Bogorad, 1991; Igloi and Kossel, 1992; Gruissem and Tonkyn, 1993 for reviews). The highly purified enzyme from maize chloroplasts has the structure α_2, β, β', β'', which is identical to the bacterial enzyme except that the β' and β'' subunits of the chloroplast enzyme correspond, respectively, to the N- and C-terminal regions of the bacterial β' subunit. In cyanobacteria β' is similarly split into two subunits consistent with the derivation of chloroplasts from this group of eubacteria (Chapter 2). All four subunits of this chloroplast enzyme are encoded by the organelle genome (but see the following). Highly purified preparations from spinach and pea chloroplasts contain several additional major and minor polypeptides when compared to the purest enzyme from maize. Whether these are genuine components of the core enzyme remains to be established.

The *rpoA* gene encodes the 38 kDa α subunit of chloroplast RNA polymerase and, aside from various deletions and insertions, the derived polypeptide has 26–28% identity with the *E. coli* α subunit. The maize *rpoB* gene encodes a basic protein of 122 kDa with ca. 38% homology to the *E. coli* β subunit. The protein specified by this and other plastid *rpoB* genes lacks a block of 103–105 amino acids at the carboxy terminus, which can be deleted without effect from the *E. coli* enzyme. The $rpoC_1$ and $rpoC_2$ genes encode polypeptides homologous to parts of the 155-kDa *E. coli* β' subunit. The $rpoC_1$ genes of different plastid genomes specify proteins of 677–684 amino acids corresponding to the N-terminal region of the *E. coli* β' subunit. The $rpoC_2$ gene encodes sequences homologous to the 829 amino acid C-terminal region of the *E. coli* subunit although the C-terminal regions encoded by the plastid genes are much larger than their *E. coli* counterparts (tobacco, 1361; liverwort, 1391; maize, 1527; and rice 1513 amino acids). This can be accounted for by the presence in the plant genes of a sequence encoding 626 amino acids and deletion of a sequence found in the *E. coli* gene, which specifies a block of 40 amino acids (Bogorad, 1991). The $rpoC_2$ genes of maize and rice are larger than those of tobacco and *Marchantia* because of insertion of a sequence of approximately 450 bp encoding about 150 amino acids. The latter sequence can be broken down into a set of 13 almost identical heptameric repeats composed of acidic amino acids.

In *E. coli* the *rpoB* and *rpoC* genes are adjacent and part of the *rif* operon, which also contains genes encoding ribosomal proteins L10 and L7/L12. In the chloroplast genomes of land plants, *rpoB*, $rpoC_1$, and $rpoC_2$ also form an operon, but the ribosomal protein genes are missing and nuclear in location (see Igloi and Kossel, 1992). The *rpoA* gene is part of a vestige of the α operon, which includes the gene encoding protein S11 (Table 3–5).

Whether plastids contain more than one species of DNA-dependent RNA polymerase is a long standing question. In run-on transcription assays with gently lysed chloroplasts virtually all RNA polymerase molecules appear to be bound within the chloroplast to multiple copies of DNA template. This leaves no soluble enzyme pool to catalyze *de novo* transcription from exogenously added DNA templates. Much of this activity can be released from the DNA by high salt treatment, but a minor portion remains tightly bound and transcriptionally active. The TAC (*t*ranscriptionally *a*ctive *c*hromosome) complex, originally isolated from *Euglena* chloroplasts, demonstrated a strong specificity for the rRNA genes. This led to the notion that the there were at least two different chloroplast RNA polymerases. The tightly bound form in the TAC complex was thought to be responsible for transcription of the rRNA genes and the soluble form for mRNA and tRNA transcription. However, this distinction has become blurred by experiments with flowering plant chloroplast extracts. Thus, the TACs obtained from mustard and spinach chloroplasts can transcribe genes encoding mRNAs in addition to the rRNA genes. As Igloi and Kossel (1992) point out, this lends support to the notion that differences between the soluble and TAC activities reflect modulation of a common core enzyme rather than the existence of two distinct enzymes. Furthermore, because uncertainty exists with respect to transcriptional initiation by the DNA-associated enzyme, the current view according to Igloi and Kossel is that the enzyme in the TAC is simply RNA polymerase engaged in the process of elongation. Once this RNA polymerase dissociates from its DNA template, the enzyme lacks the factor(s) necessary for reinitiation.

Evidence suggesting the presence of a second chloroplast RNA polymerase comes from a completely different set of results that supports the existence of an additional nuclear-encoded enzyme. Ribosome-deficient plastids obtained either from the nuclear "albostrians" mutant of barley or heat bleached rye seedlings possess transcription activity, though this is somewhat reduced with respect to normal plastids (Igloi and Kossel, 1992). Recently, Hess *et al.* (1993) analyzed transcription in the "albostrians" mutant and heat bleached rye seedlings having established by immunoblotting that an essential chloroplast ribosomal protein was absent from the tissues examined. This means that chloroplast ribosomes themselves and the chloroplast-encoded subunits of

RNA polymerase must also have been absent. Hess *et al.* found that transcripts of the *rpo* genes and *rps15* were synthesized although transcripts of several genes involved in photosynthesis were accumulated at extremely low levels. The rRNA genes are also transcribed, but the rRNA is rapidly degraded probably because of the absence of ribosomal proteins. Hess *et al.* favor the notion that chloroplasts may contain two species of RNA polymerase. The plastid-encoded enzyme would have preference for chloroplast genes encoding photosynthetic or bioenergetic componenents while nuclear-encoded enzyme would be responsible for transcribing genes involved in chloroplast gene expression.

The existence of a nuclear-encoded chloroplast RNA polymerase that can transcribe chloroplast genes involved in chloroplast gene expression is supported by observations made on the colorless parasitic plant "Beechdrops" *(Epifagus)*. This plastid genome has lost all four RNA polymerase subunit genes found in flowering plant chloroplast genomes as well as all of the chloroplast genes required for photosynthesis, but retains chloroplast genes encoding ribosomal proteins, tRNAs, rRNAs, and a few other proteins (Wolfe *et al.*, 1992a,b; also see Chapter 3). As these genes are actively transcribed, the polymerase responsible must be encoded in the nucleus. However, comparison of promoter regions in *Epifagus* and flowering plants reveals that for four genes found in both plastid genomes the typical eubacterial -35 and -10 regions found in flowering plants have either diverged or been deleted in *Epifagus*. Only in the inverted repeat region can chloroplast promoters be identified and even there the rRNA promoter has been disrupted. This raises the possibility that the existence in *Epifagus* of a chloroplast RNA polymerase encoded in the nucleus is a unique solution to a specific problem rather than of general significance. The case for the existence of a second chloroplast enzyme has recently received direct support (Lerbs-Mache, 1993). A single subunit (110-kDa) RNA polymerase has been discovered in spinach chloroplasts that seems to be of the bacteriophage T7 type. This enzyme may be responsible for transcribing the chloroplast genome early in development.

If the *E. coli*-like chloroplast RNA polymerase is as closely related to the bacterial enzyme as it seems to be, one would expect σ-factor-like proteins to be required for correct promoter recognition and initiation by the core enzyme. One such protein, the 27.5-kDa S-factor, was reported from maize by Jolly and Bogorad (1980). Compared to vector sequences, this factor increased several fold the amounts of chloroplast specific transcripts for genes encoding both mRNAs and rRNAs. However, this effect was largely dependent on the presence of supercoiled forms of the DNA template and the S factor did not affect transcription by *E. coli* RNA polymerase. Therefore, the S-factor is not considered equivalent to an *E. coli* σ factor.

There are several reports of σ-factor-like proteins from plastids. Surzycki and Schellenbarger (1976) described a 51-kDa polypeptide from *Chlamydomonas reinhardtii* that appeared to have σ'-like activity with both *E. coli* RNA polymerase and the core polymerase from *Chlamydomonas*. However, these experiments have not been followed up nor has the core polymerase been characterized from *C. reinhardtii*. Sigma-like factors have also been reported from mustard and spinach. The mustard protein fraction did not bind DNA itself, but enhanced formation of complexes between *E. coli* RNA polymerase and the mustard *psbA* promoter sequence (Bulow and Link, 1988). Tiller *et al.* (1991) characterized the mustard chloroplast fraction further and identified three σ-like factors (SLF). None of these polypeptides binds to DNA itself, but each one confers enhanced binding and transcriptional activity when added to the *E. coli* RNA polymerase core enzyme and DNA fragments carrying a chloroplast promoter. A protein fraction from spinach with similar properties contained proteins of 33 and 90 kDa that cross-reacted with antiserum against *E. coli* σ-factor (Lerbs *et al.*, 1988).

TRANSCRIPTION OF MITOCHONDRIAL GENOMES

Mitochondrial transcription has been best characterized in mammals and yeast. There are some similarities, but striking differences as well.

Mammals. Mammalian mtDNA transcription has been regularly reviewed (Attardi, 1985; Attardi and Schatz, 1988; Cantatore and Saccone, 1987; Clayton, 1984, 1991a,b; 1992). As discussed in Chapter 3, the mammalian mitochondrial genome is extremely compact with most genes being butt jointed to each other or separated by only a few nucleotides. The majority of these genes is transcribed from the H-strand including the two rRNA genes, 14 tRNA genes, and 12 protein coding genes. Transcription of the mammalian mitochondrial genome is unusual in several respects. First, the H- and L-strands are transcribed symmetrically and in their entirety. Second, there are two overlapping H-strand transcripts (see below). Third, one or more tRNA genes separate nearly all of the other coding sequences. The tRNA sequences appear to function as signals for RNA processing enzymes that generate the mature RNA species. Fourth, most reading frames lack significant untranslated regions and most of them are missing complete termination codons, instead ending with either T or TA following the last sense codon. These sequences are converted to TAA termination codons by posttranscriptional polyadenylation of the mRNAs.

Initiation of transcription and its relationship to mtDNA replication has been the focus of much attention (Chapter 5; Clayton, 1991a,b, 1992; Schinkel and Tabak, 1989). Each strand of the mitochondrial genome is transcribed from a single major promoter in the D-loop region (Fig. 5–4). The H-strand (HSP) and L-strand (LSP) promoters are usually within 150 bp or less of each other. HSP and LSP each consist of approximately 50 bp sequences that are bipartite and comprise two functionally distinct elements. The mTF1 transcription factor (see above) activates transcription by the mitochondrial RNA polymerase by binding to the upstream element in each promoter.

Following elongation of newly initiated transcripts, termination occurs. Termination has been studied in some detail for human mtDNA. The two mitochondrial rRNAs are the most abundant species in total mitochondrial RNA. This reflects a greater overall rate of synthesis of transcripts containing these RNAs alone and does not result from processing of much longer H-strand transcripts. The termination sequence responsible for preferential production of these rRNA transcripts appears to be a tridecamer sequence located at the 5'-end of the tRNA^{Leu} gene immediately adjacent to the 16 S rRNA gene (Fig. 12–1). A protein factor has been identified that binds to a 28-bp region that includes this tridecamer sequence (Kruse *et al.*, 1989). The active termination factor is a protein of ca. 34 kDa (Hess *et al.*, 1991) called mTERF or mtTERM. This factor may bind to the template and interfere with further elongation by the mitochondrial RNA polymerase (Clayton, 1992). Transcription termination of the 16S rRNA gene is severely impaired in individuals carrying the MELAS mitochondrial mutation (Hess *et al.*, 1991). The MELAS syndrome, mitochondrial encephalomyopathy, results from a

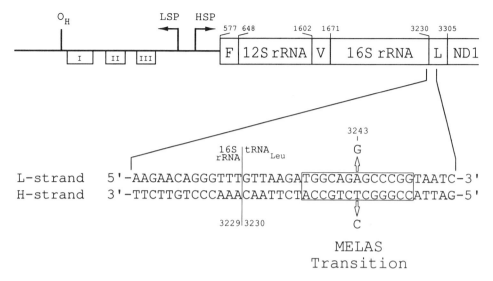

Fig. 12–1. The site of human mitochondrial DNA transcription termination at the 3'-end of the 16S rRNA gene is shown. F, V, and L are the genes for tRNAs phenylalanine, valine, and leucine, respectively. A portion of the *ndh1* (ND1) gene is shown. O_H is the origin of H-strand synthesis, HSP is the heavy strand promoter, LSP represents the light strand promoter; and I, II, and III represent the conserved sequence blocks (CSBs) (see also Fig. 5–4A). The tridecamer termination sequence is boxed and the MELAS (mitochondrial myopathy, encephalopathy, lactic acidosis, and stroke-like episodes) mutation at position 3243 is indicated. From Clayton (1991a).

single base pair transition mutation in the tridecamer sequence (Fig. 12–1, Chapter 11). This correlates with a decreased affinity of partially purified mtTERM for the MELAS template.

The structure and processing of mammalian mtDNA transcripts are discussed in several reviews (Attardi, 1985; Cantatore and Saccone, 1987; Gruissem and Schuster, 1993). There are two overlapping, primary transcripts of the mammalian H-strand (Fig. 3–1). The shortest and most abundant of these encodes the two rRNAs plus the tRNA^{Val} in between and terminates at the tRNA^{Leu} gene next to the 16S rRNA gene (see above). The second primary transcript includes the entire H-strand sequence. Similarly, the L-strand is transcribed over its entire length. The L-strand is transcribed at a rate that is two to three times higher than the H-strand, but most L-strand transcripts have half lives much shorter than the H-strand transcripts. There are 18 discrete poly(A)-containing RNAs ranging in size from ca. 215 nt to 10,400 nt and they are numbered in order of decreasing size. The three largest species (RNAs 1, 2, and 3) and the smallest (RNA 18) are encoded by the L-strand. The rest are H-strand transcripts. With the exception of RNAs 4, 6, and 10, the polyadenylated H-strand transcripts correspond to protein-encoding mitochondrial genes. Unlike eukaryotic cytoplasmic mRNAs, these mitochondrial mRNAs lack both a ''cap'' structure at the 5'-end and a 5' noncoding stretch upstream of the initiator codon. All mRNAs except RNAs 5, 9, and 16 have incomplete stop codons that are converted to UAA by polyadenylation. RNA 6 is probably a precursor of RNA 9 encoding *cox1*, RNA 10 represents a small fraction of polyadenylated 16S rRNA, and RNA 4 is the abundant rRNA transcript already discussed that terminates distal to the 16S rRNA gene. The polyadenylated L-strand transcripts 1, 2, and 3 have a common 5'-end, but different 3' ends corresponding to different tRNA genes. RNA 18, also called 7S RNA, is a ca. 215 nt L-strand transcript located near the origin of H-strand replication. This species is found in human mitochondria, has a stability comparable to the H-strand transcripts; contains an ORF that could encode a polypeptide of 23 to 24 amino acids, and has been found associated with polysomes (see Nelson, 1987). The function of 7S RNA is unclear and the species is missing from rat and mouse.

The principal mechanism by which functional mRNAs are cut out of primary transcripts is tRNA removal followed by polyadenylation to yield termination codons (Fig. 12–2). At least four enzymes are thought to be involved in these processes. One or probably two enzymes would be required to recognize the 5'- and 3'-ends of each tRNA. A third enzyme would be needed for polyadenylation and the fourth to add the CCA sequence to the 3'-end of each tRNA (see Clayton, 1991a, for discussion). One potential processing enzyme seems to be analogous to ribonuclease P (RNase P). This enzyme, an RNA–protein complex in bacteria, yeasts, and mammals, removes the 5' leaders of tRNA precursors (Altman et al., 1993). The endoribonuclease from HeLa cells has been partially purified (Doersen et al., 1985). This enzyme correctly processes the precursor of E. coli suppressor tRNA^{Tyr} with the same specificity as E. coli RNase P and has an RNA component. Both the RNA and protein portions of this enzyme are coded in the nucleus (Clayton, 1991a). Attardi (1985) remarks that a few processing sites in the H-strand transcript lack tRNA sequences. He speculates that the processing enzymes may recognize secondary structures that share some critical features with the cloverleaf structure of tRNA.

Gruissem and Schuster (1993) point out that changes in stability and turnover of mammalian mitochondrial mRNAs have been largely unexplored. While it is known that HeLa cell mRNAs are metabolically unstable with half-lives ranging between 25 and 90 min other results clearly demonstrate that inhibition of transcription does not result in noticeable diminution of continued high rates of mitochondrial protein synthesis. This increase in stability, discussed in more detail in Chapter 16, implies that mitochondrial RNA accumulation is probably partially controlled at the level of decay and/or that mitochondrial mRNA is not limiting for mitochondrial protein synthesis.

The organization of other animal mitochondrial genomes (Chapter 3) strongly suggests that transcript processing also involves tRNA excision, but this cannot be the exclusive mechanism of processing. For example, an examination of the gene map of the sea anemone *Metridium* reveals the presence of only two tRNA genes with most protein coding genes being butt jointed to each other (Wolstenholme, 1992). Other examples of individual pairs of protein coding genes not separated by tRNA sequences are to be found in nematodes and sea urchins (Wolstenholme, 1992).

In *Drosophila* termination codons are often incomplete (T or TA) as in mammals. They are likely converted to intact termination codons by polyadenylation (Clary and Wolstenholme, 1984). In nematodes, in contrast, complete termination codons are frequently observed (Okimoto et al., 1992). This is always the case for the completely sequenced mitochondrial genome of the sea anemone *Metridium senile* (Wolstenholme, 1992). These observations prompted Wolstenholme (1992) to observe that "a cleavage-

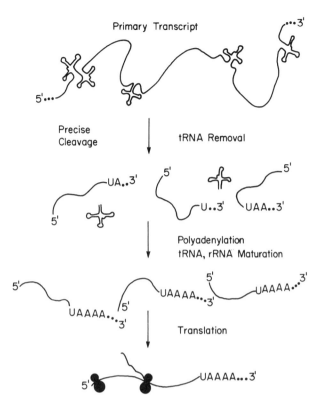

Fig. 12–2. Basic mode of primary transcript processing in mammals and other animals. The central features of processing are recognition and resection of tRNAs and polyadenylation of mRNAs. From Clayton (1984).

polyadenylation mechanism developed very early in the evolution of metazoan mitochondria."

Fungi. Attardi and Schatz (1988) have compared mtDNA transcription in yeast and mammals. In contrast to the mammalian mitochondrial genome where both DNA strands are completely transcribed, only one strand is transcribed in *S. cerevisiae* save for a short transcript of a tRNA^{Thr} gene in its complement. This reflects the fact that the distribution of genes in the mitochondrial genome of yeast is even more asymmetric than in mammalian mtDNA, with all genes except the aforementioned tRNA^{Thr} gene being localized on one strand. The mitochondrial genome of *S. cerevisiae* is also far more spacious than the mammalian mitochondrial genome and many genes are transcribed as parts of multicistronic units rather than as single transcripts of entire mtDNA strands (Grivell, 1989). There are at least 20 different promoters (Christianson and Rabinowitz, 1983; Schinkel *et al.* 1988) in contrast to the single pair of promoters (HSP and LSP) known in mammalian mitochondrial genomes. In all cases, transcription initiates within a 9 nucleotide consensus sequence 5'-TATAAGTA-3' at the *A* residue (Shadel and Clayton, 1993). Processing of these transcripts is quite involved. A conserved dodecamer motif downstream of nearly all protein coding genes has been implicated in the endonucleolytic event associated with 3'-end processing (Osinga et al., 1984). Also, tRNA genes are part of large primary transcripts and their release as mature tRNAs requires processing at both their 5'- and 3'-ends (Wolf and Del Giudice, 1988). Processing of the transcripts of the three "mosaic" genes *cox1, cytb,* and *rnl* leading to intron excision is a complex event (Chapter 10).

The yeast, *Schizosaccharomyces pombe*, has a small (ca. 19 kb) mitochondrial genome similar in size to animal mitochondrial genomes (Chapter 3). Despite its small size, this mitochondrial genome unlike those of animals contains introns in two genes (*cox1* and *cytb*) that account for ca. 25% of the genome size. All mitochondrial genes in *S. pombe* are coded on the same strand and most genes are separated by tRNA sequences that act as processing signals as in mammals (Lang et al., 1983). Only upstream of the large rRNA gene and between *cox1* and *cox3* are tRNA genes absent.

In *Neurospora crassa* processing of the single large precursor RNA containing segments corresponding to *cytb, tRNA^{Cys}, cox1, tRNA^{Arg},* and *ndh1* genes involves the two tRNA sequences. They serve as signals for the cleavage of this transcript yielding the *cox1* mRNA as well as the 3'-end of the *cytb* transcript (Wolf and Del Giudice, 1988). It was once believed that *Pst*I palindromes could serve as processing sites since six of them are located between *cytb* and *cox1* (cf. Yin et al., 1981). This possibility has since been ruled out. There is now general agreement that tRNA sequences act as primary signals for RNA processing in *N. crassa* mitochondria (Breitenberger et al., 1985; Agsteribbe and Hartog, 1987).

Flowering plants. Progress is being made in understanding transcription of the giant mitochondrial genomes of flowering plants (see Gray et al., 1992; Lonsdale, 1989; Newton, 1988). These mtDNA molecules have large intergenic regions, so one might expect that only a small portion of the genome would be transcribed. Makaroff and Palmer (1987) identified 24 abundant and unique transcripts in *Brassica campestris*. They calculated that these amounted to ca. 30% of the 218-kb mitochondrial genome of this plant. Bendich (1985) estimated that 50–70% of the 330-kb watermelon mitochondrial genome was expressed and ca. 30% of the 2400-kb muskmelon mitochondrial genome assuming completely asymmetric transcription. In reviewing these results Gray et al. (1992) conclude that "these values are probably conservative because they do not take into account low-abundance transcripts in the steady-state RNA population. Even so, transcription appears to be in considerable excess of that required to specify 20–30 polypeptides, 3 rRNA species, and 15–20 tRNAs synthesized in plant mitochondria."

The lack of clustering of plant mitochondrial genes would suggest that most genes are within their own transcriptional domains. Accordingly, many plant mitochondrial genes are expressed as monocistronic transcripts (Gray et al., 1992). But both simple and complex transcripts are found. Thus, simple transcripts are observed for most mitochondrial genes in *Brassica campestris* while in maize complex transcription patterns are seen for most genes. These complex patterns generally arise because of variability in the 3' and 5' untranslated sequences of the mRNAs and not from intron splicing, as most plant mitochondrial genes lack introns. Thus multiple transcription and termination events seem to be responsible. Dicistronic transcripts for adjacent genes have also been observed in a variety of species. Lonsdale (1989) inferred a consensus promoter sequence for plant mitochondrial genes (5'-TAAT/AGA-3') by comparison of presumptive 5' transcript termini based on nuclease protection and primer extension experiments. He reported that transcripts appeared to initiate on either of the first two A residues within the consensus sequence. However, Gray *et al.* (1992) point out that 5'-termini resulting from transcription (i.e., primary 5'-ends) must be distinguished from 5'-termini formed as a consequence of mRNA processing for meaningful comparisons to be made. This distinction can be achieved by determining whether transcripts serve as substrates for the enzyme guanyltransferase. This enzyme catalyzes incorporation of guanosine monophosphate into the 5'-terminal cap structure using as substrates only RNA species with di- or triphosphorylated termini (i.e., primary transcripts).

When this sort of analysis is done for mitochondrial transcripts in maize, wheat, and soybean, different consensus sequences emerge in which the motif CRTA (R is a purine) is the most highly conserved feature. In maize this motif is part of an 11 nucleotide conserved sequence found at the transcription initiation sites of those mitochondrial genes most strongly expressed *in vivo* (Mulligan *et al.*, 1991). *In vitro* experiments with a partially purified maize mitochondrial extract that correctly initiates transcription have identified essential promoter elements within a 19-bp sequence immediately upstream of the transcription initiation site that includes the 11-bp sequence motif identified *in vivo* (Rapp and Stern, 1992). Comparison of mitochondrial promoters in soybean and *Oenothera berteriana* indicate that CRTA belongs to a nonanucleotide consensus sequence (Binder and Brennicke, 1993). Interestingly, when the consensus sequences for 15 transcription units in the *O. berteriana* mitochondrial genome are compared, this sequence lengthens to 29 nucleotides.

Having analyzed the available data on transcription initiation in plant mitochondria Gray *et al.* (1992) conclude that

> several general characteristics of transcription in plant mitochondria are already evident: (*a*) In contrast to the situation in human or mouse mitochondria, but like that in yeast, transcription of plant mtDNA is initiated at multiple sites; however, unlike the yeast case, there may be a number of separate initiation sites for a single gene. (*b*) Contrary to the situation in yeast, there is a relatively low degree of primary sequence conservation among different sites of transcription initiation within the same plant species (i.e., no highly conserved consensus such as the yeast nonanucleotide motif is apparent). (*c*) Whereas the *S. cerevisiae* nonanucleotide consensus sequence is conserved in other yeast species, there is only limited correspondence between mitochondrial transcription initiation sites in different monocots (wheat vs. maize), and even less between monocots and dicots (wheat/maize vs. soybean). (*d*) Whereas transcription is always initiated at the final A residue in the yeast nonanucleotide motif, there is somewhat more variation in the position of the inferred transcriptional start site within the consensus sequences proposed for plant mtDNA.

Stem-loop structures are invariably identifiable in the 3' noncoding regions of plant mitochondrial genes. Whether these structures act as RNA processing signals or are involved in transcription termination is not known. Similar structures in chloroplast mRNAs are known to play a role in stabilizing transcripts against degradation as discussed in the next section.

Gruissem and Schuster (1993) review available data relating to half-lives of plant mitochondrial mRNAs and the mechanisms that control their stability. They conclude that considerably less information is available on these topics than is the case for mammals. Nevertheless, recent observations show that the rRNA genes are the most highly transcribed as in animal and yeast mitochondria and that mRNAs are transcribed at different rates depending on the gene. They also point out that these mRNAs accumulate to different levels that are not reflected by their relative rates of synthesis so these processes can be uncoupled as is true for plant chloroplast mRNAs (see Chapter 16).

TRANSCRIPTION OF CHLOROPLAST GENOMES

Most chloroplast gene clusters are transcribed polycistronically. The transcription process itself is considered in reviews by Bogorad (1991) and Igloi and Kossel (1992) while transcript processing is the subject of a separate review (Sugiura, 1991). Both subjects have been reviewed by Gruissem (1989) and Gruissem and Tonkyn (1993) while the control of plastid RNA degradation is considered in the article by Gruissem and Schuster (1993).

Unlike mitochondrial promoter sequences putative chloroplast promoter sequences are strikingly similar to the -10 and -35 promoter regions of prokaryotic genes. The proposed consensus sequence for *E. coli* genes transcribed by an RNA polymerase holoenzyme with a σ 70 factor is

$$-35 \qquad\qquad -10$$

$$5'\text{-TTGACA----17 nt----TATAAT-}3'$$

with the spacing between the -10 and -35 regions being somewhat variable. Based on comparisons of sequences 5' to the probable transcription start sites, Steinmetz *et al.* (1983) identified apparent consensus sequences for the -35 and -10 elements by analogy with *E. coli*. However, as Gruissem (1989) cautioned "structural similarities to prokaryotic promoters are of only limited applicability in locating chloroplast promoter regions. Rather such regions can be identified unequivocally only by a functional assay in homologous chloroplast transcription systems."

The requisite *in vitro* experiments have since been carried out in several plastid systems and for several chloroplast genes, notably *psbA*, using partially purified chloroplast RNA polymerase (Gruissem and Tonkyn, 1993). Both wild-type and mutant promoters have been studied. These experiments confirm that most plastid promoters consist of two regions (ctp1 = "-35" and ctp2 = "-10") separated by ca. 17 to 19 bp. Certain bases in these sequences are invariant and their deletion or mutation reduces transcription initiation efficiency, but none abolishes transcription completely. However, not all plastid genes that are transcribed efficiently *in vitro* possess the conserved ctp1–ctp2 organization. Thus, the promoters for the *atpBE* and *rps16* primary transcripts are less conserved and in the *trnR1* and *trnS1* genes internal DNA sequences seem to function as promoters (Gruissem and Tonkyn, 1993). Other plastid tRNA genes can be transcribed from a combination of upstream and internal promoters.

The technique of chloroplast transformation (Chapter 7) has made possible the introduction of genes having deleted or altered chloroplast promoters into the chloroplast genome. Using this method Blowers *et al.* (1990) showed that deleting all sequences 5' to position -24 relative to the transcription start site of the *Chlamydomonas atpB* gene had no effect on the relative rate of transcription or on the transcription initiation site for this promoter. Since this deletion eliminates any possible -35 sequence, the results show that this region is not essential for transcription of the *Chlamydomonas atpB* gene. Continuation of experiments of this type has led to the identification of a second type of chloroplast promoter that does include a -35 sequence (Klein *et al.*, 1992). Thus, deletion analysis shows that the promoter proximal to the 16 S rRNA gene of *Chlamydomonas* resembles a typical bacterial promoter with both -10 and -35 elements. Deletion analysis and transformation in each case were carried out with a foreign reporter gene downstream of each chloroplast promoter sequence. A comparison of putative promoter regions of eight *Chlamydomonas* chloroplast genes including the ones just described indicated that six of them lacked the conserved -35 element. Seven promoters contained palindromic TATAATAT in the -10 region, consisting of two overlapping -10-like elements, TATAAT and TAATAT. Often there are other like sequences in the vicinity of the -10 region.

A general feature of plastid mRNAs, and also a number of tRNA and rRNA primary transcripts, is the presence of inverted repeat (IR) sequences located in the 3'-flanking regions that can fold into stable stem-loop structures (see Gruissem and Schuster, 1993). The expectation that these IRs might form stem–loop structures that could function in a manner analogous to rho-independent bacterial terminators has been disputed. Stern and Gruissem (1987) argued that the principal function of these 3' IRs is to act as mRNA processing and stabilizing elements. They observed, using a strong chloroplast promoter, a reporter tRNA gene, and several different chloroplast 3' IRs that chloroplast RNA polymerase read through these regions. This appeared to show that the 3' IRs did not have a general function as transcription terminators. Stern and Gruissem also found, using a chloroplast transcription extract from spinach, that synthetic RNAs were processed in a 3'-5' direction by a nuclease activity present in the transcription extract that generated nearly homogeneous ends distal to the stem–loop. The 3'-ends produced *in vitro* appeared to correspond to those found for plastid mRNAs *in vivo*. Molecules possessing the 3' stem–loops also exhibited enhanced stability relative to contol RNAs *in vitro*.

Gruissem and Schuster (1993) note that two specific ribonuclease activities affecting the processing and stability of the 3'-end of *petD* mRNA have been identified in spinach chloroplast extracts. One is the aforementioned exonuclease that converts the 3'-end of the precursor mRNA to the mature form ending with the inverted repeat. The second is an endonuclease that cleaves the *petD* mRNA at the termination codon and at a second site just 5' to the IR. This results in removal of the 3' IR and rapid degradation of the upstream RNA. Nickelsen and Link (1993) have reported that a 54-kDa RNA-binding protein from mustard is an endonuclease that mediates 3'-end formation *in vitro* downstream of the *trnK* and *rps16* genes. The 3'-ends created correspond to those seen *in vivo*.

Using the *Chlamydomonas* chloroplast transformation system (Chapter 7) and deletion mutants generated *in vitro*, Stern *et al.* (1991) reported that the presence of an intact 3' IR adjacent to the chloroplast *atpB* gene is required for maximum stability of this monocistronic mRNA *in vivo*. Deletion of all or part of the IR led to decreased mRNA accumulation without affecting transcription rates.

Gruissem and Schuster (1993) discuss the accumulating evidence that the 3' stem–loop structures also bind proteins. They state that

> it is reasonable to argue that the effectiveness of the 3' IR as a stabilizing element may be modulated during plant development or in response to environmental signals. Such modulation could be achieved through participation of additional trans-acting factors such as nucleases or other RNA-binding proteins. This would be consistent with the observation that the accumulation of individual plastid mRNAs *in vivo* and the relative stabilities of plastid mRNAs *in vitro* do not necessarily reflect the theoretical free energy of the 3' inverted repeats *in vivo* (Stern and Gruissem, 1987).

The first evidence that the 3' stem–loop structures do form RNA protein complexes was published by Stern et al. (1989). Comparison of proteins bound to the *petD* 3' IR with proteins binding to transcripts of the *rbcL* and *psbA* genes revealed that certain proteins bound in a gene-specific fashion whereas others did not. Reviewing these and other results, Gruissem and Schuster (1993) note that all three inverted repeats interact with a protein (or group of proteins) of 28 kDa that, therefore, have a general affinity for different 3' ends. A 24-kDa protein also falls in this category although its affinity for different 3'-ends seems to vary. In contrast, proteins of 55 and 100 kDa bind specifically to the *petD* 3' IR RNA. Hsu-Ching and Stern (1991) have examined the regions of the stem–loop structure of the *petD* 3' IR to which different proteins bind. They report that a 33-kDa protein (previously identified as a 29-kDa protein, Stern et al., 1989) recognizes specifically the double-stranded stem of the hairpin structure. A 57-kDa protein (previously the 55-kDa protein) binds to an AU-rich sequence motif that is highly conserved in *petD* genes of land plant species. In the absence of the 55-kDa protein the 33-kDa protein fails to bind to the double–stranded stem suggesting that the former protein may be critical to stabilizing the stem–loop structure. In contrast, the 24- and 28-kDa proteins (designated 28 and 32 kDa by Hsu-Ching and Stern, 1991) possess little sequence or structural binding specificity. This is also true of the 100-kDa protein, which is somewhat puzzling if this protein does, indeed, bind specifically to the *petD* 3' IR as indicated above.

Like the 54-kDa protein of Nickelsen and Link (1993) the 28-kDa protein also seems to have a role in RNA processing (Gruissem and Schuster, 1993). Consistent with the ability of the 28-kDa protein to bind generally to 3'-ends is the finding that in the absence of this protein, the processing reaction is inhibited and the decay of the 3'-end precursor RNA is accelerated. These results suggest that the 28-kDa protein plays a direct role in 3'-end processing possibly by interacting with a specific nuclease to direct correct mRNA 3'-end processing. Gruissem and Schuster (1993) think it likely that the other chloroplast RNA-binding proteins are also involved in RNA processing events. For instance, Li and colleagues (Li and Sugiura, 1990; Ye et al., 1991) have suggested that the chloroplast RNA-binding proteins may participate in the formation of complexes that are required for intron removal much as spliceosomes are necessary for the excision of introns from nuclear-encoded mRNAs (Chapter 10).

Although the evidence summarized above strongly supports the role of the 3' IR in stabilizing chloroplast mRNAs, a discordant note has recently been heard (Blowers et al., 1993). Chimeric constructs with the *atpB* promoter fused to the *E. coli uidA* gene encoding β-glucuronidase (GUS) and followed by the 3' *rbcL* or *psaB* flanking regions were transformed into the *Chlamydomonas* chloroplast. Transformants with the 3'-flanking regions in normal, forward orientation accumulated GUS transcripts of a single size, but in transformants lacking the 3' IR the GUS transcripts proved to be heterogeneous in size. Moreover, decay rates of GUS transcripts were found to be independent of the presence of a 3' IR. Hence, the results of these experiments are at odds with those described above in suggesting that the 3' IR *is* important in defining the 3' terminus of transcript, but *is not* required for mRNA stabilization. Experiments with yet other *Chlamydomonas* chloroplast transformants indicate that 5' sequences are also important determinants of stability in this alga. When Sakamoto *et al.* (1993) analyzed transformants in which the 5' untranslated region of the *petD* gene was fused to a GUS reporter, they found that deletion of ca. 70% of the 5' untranslated region caused declines in both mRNA stability and GUS synthesis. Salvador *et al.* (1993) reported that the first 63 nts at the 5' end of the *C. reinhardtii rbcL* gene fused to the GUS gene rendered the resulting mRNA susceptible to rapid degradation in the light. Addition of the adjacent 252 nts from the coding region of the *rbcL* gene stabilizes the mRNA. Salvador *et al.* (1993) also found that rates of degradation of a reporter gene transcript were the same whether or not it was terminated by 3' flanking sequences of the *Chlamydomonas* chloroplast *rbcL* or *psbA* genes in accord with the findings of Blowers *et al.* (1993). They speculate that the 3' IR elements of these chloroplast genes may not function in mRNA stabilization *in vivo,* but rather act as RNA processing or transcription termination sites. At this point a prudent conclusion would seem to be that both the 3'- and 5'-flanking regions of chloroplast mRNAs may contain sequences important for mRNA stabilization, but that further work will have to be done before definitive conclusions are drawn. The experiments of Blowers *et al.* (1993) also raise the possibility that the 3' IRs downstream of chloroplast genes may, after all, be important in transcript termination.

Most chloroplast genes are cotranscribed to yield polycistronic transcripts and Sugiura (1991) succinctly reviewed the steps involved in processing such transcripts. The primary transcripts are extensively processed to yield complex sets of overlapping transcription products. Some of these are monocistronic, but others are not. Cutting of chloroplast mRNA precursors generally occurs in the 5'-untranslated and intercistronic regions and multiple 5'-ends are found for certain transcripts (e.g., the *rbcL* and *atpB* genes of land plants, Mullet *et al.,* 1985). *In vitro* capping experiments have established that the longest RNAs possess 5'-triphosphates and represent the primary transcripts while the shorter RNAs are processed forms of these primary transcripts.

Two examples will illustrate the complexity of transcript processing patterns encountered for polycistronic mRNAs from flowering plant chloroplasts. The *psbB* operon includes in order the genes encoding the 47-kDa chlorophyll *b* apoprotein *(psbB),* the 10-kDa phosphoprotein of PSII *(psbH),* cytochrome b_6 *(petB),* and subunit IV of the cytochrome *b/f* complex *(petD)* in that order (Fig. 12–3A). Westhoff and Herrmann (1988) found that in spinach 18 major RNA species were derived from this operon. The final products of the processing pathway were monocistronic mRNAs for the *psbB* and *psbH* genes and a dicistronic mRNA for the *petB* and *petD* genes. Single introns also had to be removed from the mRNAs encoded by each of the latter two genes, but the sequence of intron removal was not fixed. Similarly, creation of the *psbB* and *psbH* monocistronic transcripts could proceed in any of several ways (Fig. 12–3A). Several cleavage sites have been identified and comparison of DNA sequences in the

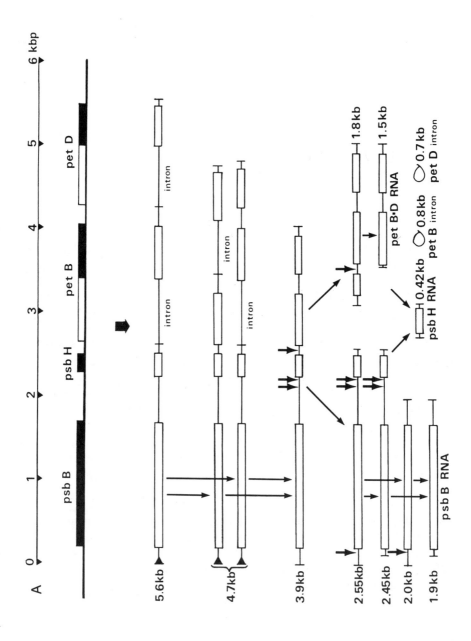

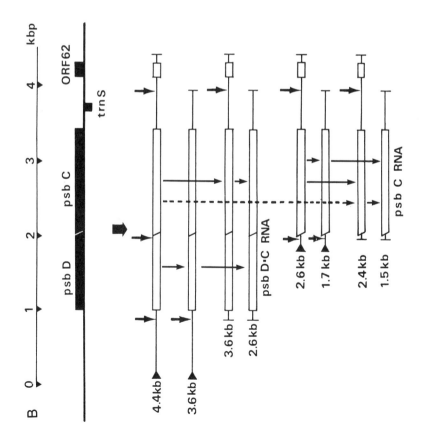

Fig. 12-3. Processing of polycistronic chloroplast mRNAs. (A) Cutting and splicing of the primary transcript of the *psbB* operon of spinach. (B) Processing the primary transcript of the *psbD/C* operon of tobacco. Triangles indicate transcription start sites and bold arrows mark cut sites. From Sugiura (1991).

vicinity of these sites revealed two conserved hexanucleotide motifs occurring in tandem. The cleaved sites were located in the first (before *petB*) or both (before *psbH*) hexanucleotide sequences. These sequence elements are conserved in other plants suggesting the involvement of a sequence-specific endonuclease that makes cuts at these hexanucleotide motifs.

Logic would dictate that only the processed transcripts would be associated with chloroplast polysomes, but Barkan (1988) found that chloroplast translation systems do not recognize such logic. The *psbB* operon in maize also yields a complex set of 20 transcripts virtually all of which cosediment with chloroplast polysomes. Barkan observed that all transcripts containing spliced *petB* or *petD* sequences were translated to yield these proteins regardless of upstream or downstream sequences and that the uninterrupted *psbB* gene was translated from all transcripts encoding it. Therefore, intercistronic processing does not seem to be required for translation of these RNAs although certain processing steps may enhance translational efficiency.

Transcripts from the *psbDC* operon of tobacco have also been analyzed (Yao et al., 1989). This operon contains in linear sequence the overlapping genes encoding the D2 *(psbD)* and 43-kDa *(psbC)* proteins of photosystem II plus ORF62 (Fig. 12–3B). Eight major RNA species were detected, four of which contained the *psbD* and *psbC* sequences with the remaining four possessing the *psbC* sequence with or without the downstream ORF62. In this operon transcription can initiate either upstream of the *psbD* gene or within the gene. Thus, two of the *psbC* mRNAs result from transcription starts within the *psbD* gene while the other two are products of the processing of the longer *psbD*-containing transcript. No conserved sequences were detected near or at the cleavage sites suggesting the existence of gene-specific endonucleases. Two other aspects of RNA processing, intron splicing and RNA editing, are discussed respectively in Chapters 10 and 14 while the regulation of transcript abundance is considered in Chapter 16.

REFERENCES

Agsteribbe, E., and M. Hartog (1987). Processing of precursor RNAs from mitochondria of *Neurospora crassa*. *Nucl. Acids Res.* 15: 7249–7263.

Altman, S., Kirsebom, L., and S. Talbot (1993). Recent studies of ribonuclease P. *FASEB J.* 7: 7–14.

Attardi, G. (1985). Animal mitochondrial DNA: An extreme example of genetic economy. *Int. Rev. Cytol.* 93: 93–145.

Attardi, G., and G. Schatz (1988). Biogenesis of mitochondria. *Annu. Rev. Cell Biol.* 4: 289–333.

Barkan, A. (1988). Proteins encoded by a complex chloroplast transcription unit are each translated from both monocistronic and polycistronic mRNAs. *EMBO J.* 7: 2637–2644.

Bendich, A.J. (1985). Plant mitochondrial DNA: unusual variation on a common theme. In *Genetic Flux in Plants* (B. Hohn and E.S. Dennis, eds.), pp. 111–138. Springer-Verlag, Vienna.

Binder, S., and A. Brennicke (1993). Transcription initiation sites in mitochondria of *Oenothera berteriana*. *J. Biol. Chem.* 268 7849–7855.

Blowers, A.D., Ellmore, G.S., Klein, U., and L. Bogorad (1990). Transcriptional analysis of endogenous and foreign genes in chloroplast transformants of *Chlamydomonas reinhardtii*. *Plant Cell* 2: 1059–1070.

Blowers, A.D., Klein, U., Ellmore, G.S., and L. Bogorad (1993). Functional *in vivo* analyses of the 3′ flanking sequences of the *Chlamydomonas* chloroplast *rbcL* and *psaB* genes. *Mol. Gen. Genet.* 238 339–349.

Bogorad, L. (1991). Replication and transcription of plastid DNA. In *The Molecular Biology of Plastids* (L. Bogorad and I.K. Vasil, eds.), pp. 93–124. Academic Press, San Diego.

Breitenberger, C.A., Browning, K.S., Alzner-DeWeerd, B., and U.L. RajBhandary (1985). RNA processing in *Neurospora crassa* mitochondria: Use of transfer RNA sequences as signals. *EMBO J.* 4: 185–196.

Bulow, S., and G. Link (1988). Sigma-like activity from mustard (*Sinapis alba* L.) chloroplasts conferring DNA-binding and transcription specificity to *E. coli* core RNA polymerase. *Plant Mol. Biol.* 10: 349–357.

Cantatore, P., and C. Saccone (1987). Organization, structure, and evolution of mammalian mitochondrial genes. *Int. Rev. Cytol.* 108: 149–208.

Christianson, T., and M. Rabinowitz (1983). Identification of multiple transcriptional initiation sites on the yeast mitochondrial genome by in vitro capping with guanylyltransferase. *J. Biol. Chem.* 258: 14025–14033.

Clary, D.O., and D.R. Wolstenholme (1984). The *Drosophila* mitochondrial genome. In *Oxford Surveys on Eukaryotic Genes*, Vol. 1 (N. Maclean, ed.), pp. 1–35. Oxford University Press, New York.

Clayton, D.A. (1984). Transcription of the mammalian mitochondrial genome. *Annu. Rev. Biochem.* 53: 573–594.

Clayton, D.A. (1991a). Replication and transcription of vertebrate mitochondrial DNA. *Annu. Rev. Cell Biol.* 7: 453–478.

Clayton, D.A. (1991b). Nuclear gadgets in mitochondrial DNA replication and transcription. *Trends Biochem. Sci.* 16: 107–111.

Clayton, D.A. (1992). Transcription and replication of animal mitochondrial DNAs. *Int. Rev. Cytol.* 141: 217–232.

Doersen, C.-J., Guerrier-Takada, C., Altman, S., and G. Attardi (1985). Characterization of an RNase P activity from HeLa cell mitochondria: comparison with the cytosol RNase P activity. *J. Biol. Chem.* 260: 5942–5949.

Gillham, N.W. (1978). *Organelle Heredity*. Raven Press, New York.

Gray, M.W., Hanic-Joyce, P.J., and P.S. Covello (1992). Transcription, processing and editing in plant mitochondria. *Annu. Rev. Plant Physiol. Plant Mol. Biol.* 43: 145–175.

Grivell, L.A. (1989). Nucleo-mitochondrial interactions in yeast mitochondrial biogenesis. *Eur. J. Biochem.* 182: 477–493.

Gruissem, W. (1989). Chloroplast RNA: Transcription and processing. In *Biochemistry of Plants*, Vol. 15 (A. Marcus, ed.), pp. 151–191. Academic Press, San Diego.

Gruissem, W., and G. Schuster (1993). Control of mRNA degradation in organelles. In *Control of Messenger RNA Stability* (Brawerman G. and J. Belasco, eds.), pp. 329–365. Academic Press, Orlando, FL.

Gruissem, W., and J.C. Tonkyn (1993). Control mechanisms of plastid gene expression. *CRC Rev. Plant Sci.* 12: 19–55.

Hess, J.F., Parisi, M.A., Bennett, J.L., and D.A. Clayton (1991). Impairment of mitochondrial transcription termination by a point mutation associated with the MELAS subgroup of mitochondrial encephalomyopathies. *Nature (London)* 351: 236–239.

Hess, W.R., Prombona, A., Fieder, B., Subramanian, A.R., and T. Borner (1993). Chloroplast *rps*15 and the *rpo*B/C1/C2 gene cluster are strongly transcribed in ribosome-deficient plastids: Evidence for a functioning non-chloroplast-encoded RNA polymerase. *EMBO J.* 12: 563–571.

Hsu-Ching, C., and D.B. Stern (1991). Specific binding of chloroplast proteins in vitro to the 3' untranslated region of spinach chloroplast *petD* mRNA. *Mol. Cell. Biol.* 11: 4380–4388.

Igloi, G.L., and H. Kossel (1992). The transcriptional apparatus of chloroplasts. *Crit. Rev. Plant Sci.* 10: 525–558.

Jaehning, J. (1993). Mitochondrial transcriptions: Is a pattern emerging? *Mol. Microbiol.* 8: 1–4.

Jolly, S.O., and L. Bogorad (1980). Preferential transcription of cloned maize chloroplast DNA sequences by maize chloroplast RNA polymerase. *Proc. Natl. Acad. Sci. U.S.A.* 77: 822–826.

Kirk, J.T.O. (1964a). DNA-dependent RNA synthesis in chloroplast preparations. *Biochem. Biophys. Res. Commun.* 14: 393–397.

Kirk, J.T.O. (1964b). Studies on RNA synthesis in chloroplast preparations. *Biochem. Biophys. Res. Commun.* 16: 233–238.

Klein, U., De Camp, J.D., and L. Bogorad (1992). Two types of chloroplast gene promoters in *Chlamydomonas reinhardtii*. *Proc. Natl. Acad. Sci. U.S.A.* 89: 3453–3457.

Kruse, B., Narasimhan, N., and G. Attardi (1989). Termination of transcription in human mitochondria: Identification and purification of a DNA binding protein factor that promotes termination. *Cell* 58: 391–397.

Kuntzel, H., and H. Noll (1967). Mitochondrial and cytoplasmic polysomes from *Neurospora crassa*. *Nature (London)* 215: 1340–1345.

Lang, B.F., Ahne, F., Distler, S., Trinkl, H., Kaudewitz, F., and K. Wolf (1983). Sequence of the mitochondrial DNA, arrangement of genes and processing of their transcripts in *Schizosaccharomyces pombe*. In *Mitochondria 1983* (R.J. Schweyen, K. Wolf, and F. Kaudewitz, eds.), pp. 313–329. De Gruyter, Berlin.

Lerbs, S., Brautigam, E., and R. Mache (1988). DNA-dependent RNA polymerase of spinach chloroplasts: characterization of α-like and σ-like polypeptides. *Mol. Gen. Genet.* 211: 459–464.

Lerbs-Mache, S. (1993). The 110-kDa polypeptide of spinach plastid DNA-dependent RNA polymerase: single subunit enzyme or catalytic core of multimeric enzyme complexes? *Proc. Natl. Acad. Sci. USA* 90: 5509–5513.

Li, Y., and M. Sugiura (1990). Three distinct ribonucleoproteins from tobacco chloroplasts: Each contains a unique amino terminal acidic domain and two ribonucleoprotein consensus motifs. *EMBO J.* 9: 3059–3066.

Lonsdale, D.M. (1989). The plant mitochondrial genome. In *The Biochemistry of Plants*, Vol. 15 (A. Marcus, ed.), pp. 229–295. Academic Press, San Diego.

Luck, D.J.L., and E. Reich (1964). DNA in mitochondria of *Neurospora crassa*. *Proc. Natl. Acad. Sci. U.S.A.* 52: 931–938.

Lyttleton, J.W. (1962). Isolation of ribosomes from spinach chloroplasts. *Exp. Cell Res.* 26: 312–317.

Makaroff, C.A., and J.D. Palmer (1987). Extensive mitochondrial specific transcription of the *Brassica campestris* mitochondrial genome. *Nucl. Acids Res.* 15: 5141–5156.

Masters, B.S., Stohl, L.L., and D.A. Clayton (1987). Yeast mitochondrial RNA polymerase is homologous to those encoded by bacteriophages T3 and T7. *Cell* 51: 89–99.

Mullet, J.E., Orozco, E.M. Jr., and N.-H. Chua (1985). Multiple transcripts for higher plant *rbcL* and *atpB* genes and localization of the transcription initiation site of the *rbcL* gene. *Plant Mol. Biol.* 4: 39–54.

Mulligan, R.M., Leon, P., and V. Walbot (1991). Transcriptional and post-transcriptional regulation of maize mitochondrial gene expression. *Nucl. Acids Res.* 11: 533–543.

Nelson, B.D. (1987). Biogenesis of mammalian mitochondria. *Curr. Topics in Bioenergetics* 15: 221–272.

Newton, K.J. (1988). Plant mitochondrial genomes: Organization, expression and variation. *Annu. Rev. Plant Physiol. Plant Mol. Biol.* 39: 503–532.

Nickelsen, J., and G. Link (1993). The 54 kDa RNA-binding protein from mustard chloroplasts mediates endonucleolytic transcript 3' end formation *in vivo. Plant J.* 3: 537–544.

O'Brien, T.W., and G.F. Kalf (1967a). Ribosomes from rat liver mitochondria. I. Isolation procedure and contamination studies. *J. Biol. Chem.* 242: 2172–2179.

O'Brien, T.W., and G.F. Kalf (1967b). Ribosomes from rat liver mitochondria. II. Partial characterization. *J. Biol. Chem.* 242: 2180–2185.

Okimoto, R., MacFarlane, J.L., Clary, D.O., and D.R. Wolstenholme (1992). The mitochondrial genomes of two nematodes, *Caenorhabdities elegans* and *Ascaris suum. Genetics* 130: 471–498.

Osinga, K.A., De Vries, E., Van der Horst, G., and H.F. Tabak (1984). Processing of yeast mitochondrial messenger RNAs at conserved dodecamer sequence. *EMBO J.* 3: 829–834.

Parisi, M.A., Xu, B., and D.A. Clayton (1993). A human mitochondrial transcription activator can functionally replace a yeast mitochondrial HMG-Box protein both in *in vivo* and *in vitro. Mol. Cell. Biol.* 13: 1951–1961.

Rapp, W.D., and D.B. Stern (1992). A conserved eleven nucleotide sequence contains an essential promoter element of the maize mitochondrial *atp1* gene. *EMBO J.* 11: 1065–1073.

Sakamoto, W., Kindle, L.L., and D.B Stern (1993). *In vivo* analysis of *Chlamydomonas* chloroplast *petD* gene expression using stable transformation of β-glucuronidase translational fusions. *Proc. Natl. Acad. Sci. USA* 90: 497–501.

Salvador, M.L., Klein, U., and L. Bogorad (1993). 5' sequences are important positive and negative determinants of the longevity of *Chlamydomonas* chloroplast gene transcripts. *Proc. Natl. Acad. Sci. USA* 90: 1556–1560.

Schinkel, A.H., and H.F. Tabak (1990). Mitochondrial RNA polymerase: Dual role in transcription and replication. *Trends Genet.* 5: 149–154.

Shadel, G.S., and D.A. Clayton (1993). Mitochondrial transcription initiation. *J. Biol. Chem.* 268: 16083–16086.

Steinmetz, A., Krebbers, E.T., Schwartz, Z., Gubbins, E.J., and L. Bogorad (1983). Nucleotide sequences of five maize chloroplast transfer RNAs and their flanking regions. *J. Biol. Chem.* 258: 5503–5511.

Stern, D.B., and W. Gruissem (1987). Control of plastid gene expression: 3' inverted repeats act as mRNA processing and stabilizing elements, but do not terminate transcription. *Cell* 51: 1145–1157.

Stern, D.B., Jones, H., and W. Gruissem (1989). Function of plastid mRNA 3' inverted repeats: RNA stabilization and gene-specific protein binding. *J. Biol. Chem.* 264: 18742–18750.

Stern, D.B., Radwanski, E.R., and K.L. Kindle (1991). A 3' stem/loop structure of the *Chlamydomonas* chloroplast *atpB* gene regulates mRNA accumulation in vivo. *Plant Cell* 3: 285–297.

Sugiura, M. (1991). Transcript processing in plastids: Trimming, cutting, splicing. In *The Molecular Biology of Plastids* (L. Bogorad and I.K. Vasil, eds.), pp. 125–137. Academic Press, San Diego.

Surzycki, S.J., and D.L. Schellenbarger (1976). Purification and characterization of a putative sigma factor from *Chlamydomonas reinhardi. Proc. Natl. Acad. Sci. U.S.A.* 73: 3961–3965.

Tiller, K., Eisermann, A., and G. Link (1991). The chloroplast transcription apparatus from mustard (*Sinapis alba* L.): Evidence for three different transcription factors which resemble bacterial σ factors. *Eur. J. Biochem.* 198: 93–99.

Westhoff, P., and R.G. Herrmann (1988). Complex RNA maturation in chloroplasts: The *psb*B operon from spinach. *Eur. J. Biochem.* 171: 551–564.

Wolf, K., and L. Del Giudice (1988). The variable mitochondrial genome of ascomycetes: Organization, mutational alterations, and expression. *Adv. in Genet.* 25: 185–308.

Wolfe, K.H., Morden, C.W., and J.D. Palmer (1992a). Function and evolution of a minimal plastid genome from a nonphotosynthetic parasitic plant. *Proc. Natl. Acad. Sci. U.S.A.* 89: 10648–10652.

Wolfe, K.H., Morden, C.W., Ems, S.C., and J.D. Palmer (1992b). Rapid evolution of the plastid translational apparatus in a nonphotosynthetic plant: Loss or accelerated sequence evolution of tRNA and ribosomal protein genes. *J. Mol. Evol.* 35: 304–317.

Wolstenholme, D.R. (1992). Animal mitochondrial DNA: Structure and evolution. *Int. Rev. Cytol.* 141: 173–216.

Xu, B., and D.A. Clayton (1992). Assignment of a yeast protein necessary for mitochondrial transcription initiation. *Nucl. Acids Res.* 20: 1053–1059.

Yao, W.B., Meng, B.Y., Tanaka, M., and M. Sugiura

(1989). An additional promoter within the protein-coding region of the *psbD-psbC* gene cluster in tobacco chloroplast DNA. *Nucl. Acids Res.* 17: 9583–9591.

Ye, L., Li, Y., Fukami-Kobayashi, K., Go, M., Konishi, T., Watanabe, A., and M. Sugiura (1991). Diversity of a ribonucleoprotein family in tobacco chloroplasts: Two new chloroplast ribonucleoproteins and a phylogenetic tree of ten chloroplast RNA-binding domains. *Nucl. Acids Res.* 19: 6485–6490.

Yin, S., Heckman, J., and U.L. RajBhandary (1981). Highly conserved GC rich palindromic DNA sequences which flank the tRNA in *Neurospora crassa* mitochondria. *Cell* 26: 326–332.

13

Protein Synthesis in Organelles

The steps involved in protein synthesis in prokaryotes have been reviewed by Hershey (1987) and serve as a useful context in which to discuss this process in organelles.

Initiation is the first step in the process of peptide bond formation. In prokaryotes a preinitiation complex made up of the 30S ribosomal subunit and tRNA$^{\text{f-Met}}_{\text{UAC}}$ is formed with the 30S subunit binding to the purine-rich Shine–Dalgarno sequence 7 ± 2 nucleotides upstream of the initiator AUG (Kozak, 1983). The canonical S-D sequence, GGAGG, or a variant, pairs with a pyrimidine-rich complementary sequence, the anti-Shine–Dalgarno (AS-D) sequence near the 3'-end of the 16S rRNA molecule. Addition of a 50S ribosomal subunit converts the preinitiation complex to an initiation complex that can enter the elongation phase of protein synthesis. These reactions are promoted by the three initiation factors, IF1, IF2, and IF3. IF2 is involved in initiator tRNA binding and GTP hydrolysis while IF3 prevents ribosomal subunit association in the absence of mRNA and is required for binding natural mRNAs. IF1 enhances the rate of ribosome dissociation and association and the activities of the other initiation factors.

A DNA sequence with homology to the bacterial *infA* gene encoding bacterial IF1 has been identified in the chloroplast genomes of land plants including the colorless parasite *Epifagus* (Shimada and Sugiura, 1991; Wolfe *et al.*, 1992). However, the tobacco *infA* gene in contrast to spinach, for example, lacks the ATG translation initiation codon and thus may be a pseudogene (Shimada and Sugiura, 1991). The *infA* gene is absent from the completely sequenced chloroplast genome of *Euglena* (Hallick *et al.*, 1993). Reading frames with homology to the genes encoding IF-2 and IF-3 have not been detected in the three sequenced chloroplast genomes of flowering plants or in *Epifagus*. In *Euglena* biochemical experiments have led to the identification of initiation factors IF-2$_{\text{chl}}$ and IF-3$_{\text{chl}}$ (Roney *et al.*, 1991), and inhibitor experiments indicate that the genes specifying these factors are nuclear in location. This is consistent with the absence of reading frames corresponding to *infB* and *infC* in the chloroplast genome of *Euglena* (Hallick *et al.*, 1993). However, a reading frame corresponding to *infB* has been reported from the chloroplast genome of the red alga *Porphyra* (Reith and Munholland, 1993).

IF-2$_{\text{chl}}$ of *Euglena* contains subunits ranging in size from 97 to >200 kDa and is present in three distinct forms ranging in molecular mass from 200 to 700–800 kDa

(Ma and Spremulli, 1992). This factor is required for binding of tRNA$^{F\text{-met}}$ to chloroplast 30S subunits so it is functionally related to prokaryotic IF-2 (Roney et al., 1991). IF-3_{chl} promotes initiation complex formation in the presence of IF-2_{chl}. Although IF-3_{chl} will replace *E. coli* IF-3 in initiation complex formation, it may not do so in a manner equivalent to that observed with the prokaryotic factor. Genes specifying initiation factors have not been found in any of the completely sequenced mitochondrial genomes (Chapter 3), but a nuclear gene *(MTF2)* encoding the mitochondrial IF2 has been identified in yeast (Tzagoloff and Dieckmann, 1990) and the factor itself has been isolated from bovine liver and proves to be a monomeric 58-kDa protein (Liao and Spremulli, 1991).

Many land plant chloroplast genes possess an upstream sequence similar to the bacterial S-D sequence that is capable of binding a sequence near the 3'-end of the chloroplast 16S rRNA (Bonham-Smith and Bourque, 1989; Ruf and Kossel, 1988; Steinmetz and Weil, 1989; Zurawski and Clegg, 1987). Ruf and Kossel (1988) report that 37 of 41 chloroplast genes examined in tobacco have such sequences if one extends the AS-D sequence in the 16S rRNA beyond the canonical CCUCC sequence to include the adjacent unpaired ACUAG sequence. Bonham-Smith and Bourque (1989) examined 196 chloroplast-encoded transcripts and observed that 92% possessed an S-D sequence within 100 bp 5' to the initiation codon. However, spacing of S-D sequences is less uniform than in bacteria. Frequency distributions of individual positions potentially involved in base pairing with 16S rRNA ranged from -2 to -29, with a major peak at -7 to -8, a second smaller peak at -15 to -16, and a third small peak at -21 to -23. Thus, chloroplast ribosomes may be able to accommodate larger distances between the ribosome recognition and translational start sites than bacterial ribosomes. Those mRNAs lacking S-D sequences may either contain out-of-frame initiator codons between the potential recognition sites and the respective in-frame start codons (e.g., *rps16*, *rpoB*, and *petD* in tobacco) or make use of highly conserved sequences just distal to the initiation codon (e.g., *atpB*) (Ruf and Kossel, 1988; Zurawski and Clegg, 1987).

In the *Euglena* chloroplast recognition seems to proceed by somewhat different rules. The putative AS-D sequence CUCCC differs from the canonical CCUCC sequence and actually forms the 3'-terminus of the 16S rRNA rather than being located several bases from the end (Steege et al., 1982). Since only about half of the *Euglena* chloroplast mRNAs contain S-D sequences, a model has been proposed which postulates that there are two modes of initiation complex formation (Roney et al., 1991; Wang et al., 1989). In one class of mRNAs, complex formation is facilitated by an S-D like sequence. However, in the second class the A/U content of the region 5' to the initiator AUG is 90% or greater and this portion of the mRNA is relatively unstructured so potential start sites in this region may be readily accessible to small subunits.

Mammalian mitochondrial mRNAs lack both an S-D sequence, required for prokaryotic ribosome binding, and the m^{7}G(5')ppp(5')N cap, which participates in binding of eukaryotic ribosomes. These mRNAs start very near the AUG, AUA, or AUU initiation codons (Cantatore and Saccone, 1987). However, a complementarity region does exist between a 3' proximal region of the small subunit rRNA and different positions of several mRNAs which might act as a variant of the S-D recognition system. Experiments with purified mammalian mitochondrial ribosomal subunits and mRNAs suggest that mitochondrial initiation factors are necessary for proper recognition and melting of the secondary structure at the 5'-end of mitochondrial mRNAs. Pietromonaco *et al.*

(1991), speculating on why mammalian mitochondrial ribosomes have so many r-proteins, raise the possibility that these "extra" proteins may be playing structural or functional roles associated with rRNA in the bacterial ribosome. They also review experiments showing that r-protein S5 has a GTP-binding site suggesting that this protein may participate directly in the formation of initiation complexes. Since neither bacterial nor eukaryotic ribosomes bind GTP, this could represent a deviation in the mechanism of protein synthesis unique to mammalian mitochondria and perhaps other animals.

The S-D sequence is also unrelated to the mature 3'-end of mitochondrial 18S rRNA from wheat (Schnare and Gray, 1982). However, analysis of sequences immediately preceding the proposed AUG initiation codon for the maize *cytb* gene revealed a limited homology to the 3'-end of the 18S rRNA (see Lonsdale, 1989). Similar sequences are found at -10 to -30 upstream of other, but not all, plant mitochondrial genes, but complementarity is often less than observed for *E. coli* genes and the position is variable (Hanson and Folkerts, 1992). Hence, ribosome binding to land plant mitochondrial mRNAs may occur by a yet to be identified mechanism.

Costanzo and Fox (1990) and Grivell (1989) have summarized the existing information on translation initiation in yeast mitochondria. Most major mRNAs have 5' leaders and 3' trailers that are A + U rich and very long. The leader of the *cytb* mRNA, for example, is 954 nucleotides long and the trailer 108 nucleotides. All of these long leaders have at least one AUG upstream of the initiator AUG codon. The only exception is *cox2* mRNA whose leader is short (54 bases) and lacks upstream AUGs. Yeast mitochondrial mRNAs are cappable indicating that they are uncapped. The small rRNA lacks the equivalent of an AS-D sequence although sites in 5' leaders complementary to the 3'-end of the small rRNA have been noted. Their positions vary between -8 and -107 relative to the initiator codon. Since a chimeric mRNA lacking any of these sites was translated efficiently (Costanzo and Fox, 1990), their significance is open to question. As in many eukaryotic cytoplasmic mRNAs, the 5' leader sometimes contains one or more short reading frames that may possibly encode proteins mediating translational control of downstream coding sequences. Proteins interacting with the 5' leader regions of different yeast mitochondrial mRNAs play prominent roles in translational regulation in this organism, a topic discussed in more detail in Chapter 17. In summary, ribosome binding to mitochondrial mRNAs remains a murky area of research.

Elongation consists of three steps (see Hershey, 1987): aminoacyl-tRNA binding, peptide bond formation, and translocation. Until recently the ribosome was thought to have only two tRNA binding sites (A and P sites), but a third tRNA binding site has now been identified (the E site) (see Rheinberger *et al.,* 1990). The aminoacylated tRNA combines with elongation factor EF-Tu and GTP to form a ternary complex that then associates with a ribosome complexed to mRNA and peptidyl-tRNA. The specific ternary complex is selected on the basis of codon–anticodon recognition at the A site following GTP hydrolysis with release of EF-Tu · GDP. Peptide bond formation then takes place with the transfer of the growing peptide chain to the aminoacyl-tRNA in the A site. Translocation is promoted by EF-G and GTP hydrolysis with the movement of the peptidyl-tRNA–mRNA complex from the A to the P site. The process is then repeated and the deacylated tRNA moves from the P to the E site. The A and E sites themselves are allosterically linked in a negative sense so that occupation of the A site by aminoacylated tRNA reduces the affinity of the E site for deacylated tRNA and vice versa. Regeneration of the active EF-Tu · GTP complex from EF-Tu · GDP is mediated by elongation factor EF-Ts.

All three elongation factors have been characterized from *Euglena* chloroplasts (Breitenberger et al., 1979; Breitenberger and Spremulli, 1980; Fox et al., 1980; Sreedharan et al., 1985). Inhibitor experiments indicated that EF-Ts and EF-G were nuclear gene products, but that EF-Tu might be encoded in the chloroplast. The latter assignment was confirmed by Montandon and Stutz (1983) who showed that the *Euglena tufA* gene was split into three exons separated by two introns. The *tufA* gene is also found in the chloroplast genome of *Chlamydomonas* and certain other green algae, but this gene is nuclear in other green algal groups and land plants (see Chapter 4).

Consistent with the sites of synthesis experiments of Spremulli and colleagues, the other two *Euglena* elongation factors must be of nuclear origin since reading frames corresponding to them are absent from the completely sequenced chloroplast genome of *Euglena* (Hallick et al., 1993). The same is true in land plants (Palmer, 1991; Shimada and Sugiura, 1991), but Reith and Munholland (1993) find that the chloroplast genome of *Porphyra* contains a reading frame corresponding to the *tsf* gene, which encodes elongation factor EF-Ts in prokaryotes. The nuclear genes encoding mitochondrial elongation factors EF-G and EF-Tu have been identified in yeast (Costanzo and Fox, 1990). No elongation factor genes have been reported in any of the completely sequenced mitochondrial genomes (Chapter 3).

Termination in bacteria involves the hydrolysis of peptidyl-tRNA and release of the completed protein from the ribosome when the ribosome reaches one of the three termination codons. Termination requires the action of two release factors, RF1, which is specific for UAA and UAG, and RF-2, which is required for UAA and UGA. A third release factor, RF-3, stimulates the activities of the other two factors. The same three codons are used for translation termination in chloroplasts (Chapter 14), but no reading frame with homology to the genes encoding bacterial termination factors has been identified in the completely sequenced chloroplast genomes (Chapter 3) nor has the isolation of these factors from chloroplasts been reported. Pel et al. (1992) report isolation of a mitochondrial release factor (mRF-1) from yeast and the cloning of this gene. mRF1 is probably equivalent to a bacterial RF1 type release factor recognizing both UAA and UAG codons. Since UGA specifies tryptophan in yeast mitochondria (Chapter 14), an RF2 type release factor would not be required to facilitate termination at this codon.

COMPONENTS OF THE TRANSLATIONAL APPARATUS

Ribosomes

Marked similarities exist between chloroplast and prokaryotic ribosomes in terms of sedimentation velocity, proteins per subunit (Table 13–1) and particularly with respect to rRNA structure and sequence (Chapter 2). About one-third of the chloroplast ribosomal (r-) proteins are coded in the chloroplast genome of land plants and substantially more in certain algae (e.g., *Cyanophora*). Most of these r-proteins have clearly identifiable homologs in in *E. coli*. In contrast, mitochondrial ribosomes differ substantially not only from prokaryotic and eukaryotic ribosomes, but from each other. For instance, all mitochondrial ribosomes except those from land plants lack the equivalent of 5S

Table 13–1. Sedimentation (S) Velocity and Protein Composition of Prokaryotic, Eukaryotic, and Organelle Ribosomes

Ribosome Source	S Value			Number of Proteins	
	Small Subunit	Large Subunit	Intact Ribosome	Small Subunit	Large Subunit
E. coli	30	50	70	21	34
Chloroplasts					
Flowering plants	30	50	70	22–26	35–39
Chlamydomonas	37–41	50	70	31	33
Mitochondria					
Saccharomyces	37	50	74	33	38
Neurospora	37	50	73	23	30
Tetrahymena	55	55	80	?	?
Maize	44	60	78	?	?
Xenopus	32	43	60	32	43
Bovine	28	39	55	33	52
Eukaryotes					
Saccharomyces	40	60	80	34	48
Rat	40	60	80	31	49

Data compiled from Gillham (1978), Kitakawa and Isono (1991), Kozak (1983), Subramanian *et al.* (1991), and Tzagoloff (1982).

rRNA. As mentioned earlier, mammalian mitochondrial ribosomes are extremely protein rich (Table 13–1). The mitochondrial rRNAs of some organisms (e.g., *Chlamydomonas,* Trypanosomes, *Plasmodium*) are minimal in structure. Mitochondrial ribosomal proteins in animals and fungi are encoded entirely, or nearly so, in the nuclear genome. However, sequencing of mitochondrial genomes from plants and protists has revealed the presence of genes encoding ribosomal proteins with sequence similarity to *E. coli* ribosomal proteins.

rRNA. Delp and Kossel (1991) have reviewed the structure, organization, and expression of chloroplast rRNA genes. As discussed in Chapter 3, land plants and algae with inverted repeats contain two sets of rRNA genes as these genes are always located in the inverted repeat. Exceptions are conifers and a few angiosperms (e.g., some leguminoseae), which lack the inverted repeat and possess one set of rRNA genes. In *Euglena* the rRNA genes are arranged in tandem repeats fluctuating from one to five in number.

Chloroplast rRNA genes are arranged in operons reminiscent of those seen in *E. coli* and cyanobacteria in the order 16S, 23S, 5S (Fig. 13–1). Within the spacer between the 16S and 23S genes are genes encoding tRNA^{Ile} and tRNA^{Ala}. In land plants and related green algae, these tRNA genes have long group II introns inserted into them, but not in algae such as *Euglena, Chlorella,* and *Chlamydomonas* or in prokaryotes. Land plants have a tRNA^{Val} gene proximal to the 16S rDNA gene and a tRNA^{Arg} distal to the 5S rDNA gene. In flowering plants and *Marchantia* the 3′-terminus of the 23S rDNA is split off as a separate gene. The 95 nucleotide rRNA encoded by this gene is referred to as 4.5S rRNA. The 23S rRNA gene is continuous in a fern of the genus *Adiantum,* but another fern *Dryopteris acuminata* has a spacer and 4.5S rRNA typical

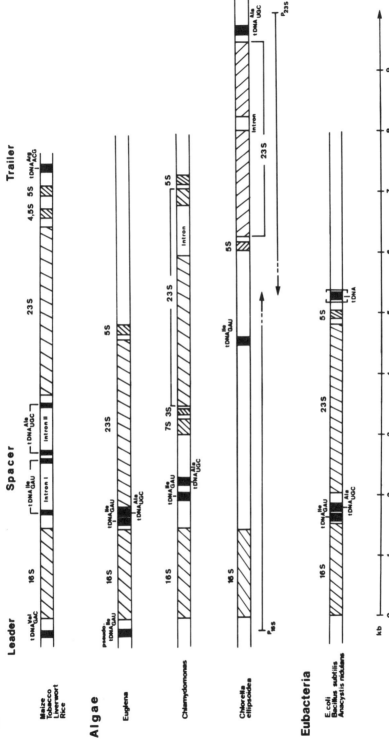

Fig. 13–1. Alignment of rRNA operons from higher plants, several algae, and eubacteria. Genes coding for tRNAs and rRNAs are marked by shading and black segments, respectively. In the case of *Chlorella ellipsoidea*, the two operons and their polarities are indicated by the long arrows starting and the respective promoter sites P_{16S} and P_{23S}. From Delp and Kossel (1991).

of flowering plants (see Bohnert et al., 1982). In contrast, in *Chlamydomonas reinhardtii* the 5'-end of the 23S rDNA is fragmented into two sequences that encode 7S and 3S rRNA. The 23S rDNA gene of this alga also contains a self-splicing group I intron (Chapter 10). The large subunit rRNA of *C. eugametos* comprises species (α and β) equivalent to the *C. reinhardtii* 7S and 3S RNAs plus two larger species (γ and δ) that together represent the remainder of the 23S rRNA molecule. The *C. eugametos* gene contains six group I introns, one of which catalyzes its own transposition (Chapter 10). The small rRNA species are part of the primary transcript of the rRNA operon in *Chlamydomonas* and are cleaved out posttranscriptionally. In *Chlorella* a 5-kb inversion has resulted in splitting of the rRNA operon into two parts oriented in opposition (Fig. 13–1). The first contains the 16S rDNA and tRNA^{Ile} genes and the second tRNA^{Ala}, the 23S rDNA and the 5S rDNA.

Whether the primary transcript contains the 5S rRNA sequence has been disputed. The apparent absence of 5S rRNA sequences from the primary transcript and the presence of promoter-like structures in the 4.5S/5S intergenic region of *Pelargonium* and spinach have been used to argue for the separate transcription of the 5S rDNA gene. However, Delp and Kossel (1991) point out that these sequences are not conserved in several other species and that the putative promoter sequences are inactive in homologous transcription systems. A full length intergenic RNA connecting the 4.5S and 5S sequences has also been found. Therefore, Delp and Kossel believe that the failure to detect 5S rRNA sequences in the primary transcript may be explained by rapid cleavage out of these sequences. Whether the primary transcript includes the tRNA^{Arg} distal to the 5S rDNA has not been settled conclusively, although Delp and Kossel cite arguments that favor its inclusion in the transcript.

Structural genes encoding the rRNAs show sequence similarity among land plants (>90%), somewhat less similarity between land plants and algae (80–90%), and a bit less similarity between plants and bacteria (65–77%). Secondary structural models are even more informative and they reveal striking similarities between chloroplast and eubacterial rRNAs (Table 13–2). These include sites of base modification, r-protein binding sites, regions of interaction between ribosomal subunits, a sequence at the 3'-end of the 16S rRNA that is complementary to the Shine–Dalgarno sequence, and a region of the 23S rRNA related to the peptidyl transferase center. Secondary structure models also permit the division of rRNAs into core and variable regions. For example, eight core regions can be identified in the small subunit RNA of prokaryotes, eukaryotes, and organelles (Fig. 2–2). When homologies in rRNA core regions are compared for chloroplast rRNA, the similarities to prokaryotic rRNAs are even more pronounced (Fig. 2–3). These secondary structural comparisons are of particular importance in deducing phylogenies of chloroplast and mitochondrial DNA (see Chapter 2).

Mitochondrial rRNAs are much more variable in size and secondary structure conservation (Table 3–2) than chloroplast rRNAs. For example, in the minimal small subunit mitochondrial rRNA of the *Trypanosoma brucei* kinetoplast, Benne (1985) finds that only three conserved elements can be identified with respect to *E. coli*. However, sequences flanking these regions can be folded into very similar structures in both rRNAs.

The organization of mitochondrial rRNAs is also variable. Vertebrate mitochondrial rDNA genes are close together, separated only by a tRNA gene and cotranscribed. They encode small rRNAs with short variable regions. The yeast and *Neurospora* mito-

Table 13-2. Secondary Structure Conservation in Organellar rRNAs[a]

		LSU rRNA		
Organelle	SSU rRNA	Complete	5'-half	3'-half
Chloroplast				
Maize	94	91	90	93
Mitochondrion				
Wheat	84	78	72	86
Paramecium	67	73	64	84
Yeast	66	60	53	70
Human	43	34	26	44
Drosophila	40	30	21	43

Modified slightly from Gray (1988).

[a] Percentage of *E. coli* 16S or 23S rRNA secondary structure that can be reconstructed as a precise replica using the sequence in question.

chondrial rDNA genes are larger, well separated on the genomic map, and not cotranscribed. The large subunit gene of *Neurospora* contains an intron as do ω[+] strains of *S. cerevisiae*. The latter intron promotes its own transfer to intronless rDNA genes in yeast (Chapter 10). In contrast, the rDNA genes of *Schizosaccharomyces pombe* are compact and close together like mitochondrial rDNA genes of metazoans. In the ciliate protozoan *Tetrahymena pyriformis,* the linear mitochondrial genome contains only one copy of the small rRNA gene, but two copies of the large rRNA genes, which are located in the subterminal inverted repeat (also see Chapter 5). All are separately transcribed. Gray (1988) remarks that this is still not the complete story since the 5'-terminal 200–300 nucleotides of both small and large subunit rRNA genes of *Tetrahymena* are encoded by additional modules called SSU α and LSU α. This is also true in *Paramecium* (Cummings, 1992). Finally, in land plants a 5S rDNA gene is present. This RNA species is not found in other mitochondrial ribosomes. As if this were not enough, the mitochondrial rRNA genes of *Chlamydomonas* (Boer and Gray, 1988) and *Plasmodium* (Feagin *et al.,* 1992) are fragmented and scrambled. Thus, in *Chlamydomonas* the small rRNA is encoded in four modules and the large rRNA in eight. In such cases, secondary structure is assumed to be responsible for bringing together the bits and pieces of transcribed rRNA molecules to form the intact molecules. Fragmentation has also been reported for the large cytoplasmic rRNAs of *Euglena* (14 rRNA species) and *Crithidia* (7 rRNA species) (Gray and Schnare, 1990).

Ribosomal proteins. Chloroplast r-proteins have been extensively characterized (see reviews by Mache, 1990; Subramanian *et al.,* 1990, 1991; Subramanian, 1993).

Subramanian *et al.* (1991) summarized the number of r-proteins per subunit for several land plants, *Chlamydomonas* and *Euglena* and arrive at an average of 59 total proteins per ribosome, which is slightly higher than the 54 recognized for *E. coli*. Of these 24 are in the small subunit and 35 are in the large subunit. Open reading frames accounting for about a third of these proteins are found in land plant chloroplast genomes and expression of seven of these putative genes has been directly demonstrated by isolating their protein products from chloroplast ribosomes of spinach and performing N-terminal sequencing (Schmidt *et al.,* 1992). Many chloroplast r-proteins with homology to *E. coli* r-proteins have been identified (Table 13–3). Similarities run from 25% (L23) amino acid sequence identity to 62% (L36). With respect to *E. coli*, polypeptide chain lengths also vary for most chloroplast r-proteins. While most differ by only a few amino acids, several (e.g., S18, L13, and L24) are considerably larger in plastids. The most conserved small subunit protein, S12, can be substituted *in vivo* in *E. coli* by its equivalent from *Chlamydomonas* (Liu et al., 1989a).

While all of the r-proteins coded in land plant chloroplasts are known because several complete genomic sequences exist (Chapter 3), identification of nuclear r-protein genes has depended on individual gene isolation. This accounts for the fact that many fewer nuclear genes than plastid genes are shown in Table 13–3. In all probability those r-proteins missing in Table 13–3 are specified by nuclear genes. Genes encoding other r-proteins not found in land plant chloroplast genomes have been identified in the chloroplast genomes of nongreen algae such as *Cyanophora* and *Porphyra* (Chapter 3). Many of the chloroplast r-proteins encoded by nuclear genes have a central core homologous to an *E. coli* r-protein surrounded by N- and C-terminal extensions with no significant homology to any bacterial r-protein (Lagrange et al., 1991). Based on amino acid sequence, Lagrange et al. (1991) argue that the N- and C-terminal extensions were

Table 13–3. Homology Relationships between *E. coli* and Plastid r-Proteins

r-Protein	Protein Sequence Identity (%)	Location of Gene
S2	39	Plastid
S3	38	Plastid
S4	39	Plastid
S7	43	Plastid
S8	38	Plastid
S11	48	Plastid
S12	66	Plastid
S14	39	Plastid
S15	34	Plastid
S16	32	Plastid
S18	32	Plastid
S19	42	Plastid
L2	43	Plastid
L9	30	Nucleus
L12	48	Nucleus
L13	54	Nucleus
L14	52	Plastid
L16	53	Plastid
L20	41	Plastid
L22	27	Plastid
L23	25	Plastid
L24	35	Nucleus
L33	35	Plastid
L35	41	Nucleus
L36	62	Plastid

Modified from Subramanian et al. (1991). The data are taken from maize except S15, S16, L2, L20, and L33, which are from rice.

added to chloroplast r-protein genes at the time of transfer of these genes from the chloroplast genome to the nucleus.

There are also certain nuclear-encoded chloroplast r-proteins with no similarity to *E. coli* r-proteins. Thus, nuclear genes specifying two proteins from the large subunit of pea, called PsCL18 and PsCL25, appear unrelated to any *E. coli* r-proteins (Gantt, 1988). In spinach a 26-kDa protein referred to as "CS-S5" (Zhou and Mache, 1989) or "PSrp-1" (Johnson *et al.*, 1990) is encoded by a nuclear gene and has no similarity to any *E. coli* r-proteins. Finally, some chloroplast r-proteins may have taken on the functions of several *E. coli* r-proteins. Thus, in spinach the *rpl22* gene, which is found in the chloroplast genomes of all angiosperms except legumes and *Epifagus* (Palmer, 1991; Wolfe *et al.*, 1992), encodes a protein in which the central region is homologous to other L22 proteins (Carol *et al.*, 1993). However, the chloroplast protein has N- and C-terminal extensions with similarity to *E. coli* L18 and L25 that bind to 5S rRNA. Like the cognate *E. coli* protein, spinach L22 binds erythromycin, but unlike *E. coli* L22 the spinach protein also binds to 5S rRNA protecting three nonoverlapping binding sites (Carol *et al.*, 1993; Toukifimpa *et al.*, 1989). Furthermore, the central region of the spinach protein is required for 5S rRNA binding showing that its L22-like region has acquired a function lacking in the *E. coli* protein. Carol *et al.* speculate that in spinach, at least, the L22 protein also embraces the roles of *E. coli* L18 and L25. Cognate genes for the latter two proteins have not yet been identified either in the chloroplast or nuclear genomes of land plants. N-terminal extensions are lacking in tobacco L22, but maize and rice both have 29-residue N-terminal extensions with mutual homology, but little similarity to the spinach extension.

Many chloroplast and bacterial r-proteins are related immunologically as would be expected on the basis of amino acid sequence homology (e.g., Randolph-Anderson *et al.*, 1989).

Most plastid-encoded r-protein genes are organized in operons. The largest of these operons includes some of the genes present in the α, *spc*, and *sr* operons (Table 3–5) as discussed in Chapter 3. Similarly, a remnant of the *S12* operon containing the *rps7* and *rps12* genes is found in the chloroplasts of land plants (Table 3–5), but this organization is disrupted in *Chlamydomonas* with the *rps7* and *rps12* genes being in different unique sequence regions. The *rps7* gene in *C. reinhardtii* is cotranscribed with the *atpE* gene (Robertson *et al.*, 1989; Hauser *et al.*, 1993) and the *rps12* gene with the *psbJ* and *atpI* genes (Hauser *et al.*, 1993). Such "mixed" operons including both r-proteins and photosynthetic proteins have also been found in land plants. Thus, the *psaA–psaB–rps14* operon includes an r-protein gene and two genes encoding photosynthetic proteins (Subramanian *et al.*, 1991). These "mixed" operons are of particular interest because the translational and photosynthetic systems differ considerably in the quantitative requirements for their protein components, especially during plastid development. This would suggest the existence of translational control mechanisms that would permit preferential synthesis of one or the other class of proteins as appropriate. Evidence has been obtained in *Chlamydomonas* for such translational control. When chloroplast protein synthesis is reduced, chloroplast r-proteins are synthesized in preference to photosynthetic proteins (Liu *et al.*, 1989b; Chapter 16).

Largely because of the work of O'Brien and his colleagues (see reviews by O'Brien and Matthews, 1976; Pietromonaco *et al.*, 1991) more is known about the structure of mammalian mitochondrial ribosomes than any other such ribosomes. These ribosomes contain nearly twice as much protein and only half the rRNA found in bacterial ribosomes. Consequently, the notion, set forth some years ago, that these ribosomes were "miniribosomes" because of their small rRNA molecules has proved to be wrong. The molecular weights of mammalian mitochondrial ribosomes are actually greater than the ribosomes of *E. coli* (see Gillham, 1978; Pietromonaco et al., 1991). The total number of mitochondrial r-proteins in the cow is about 85, with 52 residing in the large subunit and 33 in the small subunit.

Mammalian mitochondrial r-proteins have been grouped in three classes depending on their relative exposure at the surface of each subunit as determined by a double iodine labeling technique. Both subunits have a few proteins in the highly exposed group and a large number in the intermediate exposure group. Only the large subunit has an appreciable number of buried r-proteins. Pietromonaco et al. (1991) speculate that these buried polypeptides may behave as structural proteins important in the assembly process. This conjecture is supported by the finding that among the buried large subunit proteins, six are strong rRNA binding species. Seven other r-proteins have been identified as secondary rRNA binding proteins.

All mammalian mitochondrial r-proteins are encoded in the nuclear genome, consistent with their exclusive cytoplasmic site of synthesis (Schieber and O'Brien, 1985). The question, therefore, arises as to whether they represent a completely distinct set from the cytoplasmic r-proteins. Obviously, certain of these proteins might be encoded by single genes and shared between cytoplasmic and mitochondrial ribosomes. This seems unlikely based on electrophoretic and immunological comparisons showing that mammalian mitochondrial r-proteins are distinct from their counterparts in the cytoplasm (Pietromonaco et al., 1991).

In *S. cerevisiae* and *Neurospora crassa* one protein of the small subunit of the mitochondrial ribosome is encoded in the mitochondrial genome and all others are nuclear gene products (see Kitakawa and Isono, 1991; Wolf and Del Giudice, 1988). The VAR1 protein, specified by the yeast mitochondrial genome, is polymorphic (Chapter 10). There are over 80 r-proteins in the yeast mitochondrial ribosome, of which 48 have been confirmed for the large subunit by N-terminal sequencing (Kitakawa and Isono, 1991). The genes encoding 20 of these proteins have so far been cloned. The deduced amino acid sequences of the proteins coded by these genes have revealed similarities to *E. coli* r-proteins in some cases. Thus the r-protein encoded by *MRP2* is equivalent to S14, *MRP3* protein to S19, *MRP4* to S2, *MRP#A* to S21, *MRP-L9* to L3, and *MRP-L33* to L30. However, 11 nuclear encoded mitochondrial r-proteins have no detectable homology to any other r-protein (Kitakawa and Isono, 1991).

Studies of mitochondrial gene regulation in yeast have led to the identification of suppressors of a nuclear gene mutation, *pet 122*, whose product is unable to unable to activate translation of the *cox3* mRNA (see Costanzo and Fox, 1990 and Chapter 17). One suppressor affects *MRP1* and the others, two new genes *(MRP17, PET123)* encoding other small subunit r-proteins (Chapter 17, Haffter et al., 1991; Haffter and Fox, 1992). Mutations in yet another nuclear gene, *NAM9*, suppress certain mitochondrial mutations having a UAA termination codon (Boguta et al., 1992). The *NAM9* gene encodes a protein similar to the S4 proteins of eubacteria and chloroplasts.

Expression of all yeast nuclear genes coding for mitochondrial r-proteins save one is affected by glucose repression at the level of transcription (Kitakawa and Isono, 1991). The exception is *MRP7*. When the dosage of *MRP7* was increased by introducing the gene on a plasmid, its expression was found to be posttranscriptionally regulated probably at the translational level.

Compared to chloroplast r-proteins, mitochondrial r-proteins of plants seem to have received little attention. Pinel *et al.* (1986) published a preliminary characterization of the mitochondrial r-proteins from potato tuber mitochondria. They reported that the two mitochondrial ribosomal subunits had sedimentation velocities of 33S and 50S and contained 35 and 33 proteins, respectively. Gene sequencing studies on flowering plant mtDNA have revealed ORFs with homology to *E. coli rps12, rps13, rps14,* and *rps19* (Hanson and Folkerts, 1992; Lonsdale, 1989). Numerous genes encoding r-proteins have been identified in the mitochondrial genome of liverwort (Oda *et al.,* 1992a). They are localized in operon-like units reminiscent of the chloroplast genome (Table 3–5). Three ORFs with homology to *E. coli* r-protein genes (*rpl2, rps12,* and *rpl14*) have also been reported from the completely sequenced mitochondrial genome of *Paramecium* (Cummings, 1992; Pritchard *et al.,* 1990).

Antibiotics and organelle ribosome structure. Organelle ribosomes have long been known to be sensitive to inhibitors of prokaryotic ribosomes, but not to antibiotics affecting eukaryotic ribosomes (see Gillham, 1978 and Table 13–4). Mitochondrial mutations to chloramphenicol resistance in yeast and mammalian cells (Chapter 7) can be traced to specific nucleotide substitutions within the "peptidyl transferase" loop of the large subunit rRNA. In yeast mitochondria and *E. coli* mutations to erythromycin resistance have also been isolated that map within this loop. This is also true of the chloroplast genomes of *Chlamydomonas* and tobacco. In each case the nucleotide substitutions involve residues that are conserved both in *E. coli* and the chloroplast and mitochondrial genomes (Table 13–5). The same is also true of chloroplast mutations affecting the 16S rRNA molecule of the small subunit (see Harris *et al.,* 1989 for a summary). These include mutations to streptomycin and spectinomycin resistance. Sim-

Table 13–4. Inhibition of Protein Synthesis in Chloroplasts and Mitochondria by Compounds Acting on Prokaryotic and Eukaryotic Cytoplasmic Ribosomes

Ribosomal Subunit Affected	Inhibitor	Organelle	
		Mitochondria	Chloroplasts
Prokaryotic Large (50 S) subunit	Chloramphenicol	+	+
	Macrolide antibiotics (erythromycin, lincomycin etc.)	+	+
Prokaryotic small (30 S) subunit	Aminoglycoside antibiotics (streptomycin, spectinomycin, etc.)	– or ?	+
Eukaryotic large (60 S) subunit	Anisomycin	–	–
	Cycloheximide	–	–
50 S or 60 S subunits	Puromycin	+	+
30 S or 40 S subunits	Tetracycline antibiotics	+	?

Adapted from Gillham (1978).

Table 13–5. Alterations in *E. coli* and Organelle Large rRNA Genes Resulting in Erythromycin or Lincomycin Resistance

Organism and Strain	E. coli Wild-Type Sequence 5' (2050) caaGACgGAAGACCC 3' (2065)[a]
E. coli	
Mutant	A
Mutant	T
C. reinhardtii (chloroplast)	
er-u-37	A
clr-u-1	G
Tobacco (chloroplast)	
Mutant	G
Mutant	G
Yeast (mitochondria)	
rib3 mutant	G
	5' (2601) CAGTtcGGTcCcTATC 3' (2616)
C. reinhardtii (chloroplast)	
er-u-1a	T
er-u-11	G
C. moewusii	
Mutant	G
Yeast (mitochondrion)	
rib2 mutant	G

Modified from Harris *et al.* (1989)

[a] Nucleotide numbers are given for the *E. coli* sequence and small letters indicate nucleotides that differ in one or more of the corresponding organelle sequences.

ilar mutations have not been isolated for mitochondria possibly because these antibiotics are not taken up by the mitochondria or the ribosomes are naturally resistant to them (see Gillham, 1978). In *Chlamydomonas reinhardtii,* mutations to streptomycin resistance and dependence also arise as the result of specific amino acid substitutions in the S12 protein encoded by the *rps12* gene (Liu *et al.,* 1989a). Precisely the same amino acid substitutions cause resistant and dependent phenotypes in *E. coli.* Thus, comparative studies of antibiotic resistance mutations affecting *E. coli* and organelle ribosomes have helped to identify important functional similarities between these different classes of ribosomes.

Membrane-bound organelle polysomes. Membrane-bound polysomes have been observed both in plant chloroplasts and in mitochondria of *Neurospora* (see Gillham, 1978), but only in chloroplasts has the function of these polysomes been actively investigated (see Boschetti *et al.,* 1990; Jagendorf and Michaels, 1990 for reviews). High salt washes alone remove 30–45% of the thylakoid-bound RNA while addition of puromycin releases up to 80% of the bound RNA. By analogy with the rough endoplasmic reticulum these results suggest that between a third and a half of the polysomes are attached electrostatically to the membranes and the rest by both electrostatic forces and nascent polypeptide chains. The electrostatic binding of chloroplast polysomes predicts the presence of a ribosome receptor similar to the ribophorin-containing receptor found on the rough endoplasmic reticulum but missing from the bacterial cytoplasmic membrane. Other components of the system for synthesizing eukaryotic secretory proteins such as a signal recognition particle and a docking protein have not been demonstrated in chloroplasts. The fraction of membrane-bound polysomes can be enhanced markedly through the use of antibiotics such as chloramphenicol and erythromycin that inhibit transpeptidation.

Freimann and Hachtel (1988) examined the distribution of mRNAs on free and membrane-bound chloroplast polysomes of Broadbean *(Vicia faba)*. They used the criteria of release of the associated mRNA by high salt alone or high salt plus puromycin together with gene-specific probes to distinguish mRNAs electrostatically bound to thylakoids from those engaged in cotranslational protein synthesis. They recognized three classes of mRNA. The *rbcL* mRNA encoding the large subunit of RUBISCO was the only mRNA solely associated with stromal polysomes. However, other authors have reported that large subunit mRNA is also found associated with thylakoid membranes (e.g., Hattori and Margulies, 1986). Thylakoid polysomes containing mRNAs for six genes encoding integral membrane polypeptides appeared to synthesize their products in a cotranslational fashion. These mRNAs were released only by high salt plus puromycin. Thylakoid polypeptides encoded by seven other genes were assumed to be made posttranslationally as their mRNAs were found on stromal polysomes or polysomes bound electrostatically to the thylakoid membranes. There is also other supporting evidence that such chloroplast-synthesized proteins as D1, the reaction center polypeptide encoded by the *psbA* gene, and the α and β subunits of CF_1 are made, at least in part, on thylakoid-bound polysomes (see Jagendorf and Michaels, 1990 for a summary).

Membrane-binding of polysomes may play an additional role in translational regulation of chloroplast gene expression. In *Chlamydomonas* the distribution of chloroplast mRNAs varies between the thylakoid and soluble fractions in cells growing synchronously on a light–dark cycle (Boschetti *et al.*, 1990). Thus, a striking increase in the fraction of membrane-bound polysomes was observed for both *rbcL* and *psbA* mRNAs in the light period. Thylakoid binding may occur in the light phase because translation is initiated then. In contrast, Klein *et al.* (1988) found that the *psaA, psaB,* and *psbA* transcripts are primarily membrane-associated in dark-grown barley plants. The protein products of these genes are made in the dark, but are unstable in the absence of chlorophyll (see Chapter 16). Jagendorf and Michaels (1990) correctly point out that the possible role of the thylakoid membrane itself in translational regulation needs further investigation. This is equally true of the inner membrane of the mitochondrion where very little is known about membrane-bound polysomes and nothing concerning their gene products.

Transfer RNAs

Chloroplast genomes studied in detail specify enough tRNAs so that tRNA import need not be invoked to effect protein synthesis except in the colorless, parasitic flowering plant *Epifagus* (Chapter 3). Similarly, mitochondrial genomes in animals (except cnidarians) and fungi encode sufficient tRNAs for protein synthesis. However, the mitochondrial genomes of land plants, *Chlamydomonas,* and protists such *Paramecium* and *Tetrahymena,* and Trypanosomes are deficient in genes specifying tRNAs so tRNA import must occur.

Genes encoding organelle tRNAs lack the 3′-terminal CCA sequence unlike many

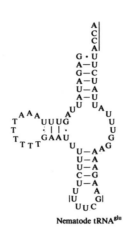

Nematode tRNA^{glu}

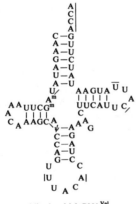

Mitochondrial tRNA^{Val}
(bovine)

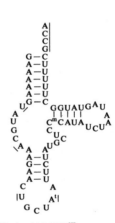

Mitochondrial tRNA^{ser}_{AGY}
(bovine)

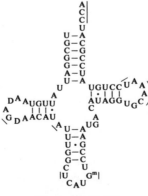

Mitochondrial tRNA_i^{met}
(Neurospora)

PROTEIN SYNTHESIS IN ORGANELLES

of the comparable genes in *E. coli*. This terminus is added posttranscriptionally as it is in eukaryotes. Interestingly, in the cyanobacterium *Anacystis nidulans* a gene coding for tRNA^{Ile}, which is highly homologous to a corresponding chloroplast gene, also does not encode the 3'-terminal CCA. However, the CCA terminus is encoded by a tRNA^{Ala} gene, which is also strongly homologous to its chloroplast counterpart (Palmer, 1985). The universal genetic code is read in most cases during translation of chloroplast mRNAs, but a modified two-letter code is used to decipher mitochondrial genes (Chapter 14). This difference is reflected in the reduced number of tRNA species present in mitochondria as compared to chloroplasts.

Besides their role in mitochondrial transcript processing, three areas of research on mitochondrial tRNAs seem particularly noteworthy. First, systematic structural stud-

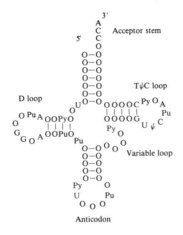

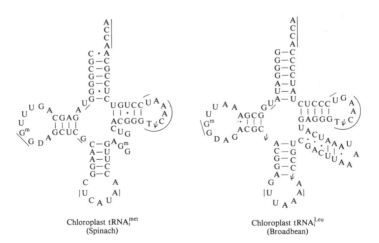

Fig. 13–2. Structures of some chloroplast and mitochondrial tRNAs compared to a generalized tRNA. Positions of invariant bases are shown in the generalized tRNA and those that are conserved in the organelle tRNAs are underlined. Note that in animal tRNA$^{Ser}_{AGY}$ species the D-loop is replaced by a loop of between 5 and 11 nucleotides. In nematodes the T ψ C and variable loops are replaced by a loop of between 4 and 12 nucleotides (TV-replacement loop).

ies of tRNAs and the genes encoding them indicate the limits to which these molecules can be varied and still function. Second, the tRNAs of higher plant mitochondria represent a heterogeneous collection of molecules gathered from the chloroplast and the cytoplasm as well as the mitochondrion itself. Third, phylogenetic investigations show that mitochondrial tRNA import has evolved independently several times, yet the mechanisms involved are still almost completely unknown.

tRNAs of animal mitochondria. The structure of animal mitochondrial tRNAs has been reviewed comprehensively by Wolstenholme (1992). Mammalian mitochondrial tRNA genes in particular are dealt with in reviews by Attardi (1985), Cantatore and Saccone (1987), and Chomyn and Attardi (1987). There are 22 tRNA coding sequences in all animal mitochondrial genomes except cnidarians, where only two tRNA genes have been reported (Chapter 3). The 22 tRNAs are sufficient to read all nontermination codons. In mammmals 14 are encoded in the H-strand and eight on the L-strand. The structures of mammalian and other metazoan tRNAs are distinct from prokaryotic, eukaryotic cytoplasmic and chloroplast tRNAs. They are generally smaller than their cytoplasmic and prokaryotic counterparts and lack many universal or common features characterizing these tRNAs. Most vertebrate and invertebrate mitochondrial tRNAs can be folded into the basic cloverleaf secondary structure (Fig. 13–2). Conserved are the 7-bp acceptor stem, the 5-bp anticodon stem, and the 7 base anticodon loop. Mammalian mitochondrial tRNAs deviate from the invariant tRNA structure as follows. They lack the TψCPuA sequence in the constant "TψC" loop, the sequence G_{18}--G_{19} in the dihydrouridine (D)-loop found so far in all tRNAs active in protein synthesis, and the constant G_{15} residue. G_{18} and G_{19} are also poorly conserved in other animals as is C_{56} in the TψC arm. The G_{18}–ψ_{55} and G_{19}–C_{56} interactions are essential for folding standard tRNAs into a functional tertiary configuration, but this is apparently not true for animal mitochondrial tRNAs (Wolstenholme, 1992). The mitochondrial tRNA$^{Ser}_{AGY}$ species of mammals and most other animals is particularly bizarre in that it lacks the entire D-loop (Fig. 13–2).

Cantatore and Saccone (1987) point out that mammalian mitochondrial tRNAs can be divided into three classes taking into account the regularity of the cloverleaf structure and conservation of invariant nucleotides. The first includes five tRNAs that have features common to other tRNAs (tRNA^{Gln}, tRNA$^{Ser}_{UCN}$, tRNA$^{Leu}_{UUR}$, tRNA^{Val}, and tRNA^{f-Met}). The tRNAs that differ strikingly in structure from normal besides tRNA$^{Ser}_{AGY}$ are tRNA^{Lys}, tRNA^{Cys}, and tRNA^{Trp}. They either have a small D loop or lack the TψC sequence and many of the invariant nucleotides. Carrying this a step further, Wolstenholme (1992) remarks that animal tRNAs that recognize AGY, AGY/A, or AGN codons have the D arm replaced by a single loop of between 2 and 11 nucleotides. The remaining tRNAs are in between and have a regular secondary structure, but lack many invariant nucleotides and sometimes the TψC sequence.

In *Drosophila* the structures predicted for the 22 tRNAs are generally similar to their mammalian counterparts including the unusual tRNA$^{Ser}_{AGY}$ gene that lacks the D arm (Clary and Wolstenholme, 1984). Sequencing of nematode mitochondrial genomes has revealed yet another unexpected variation on tRNA structure (Fig. 13–2). In 20 of the 22 mitochondrial tRNAs of nematodes the TψC arm and variable loop are replaced by a single, simple loop of between 6 and 12 nucleotides called the TV replacement loop (Okimoto and Wolstenholme, 1990; Okimoto et al., 1992; Wolstenholme et al., 1987). In the two nematode mitochondrial serine tRNAs the predicted secondary structures contain a variable loop (always four nucleotides) and TψC arm, but the D arm is replaced by a loop of five to eight nucleotides. One of these is the same tRNA$^{Ser}_{AGY}$ in which the D arm is virtually missing in mammals. In flat worms (Platyhelminthes) the same bizarre structure is seen for tRNA$^{Ser}_{AGY}$ while other tRNAs have four arms similar to other animal mitochondrial tRNAs with the exception of nematodes (Garey and Wolstenholme, 1989). This leads to a picture of animal tRNA evolution in which the tRNA$^{Ser}_{AGY}$ lacking the D-loop is a primitive feature and the TV-replacement loop of nematodes is a feature distinct to these organisms (Wolstenholme, 1992).

Heterogeneity of plant mitochondrial tRNAs. Systematic investigations of flowering plant mitochondrial tRNAs (see Dietrich *et al.,* 1992; Hanson and Folkerts, 1992; Lonsdale, 1989; Marechal-Drouard and Guillemaut, 1988; Marechal-Drouard *et al.,* 1990; Sangare *et al.,* 1990) have revealed that they fall in three groups. The first includes "native" tRNAs encoded by the mitochondrial genome. In wheat 10 such tRNAs have been identified and in potato 15. They have 60–80% sequence similarity to eubacterial tRNAs. The second is made up of "chloroplast-like" tRNAs with 90–100% homology to specific chloroplast tRNAs. Six of these tRNAs have been found in wheat and five in potato. Certain of them are the same in a variety of plants. Thus, a chloroplast-like tRNA$^{Trp}_{CCA}$ has been reported from mitochondria of French Bean *(Phaseolus vulgaris),* evening primrose, wheat, maize, and potato. Others differ between species (e.g., a chloroplast-like tRNA$^{Cys}_{GCA}$ has been found in maize and wheat, but not potato). The final class of tRNAs is "cytosolic like." These tRNAs are encoded by nuclear genes. Eleven of these tRNAs have been detected in potato and five in wheat.

Marechal-Drouard and Guillemaut (1988) found four nuclear-encoded tRNA^{Leu} species in mitochondria of *P. vulgaris.* Each species could be aminoacylated either by mitochondrial or cytoplasmic extracts. These leucine-charged tRNA molecules were competent in a wheat germ *in vitro* translation system. Each mitochondrial tRNA^{Leu} species was identical in sequence to a cytoplasmic counterpart, except for the presence of a methylated G residue at position 18 (G_m18) instead of an unmethylated G. This same modification seems to be of significance for other imported tRNAs (Dietrich *et al.,* 1992). If this methylation took place in the cytoplasm, it might act as a signal for import. However, partial sequence data have shown that some cytosolic tRNAs also possess G_m18. Small *et al.* (1992) demonstrated in transgenic potato plants carrying a leucine tRNA gene from bean that transcripts of the introduced gene can be detected in both the cytoplasm and mitochondrion. In the completely sequenced mitochondrial genome of liverwort, in contrast, 29 genes encoding 27 tRNAs have been detected, none of which is chloroplast-like (Oda *et al.,* 1992b). However, even after allowing for wobble and modification of the anticodons, two species of tRNAs (tRNA^{Ile} and tRNA^{Thr}) are missing in the liverwort mitochondrial genome so they most likely are imported.

tRNA processing and modification in yeast mitochondria. N.C. Martin and colleagues have made a detailed study of mitochondrial tRNA precursor processing in yeast (e.g., Chen and Martin, 1988; Dang and Martin, 1993; Hollingsworth and Martin, 1986; Martin *et al.,* 1985; Miller *et al.,* 1983; Najarian *et al.,* 1987; Wise and Martin, 1991). Yeast mitochondrial tRNAs are transcribed as larger precursors either in monomeric or multimeric form. These precursors must be processed to their mature forms by removal of extra bases external to both the 5'- and 3'-termini of the mature tRNA followed by the addition of the CCA residue to the processed 3'-terminus and specific nucleotide modifications. Removal of nucleotides from the 5'-end is carried out by RNase P. This RNA–protein complex is known to be required for processing 5' tRNA transcript leaders in bacteria, yeasts, and mammals. While both protein and RNA are required *in vivo*, the prokaryotic RNAs are catalytic by themselves *in vitro* (Altman *et al.*, 1993; Guerrier-Takeda *et al.*, 1983). In the case of yeast mitochondria the 490 base RNA moiety is encoded by a mitochondrial gene *(RPM1)* and the 105-kDa protein by a nuclear gene *(RPM2)* (Morales *et al.*, 1992). The *RPM2* gene has been cloned, sequenced, and found to specify a protein with 1202 residues of which the N-terminal 122 amino acids predicted by the gene are not found in the protein (Dang and Martin, 1993). At least some of the missing amino acids very likely form part of the presequence of the protein which is cleaved off following entry into the mitochondrion (Chapter 15). In another yeast *Candida glabrata* the RNA moiety is 227 bases and in *Saccharomycopsis fibuligera* the RNA is only 140 nucleotides in length (Wise and Martin, 1991). By analogy with rRNAs, which in mitochondria of certain organisms have become highly streamlined in structure (see above), one assumes that the minimal *S. fibuligera* RNase P RNA molecule has retained only those core sequences required for its function. Chen and Martin (1988) also identified a single endonucleolytic activity required for 3' processing of mitochondrial tRNA precursors at the CCA addition site.

N.C. Martin *et al.* also identified several genes required for mitochondrial tRNA modification in *S. cerevisiae* (e.g., Dihanich *et al.*, 1987; Ellis *et al.*, 1986, 1987; Najarian *et al.*, 1987). The *MOD5* gene product is required for isopentylation of a specific adenine residue. The modified base, i^6A, occurs in tRNAs at position 37 adjacent to the 3'-end of the anticodon and is found only in tRNAs that recognize codons starting with U. The yeast *MOD5* locus is the structural gene for the transferase that adds the isopentyl group both to cytoplasmic and mitochondrial tRNAs. Gene and transcript mapping indicate that there may be two translational start sites suggesting two forms of the *MOD5* gene product, one of which would have an amino terminal extension to permit mitochondrial targeting (Najarian *et al.*, 1987).

Two other dual function genes are known to be involved in modification of yeast cytoplasmic and mitochondrial tRNAs. The *TRM1* and *TRM2* gene products are required for dimethylation of guanosine and 5-methylation of specific nucleotides, respectively. The *TRM1* gene product seems to target the mitochondrial and cytoplasmic compartments by an unusual mechanism (Ellis *et al.*, 1986, 1987). There are two in frame ATGs at positions +1 and +49. The majority of the mRNAs initiate at a start site in the large open reading frame between these two ATGs although a small fraction have their 5'-ends upstream of the first ATG. The corresponding AUGs in the transcripts are used to initiate translation resulting in long and short forms of the enzyme. Which AUG is used is determined by the location of the 5'-end of the transcript. The mitochondrial targeting sequences appear to reside in a region shared by the two forms of the enzyme that is located downstream of the second in frame ATG of the *TRM1* gene. The presequence encoded therein is likely used to partition a portion of both the long and short forms of the enzyme to the mitochondria. Enzyme that is not partitioned to the mitochondrion is transported to the nucleus to be used to modify cytoplasmic tRNAs (Li *et al.*, 1989). The first 48 amino acids are necessary and sufficient for mitochondrial import while amino acids 95 to 102 contain the nuclear targeting signals (Ellis *et al.*, 1989; Rose *et al.*, 1992).

Several other cases of genes whose products can travel either to the cytoplasm or the mitochondrion have been described. Thus, there are four major transcripts for the yeast *LEU4* gene encoding α-isopropylmalate synthase. Two of these initiate 90 nucleotides downstream of the ATG at the beginning of the long open reading frame (Beltzer et al., 1986). The protein thus generated by transcription from the downstream ATG would be 30 amino acids shorter than its longer counterpart and would, presumably, go into the cytoplasm while the larger protein should target the mitochondrion. The *FUM1* gene encodes transcripts of different size for the mitochondrial and cytoplasmic fumarases. The longer transcript is assumed to encode the protein destined for the mitochondrion and the smaller the cytoplasmic polypeptide (Wu and Tzagoloff, 1987). Other instances involve single nuclear genes encoding cytoplasmic and mitochondrial forms of two aminoacyl-tRNA synthetases (see Aminoacyl-tRNA synthetases).

tRNA import into mitochondria. tRNA import has been documented for flowering plants (see above), the ciliate protozoan *Tetrahymena* (e.g., Suyama, 1986) and in one instance for yeast (Martin et al., 1979). tRNA import has also been inferred for the trypanosome kinetoplast, *Paramecium, Chlamydomonas,* and certain cnidarians. The mitochondrial genomes of these organisms are grossly deficient in genes encoding tRNAs or lack them completely in the case of the kinetoplast maxicircle mitochondrial genome (Chapter 3).

Although mitochondrial ancestry traces to a single origin in the α-purple bacteria (Chapter 2), the organisms in which loss of mitochondrial tRNA genes has occurred occupy different phylogenetic branches (Fig. 2–1). tRNA gene loss from mitochondria in each of these cases must have been accompanied by acquisition of tRNAs from the cytoplasm. Whether similar import mechanisms evolved each time remains to be seen.

How RNA is imported into mitochondria is a mystery. As Nagley (1989) points out in a review, two formal possibilities are that the RNA is naked during import or that it enters the organelle in a protein complex. Transit peptides of mitochondrially targeted proteins have a net positive charge that is required for their binding (Chapter 15) while naked RNA molecules are negatively charged. Hence, it is difficult to see how the same transport pathway could be used for both RNA and proteins. Consequently, prior binding of the RNA to a carrier protein has been the most popular hypothesis for transport. One obvious possibility is that cognate tRNA synthetases are the carriers.

Tarassov and Entelis (1992) reported what may be the first progress toward understanding tRNA import. They found that import of the cytoplasmic $tRNA^{Lys}_{CUU}$ into mitochondria of yeast *in vitro* was an ATP-dependent process that occurred in the presence of soluble cellular proteins. One of the proteins in this mixture possessed an RNA binding capacity. However, the mitochondrial lysyl-tRNA synthetase seems to be ruled out since this enzyme cannot aminoacylate the imported cytoplasmic tRNA (Martin et al., 1979).

Chloroplast tRNAs. The structure and codon recognition patterns of chloroplast tRNAs have been extensively reviewed (e.g., Marechal-Drouard et al., 1991; Steinmetz and Weil, 1989; Sugiura and Wakasugi, 1989). In tobacco the chloroplast genome specifies 30 tRNA species of which 23 are encoded by single genes and seven by genes found in duplicate in the inverted repeat. Rice has the same set of genes except that the $tRNA^{His}$ gene, found adjacent to the inverted repeat in tobacco, is within the rice repeat. In liverwort there are 31 chloroplast-encoded tRNA genes. The extra gene with respect to rice and tobacco is tRNA$^{Arg}_{CCG}$. The main features of chloroplast tRNAs from land plants are as follow:

1. Their gene sequences are highly conserved in different species.
2. They can be folded into typical cloverleaf structures unlike most mitochondrial tRNAs (Fig. 13–2).
3. They are similar in structure and sequence to prokaryotic tRNAs (ca. 70% sequence identity), but have low sequence similarity to eukaryotic tRNAs.
4. They can be aminoacylated *in vitro* with chloroplast or prokaryotic enzymes, but not with the cytoplasmic synthetases from the same organism.
5. Isoaccepting tRNAs for a given amino acid are encoded by different chloroplast genes, but these tRNAs are charged by the same chloroplast tRNA synthetases.
6. As mentioned above, the 3'-terminal CCA triplet is added posttranscriptionally.
7. The presence of certain unique modified nucleotides in chloroplast tRNAs raises the possibility that there are modification enzymes specific for chloroplast tRNAs.

Several chloroplast tRNAs have unusual feature. Two tRNAs^{Ile} are found in chloroplasts of higher plants. The major species (tRNA$^{Ile}_1$,) recognizes the codons AUU and AUC while a minor species (tRNA$^{Ile}_2$) recognizes AUA. However, the gene encoding the latter tRNA contains a CAU anticodon that normally would recognize AUG for methionine. One possible explanation revolves around the C residue, which is modified in some way posttranscriptionally. In *E. coli* the C of the homologous tRNA is modified to lysidine, a novel type of cytidine with a lysine residue, which allows it to recognize the AUA codon (Muramatsu *et al.*, 1988). Perhaps the same modification occurs with the chloroplast tRNA (Marechal-Drouard *et al.*, 1991).

The chloroplast tRNA$^{Glu}_{UUC}$ has a unique role. This tRNA participates in the metabolic conversion of glutamic acid into δ-aminolevulinic acid (DALA), an early precursor in the chlorophyll biosynthetic pathway (Schon *et al.*, 1986). Glutamate is ligated to tRNA$^{Glu}_{UUC}$ by a nuclear encoded ligase (von Wettstein, 1991). Removal of the 3'-terminal CCA sequence prevents ligation of the tRNA to glutamic acid. The nuclear-encoded glutamyl-tRNA dehydrogenase (reductase) then reduces the aminoacyl ester linkage with NADPH to generate glutamate-1-semialdehyde. Finally, the formation of 5-aminolevulinate from glutamate 1-semialdehyde is catalyzed by glutamate 1-semialdehyde aminotransferase. The synthesis of DALA from glutamate is specific for plant chloroplasts and cyanobacteria. In animals, yeast, and some bacteria, DALA synthesis occurs via condensation of glycine and succinyl-CoA.

Lastly, tRNA^{Glu} misaminoacylation is required for protein synthesis in chloroplasts (Schon *et al.*, 1988). There are two other tRNA^{Glu} species in chloroplasts besides tRNA$^{Glu}_{UUC}$. They differ only by a ribose methylation of G_{18} and both have a U*UG anticodon that is specific for glutamine. These mischarged Glu-tRNA^{Gln} species are converted to Gln-tRNA^{Gln} by a crude chloroplast extract. Since chloroplasts, cyanobacteria, and plant and animal mitochondria lack Gln-tRNA synthetase activity, the glutamine residues required for protein synthesis are provided by conversion of glutamate to glutamine following attachment of the amino acid to tRNA^{Gln}. This mischarging mechanism has been described in several species of gram-positive bacteria, but direct aminoacylation of a tRNA^{Gln} by glutamine occurs in the cytoplasm of gram-negative bacteria and eukaryotic cells.

Aminoacyl-tRNA Synthetases

The general characteristics of organelle tRNA synthetases have been known for some time (see Gillham, 1978; Tzagoloff, 1982; Hoober, 1984; Tzagoloff et. al., 1990; Weil and Parthier, 1982). These enzymes are specific for chloroplast or mitochondrial tRNAs, they do not charge cytoplasmic tRNAs, and they often (at least the chloroplast synthetases) charge prokaryotic tRNAs. No synthetase has yet been identified as an organelle gene product. Investigation of specific mitochondrial synthetases from *Neurospora* and yeast indicate (1) that the enzymes charging cognate tRNAs in the mitochondrion and cytoplasm may be encoded by the same or different genes and (2) that mitochondrial tRNA synthetases can also function to splice mitochondrial introns (see Chapter 10).

The data on mitochondrial tRNA synthetases in yeast have been summarized in two reviews (Costanzo and Fox, 1990; Tzagoloff *et al.*, 1990). Ten genes have been identified of which eight encode enzymes that function only in mitochondria. This is not the case for the histidyl or valyl tRNA-synthetases, where the cytoplasmic and mitochondrial forms of these enzymes are specified by the same gene. Both forms of the histidyl tRNA-synthetase are encoded by the *HTS1* gene (Natsoulis *et al.*, 1986). The mechanism involved in differential synthesis of the two forms of histidyl tRNA-synthetase involves pairs of ATG initiator codons and different transcriptional start sites. The long set of transcripts initiates upstream of the first ATG and the short set between the two ATGs. A mutation that destroys the first AUG on the long mRNA results in a respiratory-deficient phenotype, but does not affect the level of cytoplasmic histidine-tRNA synthetase activity or viability. However, mutations distal to the second ATG cause loss of cytoplasmic synthetase function and lethality. The longer mRNA is thought to encode the mitochondrial enzyme and the shorter mRNA the cytoplasmic synthetase. Mutational analysis of the amino-terminal presequence region of *HTS1* demonstrates that efficient targeting to mitochondria not only requires the amino-terminal presequence, which is probably cleaved on entry into the mitochondria, but a second amino-terminal sequence found in both the precursor and cytoplasmic forms of the enzyme (Chiu *et al.*, 1992). This is reminiscent of the cytoplasmic and mitochondrial forms of the tRNA dimethyltransferase encoded by the *TRM1* gene discussed earlier. The *VAS1* gene encodes both the cytoplasmic and mitochondrial forms of valyl-tRNA synthetase (Chatton *et al.*, 1988).

REFERENCES

Altman, S., Kirsebom, L., and S. Talbot (1993). Recent studies of ribonuclease P. *FASEB J.* 7: 7–14.

Attardi, G. (1985). Animal mitochondrial DNA: An extreme example of genetic economy. *Int. Rev. Cytol.* 93: 93–145.

Beltzer, J.P., Chang, L.-F. L., Hinkkanen, A.E., and G.B. Khohlhaw (1986). Structure of yeast *LEU4*: The 5' flanking region contains features that predict two modes of control and two productive translation starts. *J. Biol. Chem.* 261: 5160–5167.

Benne, R. (1985). Mitochondrial genes in trypanosomes. *Trends Genet.* 1: 117–121.

Boer, P.H., and M.W. Gray (1988). Scrambled ribosomal RNA gene pieces in Chlamydomonas reinhardtii mitochondrial DNA. *Cell* 55: 399–411.

Boguta, M., Dmochowska, A., Borsuk, P., Wrobel, K., Gargouri, A., Lazowska, J., Slonimski, P.P., Szczesniak, B., and A. Kurszewska (1992). *NAM9* nuclear suppressor of mitochondrial ochre mutation in *Saccharomyces cerevisiae* codes for a protein homologous to S4 ribosomal proteins from chloroplasts, bacteria and eucaryotes. *Mol. Cell. Biol.* 12: 402–412.

Bohnert, H.J., Crouse, E.J., and J.M. Schmitt (1982). Organization and expression of plastid genomes. *Encyclopedia Plant Physiol.* 14B: 475–530.

Bonham-Smith, P.C., and D.P. Bourque (1989). Translation of chloroplast-encoded mRNA: Potential initiation and termination sites. *Nucl. Acids. Res.* 17: 2057–2080.

Boschetti, A., Breidenbach, E., and R. Blattler (1990). Control of protein formation in chloroplasts. *Plant Sci.* 68: 131–149.

Breitenberger, C.A., and L. Spremulli (1980). Purification of *Euglena gracilis* chloroplast elongation factor G and comparison with other prokaryotic and eukaryotic translocases. *J. Biol. Chem.* 255: 9814–9820.

Breitenberger, C.A., Graves, M.C., and L.L. Spremulli (1979). Evidence for the nuclear location of the gene for chloroplast elongation factor G. *Arch. Biochem. Biophys.* 194: 265–270.

Cantatore, P., and C. Saccone (1987). Organization, structure, and evolution of mammalian mitochondrial genes. *Int. Rev. Cytol.* 108: 149–208.

Carol, P., Rozier, C., Lazaro, E., Ballesta, J.P.G., and R. Mache (1993). Erythromycin and 5S rRNA binding properties of the spinach chloroplast ribosomal protein CL22. *Nucl. Acids Res.* 21: 635–639.

Chatton, B., Walter, P., Ebel, J.-P., Lacroute, F., and Fasiolo, F. (1988). The yeast *VAS1* gene encodes both mitochondrial and cytoplasmic valyl-tRNA synthetases. *J. Biol. Chem.* 263: 52–57.

Chen, J.-Y., and N.C. Martin (1988). Biosynthesis of tRNA in yeast mitochondria: An endonuclease is responsible for the 3′-processing of tRNA precursors. *J. Biol. Chem.* 263: 13677–13682.

Chiu, I.M., Mason, T.L., and G.R. Fink (1992). *HTS1* encodes both the cytoplasmic and mitochondrial histidyl-tRNA synthetases of *Saccharomyces cerevisiae*: Mutations alter the specificity of compartmentation. *Genetics* 132: 987–1001.

Chomyn, A., and G. Attardi (1987). Mitochondrial gene products. *Curr. Topics Bioenerget.* 15: 295–329.

Clary, D.O., and D.R. Wolstenholme (1984). The *Drosophila* mitochondrial genome. In *Oxford Surveys on Eukaryotic Genes,* Vol. 1 (N. Maclean, ed.), pp. 1–35. Oxford University Press, New York.

Costanzo, M.C., and T.D. Fox (1990). Control of mitochondrial gene expression in *Saccharomyces cerevisiae*. *Annu. Rev. Genet.* 24: 91–113.

Cummings, D.J. (1992). Mitochondrial genomes of ciliates. *Int. Rev. Cytol.* 141: 1–64.

Dang, Y.L., and N.C. Martin (1993). Yeast mitochondrial RNase P: sequence of the *RMP2* gene and demonstration that its product is a protein subunit of the enzyme. *J. Biol. Chem.* 268: 19791–19796.

Delp, G., and H. Kossel (1991). rRNAs and rRNA genes in plastids. In *The Molecular Biology of Plastids* (L. Bogorad and I.K. Vasil, eds.), pp. 139–167. Academic Press, San Diego.

Dietrich, A., Weil, J.H., and L. Marechal Drouard (1992). Nuclear-encoded transfer RNAs in plant mitochondria. *Annu. Rev. Cell Biol.* 8: 115–131.

Dihanich, M.E., Najarian, D., Clark, R., Gillman, E.C., Martin, N.C., and A.K. Hopper (1987). Isolation and characterization of *MOD5*, a gene required for isopentenylation of cytoplasmic and mitochondrial tRNAs of *Saccharomyces cereviseae*. *Mol. Cell. Biol.* 7: 177–184.

Ellis, S.R., Morales, M.J., Li, J.-M., Hopper, A.K., and N.C. Martin (1986). Isolation and characterization of the *TRM*1 locus, a gene essential for the N^2,N^2-dimethylguanosine modification of both mitochondrial and cytoplasmic tRNA in *Saccharomyces cerevisiae*. *J. Biol. Chem.* 261: 9703–9709.

Ellis, S.R., Hopper, A.K., and N.C. Martin (1987). Amino-terminal extension generated from an upstream AUG codon is not required for mitochondrial import of yeast N^2,N^2-dimethylguanosine-specific tRNA methyl-transferase. *Proc. Natl. Acad. Sci. U.S.A.* 84: 5172–5176.

Ellis, S.R., Hopper, A.K., and N.C. Martin (1989). Aminoterminal extension generated from an upstream AUG codon increases the efficiency of mitochondrial import of N^2,N^2-dimethylguanosine-specific tRNA methyltransferase. *Mol. Cell. Biol.* 9: 1611–1620.

Feagin, J.E., Werner, E., Gardner, M.J., Williamson, D.H., and R.J.M. Wilson (1992). Homologies between the contiguous and fragmented rRNAs of the two *Plasmodium falciparum* extrachromosomal DNAs are limited to core sequences. *Nucl. Acids Res.* 20: 879–887.

Fox, L., Erion, J., Tarnowski, J., Spremulli, L. Brot, N., and H. Weissbach (1980). *Euglena gracilis* chloroplast EF-Ts: Evidence that it is a nuclear-coded gene product. *J. Biol. Chem.* 255: 6018–6019.

Freimann, A., and W. Hachtel (1988). Chloroplast messenger RNAs of free and thylakoid-bound polysomes from *Vicia faba* L. *Planta* 175: 50–59.

Gantt, J.S. (1988). Nucleotide sequences of cDNAs encoding four complete nuclear-encoded plastid ribosomal proteins. *Curr. Genet.* 14: 519–528.

Garey, J.R., and D.R. Wolstenholme (1989). Platyhelminth mitochondrial DNA: evidence for early origin of a tRNA-ser(AGN) that contains a dihydrouridine arm replacement-loop, and or serine-specifying AGA and AGG codons. *J. Mol. Evol.* 28: 374–387.

Gillham, N.W. (1978). *Organelle Heredity* Raven Press, New York.

Gray, M.W. (1988). Organelle origins and ribosomal RNA. *Biochem. Cell Biol.* 66: 325–348.

Gray, M.W., and M.N. Schnare (1990). Evolution of the modular structure of rRNA. In *The Ribosome. Structure, Function and Evolution* (W.E. Hill, A. Dahlberg, R.A. Garrett, P.B. Moore, D. Schlessinger, and J.R. Warner, eds.), pp. 589–597. American Society for Microbiology, Washington, D.C.

Grivell, L.A. (1989). Nucleo-mitochondrial interactions in yeast mitochondrial biogenesis. *Eur. J. Biochem.* 182: 477–493.

Guerrier-Takeda, C., Gardiner, K., Marsh, T., Pace,

N., and S. Altman (1983). The RNA moiety of ribonuclease P is the catalytic subunit of the enzyme. *Cell* 35: 849–857.

Haffter, P., and T.D. Fox (1992). Suppression of carboxy-terminal truncations of the yeast mitochondrial mRNA-specific translational activator PET122 by mutations in two new genes *MRP17* and *PET127. Mol. Gen. Genet.* 235: 64–73.

Haffter, P., McMullin, T.W., and T.D. Fox (1991). Functional interactions among two yeast mitochondrial ribosomal proteins and an mRNA-specific translational activator. *Genetics* 127: 319–326.

Hallick, R.B., Hong, L., Drager, R.G., Favreau, M.R., Monfort, A., Orsat, B., Spielmann, A., and E. Stutz (1993). Complete sequence of *Euglena gracilis* chloroplast DNA. *Nucl. Acids Res.*, 21: 3537–3544.

Hanson, M.R., and O.F. Folkerts (1992). Structure and function of the higher plant mitochondrial genome. *Int. Rev. Cytol.* 141: 129–172.

Harris, E.H., Burkhart, B.D., Gillham, N.W., and J.E. Boynton (1989). Antibiotic resistance mutations in the chloroplast 16S and 23S rRNA genes of *Chlamydomonas reinhardtii:* Correlation of genetic and physical maps of the chloroplast genome. *Genetics* 123: 281–292.

Hattori, T., and M. Margulies (1986). Synthesis of large subunit of ribulose-bisphosphate carboxylase by thylakoid-bound polyribosomes from spinach chloroplasts. *Arch. Biochem. Biophys.* 244: 630–640.

Hauser, C.R., Randolph-Anderson, B.L., Boynton, J.E., and N.W. Gillham (1993). Molecular genetics of chloroplast ribosomes in *Chlamydomonas reinhardtii.* In *The Translational Apparatus,* (K. Nierhaus *et al.* eds.), pp. 545–554. Plenum Press, New York.

Hershey, J.B. (1987). Protein synthesis. In *Escherichia coli and Salmonella typhimurium,* Vol. 1 (J.L. Ingraham, K.B. Low, B. Magasanik, M. Schaechter, and H.E., Umbarger, eds.), pp. 613–647. American Society of Microbiology, Washington, D.C.

Hollingsworth, M.J., and N.C. Martin (1986). RNase P activity in the mitochondria of *Saccharomyces cerevisiae* depends on both mitochondrion and nucleus-encoded components. *Mol. Cell. Biol.* 6: 1058–1064.

Hoober, J.K. (1984). *Chloroplasts.* Plenum Press, New York.

Jagendorf, A.T., and A. Michaels (1990). Rough thylakoids: Translation on photosynthetic membranes. *Plant Sci.* 71: 137–145.

Johnson, C.H., Kruft, V., and A.R. Subramanian (1990). Identification of a plastid-specific ribosomal protein in the 30S subunit of chloroplast ribosomes and isolation of the cDNA clone encoding its cytoplasmic precursor. *J. Biol. Chem.* 265: 12790–12795.

Kitakawa, M., and K. Isono (1991). The mitochondrial ribosomes. *Biochimie* 73: 813–825.

Klein, R.R., Mason, H.S., and J.E. Mullet (1988). Light-regulated translation of chloroplast proteins. I. Transcripts of *psaA-psaB, psbA,* and *rbcL* are associated with polysomes in dark-grown and illuminated barley seedlings. *J. Cell Biol.* 106: 289–301.

Kozak, M. (1983). Comparison of initiation of protein synthesis in procaryotes, eucaryotes, and organelles. *Microbiol. Rev.* 47: 1–45.

Lagrange, T., Carol, P., Bisanz-Seyer, C., and R. Mache (1991). Comparative analysis of four different cDNA clones encoding chloroplast ribosomal proteins. In *The Translational Apparatus of Photosynthetic Organelles* (R. Mache, E. Stutz, and A.R. Subramanian, eds.). *NATO ASI Series H,* Vol. 55, pp. 107–115. Springer-Verlag, Berlin.

Li, J., Hopper, A.K., and N.C. Martin (1989). N^2,N^2-Dimethylguanosine-specific tRNA methyltransferase contains both nuclear and mitochondrial targeting signals in *Saccharomyces cerevisiae. J. Cell Biol.* 109: 1411–1419.

Liao, H.-X., and L.L. Spremulli (1991). Initiation of protein synthesis in animal mitochondria: purification and characterization of translational initiation factor 2. *J. Biol. Chem.* 266: 20714–20719.

Liu, X.-Q., Gillham, N.W., and J.E. Boynton (1989a). Chloroplast ribosomal protein gene *rps12* of *Chlamydomonas reinhardtii:* Wild-type sequence, mutations to streptomycin resistance and dependence, and function in *Escherichia coli. J. Biol. Chem.* 264: 16100–16108.

Liu, X.-Q., Hosler, J.P., Boynton, J.E., and N.W. Gillham (1989b). mRNAs for two ribosomal proteins are preferentially translated in the chloroplast of *Chlamydomonas reinhardtii* under conditions of reduced protein synthesis. *Plant Mol. Biol.* 12: 385–394.

Lonsdale, D.M. (1989). The plant mitochondrial genome. In *The Biochemistry of Plants,* Vol. 15 (A. Marcus, ed.), pp. 229–295. Academic Press, San Diego.

Ma, L., and L.L. Spremulli (1992). Immunological characterization of the complex forms of chloroplast translational intitiation factor 2 from *Euglena gracilis. J. Biol. Chem.* 267: 18356–18360.

Mache, R. (1990). Chloroplast ribosomal proteins and their genes. *Plant Sci.* 72: 1–12.

Marechal-Drouard, L., and P. Guillemaut (1988). Import of several tRNAs from the cytoplasm into the mitochondria in bean *Phaseolus vulgaris. Nucl. Acids Res.* 16: 4777–4788.

Marechal-Drouard, L., Guillemaut, P., Cosset, A., Arbogast, M., Weil, J.-H., and A. Dietrich (1990). Transfer RNAs of potato *(Solanum tuberosum)* have different genetic origins. *Nucl. Acids Res.* 18: 3689–3696.

Marechal-Drouard, L., Kuntz, M., and J.-H. Weil (1991). tRNAs and tRNA genes of plastids. In *The Molecular Biology of Plastids* (L. Bogorad and I.K. Vasil, eds.), pp. 169–189. Academic Press, San Diego.

Martin, N.C., Miller, D.L., Underbrink, K., and X.

Ming (1985). Structure of a precursor to the yeast mitochondrial tRNA^{fMet}: Implications for the function of the tRNA synthesis locus. *J. Biol. Chem.* 260: 1479–1483.

Martin, R.P., Scheller, J.M., Stahl, A.J.C., and G. Dirheimer (1979). Import of a nuclear deoxyribonucleic acid coded lysine-accepting transfer ribonucleic acid (anticodon CUU) into yeast mitochondria. *Biochemistry* 18: 4600–4605.

Miller, D.L., Underbrink-Lyon, K., Najarian, D.R., Krupp, J., and N.C. Martin (1983). Transcription of yeast mitochondrial tRNA genes and processing of tRNA gene transcripts. In *Mitochondria 1983* (R.J. Schweyen, K. Wolf, and F. Kaudewitz, eds.), pp. 151–164. De Gruyter, Berlin.

Montandon, P.-E., and E. Stutz (1983). Nucleotide sequence of a *Euglena gracilis* chloroplast genome region coding for the elongation factor tu; evidence for a spliced mRNA. *Nucl. Acids Res.* 11: 5877–5892.

Morales, M.J., Dang, Y.L., Lou, Y.C., Sulo, P., and N. Martin (1992). A 105-kDa protein is required for yeast mitochondrial RNase P activity. *Proc. Natl. Acad. Sci. U.S.A.* 89: 9875–9879.

Murumatsu, T., Nishihawa, K., Nemoto, F., Kuchino, Y., Nishinura, S., Miyazawa, T., and S. Yokoyama (1988). Codon and amino acid specificities of a transfer RNA are both converted by a single post-transcriptional modification. *Nature (London)* 336: 179–181.

Nagley, P. (1989). Trafficking in small mitochondrial RNA molecules. *Trends Genet.* 5: 67–69.

Najarian, D., Dihanich, M.E., Martin, N.C., and A.K. Hopper (1987). DNA sequence and transcript mapping of *MOD5:* Features of the 5′ region which suggest two translational starts. *Mol. Cell. Biol.* 7: 185–191.

Natsoulis, G., Hilger, F., and G.R. Fink (1986). The *HTS1* gene encodes both the cytoplasmic and mitochondrial histidine tRNA synthetases of S. cerevisiae. *Cell* 46: 235–243.

O'Brien, T.W., and D.E. Matthews (1976). Mitochondrial ribosomes. In *Handbook of Genetics,* Vol. 5 (R.C. King, ed.), pp. 535–580. Plenum Press, New York.

Oda, K., Yamato, K., Ohta, E., Nakamura, Y., Takemura, M., Nozato, N., Akashi, K., Kanegae, T., Ogura, Y., Kohchi, T., and K. Ohyama (1992a). Gene organization deduced from the complete sequence of liverwort *Marchantia polymorpha* mitochondrial DNA: A primitive form of plant mitochondrial genome. *J. Mol. Biol.* 223: 1–7.

Oda, K., Yamato, K., Ohta, E., Nakamura, Y., Takemura, M., Nozato, N., Akashi, K., and K. Ohyama (1992b). Transfer RNA genes in the mitochondrial genome from a liverwort, *Marchantia polymorpha:* The absence of chloroplast-like tRNAs. *Nucl. Acids Res.* 20: 3773–3777.

Okimoto, R. and D.R. Wolstenholme (1990). A set of tRNAs that lack either the T ψ C arm or the dihydrouridine arm: towards a minimal tRNA adaptor. *EMBO J.* 9: 3405-3411.

Okimoto, R., MacFarlane, J.L., Clary, D.O., and D.R. Wolstenholme (1992). The mitochondrial genomes of two nematodes, *Caenorhabditis elegans* and *Ascaris suum*. *Genetics* 130: 471–498.

Palmer, J.D. (1985). Evolution of chloroplast and mitochondrial DNA in plants and algae. In *Molecular Evolutionary Genetics* (R.J. MacIntyre, ed.), pp. 131–240. Plenum, New York.

Palmer, J.D. (1991). Plastid chromosomes: Structure and evolution. In *The Molecular Biology of Plastids* (L. Bogorad and I.K. Vasil, eds.), pp. 5–53. Academic Press, San Diego.

Pel, H.J., Rep, M., and L.A. Grivell (1992). Sequence comparison of new prokaryotic and mitochondrial members of the polypeptide chain release factor family predicts a five-domain model for release factor structure. *Nucl. Acids Res.* 20: 4423–4428.

Pietromonaco, S.F., Denslow, N.D., and T.W. O'Brien (1991). Proteins of mammalian mitochondrial ribosomes. *Biochimie* 73: 827–836.

Pinel, C., Douce, R., and R. Mache (1986). A study of mitochondrial ribosomes from the higher plant *Solanum tuberosum* L. *Mol. Biol. Rep.* 11: 93–97.

Pritchard, A.E., Seilhamer, J.J., Mahalingam, R., Sable, C.L., Venuti, S.E., and D.J. Cummings (1990). Nucleotide sequence of the mitochondrial genome of *Paramecium*. *Nucl. Acids Res.* 18: 173–180.

Randolph-Anderson, B.L., Gillham, N.W., and J.E. Boynton (1989). Electrophoretic and immunological comparisons of chloroplast and prokaryotic ribosomal proteins reveal that certain families of large subunit proteins are evolutionarily conserved. *J. Mol. Evol.* 29: 68–88.

Reith, M., and J. Munholland (1993). A high resolution gene map of the chloroplast genome of the red alga *Porphyra purpurea*. *Plant Cell* 5: 465–475.

Rheinberger, H.-J., Geigenmuller, U., Gnirke, A., Hausner, T.-P., Remme, J., Saruyama, H., and K.H. Nierhaus (1990). Allosteric three-site model for the ribosomal elongation cycle. In *The Ribosome* (W.E. Hill, P.B. Moore, A. Dahlberg, D. Schlessinger, R.A. Garrett, and J.R. Warner *eds.*), pp. 318–330. American Society for Microbiology, Washington, D.C.

Robertson, D., Boynton, J.E., and N.W. Gillham (1990). Cotranscription of the wild-type chloroplast *atp*E gene encoding the CF_1/CF_0 epsilon subunit with the 3′ half of the *rps7* gene in *Chlamydomonas reinhardtii* and characterization of frameshift mutations in *atp*E. *Mol. Gen. Genet.* 221: 155–163.

Roney, W.B., Ma, L., Wang, C.-C., and L.L. Spremulli (1991). Recent progress on understanding the initiation of translation in the chloroplasts of *Euglena gracilis*. In *The Translational Apparatus of Photosynthetic Organelles* (R. Mache, E. Stutz, and A.R. Subramanian, eds.), *NATO ASI Series H,* Vol. 55, pp. 197–205. Springer Verlag, Berlin Heidelberg.

Rose, A.M., Joyce, P.B.M., Hopper, A.K., and N.C. Martin (1992). Separate information required for nuclear and subnuclear localization: Additional complexity in localizing an enzyme shared by mitochondria and nuclei. *Mol. Cell. Biol.* 12: 5652–5658.

Ruf, M., and H. Kossel (1988). Occurrence and spacing of ribosome recognition sites in mRNAs of chloroplasts of higher plants. *FEBS Lett.* 240: 41–44.

Sangare, A., Weil, J.H., Grienenberger, J.M., Fauron, C., and D. Lonsdale (1990). Localization and organization of tRNA genes on the mitochondrial genomes of fertile and male sterile lines of maize. *Mol. Gen. Genet.* 223: 224–232.

Schieber, G.L., and T.W. O'Brien (1985). Site of synthesis of the proteins of mammalian mitochondrial ribosomes: Evidence from cultured bovine cells. *J. Biol. Chem.* 260: 6367–6372.

Schmidt, J., Herfurth, E., and A.R. Subramanian (1992). Purification and characterization of seven chloroplast ribosomal proteins: Evidence that organelle ribosomal protein genes are functional and that NH_2-terminal processing occurs via multiple pathways in chloroplasts. *Plant Mol. Biol.* 20: 459–465.

Schnare, M.N., and M.W. Gray (1982). 3'-terminal sequence of wheat mitochondrial 18S ribosomal RNA: Further evidence of a eubacterial evolutionary origin. *Nucl. Acids Res.* 10: 3921–3932.

Schon, A., Krupp, G., Gough, S., Berry-Lowe, S., Kannangara, C.G., and D. Soll (1986). The RNA required in the first step in chlorophyll biosynthesis is a chloroplast glutamate tRNA. *Nature (London)* 322: 281–284.

Schon, A., Kannangara, C.G., Gough, S., and D. Soll (1988). Protein biosynthesis in organelles requires misaminoacylation of tRNA. *Nature (London)* 331: 187–190.

Shimada, H., and M. Sugiura (1991). Fine structural features of the chloroplast genome: Comparison of the sequenced chloroplast genomes. *Nucl. Acids Res.* 19: 983–995.

Small, I., Marechal-Drouard, L., Masson, J., Pelletier, G., Cosset, A., Weil, J.-H., and A. Dietrich (1992). In vivo import of a normal or mutagenized heterologous transfer RNA into the mitochondria of transgenic plants: Towards novel ways of influencing mitochondrial gene expression? *EMBO J.* 11: 1291–1296.

Sreedharan, S.P., Beck, C.M., and L.L. Spremulli (1985). *Euglena gracilis* chloroplast elongation factor Tu: Purification and initial characterization. *J. Biol. Chem.* 257: 10430–10439.

Steege, D.A., Graves, M.C., and L.L. Spremulli (1982). *Euglena gracilis* chloroplast small subunit rRNA. Sequence and base-pairing potential of the 3' terminus, cleavage by colicin E3. *J. Biol. Chem.* 257: 10430–10439.

Steinmetz, A., and J.-H. Weil (1989). Protein synthesis in plants. In *Biochemistry of Plants*, Vol. 15 (A. Marcus, ed.), pp. 193–227. Academic Press, San Diego.

Steinmetz, A., Krebbers, E.T., Schwartz, Z., Gubbins, E.J., and L. Bogorad (1983). Nucleotide sequences of five maize chloroplast transfer RNAs and their flanking regions. *J. Biol. Chem.* 258: 5503–5511.

Subramanian, A. (1993). Molecular genetics of chloroplast ribosomal proteins. *Trends Biochem. Sci.* 18: 177–180.

Subramanian, A.R., Smooker, P.M., and K. Giese (1990). Chloroplast ribosomal proteins and their genes. In *The Ribosome. Structure, Function and Evolution* (W.E. Hill, A. Dahlberg, R.A. Garrett, P.B. Moore, D. Schlessinger, and J.R. Warner, eds.), pp. 655–663. American Society for Microbiology, Washington, D.C.

Subramanian, A.R., Stahl, D., and A. Prombona (1991). Ribosomal proteins, ribosomes, and translation in plastids. In *The Molecular Biology of Plastids* (L. Bogorad and I.K. Vasil, eds.), pp. 191–215. Academic Press, San Diego.

Sugiura, M., and T. Wakasugi (1989). Compilation of transfer RNA genes from tobacco chloroplasts. *Crit. Rev. Plant Sci.* 9: 89–101.

Suyama, Y. (1986). Two dimensional polyacrylamide gel electrophoresis analysis of *Tetrahymena* mitochondrial tRNA. *Curr. Genet.* 10: 411–420.

Tarassov, I.A., and N.S. Entelis (1992). Mitochondrially-imported cytoplasmic tRNA^{Lys}(CUU) of *Saccharomyces cerevisiae*: In vivo and in vitro targetting systems. *Nucl. Acids Res.* 20: 1277–1281.

Toukifimpa, R., Romby, P., Rozier, C., Ehresmann, C., Ehresmann, B., and R. Mache (1989). Characterization and footprint analysis of two 5S rRNA binding proteins from spinach chloroplast ribosomes. *Biochemistry* 28: 5840–5846.

Tzagoloff, A. (1982). *Mitochondria*. Plenum Press, New York.

Tzagoloff, A., and C.L. Dieckmann (1990). PET genes of *Saccharomyces cerevisiae*. *Microbiol. Rev.* 54: 211–225.

Tzagoloff, A., Gatti, D., and A. Gampel (1990). Mitochondrial aminoacyl-tRNA synthetases. *Prog. Nucle. Acids Res.* 39: 129–158.

von Wettstein, D. (1991). Chlorophyll Biosynthesis. In *Plant Molecular Biology 2* (R.G. Herrmann and B. Larkins, eds.), pp. 449–459. Plenum Press, New York.

Wang, C.C., Roney, W.B., Alston, R.L., and L.L. Spremulli (1989). Initiation complex formation on *Euglena* chloroplast 30 S subunits in the presence of natural mRNAs. *Nucl. Acids Res.* 17: 9735–9747.

Weil, J.H., and B. Parthier (1982). Transfer RNA and aminoacyl-tRNA synthetases in plants. *Encyclopedia Plant Physiol.* 14A: 65–112.

Wise, C.A., and N.C. Martin (1991). Dramatic size variation of yeast mitochondrial RNAs suggests that RNase P RNAs can be quite small. *J. Biol. Chem.* 266: 19154–19157.

Wolf, K., and L. Del Giudice (1988). The variable

mitochondrial genome of ascomycetes: Organization, mutational alterations, and expression. *Adv. Genet.* 25: 185–308.

Wolstenholme, D.R. (1992). Animal mitochondrial DNA: Structure and evolution. *Int. Rev. Cytol.* 141: 173–216.

Wolstenholme, D.R., Macfarlane, J.L., Okimoto, R., Clary, D.O., and J.A. Wahleithner (1987). Bizarre tRNAs inferred from DNA sequences of mitochondrial genomes of nematode worms. *Proc. Natl. Acad. Sci. U.S.A.* 84: 1324–1328.

Wu, M., and A. Tzagoloff (1987). Mitochondrial and cytoplasmic fumarases in *Saccharomyces cerevisiae* are encoded by a single nuclear gene *FUM1*. *J. Biol. Chem.* 262: 12275–122282.

Zhou, D.-X., and R. Mache (1989). Presence in the stroma of chloroplasts of a large pool of a ribosomal protein not structurally related to any *Escherichia coli* ribosomal protein. *Mol. Gen Genet.* 216: 439–445.

Zurawski, G., and M.T. Clegg (1987). Evolution of higher-plant chloroplast DNA-encoded genes: Implications for structure-function and phylogenetic studies. *Annu. Rev. Plant Physiol.* 38: 391–418.

14

Reading and Editing Messages in Chloroplasts and Mitochondria

The evolution of the genetic code has been comprehensively reviewed by Osawa *et al.* (1992) and earlier by Fox (1987) and for chloroplasts and mitochondria specifically by Jukes and Osawa (1990). The 64 codons of the universal code can be assigned to 16 quartets of triplets UUN, CUN, GCN, etc. in which N represents the third of 3' base in the codon (i.e., A, C, G or U). Eight of these quartets (coding for serine, leucine, proline, arginine, threonine, valine, alanine, and glycine) are "families" in which all four codons in the quartet code for the same amino acid (Table 14–1). Five quartets (coding for phenylalanine/leucine, aspartic acid/glutamic acid, histidine/glutamine, serine/arginine, and asparagine/lysine) are designated "nonfamilies" because the two codons with a 3'-pyrimidine (Py) designate one amino acid while the two with a 3'-purine (Pu) code for a second. There are also three special codon groups. Two of these include termination codons and the third the initiator codon. UAN includes two codons (UAPy) specifying tyrosine and two (UAPu) that act as terminators so a single tRNA^{Tyr} is required for this quartet. In the quartet UGN two codons (UGPy) code for cysteine, one (UGG) for tryptophan, and one (UGA) for termination. Finally, the quartet AUN includes three isoleucine codons: two (AUPy) are read by one tRNA and the third (AUA) is recognized by a second. The fourth codon (AUG) specifies methionine and is read either by the initiator tRNA or the tRNA that inserts methionine residues internally.

The universal (classical wobble) genetic code requires a minimum of 32 tRNAs including the initiator tRNA to decipher the 61 "sense" codons. Both families and nonfamilies require two tRNAs for a total of 26 different species. The codons in the special quartets account for the remaining six (Table 14–1).

In addition to classical wobble other simplified mechanisms for reading the code have been proposed that require less than 32 tRNAs. In the "two out of three" code of Lagerkvist (1978, 1981) all four codons in a family are read by a single tRNA anticodon (i.e., the base in the 3' position is ignored). Thus, only 24 tRNAs are required to read this code. In "four-way" wobble families are treated in the same way except an unmodified U in the 5' position of the anticodon pairs with U, C, A, or G in the 3' position of the codon (Osawa *et al.*, 1992). The completely sequenced chloroplast

Table 14–1. Reading the Genetic Code by the Classical Wobble (WOB) versus the "Two Out of Three" (TOT) and "Four-Way Wobble" Models

WOB	TOT	WOB	TOT	WOB	TOT	WOB	TOT
UUU ⟩ Phe ← Phe UUC		UCU ⟩ Ser1 UCC		UAU ⟩ Tyr ← Tyr UAC		UGU ⟩ Cys ← Cys UGC	
			Ser1				
UUA ⟩ Leu ← Leu UUG		UCA ⟩ Ser3 UCG		UAA ⟩ Ter ← Ter UAG		UGA ← Ter ← Ter UGG ← Trp	
CUU ⟩ Leu1 CUC		CCU ⟩ Pro1 CCC		CAU ⟩ His ← His CAC		CGU ⟩ Arg1 CGC	
	Leu		Pro				Arg1
CUA ⟩ Leu2 CUG		CCA ⟩ Pro2 CCG		CAA ⟩ Gln ← Gln CAG		CGA ⟩ Arg3 CGG	
AUU ⟩ Ile1 ← Ile1 AUC		ACU ⟩ Thr1 ACC		AAU ⟩ Asn ← Asn AAC		AGU ⟩ Ser2 ← Ser2 AGC	
			Thr				
AUA ⟩ Ile2 ← Ile2 AUG Met ← Met		ACA ⟩ Thr2 ACG		AAA ⟩ Lys ← Lys AAG		AGA ⟩ Arg2 ← Arg2 AGG	
GUU ⟩ Val1 GUC		GCU ⟩ Ala1 GCC		GAU ⟩ Asp ← Asp GAC		GGU ⟩ Gly1 GGC	
	Val		Ala				Gly
GUA ⟩ Val2 GUG		GCA ⟩ Ala2 GCG		GAA ⟩ Glu ← Glu GAG		GGA ⟩ Gly2 GGG	

genomes of three flowering plants, and *Marchantia*, encode sufficient tRNAs so that most codons can be deciphered with a few exceptions via the classical wobble. However, animal and fungal mitochondrial genomes do not specify enough tRNAs to read mRNAs via the universal mechanism, but they encode sufficient tRNAs to decode mitochondrial messages using one or the other variation of the universal code mentioned above. It is also important to remember that the discussion that follows assumes that translation is solely dependent on tRNAs encoded by the organelle genome. Yet we know that tRNA import has arisen independently several times in mitochondria and at least once in plastids (Chapter 13) so imported tRNAs could play a role in some instances.

Not only do variations exist in how the code is deciphered, particularly in mitochondria, but mRNAs themselves may be unreadable without the process of RNA editing. RNA editing, like tRNA import, seems to have evolved independently for both chloroplasts and mitochondria and several times for mitochondria alone. RNA editing is the subject of the second half of this chapter.

READING THE CODE IN ORGANELLES

Reading the code in chloroplasts. Chloroplasts employ the universal code and the 30–31 tRNA genes encoded by land plant chloroplast genomes (Chapter 13) are almost sufficient to read the code by the classical wobble method. The chloroplast genome encodes a typical initiator tRNA$^{fMet}_{CAU}$ and employs the three classical termination codons (UAA, UAG, UGA). However, tRNAs recognizing the codons CUU/C (Leu), CCU/C (Pro), GCU/C (Ala), and CGC/A/G (Arg) are absent from the chloroplast genomes of tobacco and rice, although the tRNA that recognizes CGG is encoded in the liverwort chloroplast genome (Shimada and Sugiura, 1991; Sugiura and Wakasugi, 1989). Since all 61 sense codons are used in all three genomes, this deficit in specific tRNAs requires that the tRNAs be imported or that these codons be read by the two-out-of-three- or four-way wobble mechanisms. One of the latter two mechanisms seems to be used for tRNA$^{Ala}_{U*GC}$, tRNA$^{Pro}_{U*GG}$ (U* is a modified U), and tRNA$^{Arg}_{ICG}$, which can read respectively all four alanine (GCN), proline (CCN), and arginine (CGN) codons (Pfitzinger *et al.,* 1990).

The problem of decoding the six leucine codons is solved somewhat differently. Two of the leucyl tRNAs translate the UUA and UUG codons (Pfitzinger *et al.,* 1987). The remaining tRNA$^{Leu}_{UAm7G}$ translates all four CUN codons for leucine apparently using a U:N wobble mechanism (Pfitzinger *et al.,* 1990).

The completely sequenced chloroplast genome of *Euglena gracilis* specifies only 27 tRNAs. Hallick *et al.* (1993) propose that a single tRNA is used to decipher all the members of seven of the eight codon families. For the eighth family, glycine, genes encoding two tRNAs have been identified.

Reading the code in mitochondria. The genetic code in plant mitochondria is identical to the universal code (Jukes and Osawa, 1990). Furthermore, plant mitochondria probably have enough different tRNA species to read the code by classical wobble, which would also be consistent with the fact that they contain "chloroplast-like" and "cytosolic" tRNAs in addition to the "native" species (Chapter 13).

While fungal mitochondrial genomes contain between 25 and 27 tRNA genes, sufficient to read the code by the two-out-of-three method or four-way wobble, animals make do with 22, which is two less than the required number (Chomyn and Attardi, 1987). The solution to this problem for most animals can be traced to the nonfamily containing methionine and isoleucine. In the universal code four tRNAs are required. Two of these recognize isoleucine codons (Table 14–1) and two methionine codons. In the latter case AUG codons at the beginning of the mRNA are read by the initiator tRNA^{F-Met} coded by one gene and internal AUG codons by tRNA^{Met} coded by a second gene. This is also the way it works in fungi. In animals there is a single mitochondrial methionine tRNA gene that is thought to encode both tRNA species, (Chomyn and Attardi, 1987). Secondary modifications of the primary transcripts are assumed to yield the two tRNA species both of which have been detected in mammalian mitochondria. Furthermore, not only AUG, but AUA, AUU, and AUC can serve as initiator codons in animal mitochondria. All three of the latter codons specify isoleucine in the universal code (Table 14–1). However, in the mitochondria of most animals, AUA codes for methionine (Table 14–2), which means only two tRNAs rather than four are required to decode the isoleucine/methionine nonfamily. Assuming that tRNA$^{F\text{-met}}$ is the normal initiator tRNA, its CAU anticodon must be able to read not only the normal AUG initiator codon, but also the three other initiator codons found in metazoan mitochondrial mRNAs. The assumption is that modification of the C residue or lack thereof is responsible (Chomyn and Attardi, 1987). Other initiation codons also pop up in different animal genes (Wolstenholme, 1992). GUG seems to be the initiator for a few genes and the four base sequence AUAA has been proposed as the initiator codon for the *cox1* gene of *Drosophila yakuba*. UUG is a popular initiation codon for nematode mitochondrial genes and GUU is also used in at least one instance. UUG specifies leucine in the universal code and GUU normally encodes valine (Table 14–1).

One of the more important deviations in the mitochondrial genetic code concerns UGA, which (in addition to UGG) specifies tryptophan instead of termination in animals, yeasts and filamentous ascomycetes, but may have its universal code meaning in the basidiomycete *Schizophyllum* (Clark-Walker, 1992). UGA also specifies tryptophan in *Mycoplasma* (Osawa et al., 1992). Osawa and Jukes (1989) postulated that the steps involved in this process of "codon capture" in *Mycoplasma* were (1) an increase in AT directional mutation pressure that converted all UGA stop codons to UAA; (2) an accompanying deletion of release factor 2; and (3) a change in the anticodon of tryptophan tRNA from CCA to UCA that permits pairing of the anticodon with both UGA and UGG. They assume the same series of events took place in mitochondria (Osawa et al., 1992). Utilization of UGA to encode tryptophan means that a mitochondrial analog of release factor 2 can be dispensed with, which is consistent with the finding of only release factor 1 in yeast mitochondria (Chapter 13).

In the serine/arginine nonfamily (Table 14–1) AGPy encodes serine in the universal code and AGPu specifies arginine. This is also true of ascomycetes and protozoa, but only in the most basal animals (Cnidaria) (Table 14–2). The AGPu codons are read

Table 14–2. Genetic Code Variations in Mitochondria

Organism	UGA Stop	AUA Ile	AAA Lys	AGPu Arg	CUN Leu
Vertebrates	Trp	Met	Lys	Stop[a]	Leu
Arthropods	Trp	Met	Lys	Ser[b]	Leu
Echinoderms	Trp	Ile	Asn	Ser	Leu
Molluscs	Trp	Met	Lys	Ser	Leu
Nematodes	Trp	Met	Lys	Ser	Leu
Platyhelminthes	Trp	Met	Asn	Ser	Leu
Cnidarians	Trp	Ile	Lys	Arg	Leu
Yeasts	Trp	Ile, Met	Lys	Arg	Thr
Aspergillus, Neurospora	Trp	Ile	Lys	Arg	Leu
Paramecium	Trp	Ile	Lys	Arg	Leu

Data from Clark-Walker (1992), Osawa et al. (1992), and Wolstenholme (1992).

[a] Either AGA or AGG may be missing completely from certain mitochondrial genomes. For example, AGA is not found in the mitochondrial genome of chicken while AGG is absent from the mitochondrial genome of the toad *Xenopus laevis*.

[b] AGA only codes for Ser.

as serine from Nematodes through Arthropods so the quartet of codons beginning with AG is treated as a family. However, in vertebrates AGPu means stop. In the universal code AAA means lysine, but it means asparagine in echinoderms and platyhelminths (Osawa et al., 1992). AUA specifies methionine in most yeasts as well as most animals, but has its universal code meaning of isoleucine in at least one yeast *(Brettanomyces)* (Clark-Walker, 1992), filamentous ascomycetes, ciliates, cnidarians, and echinoderms (Table 14–2). Finally, the CUN codons are reassigned from Leu to Thr in yeasts. The postulated chain of events leading to each of these codon captures is reviewed by Osawa et al. (1992).

Why has the genetic code and tRNA usage become so streamlined in mitochondria? Kurland (1992) advances the interesting thesis that mitochondrial genomes within a cell compete for replicative advantage. Replication will be faster if some genes and their products are eliminated. Kurland supposes that defective genomes, particularly those with deletions, "can take over the micro-populations of a cell line when the intracellular genomic populations are small and random fluctuations in the proportional segregation of different genomes are common." Kurland believes that as long as the deleted genes are transferred to the nuclear genome and expressed, mitochondria with deleted genes can survive. "The reassignment of codons to amino acids different from those designated in the so-called universal code is seen in part as an expression of the reduction of the number of genes used by these genomes to code for tRNA species."

RNA Editing

As if there were not enough peculiarities in reading the genetic code in organelles, RNA editing had to come along. So far, this process has been identified in organelles with "tamer" genetic codes (i.e., chloroplasts and plant mitochondria) as well as in the trypanosome kinetoplast where, like everything else associated with this organelle, the bizarre and unexpected are always the rule.

Two reviews by Cattaneo (1990, 1991) serve as a useful contextual basis from which to consider the overall problem of RNA editing. Cattaneo recognizes three general classes of editing mechanisms (Table 14–3).

1. Modification–substitution of nucleotides occurs extensively in flowering plant mitochondria and has been discovered more recently in chloroplasts. This mechanism is also employed in the case of the mRNA for the apolipoprotein B transcript of mammalian cells where a C → U posttranscriptional conversion at position 6666 in the 14-kb transcript generates a novel stop codon. This stop codon defines a major form of the protein found in the intestine.
2. Posttranscriptional insertion and deletion of nucleotides are elaborate RNA-guide-directed processes in trypanosomes. Cytidine insertions have also been discovered in mitochondrial transcripts of the slime mold *Physarum,* but whether guides are involved is unknown.
3. Cotranscriptional insertion of nucleotides, the third editing mechanism, will not be discussed. This mechanism applies to certain viruses and to the creation of stop codons in animal mitochondrial mRNAs by polyadenylation.

Modification–substitution of nucleotides. RNA editing in flowering plant mitochondria has been the source of much excitement with the percent amino acid changes occurring

Table 14-3. Mechanisms and Consequences of Different Types of mRNA Editing

Type of mRNA Editing	Base	Mechanism	Consequences for Protein Expression	Gene(s) and Organism
Modification–substitution of nucleotides	C → U	Deamination?	Shorter protein	Apolipoprotein gene of mammals
	C → U, U → C[a]	Unknown	Modified amino acid composition	Plant mitochondrial and chloroplast genes
Cotranscriptional insertion of nucleotides	A	Polyadenylation	Introduction of UAA stop codons	Mitochondrial genes of animals
	G	Polymerase stuttering	Additional protein (efficiency 20–50%)	Phosphoprotein gene of paramyxoviruses
	G			P gene of measles and mumps viruses
Posttranscriptional insertion and deletion of nucleotides	U	Correction by − strand "guide" RNAs	Limited or extensive alteration of the reading frame	Mitochondrial genes of trypanosomes
	C	Unknown	Modified amino acid composition	Mitochondrial genes of slime molds

Modified from Catteneo (1990).

[a] Rarely for plant mitochondrial genes.

varying from 0.4% to 14.6% depending on the gene and plant (see reviews by Bonnard et al., 1992; Grienenberger et al., 1991; Gray et al.; 1992 Mulligan, 1991; Schuster et al., 1991, 1993; Walbot, 1991). The discovery of RNA editing in these mitochondria is an interesting story. Fox and Leaver (1981) reported the sequence of the maize *cox2* gene and found a CGG (Arg) codon at three positions where tryptophan (UGG) was conserved in the COX2 polypeptides of other organisms. They postulated that CGG might code for tryptophan in plant mitochondria, a perfectly reasonable hypothesis at the time given the deviations from the universal code being found in animal mitochondria. However, some CGG codons were also found at positions conserved as arginine and no tRNA^{Trp} able to recognize CGG was identified in plant mitochondria (see Bonnard et al., 1992). Recognizing this anomaly, Covello and Gray (1989) and Gualberto et al. (1989) published back-to-back papers in *Nature*. They demonstrated that when cDNAs made from wheat mitochondrial mRNAs were sequenced, the anomalous CGG codons specified in the mitochondrial genome had been converted to UGG for tryptophan. Thus, the problem had been solved. C → U editing converted CGG → UGG codons and there was no deviation from the universal code in flowering plant mitochondria.

Editing is almost always from C → U and can be quite extensive. Several cases of U → C edits have also been reported (see Bonnard et al., 1992 for discussion). Editing is most pronounced in protein coding regions, but has been reported in flanking regions as well as in intron sequences. In fact, certain intron sites in conserved secondary structure domains are frequently edited and editing seems to be essential for correct folding of such introns (Schuster et al., 1993). Structural RNAs in contrast to mRNAs do not seem to be edited via modification-substitution except in *Acanthamoeba*. RNA editing seems to be correlated with transcript maturation, but translation of a heterogeneous population of transcripts cannot currently be ruled out. Most modifications (86%) result in an amino acid change, although others create stop and start codons. Silent modifications represent only 14–16% of the number of conversions although a random distribution of edited Cs in all editable codons would dictate that nearly 30% of possible modifications would be silent. The end result of RNA editing in plant mitochondria is that most mitochondrially encoded proteins are nearly identical in all plant species.

C → U editing has also been demonstrated for four positions in the *ndh1* gene of maize chloroplasts (Maier *et al.*, 1992). In each case editing restores codons for conserved amino acids in the polypeptide encoded by this gene. Editing is also responsible for converting ACG triplets to AUG initiation codons for two other chloroplast mRNAs (Hoch *et al.*, 1991; Kudla *et al.*, 1992). Editing of transcripts for two photosynthetic mRNAs in spinach has been found to be stage-specific (Bock *et al.*, 1993). Editing is complete in etioplasts as well as chloroplasts suggesting that the process is light-independent, but editing occurs with much lower efficiency in seeds and roots (Bock *et al.*, 1993). Hence, editing may serve as a regulatory mechanism in plastid gene expression.

The mechanism of modification–substitution editing is not understood. Bonnard *et al.* (1992) suggest three possible mechanisms. Deamination/amination is the simplest. Deamination of C at position 4 by cytidine deaminase would yield U. Conversely, the reverse modification U → C could be carried out by a cytosine triphosphate synthetase. Base nucleotide exchange is the second mechanism and would involve a trans-glycosylation reaction leading to the removal of a cytosine from its ribose residue without breakage of the phosphodiester backbone. A uridine would then be inserted into the apyrimidinic site. Finally, deletion and insertion of bases could occur in which case the editing process would not involve a modification–substitution process after all.

No matter which of these mechanisms prevails the question is how precise editing is achieved. Are guide (g) RNAs involved as they are in trypanosome editing? Such gRNAs have not so far been identified in hybridization experiments although small, cappable RNA species have been found in soybean mitochondria (Gray *et al.*, 1992; Schuster *et al.*, 1993). Walbot (1991) wonders in her review whether plant mitochondrial gRNAs could be the imported products of nuclear genes. After all tRNAs are transported into plant mitochondria so why not gRNAs? An alternative possibility is that the information required for editing is specified in the structure of the preedited mRNA itself (Gray *et al.*, 1992). This seems to be the case of the mammalian lipoprotein B mRNA.

Amidst all this speculation Lu and Hanson (1992) report an interesting genetic result that provides a possible glimpse into one aspect of the editing mechanism. They find that editing of the *Petunia ndh3* gene at the three sites studied seems to be controlled by a single nuclear gene that also controls transcript abundance. Other transcripts are unaffected by this gene. Lu and Hanson surmise that the gene product could either affect transcript stability and, therefore, the opportunity of the transcript to be edited or the gene could control the editing process itself with the edited transcripts being stable. If the latter interpretation is correct, it could mean that the editing process is highly specific with a 1:1 correspondence between nuclear genes and mitochondrial transcripts. Hence, editing may represent just one more example of a specific posttranscriptional control mechanism by which nuclear genes influence expression of mitochondrial genes (Chapter 17).

Lonergan and Gray (1993) have described yet another editing system in *Acanthamoeba castellani* that seems to fit the modification–substitution category. This is the first reported case of tRNA editing and consists of single nucleotide conversions (U → A, U → G, and A → G) in the tRNA acceptor stem that correct mismatched base pairs.

Posttranscriptional insertion and deletion of nucleotides. The best understood RNA editing system is that of the trypanosome kinetoplast (reviewed by Benne, 1990, 1993; Cattaneo, 1991; Hajduk et al., 1993; Simpson and Shaw, 1989; and Stuart, 1991a,b, 1993; Stuart and Feagin, 1992). Editing in trypanosomes creates translatable open reading frames by insertion or to a lesser degree deletion of U residues. Table 14–4 presents one example of RNA editing for the *ndh7* gene in three species of trypanosomes. As the table shows the degree of editing of a specific gene can vary greatly. In *Leishmania tarentolae* 25 Us are added to the mRNA by editing, but in *Trypanosoma brucei* 553 Us are added and 89 deleted. The latter case is extreme, but not atypical of *T. brucei* where editing is very extensive (Stuart, 1991a,b, 1993).

Comparison of edited sequences between flowering plant mitochondrial and kinetoplast genes is instructive (Table 14–5). A partial sequence of the wheat *cox2* gene shows that edited sites are usually not clustered and base modification rather than insertion seems to be involved. In contrast, massive insertion of U residues occurs to create new codons in the mRNAs produced by trypanosome kinetoplast genes or cryptogenes as they are frequently called. Simpson and Shaw (1989) distinguish three classes of cryptogenes. Type I cryptogenes yield RNA molecules that are edited at internal positions within protein coding regions. Type II cryptogenes yield transcripts that are edited at the 5′-ends to create initiator as well as new internal codons (Table 14–5). Some genes also have both Type I and Type II cryptogene properties. Type III cryptogenes yield transcripts that are extensively edited (pan-edited) over their entire length to generate RNA nucleotide sequences that encode completely new proteins. In fact, kinetoplast genes can be so encrypted that homology to related genes cannot be detected without knowing the edited sequence. An example of such a cryptogene is a conserved, short G-rich ''intergenic'' region that turns out not to be an intergenic region after all, but a pan-edited cryptogene encoding ribosomal protein S12 (Maslov et al., 1992). Editing also creates continuous ORFs from frameshifted genomic coding sequences. Editing is posttranscriptional and proceeds in the 3′ to 5′ direction. Unedited, partially edited, and fully edited molecules are all found as they are in plant mitochondria. On the other hand, not all trypanosome mRNAs require editing. Examples are the *ndh1, ndh4, ndh5,* and *cox1* transcripts (Stuart, 1991a,b).

Table 14–4. *ndh7* Editing in the Mitochondria of Three Trypanosome Species

Species	Added/ Deleted Us	Initiation Codon	Termination Codon	Location of gRNAs
Leishmania tarentolae	25/0	Non-AUG?	Encoded	Maxicircle
Crithidia fasciculata	27/0	Altered	Encoded	Maxicircle
Trypanosoma brucei	553/89	Created	Created	Minicircle

Modified from Cattaneo (1991).

Table 14–5. Comparison of Gene and Edited mRNA Sequences for the *cox2* Gene of Wheat and Three Edited Sequences from *Trypanosoma brucei*.[a]

Gene	Sequence	
Wheat *cox2*	DNA	ACT ACT ATC GAA ATT ATT ATT CGG ACC ATA TTT CCA AGT
	RNA	AC<u>U</u> AC<u>U</u> AC<u>U</u> GAA A<u>U</u>U A<u>U</u>U A<u>U</u>U <u>U</u>GG ACC A<u>U</u>A <u>U</u>UU CCA AG<u>U</u>
	Protein	Trp
T. brucei		
cox2 internal	DNA	AAG GTA GA G A A CCT GGA
	RNA	AAG GUA GA<u>U</u> <u>U</u>GU A<u>U</u>A CCU GGA
	Protein	Asp Cys Ile
cytb 5'	DNA	ATATAAA A G CG G AGA A A A
	RNA	A<u>U</u>A<u>U</u>AAA<u>U</u> A<u>U</u>G <u>UU</u>U CG<u>U</u> <u>U</u>G<u>U</u> AGA <u>U</u>UU <u>UU</u>A <u>UU</u>A <u>U</u>UU <u>U</u>UU <u>UU</u>A
	Protein	Met Phe Arg Cys Phe Leu Leu Phe Phe Leu
cox3 pan	DNA	A CG GG A GA G G A A CG G
	RNA	<u>UU</u>A CG<u>U</u> GG<u>U</u> <u>UU</u>A <u>U</u>UU GA<u>U</u> <u>U</u>UU <u>U</u>G<u>U</u> G<u>U</u>U <u>UU</u>A CG<u>U</u> <u>U</u>G<u>U</u>
	Protein	Leu Arg Gly Leu Phe Asp Phe Cys Val Leu Arg Cys

Partial sequence data for wheat *cox2* taken from Covello and Gray (1989) and *T. brucei* data from Simpson and Shaw (1989). Note that there seems to be a sequencing error for *cox2* in Fig. 4 of the Simpson and Shaw paper. An extra A is inserted after the second codon for Val which brings the rest of the sequence shown out of frame.

[a] Edited Us are underlined and only those amino acids changed as a result of editing are indicated in the protein sequence. Internal, internal editing; 5', 5' edited; pan, pan edited (see Simpson and Shaw, 1989).

Guide RNAs, discovered by Simpson and colleagues (Blum and Simpson, 1990), are small (50–80 nucleotide) molecules that carry the editing information. These gRNAs also possess on average 15 noncoded U residues at their 3'-termini that are probably added by a 3'-terminal uridyl transferase (TUTase). The gRNAs contain sequences of perfect complementarity to the edited regions. This complementarity of the 5'-end of the gRNA with mRNA involves standard Watson–Crick base pairing plus some wobble (G:U) base pairing. The wobble base pairing implies that gRNA is not functioning as a conventional template because gRNA G residues would specify Cs while gRNA U residues would be complementary to As. Arts *et al.* (1993) report for the *ndh7* gene of *Crithidia fasciculata* that there is an 8-fold difference in the molar ratio of the two gRNAs required to edit the two domains of the mRNA, but that both domains are edited with frequencies of around 50%. These investigators also find that many gRNAs exist with the same 5'-end, but many different 3'-uridylation sites. About 20% of these gRNAs are short and lack the information required to edit a complete domain. Arts *et al.* speculate that participation of these truncated gRNAs in the editing process could explain the high incidence of partially edited mRNAs.

The original model to explain the mechanism by which gRNAs cause insertion of U residues proposed by Blum *et al.* (1990) assumed cycles of mRNA cleavage, U addition (or deletion), and religation. This model has been since revised so that editing is now thought to involve a series of transesterification reactions similar to those involved in intron removal (see Chapter 10; Blum *et al.,* 1991; Cech, 1991). In the transesterification model, (see reviews by Cattaneo, 1991; Sollner-Webb, 1992, and Stuart, 1993) the 5'-end of the gRNA binds to its complementary 3' "anchor" in the mRNA prior to editing as it would in the cleavage-religation model (Fig. 14–1A). After base pairing, the editing reaction begins with an attack at the 3'-phosphate of the first mismatched base in the mRNA by the terminal hydroxyl (—OH) group of the poly(U) tail of the gRNA (thick arrow, Fig. 14–1A). This would lead to the formation of a chimeric intermediate in which the gRNA is transiently linked covalently to the 3' portion of the mRNA. This chimeric intermediate has been identified (Blum *et al.,* 1991; Koslowsky *et al.,* 1991, Read *et al.,* 1992). The first block of seven U residues from the poly(U) tail inserted into the mRNA shown in Fig. 14–1B then pair with the complementary sequence of seven A and G residues in the guide. In a second transesterification reaction the 3' OH— end of the released mRNA fragment mounts a counterattack within the poly(U) tail of the chimeric molecules (thick arrow, Fig. 14–1B), liberating the gRNA and rejoining the now partially edited mRNA. For transfer of each additional block of Us, at least one transesterification cycle is required. However, the poly(U) tail of the gRNA shown in Fig. 14–1 is initially not long enough to provide the 20 U residues required for editing, so elongation of the tail is assumed to occur during editing.

In vitro systems have been developed to study the details of the editing process, chimeric molecules have been detected as mentioned above (see Sollner-Webb, 1992, for review) and a specific endonuclease has been found that cleaves the pre-edited mRNA near the 3'-most uridine addition sites (Harris *et al.,* 1992). Characterization of the editing machinery or "editosome" particularly in an efficient *in vitro* system will be important because there are differences in opinion as to the mechanism involved (Arts *et al.,* 1993; Benne, 1993).

How is the editing machinery organized? Pollard *et al.* (1992) report the isolation

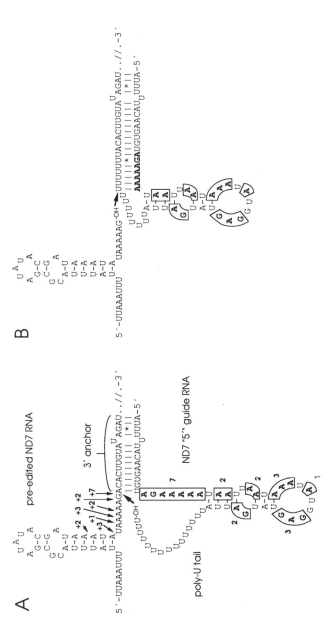

Fig. 14–1. Model for RNA editing in trypanosomes by transesterification as exemplified by the 5' portion of the *ndh7* (ND7) transcript. (A) The *ndh7* preedited RNA is represented on top and the *ndh7* "5'" guide RNA (gRNA) on the bottom. A region of complementarity between these two RNAs is indicated as the 3' anchor (right). Standard complementarity is indicated by dashes and G:U base pairs by asterisks. Residues (As or Gs) guiding U insertion are bold and boxed. The poly(U) tail, added posttranscriptionally to the gRNA, is also shown. (B) Same as A, but after the first transesterification and base pairing of the gRNA 3' seven U residues with the "guide" A and G residues of the first block. From Cattaneo (1991).

of two complexes from *T. brucei* that are involved in editing. Complex I (19S) contains a gRNA complexed with TUTase, RNA ligase and chimera-forming activities. Complex II (35-40S) possesses pre-edited mRNA, gRNA, RNA ligase and chimera-forming activities. Pollard *et al.* speculate that complex I forms first and that TUTase in this complex adds the nonencoded poly (U) tail to the gRNA. Following this, TUTase is released and the pre-edited mRNA is bound to form complex II which would be the active editosome. Gel-retardation and UV-crosslinking experiments have also led to the identification of several complexes and polypeptides that seem to be associated with the editing process (Stuart, 1993). Thus, five proteins with molecular weights varying from 38 to 90 kDa appear to be associated with gRNA.

The requirement for numerous different species of gRNAs encoded by minicircle as well as maxicircle DNA has provided a functional role for the minicircles beyond their probable requirement for orderly segregation of kinetoplast DNA complexes during cell division (Chapter 5).

Is editing a conservative force in trypanosome evolution? If anything, the opposite appears to be true according to a paper by Landweber and Gilbert (1993). They examined RNA editing in the mitochondria of four species of insect parasites of the genus *Herpetomonas* and found that editing was extensive. Landweber and Gilbert observed that RNA editing was a novel source of frameshift mutations over time and that edited proteins accumulate mutations at nearly twice the rate of unedited proteins.

A distinct insertional editing system has been discovered in the mitochondrion of the slime mold *Physarum polycephalum* that applies both to mRNAs and structural RNAs (Mahendran *et al.*, 1991; Miller *et al.*, 1993). So far, 297 nucleotide insertions have been identified at 285 sites in eleven RNAs which include five mRNAs, both rRNAs, and four tRNAs. About 92% of these insertions are single cytidines, with the rest being single uridine insertions (4%) and various dinucleotide insertions (4%).

REFERENCES

Arts, G.J., van der Spek, H., Speijer, D., van den Burg, J., van Steeg, H., Sloof, P., and R. Benne (1993). Implications of novel guide RNA features for the mechanism of RNA editing in *Crithidia fasciculata*. *EMBO J.* 12: 1523–1532.

Benne, R. (1990). RNA editing in trypanosomes: Is there a message? *Trends Genet.* 6: 177–181.

Benne, R. (1993). RNA editing in mitochondria of *Leishmania tarentolae* and *Crithidia fasciculata*. *Seminars in Cell Biology* 4: 241–249.

Blum, B., and L. Simpson (1990). Guide RNAs in kinetoplastid mitochondria have a nonencoded 3′ oligo(U) tail involved in recognition of the preedited region. *Cell* 62: 391–397.

Blum, B., Bakalara, N., and L. Simpson (1990). A model for RNA editing in kinetoplastid mitochondria: "Guide" RNA molecules transcribed from maxicircle DNA provide the edited information. *Cell* 60: 189–198.

Blum, B., Sturm, N.R., Simpson, A.M., and L. Simpson (1991). Chimeric gRNA-mRNA molecules with oligo(U) tails covalently linked at sites of RNA editing suggest that U addition occurs by transesterification. *Cell* 65: 543–550.

Bock, R., Hagemann, R., Kossel, H., and J. Kudla (1993). Tissue- and stage-specific modification of RNA editing of the *psbF* and *psbL* transcripts from spinach plastids—a new regulatory mechanism? *Mol. Gen. Genet.* 240: 238–244.

Bonnard, G., Gualberto, J.M., Lamattina, L., and J.M. Grienenberger (1992). RNA editing in plant mitochondria. *Crit. Rev. Plant Sci.* 10: 503–524.

Cattaneo, R. (1990). Messenger RNA editing and the genetic code. *Experientia* 46: 1142–1148.

Cattaneo, R. (1991). Different types of messenger RNA editing. *Annu. Rev. Genet.* 25: 71–88.

Cech, T.R. (1991). RNA editing: World's smallest introns? *Cell* 64: 667–669.

Chomyn, A., and G. Attardi (1987). Mitochondrial gene products. *Curr. Topics Bioenerget.* 15: 295–329.

Clark-Walker, G.D. (1992). Evolution of mitochondrial genomes in fungi. *Int. Rev. Cytol.* 141: 89–127.

Costanzo, M.C., and T.D. Fox (1990). Control of mitochondrial gene expression in *Saccharomyces cerevisiae*. *Annu. Rev. Genet.* 24: 91–113.

Covello, P.S., and M.W. Gray (1989). RNA editing in plant mitochondria. *Nature (London)* 341: 662–666.

Fox, T.D. (1987). Natural variation in the genetic code. *Annu. Rev. Genet.* 21: 67–91.

Fox, T.D., and C.J. Leaver (1981). The *Zea mays* mitochondrial gene coding cytochrome oxidases subunit II has an intervening sequence and does not contain TGA codons. *Cell* 26: 315–323.

Gray, M.W., Hanic-Joyce, P.J., and P.S. Covello (1992). Transcription, processing and editing in plant mitochondria. *Annu. Rev. Plant Physiol. Plant Mol. Biol.* 43: 145–175.

Grienenberger, J.-M., Lamattina, L., Weil, J.-H., Bonnard, G., and J. Gualberto (1991). RNA editing in wheat mitochondria: A new mechanism for the modulation of gene expression. In *Plant Molecular Biology*, Vol. 2 (R.G. Herrmann and B. Larkins, eds.), pp. 365–373. Springer-Verlag, Wien, New York.

Gualberto, J.M., Lamattina, L., Bonnard, G., Weil, J.-H., and J.-M. Grienenberger (1989). RNA editing in wheat mitochondria results in the conservation of protein sequences. *Nature (London)* 341: 660–662.

Hajduk, S.L., Harris, M.E., and V.W. Pollard (1993). RNA editing in kinetoplastid mitochondria. *FASEB J.* 7: 54–63.

Hallick, R.B., Hong, L., Drager, R.G., Favreau, M.R., Monfort, A., Orsat, B., Spielmann, A., and E. Stutz (1993). Complete sequence of *Euglena gracilis* chloroplast DNA. *Nucl. Acids Res.* 21: 3537–3544.

Harris, M., Decker, C., Sollner-Webb, B., and S. Hajduk (1992). Specific cleavage of pre-edited mRNAs in Trypanosome mitochondrial extracts. *Mol. Cell. Biol.* 12: 2591–2598.

Hoch, B., Maier, R.M., Appel, K., Igloi, G.L., and H. Kossel (1991). Editing of a chloroplast mRNA by creation of an initiation codon. *Nature (London)* 353: 178–180.

Jukes, T.H., and S. Osawa (1990). The genetic code in mitochondria and chloroplasts. *Experientia* 46: 1117–1126.

Koslowsky, D.J., Bhat, G.J., Read, L.K., and K. Stuart (1991). Cycles of progressive realignment of gRNA with mRNA in RNA editing. *Cell* 67: 537–546.

Koslowsky, D.J., Goringer, H.U., Morales, T.H., and K. Stuart (1992). In vitro guide RNA/mRNA chimaera formation in *Trypanosoma brucei* RNA editing. *Nature* 356: 807–809.

Kudla, J., Igloi, G.L., Metzlaff, M., Hagemann, R., and H. Kossel (1992). RNA editing in tobacco chloroplasts leads to the formation of a translatable *psbL* mRNA by a C to U substitution within the initiation codon. *EMBO J.* 11: 1099–1103.

Kurland, C.G. (1992). Evolution of mitochondrial genomes and the genetic code. *BioEssays* 14: 709–714.

Lagerkvist, U. (1978). "Two out of three": An alternative method for codon reading. *Proc. Natl. Acad. Sci. U.S.A.* 75: 1759–1762.

Lagerkvist, U. (1981). Unorthodox codon reading and the evolution of the genetic code. *Cell* 23: 305–306.

Landweber, L.F., and W. Gilbert (1993). RNA editing as a source of genetic variation. *Nature (London)* 363: 179–182.

Lonergan, K.M., and M.W. Gray (1993). Editing of transfer RNAs in *Acanthamoeba castellanii* mitochondria. *Science* 259: 812–816.

Lu, B., and M.R. Hanson (1992). A single nuclear gene specifies the abundance and extent of RNA editing of a plant mitochondrial transcript. *Nucl. Acids Res.* 20: 5699–5703.

Mahendran, R., Spottswood, M.R., and D.L. Miller (1991). RNA editing by cytidine insertion in mitochondria of *Physarum polycephalum*. *Nature (London)* 349: 434–438.

Maier, R.M., Hoch, B., Zeitz, P., and H. Kossel (1992). Internal editing of the maize chloroplast *ndhA* transcript restores codons for conserved amino acids. *Plant Cell* 4: 609–616.

Maslov, D.A., Sturm, N.R., Niner, B.M., Gruszynski, E.S., Peris, M., and L. Simpson (1992). An intergenic G-rich region in *Leishmania tarentolae* kinetoplast maxicircle DNA is a pan-edited cryptogene encoding ribosomal protein S12. *Mol. Cell. Biol.* 12: 56–67.

Miller, D., Mahendran, R., Spottswood, M., Costandy, H., Wang, S., Ling, M.-I., and N. Yang (1993). Insertional editing in mitochondria of *Physarum*. *Seminars in Cell Biology* 4: 261–266.

Mulligan, R.M. (1991). RNA editing: When transcript sequences change. *Plant Cell* 3: 327–330.

Osawa, S., and T.H. Jukes (1989). Codon reassignment (codon capture) in evolution. *J. Mol. Evol.* 28: 271–278.

Osawa, S., Jukes, T.H., Watanabe, K., and A. Muto (1992). Recent evidence for evolution of the genetic code. *Microbiol. Rev.* 56: 229–264.

Pfitzinger, H., Guillemaut, P., Weil, J.H., and D.T.N. Pillay (1987). Adjustment of the tRNA population to codon usage in chloroplasts. *Nucl. Acids Res.* 8: 1377–1386.

Pfitzinger, H., Weil, J.H., Pillay, D.T.N., and P. Guillemaut (1990). Codon recognition mechanisms in plant chloroplasts. *Plant Mol. Biol.* 14: 805–814.

Pollard, V.W., Harris, M.E., and S.L. Hajduk (1992). Native mRNA editing complexes from *Trypanosoma brucei* mitochondria. *EMBO J.* 11: 4429–4438.

Read, L.K., Corell, R.A., and K. Stuart (1992). Chimeric and truncated RNAs in *Trypanosoma brucei* suggest transesterifications at non-consecutive sites during RNA editing. *Nucl. Acids Res.* 20: 2341–2347.

Schuster, W., Hiesel, R., and A. Brennicke (1993).

RNA editing in plant mitochondria. *Seminars in Cell Biology* 4: 279–284.

Schuster, W., Wissinger, B., Hiesel, R., Unseld, M., Gerold, E., Knoop, V. Marchfelder, A., Binder, S., Schobel, W., Scheike, R., Gronger, P., Ternes, R., and A. Brennicke (1991). Between DNA and protein—RNA editing in plant mitochondria. *Physiol. Plant.* 81: 437–445.

Shimada, H., and M. Sugiura (1991). Fine structural features of the chloroplast genome: Comparison of the sequenced chloroplast genomes. *Nucl. Acids Res.* 19: 983–995.

Simpson, L., and J. Shaw (1989). RNA editing and the mitochondrial cryptogenes of kinetoplastid protozoa. *Cell* 57: 355–366.

Sollner-Webb, B. (1992). Guides to experiments. *Nature (London)* 356: 743–744.

Stuart, K. (1991a). RNA editing in mitochondrial mRNA of trypanosomatids. *Trends Biochem. Sci.* 16: 68–72.

Stuart, K. (1991b). RNA editing in trypanosomatid mitochondria. *Annu. Rev. Microbiol.* 45: 327–344.

Stuart, K. (1993). The RNA process in *Trypanosoma brucei*. *Seminars in Cell Biology* 4: 251–260.

Stuart, K., and J.E. Feagin (1992). Mitochondrial DNA of kinetoplastids. *Int. Rev. Cytol.* 141: 65–88.

Sugiura, M., and T. Wakasugi (1989). Compilation of transfer RNA genes from tobacco chloroplasts. *Crit. Rev. Plant Sci.* 9: 89–101.

Walbot, V. (1991). RNA editing fixes problems in plant mitochondrial transcripts. *Trends Genet.* 7: 37–39.

Wolstenholme, D.R. (1992). Animal mitochondrial DNA: Structure and evolution. *Int. Rev. Cytol.* 141: 173–216.

15

Protein Targeting and Import

Most chloroplast and mitochondrial proteins are nuclear gene products imported from the cytoplasm. The mechanisms involved in these processes have been a major topic of interest to many investigators and are chronicled in numerous reviews. Unfortunately, with the exception of Pugsley's (1989) excellent monograph, which reviews the entire subject of protein targeting in prokaryotic and eukaryotic cells, targeting and import in chloroplasts and mitochondria are usually reviewed separately despite their many similarities. In the account that follows, these processes are considered together. While the approaches used to study protein targeting and import to chloroplasts and mitochondria are similar in many ways, there is one striking difference. Genetics has been used to great advantage to study these processes in mitochondria because yeast serves as the model system of choice. In contrast, genetic analysis has had virtually no role in examining targeting and import of proteins into plastids because most of the work has been done with flowering plant species that are not readily amenable to this approach. The reason *Chlamydomonas* has been used so sparingly may be because chloroplast isolation is not as easy in this organism as in land plants.

The account that follows begins with one of the oldest questions concerning protein import into organelles. Is the process cotranslational or posttranslational? This discussion leads into a consideration of the nature of the routing signals used to direct proteins to chloroplasts and mitochondria, the structure of the import machinery, and the mechanisms involved in protein sorting within different organellar compartments.

IS IMPORT COTRANSLATIONAL OR POSTTRANSLATIONAL?

In cotranslational import, as the name implies, the protein precursor crosses the membrane while translation continues on membrane-associated polysomes. This is the mode of translocation of the majority of eukaryotic, secretory proteins across the rough endoplasmic reticulum membrane (Pugsley, 1989). In contrast, a few eukaryotic secretory polypeptides and the great majority of prokaryotic secretory polypeptides can be posttranslationally translocated across their respective rough endoplasmic reticulum or cytoplasmic membranes *in vitro*. This is also true *in vitro* of proteins destined for the chloroplast and mitochondrion (e.g., see reviews by Schatz and Butow, 1983; Schmidt and Mishkind, 1986). However, the question remains as to whether the same is true *in vivo*. In their review Schmidt and Mishkind (1986) point out that in *Chlamydomonas* the mRNA encoding the small subunit of RUBISCO is localized to preparations of free polysomes whereas those mRNAs for secretory proteins are enriched in membrane-associated fractions. Furthermore, they note the virtual absence of cytoplasmic ribosomes associated with chloroplasts as determined by transmission electron microscopy. In contrast, in yeast spheroplasts incubated in the presence of cycloheximide, cytoplasmic ribosomes were found to be closely packed and bound to the outer membranes of the mitochondria (Ades and Butow, 1980; Kellems and Butow, 1972, 1974; Kellems *et al.*, 1974). About a third could be released with high KCl concentrations, but removal of all of them could be achieved only with high KCl plus puromycin. The puromycin experiments suggested that most of the polysomes were bound to the outer membrane by nascent polypeptide chains. Suissa and Schatz (1982) subsequently found that *in vitro* translation of mRNA extracted from ribosomes bound to mitochondria showed that these ribosomes were enriched in mRNAs encoding mitochondrial proteins. However, only part of the total mRNA for any given mitochondrial precursor was recovered from bound polysomes. The rest of the mRNA proved to be associated with free cytoplasmic ribosomes. To explain the previous observations concerning polysomes bound to mitochondria of yeast spheroplasts *in vivo,* Schatz and Butow (1983) argued that since incubation of spheroplasts in the presence of cycloheximide reduces the size of extramitochondrial precursor pools, the number of unoccupied receptor sites will increase. This will favor binding of the amino termini of polysome-bound precursor polypeptides to receptor sites.

Hartl *et al.* (1989) review the history of posttranslational and cotranslational import into mitochondria. They make the point that while cotranslational import may be possible under appropriate conditions (e.g., after inhibition of elongation with cycloheximide), the demonstration that posttranslational import occurs proves that translation and translocation can be mechanistically independent. Hartl *et al.* also say that with respect to prokaryotes, there is now general agreement that cotranslational and posttranslational translocation reflect kinetic differences and not fundamental mechanistic differences.

Consistent with this view, experiments with yeast employing a homologous *in vitro* system coupled with *in vivo* analysis are swinging the pendulum back in favor of a cotranslational mechanism for mitochondrial import (Fujiki and Verner, 1991, 1993; Verner, 1992, 1993). Fujiki and Verner found that import of a chimeric protein *in vitro* was very inefficient in a strictly posttranslation reaction, but efficient import was achieved when precursor synthesis and import were coupled. Also, the cytoplasmic protein synthesis inhibitor cycloheximide blocked import of the bulk of mitochondrial protein *in vitro* and *in vivo* suggesting that cotranslational mechanisms play a significant role in import into mitochondria. Verner (1993), in an overview, presents a model which supposes that both co- and posttranslational import occur in mitochondria. The latter process would require heat shock protein 70 (see below) to keep the precursor in an unfolded state. In contrast, maintenance of import-competent precursor forms would be less important for cotranslational import where precursor proteins penetrate the membrane as they are synthesized.

PRESEQUENCES OF IMPORTED PROTEINS

Pugsley (1989) defines four classes of protein-targeting signals of which three will enter into the discussion that follows. *Routing signals* mediate the interaction between targeted polypeptides and the first membrane with which they come in contact. *Sorting signals* direct proteins into different branches of a targeting pathway or redirect them once they have reached their initial targets. *Stop transfer signals* halt translocation of a polypeptide through the membrane. Pugsley (1989) has also made sense out of the taxonomy that has evolved for the naming of different targeting sequences. For example, he lists as routing signals the terms signal peptide (chloroplasts and mitochondria), presequence (mitochondria), and transit peptide (chloroplasts). This classification is very helpful in that presequences and transit peptides always contain routing information directing the precursor to the outer membrane (envelope) of the appropriate organelle. However, they may also contain signals for sorting or stop transfer for proteins that are directed into the electron transfer membranes, the intermembrane space, or the thylakoid lumen. The term presequence is used throughout since it is functionally neutral.

Comprehensive summaries of known presequences currently can be found in reviews by Berry Lowe and Schmidt (1991), de Boer and Weisbeek (1991), and Keegstra *et al.* (1989) for the chloroplast and in reviews by Grivell (1988) and Hartl *et al.* (1989) for the mitochondrion.

Structure

N-Terminal presequences in chloroplasts and mitochondria are usually 20 to 80 amino acids in length (Table 15–1). They are processed in the course of translocation into the organelle. Since the processed (mature) proteins are smaller than their precursors, the site of processing can be determined by comparing the N-terminal sequence of the protein with that of the precursor. One of the most remarkable features of organelle presequences is the nearly complete lack of similarity between signals from different precusor polypeptides. This is also true of signal peptides of secretory proteins (Pugsley, 1989). The major defining features of organelle presequences appear to be the presence of basic residues, usually arginine (abundant in mitochondrial presequences, but less so in chloroplasts despite a net positive charge), a marked deficiency or complete absence of acidic residues, and a preponderance of hydroxylated (serine and threonine) residues (Hartl *et al.,* 1989; Keegstra *et al.,* 1989; Keegstra and von Heijne 1992). It was originally thought, based on comparisons of presequences of available chloroplast precursors (several CAB protein and RUBISCO small subunit presequences plus a single ferredoxin presequence), that these presequences were divisible into three conserved domains (see Berry-Lowe and Schmidt, 1991, for a review). The first was near the amino terminus, the second in the middle, and the third at the carboxyl terminus of the presequence. The amino terminal and central regions were proposed to be involved in the binding or translocation steps while the region near the carboxyl terminus was thought to be required for proteolytic removal of the presequence (Mishkind *et al.,* 1985; Karlin-Neumann and Tobin, 1986). However, as additional presequences have been determined, some have been found that do not fit the three domain model (Berry-Lowe and Schmidt, 1991; Keegstra *et al.,* 1989; Keegstra and von Heijne, 1992). Therefore, the three domain model is of questionable utility, although it may be applicable in some cases.

Keegstra *et al.* (1989) point out that the lack of sequence similarity between presequences from different polypeptides can be explained in at least two different ways. First, precursors interact with different receptors and the presequences of different pre-

Table 15–1. Amino Acid Sequences of Some Typical Presequences Located at the N-Termini of Nuclear-Encoded Mitochondrial and Chloroplast Proteins[a]

Protein	Location	Amino Acid Sequence[b]
Ornithine transcarbamylase (human mitochondrion)	Matrix	M L F N L R I L L N N A A F R N G H N F M V R N F R C G Q P L Q ∣ N K 　　　　　　　　　　　　　　　　+　　　　　+　　+　　　+　　　　+
Cytochrome oxidase subunit IV (yeast mitochondrion)	Inner membrane	M L S L R Q S I R F F K P A T R T ∣ L C S S R Y L L Q Q 　+　•　　+　•　•　+　+　•　•　+　•　•　•　+
Cytochrome cI (yeast mitochondrion)	Inner membrane	M F S N L S K R W A Q R T L S K S F Y S T A T G A ∣ A S K S G 　　•　　　　+　+　　　+　•　　+　•　•　•　•　•　•　　•　+　•　　 K L T Q K L V T A G V A A A G I T A S T L L Y A D S L T A E A ∣ M T A +　•　•　+　　•　　　　　　　　　•　•　•　　•　•　•　•　　•　　　•　•
Porin[b] (yeast mitochondrion)	Outer membrane	M S P P V V Y S D I S R N I N D L L N K D F Y H A T P A A F D V Q T T 　•　　　　　•　•　　+　　　　•　•　+　•　　　•　•　•　•　　•　•　•　• T A N G I K F S L K A K Q P V K •　　　　+　•　•　+　•　+
RUBISCO small subunit (*Chlamydomonas*)	Stroma	M A V I A K S S V S A A V A R P A R S S V R P M A ∣ A L K P A 　　　　+　•　•　•　•　　　　+　　+　•　•　+　　•　　　+

Modified slightly from Pugsley (1989).

[a] +, basic charges; •, hydroxylated residues ∣ processing cleavage sites.

[b] Sequence for yeast porin is from Hartl *et al.* (1989). Note that this protein does not possess a cleavable presequence. The routing information is in the N-terminus of the mature protein.

cursors reflect this sort of "lock and key" mechanism. If this hypothesis were correct, presequences should fall into several different classes based on the receptors with which they interact. Currently, there is little evidence for this hypothesis. Second, the essential signals in presequences are to be found in the secondary structure rather than in the primary amino acid sequence. This hypothesis is supported by studies of both signal sequences that route secretory proteins (Pugsley, 1989) and mitochondrial presequences.

Peptides that form an amphiphilic helix have intermingled polar and nonpolar residues. When these peptides fold into α-helices one face of the helix is nonpolar while the other is polar. Such sequences can interact directly with biological membranes. These interactions may be important in protein translocation. Keegstra et al. (1989) reviewed the evidence for amphiphilic α-helices in chloroplast and mitochondrial presequences. They point out that theoretical considerations are consistent with the hypothesis that mitochondrial presequences often form amphiphilic α-helices (see also Hartl et al., 1989). Experimental data for mitochondrial presequences also support the theoretical conclusions. Thus, synthetic peptides corresponding to presequences of several precursors are water soluble, but will insert into synthetic membranes. In one study Roise et al. (1988) found that all import-active presequences were amphiphilic while sequences that were not amphiphilic did not exhibit import activity. They also found one active presequence that was very amphiphilic, but unable to form an α-helix. This suggested that amphiphilicity itself is the key presequence feature, with the α-helical structure simply being one manifestation of amphiphilicity. This significant finding is supported by preliminary analyses of Keegstra et al. (1989). They conclude that an amphiphilic helix is unlikely to be an essential structure in all chloroplast transit peptides and that amphiphilicity itself is the key feature with no specific secondary structure being required.

Roise (1992) used a fluorescent-labeled peptide identical to a naturally occurring mitochondrial presequence to study the characteristics of precursor binding and import. He reached the following conclusions from his experiments:

1. The presequence can bind directly to the mitochondrial outer membrane.
2. The import rate of the presequence is strictly proportional to its concentration on the mitochondrial surface.
3. The transport apparatus did not display saturation by the bound presequence in the concentration range used.

The data on presequences summarized in Table 15–1 also call attention to another point. While an enzyme such as ornithine transcarbamylase destined for the mitochondrial matrix has a single processing site in the presequence, other proteins such as cytochrome oxidase subunit IV or cytochrome cI, targeted, respectively, to the inner membrane and inter membrane space, have two. The first prepeptide demarked by the first cleavage site contains the routing signal for the mitochondrial outer membrane. The second prepeptide contains the information for proper sorting (or stop transfer) of the mature polypeptide. Proteins destined for the thylakoid membrane also have a bipartite presequence consisting of a routing signal for the chloroplast and a short thylakoid transfer domain that has similarities to prokaryotic and eukaryotic signal sequences (see Distinguishing Routing and Sorting Sequences).

To Have or Not to Have?

More than 40 different precursor proteins destined for import into the matrix or inner membrane of the mitochondrion have been analyzed (Hartl et al., 1989). Most of these proteins have cleavable N-terminal presequences, but there are some exceptions. The yeast ADP/ATP translocator protein of the inner mitochondrial membrane lacks a cleavable presequence. This protein is made up of three homologous domains of about 100 amino acids each suggesting the protein evolved by triplication of an ancestral gene. A stretch of 20 amino acids near the C-terminus of each domain can form an amphiphilic helix and therefore possesses a feature characteristic of most mitochondrial presequences. Furthermore, a truncated ADP/ATP carrier lacking the N-terminal third of the polypeptide is import competent, showing that the carboxy-terminal two-thirds of the protein contain sufficient routing information. These results point to the C-terminal regions of the three homologous domains as being the likely locations of the routing signals for the ADP/ATP carrier protein (Fig. 15-1).

Several complex III subunits also lack N-terminal extensions in at least some cases (Hartl et al., 1989). The mitochondrial matrix protein, 3-oxoacyl-CoA thiolase, is also made without a cleavable presequence (Hartl et al., 1989). Its routing signal is contained within the first 61 N-terminal residues. The highly homologous peroxisomal enzyme is made as a larger precursor and the presequence may mask the mitochondrial routing signal in the N-terminal region of the mature protein. The only other matrix protein not known to have an N-terminal extension is 2-isopropylmalate synthetase.

Most proteins of the intermembrane space have cleavable presequences, but cytochrome c does not (see The Unconventional Routing and Sorting of Cytochrome c). The same may be true of adenylate kinase (Hartl et al., 1989). Porin, a channel-forming outer membrane protein, lacks a presequence as do three other outer membrane proteins that have been studied. However, a 35-kDa outer membrane protein from rat liver mitochondria may be exceptional, as a potential precursor of slightly higher molecular weight has been detected on SDS-polyacrylamide gels (Hartl et al., 1989).

With one possible exception, all of the many proteins whose transport into chloroplasts has been demonstrated have cleavable presequences (de Boer and Weisbeek, 1991). The possible exception is a heat shock protein from *Chlamydomonas* that lacks an N-terminal extension as deduced from the nucleotide sequence of the gene. The 30-kDa phosphate translocator and another 37-kDa inner envelope protein, studied by Flugge and colleagues (see Flugge, 1990), both have cleavable presequences. The structure of the presequence of the 30-kDa protein differs in a number of respects from the presequences of other imported chloroplast proteins. However, the protein possesses an N-terminal amphiphilic α-helix as does the 37-kDa protein. On the other hand, the three outer envelope proteins examined thus far both lack cleavable presequences (Ko et al., 1992; Salomon et al., 1990; Li et al., 1991) and resemble mitochondrial outer membrane proteins in this respect.

Distinguishing Routing and Sorting Sequences

So far we have discussed only routing signals, which are usually, but not always, in the cleavable presequence. Sorting signals may also be present depending on the final

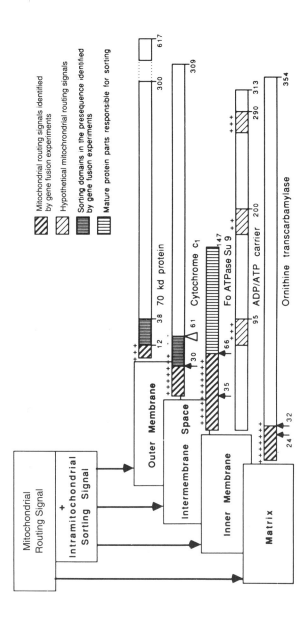

Fig. 15-1. Examples of structural domains in mitochondrial precursor proteins for routing and intramitochondrial sorting. Mitochondrial routing signals are typically hydrophilic, positively charged sequences contained in amino-terminal presequences. They are usually cleaved off during import. Routing signals can also reside in internal parts of the mature protein (e.g., ADP/ATP carrier). In addition to routing signals, sorting signals are found in the amino-terminal sequences (e.g., 70 kDa outer membrane protein, cytochrome c_1). Information for intramitochondrial sorting can also reside in the mature part of the protein (e.g., F_0 ATPase subunit 9). The respective cleavage site of the matrix processing protease is indicated by an arrow and the cleavage site for the second proteolytic processing step of cytochrome c_1 (occurring at the outer surface of the inner membrane) by an open arrowhead. Charged amino acid residues in functionally relevant domains are indicated. From Hartl et al. (1989).

destination of the protein. Gene fusions involving "passenger" proteins and *in vitro* mutagenesis have been used to distinguish routing and sorting signals as described in the next section.

The various possible combinations of routing and sorting signals that can exist for polypeptides imported into chloroplasts and mitochondria are vividly illustrated by four examples of proteins imported into mitochondria (see Hartl *et al.*, 1989 and Fig. 15–1). Matrix proteins, such as the enzyme ornithine transcarbamylase, require only a routing signal, which is normally present in the cleavable presequence. The 70-kDa outer membrane protein does not possess a cleavable presequence, but does have adjacent routing and sorting sequences at the N-terminus. The same is true of a 70-kDa heat shock protein of the spinach chloroplast outer envelope where the extreme NH_2 terminus of the protein is required for proper targeting (Wu and Ko, 1993). As discussed earlier, the inner membrane ADP/ATP carrier protein, unlike most other inner membrane proteins, lacks a cleavable presequence, but has three homologous internal sequences that are probably responsible for routing. Subunit 9 of the mitochondrial F_1/F_0 ATP synthase is a more typical inner membrane protein in that it possesses a typical cleavable presequence containing a routing signal. The cleaved, mature polypeptide also has what is either a sorting signal or a stop-transfer sequence depending on how one views intramitochondrial translocation (see Protein Sorting). Similarly, cytochrome c_1, a protein of the inter membrane space, has a cleavable presequence with a routing signal and an adjacent N-terminal sorting or stop-transfer sequence in the mature polypeptide.

Proteins destined for the thylakoid lumen of the chloroplast seem to possess two routing signals and undergo two-step processing (see Keegstra and von Heijne, 1992; de Boer and Weisbeek, 1991; Smeekens *et al.*, 1990). Thus, plastocyanin possesses a normal cleavable presequence with a routing signal for entry into the stroma. On entry into the stroma a second routing signal (the thylakoid transfer domain) targets the protein to the thylakoid membrane. Once the protein has passed through the thylakoid this part of the presequence is removed. This two-step translocation mechanism also applies to other nuclear-encoded thylakoid lumen polypeptides such as the 16-, 23-, and 33-kDa subunits of the water splitting complex and very likely to subunit III of PSI, the putative plastocyanin-docking protein (de Boer and Weisbeek, 1991).

Routing and sorting of the the chlorophyll *a/b* (CAB) proteins of LHC II (see Chapter 1) into thylakoid membranes have been scrutinized (see Berry-Lowe and Schmidt, 1991; de Boer and Weisbeek, 1991). The CAB protein presequence has no function in thylakoid membrane insertion and can be replaced by another presequence. Hence, CAB II proteins first integrate in the precursor form following which the presequence is removed. Integration into the thylakoid membrane is dependent on the mature sequence of the protein. Specifically, there are three membrane-spanning domains, with the region located closest to the C-terminus being essential for initial integration (Kohorn and Tobin, 1989). Integration of these CAB II proteins occurs into stromal lamellae followed by transport to the grana stacks.

A unique role has now been reported for one presequence. The 78 amino acid-targeting sequence of the bovine Rieske iron–sulfur protein associated with the cytochrome b,c_1 complex is actually retained as a subunit of the complex (Brandt *et al.*, 1993).

Use of Passenger Proteins to Study Presequences

The properties of different presequences have been largely defined by making chimeric gene constructs in which the DNA sequence encoding the presequence of interest is fused to a reporter gene specifying a conveniently assayed foreign protein. The hybrid precursor encoded by the chimeric construct is then expressed and its uptake into the target organelle studied *in vitro*. Favorite passenger proteins for the mitochondrion are *E. coli* β-galactosidase and mouse dihydrofolate reductase (DHFR) (Pugsley, 1989). For chloroplasts they include such bacterial proteins as neomycin phosphotransferase and chloramphenicol transacetylase, brome mosaic virus coat protein, and a couple of yeast mitochondrial proteins (see de Boer and Weisbeek, 1991; Keegstra, 1989).

Experiments with chimeric genes were done initially with the mouse DHFR gene fused to the coding region for the 25 amino acid presequence of the gene specifying cytochrome oxidase subunit IV (COX4). They demonstrated that the resulting chimeric precursor could be successfully transported into the mitochondrion and processed (see Hartl *et al.*, 1989). Deletion experiments then revealed that the 12 amino terminal amino acids were sufficient to route the polypeptide into the mitochondrion, but deletion of the first seven amino acids abolished import.

Care must be taken in choosing the right precursor–passenger combination. For example, Lubben *et al.* (1989) compared the transport properties of the RUBISCO small subunit precursor with three chimerics. One had the presequence fused to the gene encoding the brome mosaic virus coat protein, the second had a small part of the mature small subunit gene sandwiched in between, and the third included almost the entire sequence of the small subunit precursor fused to the coat protein gene. The rate of import of the first two chimeric proteins was only 2% of the authentic precursor and the third was undetectable (also see Berry-Lowe and Schmidt, 1991, for a summary of experiments of this type).

Organelle Specificity Is Determined by Presequence Routing Signals

The presequence possesses the correct routing information to discriminate between plastids and mitochondria. This is illustrated by experiments in which the presequence of a protein destined for one organelle has been fused with a passenger protein that is normally routed to the other organelle. Thus, fusing the DNA sequence encoding the presequence of the COXIV subunit to the sequence encoding the mature RUBISCO small subunit yields a chimeric precursor that is directed to the mitochondrion (Hurt *et al.*, 1986). The converse is true when chloroplast presequences are coupled to mature polypeptides encoding mitochondrial proteins (Smeekens *et al.*, 1987). In fact Boutry *et al.* (1987) have shown in plant cells that mitochondrial and chloroplast targeting sequences direct the attached reporter proteins exclusively to the correct organelle *in vivo*. Despite this, chloroplast presequences can direct proteins into mitochondria with very low efficiency (see Pfanner and Neupert, 1990). However, this ''bypass'' import does not employ the normal protease-sensitive mitochondrial receptors on the mitochondrial outer surface and the precursor enters the import pathway at a later point. Such experiments are reminiscent of those done by Baker and Schatz (1987) who found that >2.7% of clones generated from *E. coli* DNA encode sequences that specify a presequence for a truncated COXIV subunit of yeast. Presumably, early in organelle evolution protein import did not involve receptors. These would likely have evolved subsequently as the process became more sophisticated.

MECHANISMS INVOLVED IN PROTEIN IMPORT

Protein import refers to the events leading to the translocation of a precursor protein from the cytoplasm to mitochondrial matrix or chloroplast stroma where it is then usually processed. Protein sorting, discussed later in this chapter, refers to the mechanisms required to move the protein to its final destination within the organelle. The details of protein import are best understood in mitochondria and have received detailed treatment in numerous reviews (e.g., Eilers *et al.,* 1988; Grivell, 1988; Geli and Glick, 1990; Glick and Schatz, 1991; Hartl *et al.,* 1989; Pfanner, 1992; Pfanner and Neupert, 1990; and Weinhaus and Neupert, 1992). While there is agreement on most aspects of the process there are areas of disagreement as well. Lest the reader gain the wrong impression, workers on chloroplast import have also been very active (see reviews by Berry-Lowe and Schmidt, 1991; de Boer and Weisbeek, 1991; Flugge, 1990; Keegstra, 1989; Keegstra and Bauerle, 1988; Keegstra *et al.,* 1989; Schmidt and Mishkind, 1986; Smeekens *et al.,* 1990; Soll and Alefsen, 1993). The account that follows summarizes these developments beginning with the role of molecular chaperones and ending with energetics.

One view of protein import into mitochondria is depicted in Fig. 15–2, and the protein components themselves together with their functions are summarized in Table 15–2. The import process is seen to involve molecular chaperones (hsp 70) and, in rabbits and rats at least, the presequence binding factor (PBF). These proteins keep the precursor unfolded while its presequence routes it to a receptor on the outer membrame (R). Actual receptor proteins include MOM19 and MAS70 (MOM72) (Table 15–2). The precursor then travels through an import channel and across the outer and inner membranes. One of the proteins involved is called ISP42 in yeast and MOM38 in *Neurospora.* Other proteins shown as X, Y, Z, and A–F in Fig. 15–2, some of which have been identified (Table 15–2), are probably involved as well. Additional chaperones inside the mitochondrion (mhsp70, hsp60) play a role in folding and assembly of the precursor protein. The matrix-processing protease consisting of two subunits (Mas1p and Mas2p) then cleaves off the presequence. Each of these steps is now examined in more detail and contrasted to the equivalent process in chloroplasts.

Molecular Chaperones

Molecular chaperones are defined as a family of unrelated classes of proteins that mediate the correct assembly of other polypeptides, but are not components of the functional assembled structures when they are performing their normal biological functions (Ellis, 1993). All molecular chaperones so far studied seem to act not by providing the steric information required for assembly, but by inhibiting improper interactions that yield nonfunctional structures during self-assembly.

Included in the molecular chaperone family are most of the heat shock (hsp) proteins. As their name implies, the hsp proteins are heat induced. Nevertheless, both hsp60 and hsp70 proteins, the only species discussed in detail here, are also abundant consti-

PROTEIN TARGETING AND IMPORT

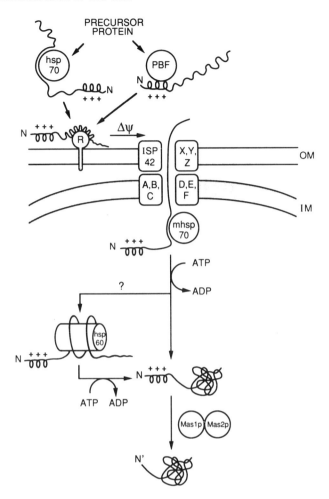

Fig. 15–2. Transport of a mitochondrial precursor protein with a cleavable amino-terminal matrix targeting signal into the matrix (see text for details and Table 15–2). In some cases, the requirements for cytosolic hsp70, PBF, and outer membrane receptors may be circumvented. The functional relationship between mitochondrial hsp70 (mhsp70) and hsp60 is not yet clear, nor is it known at which step in the matrix the routing signal is removed. R, one of several import receptors (such as MOM19 and Mas70p) on the mitochondrial surface; X, Y, Z, and A-F, hypothetical additional subunits of the import channels across the outer and inner membrane, respectively; PBF, presequence-binding factor; Mas1p and Mas2p, the two subunits of the matrix processing protease (MPP). From Glick and Schatz (1991).

tutively in the absence of heat stress. The properties of molecular chaperones have been extensively reviewed by Ellis (1990, 1993), Ellis and van der Vies (1991), Gatenby (1992), Horwich and Willison (1993), Landry and Gierasch (1991), Lorimer *et al.* (1993), Neupert and Pfanner (1993), and Zeilstra-Ryalls *et al.* (1991). The heat shock proteins are reviewed by Hendrick and Hartl (1993), Lindquist and Craig (1988) and Vierling (1991) while stress proteins generally are discussed in the volume edited by Morimoto *et al.* (1990).

Table 15–2. Components Involved in Protein Transport Across Mitochondrial Membranes

Location and Name[a]	Organism	Function
Cytosol		
PBF (presequence binding factor)	Rat, rabbit	Stabilize precursor in translocation competent form
hsp70	Yeast	Stabilize precursor in translocation competent forms
Outer membrane		
MOM19	*Neurospora*	Binding protein for several precursors and for porin
MAS70 or Mas70p (MOM72)	Yeast, *Neurospora*	Binding protein for precursor of ADP/ATP carrier
MOM22	*Neurospora*	Transfer of proteins from receptor to general insertion pore
p32	Yeast	Putative binding protein at translocation contact sites
ISP42 (MOM38)[b]	Yeast, *Neurospora*	Import channel in outer membrane
Matrix		
mhsp70 (SSC)	Yeast	Import and folding of precursors
hsp60 MIF4	Yeast, *Neurospora*	Folding and assembly of precursors
Maslp (MIFI, PEP)	Yeast, *Neurospora*, rat	Cleavage of N-terminal presequences of imported precursors
Mas2p (MIF2, MPP)	Yeast, *Neurospora*	Cooperates in presequence cleavage with Maslp
Inner membrane		
Inner membrane protease I	Yeast	Processing of cytochrome b_2 and COX2
Intermembrane space		
Heme lyase	Yeast, *Neurospora*	Import of apocytochrome c and covalent attachment of heme

Modified after Glick and Schatz (1991) and Wienhues and Neupert (1992), which cite the original references. See Kiebler et al. (1993) for MOM22 function.

[a]The yeast name is given first, if known, followed by synonyms. Numbers after gene symbols usually equal molecular weights in kDa.

[b]Sollner et al. (1992) report that MOM7, MOM8, and MOM30 crosslink to MOM38.

Molecular chaperones play a role both outside and inside mitochondria and chloroplasts. Members of the hsp70 family have molecular weights of approximately 70 kDa as their name implies and occur in both prokaryotes and eukaryotes. These proteins are thought to play a role in the ATP-dependent folding and assembly of proteins (Pelham, 1986). The DnaK protein of *E. coli* is an hsp70 protein. In yeast, there are eight HSP70 genes that comprise four nuclear gene families. The cytoplasmic SSA proteins, encoded by a four gene family, facilitate transport of proteins across the endoplasmic reticulum and also bind to newly formed mitochondrial precursor proteins, keeping them in the unfolded state required for import. Thus a yeast mutant defective in three of the four SSA genes exhibits reduced import of several proteins into mitochondria *in vivo* accompanied by accumulation of precursors of these proteins in the cytoplasm (Deshaies et al., 1988).

Two proteins purified from rabbit reticulocyte lysates may also play roles as mitochondrial chaperones (see Glick and Schatz, 1991). One is the presequence binding factor (PBF). This is an oligomer of 50 kDa subunits that binds to the presequence of ornithine transcarbamylase, and thereby prevents folding or nonspecific aggregation of the precursor. The second is a 28-kDa protein whose function *in vitro* appears to be similar to that of PBF.

The story in the case of chloroplasts is less clear (de Boer and Weisbeek, 1991). Studies with the purified precursor of a CABII protein showed that the denatured precursor could be imported into chloroplasts only in the presence of dialyzed leaf extract. The effect was incompletely compensated by hsp70, suggesting the involvement of a second factor. In contrast, cytosolic factors do not seem to be required for import of the denatured precursors of ferredoxin or plastocyanin. These experiments do not rule out the participation of hsp70 proteins in chloroplast import *in vivo*.

Chloroplasts and mitochondria themselves possess unique hsp70 proteins (see Vierling, 1991). In yeast, the SSC family, containing a single gene, encodes the mitochondrial hsp70 (SSC1). The SSC1 protein shows greater similarity to the DnaK protein of *E. coli* than to the hsp70s of the eukaryotic cytoplasm and endoplasmic reticulum. A similar protein has now been found in mitochondria of several other organisms. DnaK homologs have also been identified in chloroplasts of pea, maize, spinach, *Arabidopsis*, and *Euglena* (Vierling, 1991). They are located in the stroma. The cDNA encoding the Hsp70 of pea chloroplast stroma has been isolated and specifies a 75.5-kDa protein possessing a presequence and closely related to hsp70 from the cyanobacterium *Synechocystis* (Marshall and Keegstra, 1992). The purified chaperone from barley chloroplasts plays a major role in CABII precursor insertion into thylakoid membranes (Valovsky *et al.*, 1992).

Transfer of precursors into mitochondria is defective in a yeast mutant with an altered mitochondrial hsp70 (Kang *et al.*, 1990; Wienhues and Neupert, 1992). Precursor proteins were found to be arrested in translocation contact sites. Denaturation of precursors prior to import overcame the defect. This suggests that the mitochondrial hsp70 may be needed for unfolding and translocation of mitochondrial precursors following their insertion into mitochondrial contact sites. The mitochondrial hsp70 also appears to be involved in refolding imported precursors, since precursors imported after urea denaturation remain unfolded in the mutant. The chloroplast hsp70 is presumed to function in a similar fashion (Yalovsky *et. al.,* 1992).

The second molecular chaperone found in chloroplasts and mitochondria is chaperonin 60 or hsp60. The degree of sequence relatedness in this group of proteins is higher. Thus, the yeast mitochondrial chaperonin 60 shares 45% identical amino acid residues with wheat chloroplast chaperonin 60 and 54% identical residues with *E. coli* chaperonin 60. Ellis and van der Vies (1991) point out that the distribution of these chaperonins in prokaryotes, plastids, and mitochondria, coupled with their high degree of sequence identity, is consistent with an ancient origin for these proteins and their appearance in chloroplasts and mitochondria as a consequence of endosymbiosis. No related proteins were thought to occur in other compartments of the eukaryotic cell (Hendrick and Hartl, 1993; Vierling, 1991), but a protein, TCP-1, having some similarity to chaperonin 60, has been reported to be encoded by the *t*-complex locus on mouse chromosome 17 (Ellis and van den Vies, 1991; Hendrick and Hartl, 1993). A second chaperonin, chaperonin 10, is much smaller than chaperonin 60 (60 versus 10 kDa), but is strongly sequence related to one specific region of chaperonin 60.

Chaperonin 60 from bacteria, plastids, and mitochondria consists of 14 subunits arranged in two rings of seven subunits each termed the "double donut." This is a homooligomer in *E. coli,* but the plastid chaperonin 60 contains two closely related proteins called α and β. Bacterial chaperonin 10 forms a ring of seven subunits. In *E. coli* chaperonins 10 and 60 are encoded in the *groES* and *groEL* genes, respectively, which constitute the *groE* operon (Gatenby, 1992; Zeilstra-Ryals *et al.,* 1991; Hendrick and Hartl, 1993). Genetic and biochemical data for the bacterial chaperonins have led to a model for how these proteins facilitate folding of proteins (Zeilstra-Ryals *et al.,* 1991). The first step is the reversible binding of chaperonin 60 to the unfolded form of some polypeptide. This association is thought to prevent incorrect or premature folding of the protein. At this point the folded protein may either be released directly from chaperonin 60, a step that proceeds very slowly and requires ATP, or chaperonin 10 can act as a cogwheel to displace the folded protein from the chaperonin 60 complex. The hydrolysis of ATP by chaperonin 60 provides the energy for this process. Eventually the two chaperonins dissociate.

In mitochondria, precursors associated with the mitochondrial hsp70 are thought to be transferred to chaperonin 60 where ATP-dependent folding of these proteins and assembly can occur. The yeast chaperonin 60 is encoded by the nuclear *MIF4* gene (Table 15–2). A temperature-sensitive mutant in this gene was shown to be unable to form an enzymatically active trimer of human ornithine transcarbamylase at the restrictive temperature, although transport and processing of the precursor occurred normally (Cheng *et al.,* 1989). A number of other assembly defects were also noted in the mutant (e.g., the imported β subunit of the ATP synthetase failed to assemble into the ATPase complex). These results provide strong support for the requirement of chaperonin 60 in the folding of mitochondrial polypeptides. If the mitochondrial chaperones operate by a mechanism similar to the one described above for bacteria, one would predict either that the organelle chaperonin 60 also has a chaperonin 10-like function (e.g., a cogwheel domain) or that the mitochondrion also contains chaperonin 10. A partially purified protein with chaperonin 10-like function was first reported from liver mitochondria (Lubben *et al.,* 1990). This protein has since been purified to homogeneity from rat hepatoma cells (Hartman *et al.,* 1992). In combination with *E. coli* Hsp60, rat mitochondrial Hsp10 causes the efficient reconstitution of rat ornithine transcarbamylase from the denatured state.

The chloroplast chaperonin 60 was discovered in the course of experiments designed to study the assembly of RUBISCO (Ellis and van der Vies, 1991). In land plants and green algae, RUBISCO holoenzyme consists of eight plastid-encoded large subunits and eight small subunits imported from the cytoplasm. Binding of large subunits to this protein was hypothesized to be an obligatory step in the assembly of the RUBISCO holoenzyme and so the protein was originally called the RUBISCO subunit binding protein. The plastid chaperonin 60 contains equal amounts of two polypeptides called *alpha* and *beta* having 50% sequence identity and encoded by small families of nuclear genes. A variety of imported proteins is found associated with chloroplast chaperonin 60, but a few are not (e.g., ferredoxin and superoxide dismutase), indicating some specificity in binding. The presequence of the imported precursor may be required for binding to chaperonin 60 since the mature RUBISCO small subunit will not bind to this protein. The interaction between RUBISCO large subunit and chaperonin 60 has been used extensively to model the reactions involved in chaperonin-assisted protein folding (see Lorimer et al., 1993).

Chloroplasts, like mitochondria, contain a chaperonin 10 homolog (Gatenby, 1992; Lubben et al., 1990). Bertsch et al. (1992) isolated chloroplast chaperonin 10 from pea chloroplasts and shown that it is a 24-kDa protein. They have also isolated a spinach cDNA encoding this protein. DNA sequencing reveals that the chloroplast protein is actually a "double" chaperonin 10 accounting for its unusual molecular weight.

Receptors

So far, two mitochondrial import receptor proteins have been identified in the outer membrane of the mitochondrion (Table 15–2). One of these, called MOM72 in *Neurospora*, seems to be mainly a receptor for ADP/ATP translocator (Pfanner, 1992; Wienhues and Neupert, 1992). The equivalent protein in yeast is Mas70p (Glick and Schatz, 1991). Both proteins enhance import of the translocator at an early step, probably by binding the protein to the outer membrane in the vicinity of an import channel. Experiments with a *mas70* null mutant in yeast showed that the translocator still accumulated to nearly wild-type amounts in mitochondria, but pulse–chase experiments revealed that the rate of uptake of this and all other precursors tested was severalfold slower than wild type (Hines et al., 1990). The fact that Mas70p is not essential indicates that mitochondria also possess other import receptors. A 19-kDa outer membrane protein (MOM19), so far identified only in *Neurospora*, also seems to have receptor function. This protein binds several precursors and accelerates import of a variety of precursors into mitochondria. Immunoprecipitation from digitonin extracts of *Neurospora* mitochondria with antibodies directed against MOM19 and MOM72 led to the discovery of a receptor complex containing MOM19, MOM22, MOM38, and MOM72 (Wienhues and Neupert, 1992). MOM38 is an import channel protein in the outer membrane (Table 15–2) so the receptor complex and import channel are likely close to one another. Import of MOM22 is absolutely dependent on the presence of surface receptors and both MOM19 and MOM72 are involved (Keil and Pfanner, 1993; Keil et al., 1993). The requirement for both receptor proteins may reflect the existence of a specific control system that ensures selective targeting of proteins of the receptor complex. MOM72 itself requires MOM19 for import, but MOM9 does not require a protease-sensitive surface receptor for its import.

One other import receptor has been identified on the outer membrane using an antiidiotypic antibody directed against a mitochondrial presequence peptide (Pain et al., 1990). A 32-kDa protein recognized by this antibody forms complexes with precursors bound to the mitochondrial surface, but deletion of the gene encoding this protein in yeast, while not lethal, renders the cells respiration deficient (Murakami et al., 1990). This suggests that the protein may be required for translocation of certain polypeptides necessary either for respiratory electron transport or the synthesis of electron transport components in the mitochondrion (e.g., mitochondrial ribosomal proteins coded in the nucleus). However, caution is necessary in interpreting these results because Joyard et al. (1991) note that the 32-kDa protein also shows 40% similarity to the mammalian phosphate translocator protein. This is an even bigger problem with respect to a putative 30-kDa chloroplast receptor protein discussed below. Purified preparations of mitochondrial p32 have since been reported to bind specifically preproteins containing mitochondrial presequences (Murakami *et al.,* 1993). These data support the previous assignment of p32 as a receptor protein, but do not resolve the conflicting observations suggesting that this or a similar protein is the phosphate translocator. One possibility is that p32 is actually bifunctional (Murakami *et al.,* 1993).

Detection of translocation contact sites between the inner and outer membranes has been facilitated by using precursors whose C-termini can become tightly folded so they get stuck in the process of translocation (Wienhaus and Neupert, 1992). Immunoelectron microscopy demonstrated that the accumulated precursors were preferentially localized to sites of close contact between the mitochondrial membranes, with the minimum length of the membrane-spanning portion of the precursor being 50 amino acids. There appear to be a few thousand contact sites per mitochondrion, and competition experiments indicate that different precursors share the same translocation machinery. This observation is encouraging since it suggests that the number of different species of receptors will probably be limited.

Experiments reviewed by Berry-Lowe and Schmidt (1991) and de Boer and Weisbeek (1991) suggest there are anywhere from 1500 to 3000 precursor binding sites per plastid. However, progress toward identification of receptor proteins has been marked by a controversy summarized in detail by Joyard *et al.* (1991). The antiidiotypic antibody approach using a synthetic chloroplast presequence led to the identification of proteins with molecular weights of 52 and 30 kDa. The 52-kDa protein seems to be identical with the large subunit by RUBISCO and probably represents a specific interaction of the antibody with this protein. The 30-kDa species, an integral membrane protein, was suggested to be the receptor. However, the phosphate-triose phosphate translocator, a major envelope protein, also has a molecular weight of 30 kDa and the protein from spinach has an 84% amino acid sequence identity with the "import" protein from pea. Quantitative calculations showed that with 3000 precursor binding sites per chloroplast, one would expect a corresponding number of receptor proteins amounting to 0.04% of the total envelope protein. However, the 30-kDa phosphate translocator protein represents 20% of the total envelope protein. A solution to the controversy now seems to revolve around three possibilities. The first is that there are two 30-kDa proteins. One of these is the phosphate translocator protein and the other is a putative receptor. The second is that the phosphate translocator protein was misidentified in the first place, a possibility that seems unlikely. The third is that the antiidiotype antibody did not, after all, identify a receptor protein.

Another candidate receptor protein was detected by crosslinking the RUBISCO small subunit precursor with a 66-kDa protein, but no correlation with any outer envelope protein was observed (Joyard *et al.,* 1991; de Boer and Weisbeek, 1991). If this is an envelope protein, it would have to be present in small amounts. Alternatively, the protein might be a dimer of the aforementioned 30-kDa protein.

The possibility of progress towards identification of the chloroplast receptor proteins has brightened with the report of isolation of a receptor complex from the outer envelope of pea (Soll and Waegemann, 1992). This complex will react with imported proteins only when the presequence and ATP are present. The complex contains a limited number of proteins of which two, OEP34 and OEP86, are protease sensitive like the receptor itself.

Protein Translocation Channels

Precursors whose import into mitochondria is arrested by low temperature or other methods normally become stuck across both membranes with their N-termini protruding into the matrix. Since these stuck precursors are solubilized by urea or alkaline pH, whereas typical integral membrane proteins are not, they appear to be within a hydrophilic channel composed of proteins (Glick and Schatz, 1991). One such protein, *I*mport *S*ite *P*rotein 42 (ISP42), has been identified by crosslinking it to a stuck precursor. This is an essential protein in yeast and its depletion is followed by accumulation of uncleaved mitochondrial precursors outside the mitochondrial inner membrane and cell death. The homologous protein in *Neurospora* is MOM 38. MOM38 was originally thought of as the GIP (*g*eneral *i*mport *p*rotein), a term that was actually proposed prior to the discovery of the actual protein to explain the fact that different precursors seem to use the same translocation machinery (Pfaller *et al.*, 1988). However, since MOM38 can be immunoprecipitated with a group of receptor complex proteins (see above), it is better thought of as part of a structure, the *g*eneral *i*mport *p*ore, that mediates translocation of preproteins across the mitochondrial outer membrane. Protein crosslinking studies in *Neurospora* have networked into yet other proteins that may be associated with the receptor/translocation apparatus since MOM7, MOM8, and MOM30 form a complex with MOM38 (Sollner *et al.*, 1992). Other experiments indicate that MOM22 is also part of this complex and may play the crucial role of transferring precursor proteins from the receptors (MOM19, MOM72) to the GIP (Kiebler *et al.*, 1993). In yeast another import gene *MIP1* has been identified (Maarse *et al.*, 1992). This essential gene encodes a mitochondrial inner membrane protein (MpiIp) that may be a component of a proteinaceous import channel required for translocation of precursor proteins across the mitochondrial inner membrane. This protein is identical to a 45 kDa inner membrane protein identified as part of a four protein inner membrane protein complex that can be crosslinked to precursor proteins stuck in the process of import across the inner membrane (Horst *et al.*, 1993). Whether, or if, MpiIp is identical to any of the aforementioned *Neurospora* translocation proteins remains to be established.

Mitochondrial import channels probably do not require energy for protein transport. The precursors are likely "pulled" in by ATP-hydrolyzing chaperones in the mitochondrial matrix (Glick and Schatz, 1991). In fact the import channels seem to be flexible pores that can even accommodate short DNA sequences chemically linked to a precursor (Vestweber and Schatz, 1989). As yet, translocation channel proteins do not seem to have been identified for chloroplasts.

Presequence Cleavage

Most N-terminal presequences containing routing information for the mitochondrial outer membrane are cleaved off by a metalloprotease, or *m*itochondrial *p*rocessing *p*eptidase (MPP), in the matrix (Glick and Schatz, 1991). The protein is a soluble heterodimer whose subunits are encoded by the *MAS1* and *MAS2* genes in yeast. Both genes are essential for survival. Similarly, the chloroplast contains a stromal processing peptidase (spp) that cleaves presequences of imported precursor proteins (de Boer and Weisbeek, 1991). Oblong and Lamppa (1992) have succeeded in purifying this labile enzyme and report that spp contains two subunits of 143 and 145 kDa. The enzyme also seems to be a metalloprotease and does not require ATP for function.

Energetics

The energy requirements for mitochondrial import are reviewed in detail by Geli and Glick (1990), Glick et al. (1992a), and Hartl et al. (1989). Assembly of porin into the outer membrane is unaffected by eliminating an electrochemical potential across the inner membrane. In contrast, specific inhibitors of the respiratory chain (e.g., rotenone, antimycin A) or the ATP synthetase (e.g., oligomycin, valinomycin) block import of proteins into the matrix. These experiments demonstrate that the electrochemical potential is essential for transport of proteins into or across the inner membrane.

The electrochemical potential (total protonmotive force) consists of two components: the membrane potential $\Delta\psi$ (electrical component) and the proton gradient ΔpH (chemical component). Transport of precursors requires only the membrane potential (negative inside) since addition of protonophores does not block translocation. This potential appears to facilitate an early interaction of matrix-targeting sequences with the inner membrane. The potential is needed only for insertion of the N-terminal portion of a precursor protein into the inner membrane. Subsequent translocation of the rest of the polypeptide is potential independent. Geli and Glick (1990) point out that it is as if the positively charged presequence is "electrophoresed" across the inner membrane early in translocation.

Protein import requires nucleoside triphosphates in addition to $\Delta\psi$. In fact, it is now known that ATP is needed both for initial insertion of precursors into the outer membrane and for translocation (Pfanner et al., 1990). The ATP requirement for insertion probably has to do in part with the ATP dependency of the hsp70 unfolding proteins. Lack of dependence on cytosolic ATP was observed with precursors that had been unfolded by treatment with high urea concentrations (Wienhaus and Neupert, 1992). ATP is also required for folding and for oligomeric assembly of imported proteins catalyzed by chaperonin 60 and, possibly, to "pull" the precursor through the mitochondrial inner membrane, a chaperone-mediated process (Geli and Glick, 1990).

The energy requirements for protein import into chloroplasts have also been reviewed (de Boer and Weisbeek, 1991; Flugge, 1990; Keegstra and von Heijne, 1992; Keegstra et al., 1989; Joyard et al., 1991). Chloroplasts differ from mitochondria in that precursor import does not require a membrane potential, but the two organelles are similar in requiring ATP for import. Isolated chloroplasts can import proteins without the addition of ATP when incubated in the light because they generate their own ATP via photosynthetic phosphorylation. In the dark, however, when great care is taken to remove all external ATP from the *in vitro* import system an absolute dependence on added ATP can be demonstrated (Flugge, 1990).

There seem to be two points at which ATP is required for chloroplast import. The first is an apparent requirement for low amounts of ATP (50–100 μM) for precursor binding (Olsen et al., 1989). Mitochondria do not have a similar requirement (Keegstra et al., 1989). This binding requirement is disputed by Flugge (1990), but he allows that the experiments of Olsen et al. (1989) were done with isolated pea chloroplasts whereas his experiments were carried out with spinach chloroplasts. Nevertheless, it is hard to believe that such a fundamental difference is species-specific. Second, ATP is required for translocation. As Flugge (1990) points out, this ATP requirement could be manifested in three mutually compatible ways. First, ATP may be required by cytosolic antifolding proteins such as hsp70 and for protein refolding inside the stroma by chaperonin 60. Second, ATP could be required for a transmembrane system functioning as a protein translocase that couples the energy derived from ATP hydrolysis to the translocation of the precursor protein across the membrane. Third, ATP might serve as an energy source for the phosphorylation of a component of the translocation apparatus, thus triggering the import reaction.

PROTEIN SORTING

Precursor proteins arriving at the mitochondrial or chloroplast surfaces must be sorted so that they go to the proper compartment. In the case of the mitochondrion this means the outer envelope, the matrix, the inner membrane, or the intermembrane space. For the chloroplast choices must be made between the inner and outer envelopes, the stroma, the thylakoid, and the thylakoid lumen. Correct decision making involves sorting and stop transfer signals often followed by additional proteolytic processing. The basic strategy for routing precursors into the matrix and stroma has been described already. Here the focus will be on membranes, the intermembrane space and the thylakoid lumen. The mechanism by which sorting occurs to the intermembrane space is of particular interest because two fundamentally different explanatory hypotheses (conservative sorting and stop-transfer) have been proposed.

The Mitochondrial Outer Membrane and the Chloroplast Envelopes

One of the least understood aspects of protein import concerns the targeting and sorting of proteins destined for the bounding membranes of chloroplasts and mitochondria. As discussed earlier, these proteins seem to be made without presequences. Glick and Schatz (1991) reviewed mitochondrial outer membrane targeting. Insertion of mitochondrial outer membrane proteins does not require an electrochemical potential across the inner membrane although the transport process itself probably shares some import steps with precursors destined for internal mitochondrial compartments. Several pieces of evidence favor this. First, the 70-kDa outer membrane protein can be mistargeted to the matrix by a single, conservative amino acid replacement. Second, a matrix enzyme can be converted to an outer membrane protein by placing a hydrophobic sequence near its amino terminus. Third, a water-soluble form of porin was found to interact with the outer membrane import components required for translocation of many precursors destined for the matrix.

The most detailed analysis of outer membrane targeting involves experiments with the Mas70p receptor protein of yeast, which is specific for the ADP/ATP translocator (see Receptors). When the N-terminal 12 amino acids of Mas70p are fused to a passenger protein, they function as a matrix sorting signal. This region merges into a hydrophobic domain (residues 10–37) that seems to act as stop-transfer signal for the outer membrane. These results seem to support the straightforward hypothesis that the Mas70p protein first targets the import machinery used to transfer precursors to the matrix following which the hydrophobic sequence anchors the protein in the outer membrane. However, Glick and Schatz (1991) counsel extreme caution in arriving at this conclusion for three reasons. First, many different sequences serve as matrix-targeting signals. Second, other precursors with a matrix-targeting signal followed by a hydrophobic region are imported into the intermembrane space. Third, the N-terminus of the homologous *Neurospora* protein (MOM72) lacks a typical matrix-targeting sequence. Other outer membrane proteins also appear to lack matrix-targeting sequences.

The import of four chloroplast envelope proteins has been examined. One of these is the 30-kDa phosphate translocator discussed earlier in connection with the receptor controversy (see Receptors). This protein is made as a 42.2-kDa precursor and the energy requirements for insertion of this protein resemble those needed to import stromal and thylakoid polypeptides (Flugge, 1990). This protein is one of the major envelope membrane proteins (15–20% of the total as mentioned above) and is located in the inner envelope (Douce and Joyard, 1991). Sorting of three outer membrane proteins has also been studied. One of these is a 6.7-kDa spinach protein (Salomon *et al.*, 1990). This protein is synthesized without a cleavable transit peptide and its import into the outer membrane appears to be ATP independent. Li *et al.* (1991) examined transport of a 14-kDa outer membrane protein. Like the 6.7-kDa protein the 14-kDa protein also lacks a cleavable transit peptide and its insertion is ATP independent. Based on results obtained for these two proteins, Li *et al.* (1991) postulate that proteins destined for the outer envelope of the chloroplast use a different pathway than those targeted for chloroplast compartments. The lack of a cleavable presequence and the absence of an ATP requirement are features that seem to be shared by the outer membrane import pathways of both chloroplasts and mitochondria. However, there is evidence, discussed earlier for the 70 kDa heat shock protein of the outer envelope, indicating that the extreme amino terminus of this protein is required for sorting (Wu and Ko, 1993).

The Mitochondrial Intermembrane Space: Conservative Sorting Versus Stop Transfer

Glick and Schatz (1991) separate precursors destined for the intermembrane space into three classes. The first includes cytochrome *c* and adenylate kinase, which lack cleavable presequences and seem to be imported and sorted differently than most mitochondrial precursors (see The Unconventional Routing and Sorting of Cytochrome *c*). The iron–sulfur protein of the cytochrome bc_1 complex typifies the second precursor class. This protein faces the outer surface of the inner membrane. According to Hartl *et al.* (1989) the iron–sulfur protein is first translocated completely into the mitochondrial matrix. MPP then cleaves off part of the presequence, yielding an intermediate with an eight amino acid extension. This intermediate is then processed to the mature-sized protein, retranslocated back across the inner membrane, and finally assembled into complex III. The role of the eight amino acid extension for sorting is unclear, but it probably does not contain targeting information for retranslocation since it is cleaved in the matrix space. According to one school of thought (Hartl *et al.*, 1989) translocation of the iron–sulfur protein to the intermembrane space is an example of conservative sorting. Other polypeptides, including subunit IV of cytochrome oxidase, probably follow a similar pathway (Glick and Schatz, 1991).

The third precursor class includes cytochrome *c* peroxidase, cytochrome c_1, cytochrome b_2, and creatine kinase. Each of these proteins possesses a bipartite presequence. The amino termini exhibit the typical features of the positively charged mitochondrial targeting sequences (Hartl and Neupert, 1990). Cleavage by the MPP processing enzyme removes the N-terminal sequences, yielding an intermediate. The remaining COOH-terminal part of the presequence contains essentially uninterrupted hydrophobic stretches of ca. 20 amino acids that are preceded by one to four basic residues, a motif reminiscent of the leader sequences that target the export of proteins across the plasma membrane in bacteria (Hartl and Neupert, 1990). It is this third class of proteins whose sorting is controversial. Some interpret the hydrophobic sequence as a reexport signal for sorting the protein conservatively from the matrix into the intermembrane space. Others believe the hydrophobic sequence is a stop-transfer signal that anchors the intermediate in the inner membrane.

The conservative sorting hypothesis posits that proteins of the intermembrane space are translocated in their entirety to the matrix and then reexported through the inner membrane (Fig. 15–3A). This hypothesis is highly appealing in view of the probable endosymbiotic origin of chloroplasts and mitochondria (Chapter 2). It predicts that the reexport process should be comparable to prokaryotic protein secretion. This would explain the similarity of the hydrophobic region in the presequences of proteins such as cytochrome c_1 to the leader sequence of bacterial proteins designed for export across the plasma membrane. The mature polypeptide sequence protruding into the intermembrane space would be freed from the inner membrane by endopeptidase cleavage of the intermediate. Conservative sorting is consistent with fixed or dynamic contact sites for translocation of precursors into mitochondria.

The initial step in the stop-transfer model is similar to conservative sorting in that the precursor begins to enter the matrix space through an import channel. Cleavage of the part of the presequence containing the routing information follows, yielding an intermediate. However, the two models now diverge. The stop-transfer hypothesis supposes that the hydrophobic sequence in the intermediate is a stop-transfer signal that anchors the intermediate into the inner membrane (Fig. 15–3B). The contact site is dynamic and the inner membrane region containing the anchored intermediate moves away from the outer membrane, permitting cleavage of the intermediate to yield the free mature form of the polypeptide in the intermembrane space.

Both models agree that the intermediate forms are bound to the outer face of the inner membrane. Conversion of the intermediates to mature forms would be catalyzed by a specific protease. Such an inner membrane protease has been identified in yeast (Glick and Schatz, 1991). This enzyme cleaves the intermediate form of cytochrome b_2 to the mature polypeptide. The amino acid sequence of the enzyme possesses regions of similarity to *E. coli* signal (leader) peptidase I, an enzyme that processes most *E. coli* secretory protein precursors (Pugsley, 1989). However, the processing site for the enzyme does not resemble typical bacterial cleavage sequences, and gene disruption experiments reveal that the enzyme is not required to process the cytochrome c_1 and cytochrome *c* peroxidase precursors. Surprisingly, the cleavage sites of the two enzymes resemble closely the motif recognized by bacterial signal peptidase I.

The two models make different predictions concerning the location of intermediates. The intermediates in the conservative sorting model would be located exclusively in the matrix whereas they would be in the intermembrane space according to the stop-transfer hypothesis. Furthermore, the stop-transfer model requires dynamic contact sites

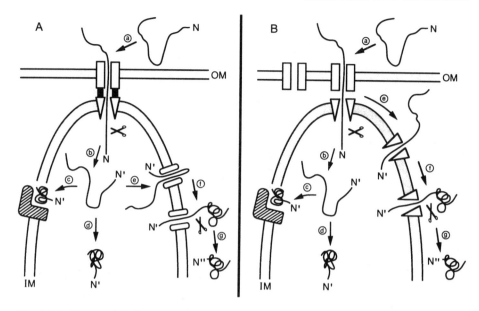

Fig. 15-3. Two models for protein sorting to the intermembrane space. (A) Fixed import sites and the conservative sorting (reexport) model. The extended line represents a loosely folded precursor protein with amino-terminus "N." (a) Precursors enter a translocation complex that contains outer membrane components (rectangles) and inner membrane components (triangles), joined tightly (black bars) at contact sites to form a continuous channel to the matrix. (b) All precursors pass completely into the matrix, where proteolytic cleavage generates a new amino terminus N'. (c) Some proteins assemble with partner subunits (cross-hatched) into complexes in the inner membrane. (d) Other proteins assume their mature configuration as soluble matrix components. (e) Some precursors are transported back across the inner membrane through a reexport machinery (ellipses). (f) After reexport, the protein spans the inner membrane facing the intermembrane space. (g) Cleavage on the outer face of the inner membrane yields soluble intermembrane space protein with mature amino terminus N'. (B) Dynamic import sites and the stop-transfer model. Translocation machineries of the outer membrane (rectangles) and inner membrane (triangles) are not tightly joined, and can diffuse into and out of contact sites. (a) Translocation across the two membranes is initiated at contact sites. (b) Precursors of matrix proteins, and some precursors destined for the inner membrane, pass through both membranes into the matrix and are proteolytically processed. (c), (d) As above. (e) A stop-transfer signal arrests translocation through the inner membrane, and the remainder of the precursor continues to cross the outer membrane into the intermembrane space. The amino-terminal domain faces the matrix and can be proteolytically processed. (f) An intermediate form is generated that is anchored in the inner membrane translocation machinery. (g) As above. OM, outer membrane; IM, inner membrane. From Glick and Schatz (1991).

whereas conservative sorting does not. The two models also make different predictions concerning the role of hsp60 and the ATP required for release of proteins from hsp60.

The question of where intermediates are located is problematical. Initially, van Loon and Schatz (1987) reported that the once cleaved intermediate form was always found outside the inner membrane. Subsequent publications reported that import intermediates of cytochromes c_1 and b_2 could be detected in the matrix if import took place at low temperatures and if cleavage by the matrix processing protease was inhibited

(Hartl et al., 1987; Nicholson et al., 1989). These experiments have been repeated by Glick et al. (1992) who claim that productive cytochrome c_1 and b_2 intermediates are found only outside the inner membrane under any conditions. However, they also report that 1% of the internalized molecules are present in the matrix, but that these are not reexported and "therefore represent a 'dead end' pathway." In contrast, Hartl et al. (1987) claimed that up to 50% of the cytochrome b_2 intermediates were in the matrix. Glick et al. (1992a) argue that the discrepancy stems from the different methods used by themselves and by Hartl et al. (1987) to assess the intramitochondrial location of imported proteins. If Glick et al. are correct, the mitochondrial fractionation procedure used by Hartl et al. (1987) must have given misleading results. However, Jensen et al. (1992) report that in yeast mutants that alter the cytochrome c_1 presequence by amino acid substitutions or deletions in the hydrophobic carboxy-terminal end of the presequence the intermediates do not localize to the intermembrane space. Instead, the intermediates are found in the mitochondrial matrix, which supports the conservative sorting hypothesis. Of particular importance is one mutant, T47N, in which the presequence is capable of slow processing. In this mutant mature protein is detectable in the intermembrane space, but all of the intermediate-sized proteins appear to be within the inner membrane.

Nevertheless, there is a potential drawback in experiments of the type reported by Jensen et al. that arises because of the selection scheme employed. In order to select mistargeting mutants, Jensen et al. attached the *COXIV* gene to the cytochrome c_1 presequence in a yeast strain with a disruption in *COXIV*. By definition, this meant that all surviving mutants had to be capable of directing the chimeric protein to the matrix so that the COXIV passenger protein could become part of a functional cytochrome oxidase complex. Beasley et al. (1993) used a similar approach to dissect the sorting signal of cytochrome b_2. These experiments revealed that the intermembrane space sorting signal was much longer than anticipated and consisted of three distinct regions. But more important from the viewpoint of stop-transfer versus conservative sorting, Beasley et al. interpreted their mutants as examples of *missorting*. That is, the chimeric proteins would normally *not* have been directed into the matrix were it not for the rare mutational alterations that made missorting possible. Schwarz et al. 1993 took a different approach to the same problem, which is free from this inherent selectional bias, but is at the same time more restricted. They believe the intermembrance space sorting sequence contains at least two typical motifs: (i) a core of hydrophobic uncharged amino acid residues preceded by (ii) one or more positive charges. Schwarz et al. focused on each of three positively charged residues by coupling mutated cytochrome b_2 presequences to a mouse dihydrofolate reductase (DHFR) reporter gene. The mutant proteins were synthesized *in vitro* and their localization into isolated mitochondria followed. Schwarz et al. believe that their results support the view that (i) there are one or more matrix components that recognize the second sorting signal and thereby trigger translocation into the intermembrane space, (ii) the mutant signals have reduced ability to interact with these recognition components and so embark on a default pathway in the mitochondrial matrix, and (iii) that their results argue against a role of the hydrophobic segment of the sorting sequence in stop transfer. Nevertheless, Schwarz et al. concede that their findings do not exclude a modified version of stop-transfer, which assumes that the positive amino acid residues at the amino-terminus of the hydrophobic domain are crucial for stop-transfer.

This rather lengthy discussion shows that the answer one gets is, to some degree, dependent on the method used. In the opinion of this reviewer none of the three aforementioned papers conclusively distinguishes between the stop-transfer and conservative sorting models although they each provide crucial information regarding the regions of the presequence that are important for localization of cytochromes c_1 and b_2 to the intermembrane space.

The second question is whether contact sites are dynamic, as required by the stop-transfer model, or static. Schatz and colleagues (Glick and Schatz, 1991; Horst et al., 1993) believe the data are consistent with the existence of dynamic contact sites. They note that the inner and outer membranes do not appear to be fused at contact sites, based on recent electron microscopic observations, despite an earlier report to the contrary. However, the most compelling evidence for dynamic contact sites comes from experiments with a chimeric, artificial precursor protein that had the pancreatic trypsin inhibitor protein as its carboxy-terminal domain (Horst et al., 1993). These experiments indicated that transport occurs in a biphasic fashion with the protein moving first across the outer membrane, then into the intermembrane space, and subsequently across the inner membrane. The same pattern of "biphasic" transport into the matrix was also noted for authentic precursors. Horst et al. (1993) also observe that import of the mitochondrial enzyme cytochrome c heme lyase from the cytoplasm to the intermembrane space seems to be completely independent of the inner membrane system. These observations lead Horst et al. (1993) to conclude that the two mitochondrial membrane systems each possess its own transport system and that these transport systems can link up when required to form a channel spanning both membranes. However, this linkage seems to be dynamic in that the two systems can become disengaged when transport across the inner membrane is interfered with by depletion of matrix ATP.

In contrast, Wienhaus and Neupert (1992) remark in passing that contact sites appear to be quite stable structures that survive even lengthy manipulations. As Pfanner and Neupert (1990) point out, translocation contact sites can be enriched by fractionation of mitochondria into submitochondrial vesicles. These observations certainly seem more consistent with stability than dynamic flux. However, Glick and Schatz (1991) report that contact regions persist in mitochondria even when no precursors are being imported. Are the contact regions and the import channels the same? Glick and Schatz say perhaps not.

The final piece of evidence that might discriminate between conservative sorting and stop-transfer revolves around chaperonin 60 and its ATP requirement. If the precursors destined for the intermembrane space are initially translocated to the matrix, one might expect chaperonin 60 to be required to help "pull" the precursor or otherwise facilitate its translocation across the inner membrane. Reexport of cytochrome b_2 from the matrix was reported to be defective under nonpermissive conditions in the *mif4* mutant of yeast, which contains a temperature-sensitive chaperonin 60 protein (Cheng et al., 1989). This would be consistent with matrix translocation of precursors destined for the intermembrane space. However, Glick et al. (1992b) state that they have been unable to verify these results despite repeated attempts. In their hands, maturation of both cytochromes b_2 and c_1 proceeded normally in the *mif4* mutant at the nonpermissive temperature. Furthermore, they report that fusion proteins with the cytochrome b_2 or c_1 presequences were transported to the intermembrane space and correctly processed under conditions of ATP depletion. This also argues against conservative sorting since

PROTEIN TARGETING AND IMPORT

ATP is needed not only for chaperonin 60 function, but also for complete transport of precursors across the inner membrane (Glick and Schatz, 1991). Additionally, Hallberg *et al.* (1993) examined *mif4* plus three additional *HSP60* mutants generated by *in vitro* mutagenesis that, like *mif4,* were all temperature sensitive. Hallberg *et al.* found that matrix proteins synthesized at the nonpermissive temperature were correctly targeted to and processed within the mitochondria whereupon they accumulated as insoluble aggregates as expected in the absence of functional hsp60. In contrast, the metabolism of cytochromes c_1 and b_2 was unaffected at the nonpermissive temperature as judged by their correct processing and the complete solubility of these proteins when synthesized at the nonpermissive temperature. These findings are consistent with the stop-transfer model.

Like all good hypotheses, conservative sorting and stop-transfer point the way to future experiments. Conservative sorting relies on the endosymbiont analogy, which has proved such an effective guidepost for delving into properties of chloroplasts and mitochondria. The investigator is encouraged to explore the possibility that protein transport across the inner membrane is analogous to protein secretion in bacteria. Thus, Hartl and Neupert (1990) use the conservative sorting hypothesis to compare the assembly pathways of the cytochrome bc_1 complex in mitochondria and prokaryotes from the α-subdivision of the purple bacteria *(Rhodobacter, Paracoccus)* from which mitochondria are supposed to derive (Chapter 2). They note that the bc_1 complex of these bacteria is highly similar to that of mitochondria and that bacterial cytochrome c_1 is sorted from the bacterial cytosol and through the plasma membrane where it faces the periplasmic space. The analogy to the proposed conservative sorting of mitochondrial cytochrome c_1 is obvious. On the other hand, stop transfer leads one to question whether this comparison is really valid. Perhaps it is fortuitous that the second part of the presequence of the mitochondrial cytochrome c_1 resembles the typical bacterial export signal found in cytochrome c_1 of *Rhodobacter.* Stop transfer also forces one to view contact sites between the inner and outer membrane where precursor transport occurs as dynamic rather than static. This is a prediction that has to be tested in more detail and in the process we will gain further insight into the mechanisms involved in protein import.

The Thylakoid Lumen

The thylakoid lumen is believed to be a space completely enclosed by the thylakoid membrane without any contact with the stroma (De Boer and Weisbeek, 1991). Proteins present in the lumen include three members of the water splitting complex, plastocyanin, and the plastocyanin docking protein. The presequences possess two domains. The N-terminal domain contains routing information directing the protein into the chloroplast import pathway. This sequence is processed off in the stroma. The second hydrophobic domain includes information that sorts the intermediate through the thylakoid membrane. This sequence is removed by the thylakoid-processing peptidase (TPP) on transfer of the intermediate to the lumen. The thylakoid transfer domains of lumen proteins strongly resemble secretory pathway signal sequences of prokaryotic and eukaryotic cells (Smeekens *et al.,* 1990). The information specifying proteolytic cleavage appears to be highly conserved in all three sequence types since TPP can cleave signal peptides of both prokaryotic and eukaryotic origin.

The two-domain structure of the presequence and the proposed mechanism of protein translocation to the thylakoid lumen is compatible with conservative sorting (Hartl and Neupert, 1990). In cyanobacteria, the ancestors of chloroplasts (Chapter 2), proteins like plastocyanin and the PSII manganese-stabilizing protein of the oxygen evolving complex are synthesized in the cytosol (equivalent to the chloroplast stroma) and translocated across the thylakoid membrane (Smeekens et al., 1990). These cyanobacterial proteins are also made as precursors containing N-terminal extensions resembling signal sequences. As Smeekens et al. (1990) point out, this is appealing in evolutionary terms since the addition of a presequence containing a chloroplast routing signal to a prokaryotic precursor protein will be sufficient to route and sort the protein correctly. In this regard it is of interest that cytochrome f, a chloroplast gene product that also targets the lumen, is made as a precursor with an N-terminal signal sequence. The protein is inserted into the thylakoid membrane such that the heme-binding domain is exposed in the lumen while a stop-transfer signal anchors the C-terminus to the thylakoid.

Other evidence makes the comparison of lumenal transport to bacterial secretion even more compelling (de Boer and Weisbeek, 1991). First, the precursor of the Mn-stabilizing subunit of the water splitting complex of cyanobacteria can be translocated into the periplasmic space of *E. coli*. Second, the thylakoid transfer domain of this protein will translocate alkaline phosphatase into the periplasmic space. Third, the cyanobacterial plastocyanin precursor and the plastocyanin intermediate from *Arabidopsis* are both transported to the periplasmic space in *E. coli*.

Inner Membrane and Thylakoid

Glick and Schatz (1991) classify inner membrane proteins into two groups: homooligomeric proteins such as the ADP/ATP carrier and heterooligomeric complexes exemplified by the F_oF_1 ATP-synthase and the respiratory chain complexes.

The ADP/ATP carrier of yeast and *Neurospora* has no cleavable presequence as discussed earlier, but instead contains internal routing information. This protein appears to be anchored to the inner membrane by three amphiphilic helices. The ADP/ATP carrier also shares several import components with precursors targeted to the matrix (e.g., Mas70p). In contrast, the bovine mitochondrial phosphate carrier is made as a precursor possessing a transient presequence. Pfanner and Neupert (1990) use the fungal ADP/ATP carrier as an example of nonconservative sorting. They remark that a prokaryotic equivalent probably does not exist. There is also no ATP requirement for its transport in contrast to proteins that sort conservatively.

Other proteins of the inner membrane are first imported completely into the matrix before assuming their final configuration in the inner membrane (Glick and Schatz, 1991). These include cytochrome oxidase subunit IV and the Rieske iron–sulfur protein of the cytochrome bc_1 complex. As mentioned earlier, the presequence of the Rieske iron–sulfur protein is processed in the matrix in two steps. Therefore, information for transfer across the inner membrane must reside in the mature sequence. Hartl and Neupert (1990) note that an iron–sulfur protein in the bacterium *Desulfovibrio vulgaris* is likewise translocated to the periplasmic space without a cleavable hydrophobic signal. Also interesting is the case of subunit 9 of the F_1/F_0 ATP-sythetase. Subunit 9 in *Neurospora* is a nuclear gene product that possesses a long presequence (66 amino acids) to target it to the mitochondrion (Hartl et al., 1989). Subunit 9 integrates into the inner membrane following complete translocation into the matrix. While Glick and Schatz (1991) state that this translocation pattern is compatible with both stop-transfer and conservative sorting, it seems to fit the latter hypothesis better. Otherwise, why would the entire protein be sorted into the matrix before entering the inner membrane? Why would it not simply insert into the inner membrane during import and stop transfer?

Approximately half of the integral membrane proteins of the thylakoid are imported from the cytoplasm (Smeekens et al., 1990). The most abundant of these are the LHCII (CAB-II) proteins whose import and sorting to the thylakoids have also been subject to the most attention. The CAB-II protein presequence has no function in membrane insertion and can be replaced by a presequence from a different protein (de Boer and Weisbeek, 1991). Also, the mature protein lacking the presequence will integrate into the thylakoid membrane without the help of a presequence. After entry into the chloroplast the CAB-II precursor associates with a chaperonin (Lubben et al., 1989). An unidentified stromal factor and ATP are required for integration (Cline, 1986; Chitnis et al., 1987). CAB-II proteins possess three membrane-spanning domains of which only the most C-terminal is essential for initial integration (Cline et al., 1989; Kohorn and Tobin, 1989). Once precursor integration occurs, the presequence is clipped off by the stromal processing peptidase (de Boer and Weisbeek, 1991).

Like the mitochondrial Rieske iron–sulfur protein associated with the cytochrome bc_1 complex, the homologous protein found in the chloroplast b_6f complex is encoded by a nuclear gene. This protein, like its mitochondrial counterpart, has a presequence routing it to the organelle and the signal that sorts it to the thylakoid membrane is contained in the mature polypeptide (de Boer and Weisbeek, 1991).

THE UNCONVENTIONAL ROUTING AND SORTING OF CYTOCHROME c

Gonzales and Neupert (1990) reviewed the biogenesis of cytochromes c and c_1. The import pathway followed by cytochrome c is quite different from those described above for either mitochondria or chloroplasts. The precursor of cytochrome c (apocytochrome c) is made in the cytoplasm, has no heme, and lacks a cleavable presequence. Furthermore, obvious mitochondrial routing signals cannot be detected from analysis of the primary structure of the protein and ATP is not required for import. Apocytochrome c binds efficiently to mitochondria. The cytochrome c binding sites differ in many ways from those of other precursors. They have a lower affinity for their substrates, they are more abundant than the receptor complexes that bind other precursors, and they are insensitive to low levels of protease. Apocytochrome c import can be blocked at the binding stage by preventing heme addition, which is catalyzed by cytochrome c heme lyase (CCHL). This enzyme transfers reduced heme to apocytochrome c whereupon the enzyme translocates directly into the intermembrane space. It is not known whether there is a special mitochondrial enzyme for heme reduction or whether reduced heme is delivered directly from ferrochelatase to CCHL. CCHL is a membrane-bound enzyme that seems to comprise all or part of the cytochrome c receptor.

The equivalent of cytochrome c in *Rhodobacter* is cytochrome c_2 (Hartl and Neupert (1990). However, the bacterial protein does have a cleavable presequence containing typical export signal.

Genetic studies with yeast (Sherman, 1990) have resulted in the identification of two structural genes for cytochrome c, *CYC1* and *CYC2*, which encode, respectively, the iso-1-cytochrome c and iso-2-cytochrome c forms of the protein. Iso-1-cytochrome c is the major form of the enzyme found under most conditions. The *CYC3* gene encodes CCHL, and *CYC2* specifies another protein required for efficient mitochondrial import of cytochrome c (Dumont et al., 1993).

REFERENCES

Ades, I.Z., and R.A. Butow (1980). The products of mitochondria-bound cytoplasmic polysomes in yeast. *J. Biol. Chem.* 225: 991–9924.

Baker, A., and G. Schatz (1987). Sequences from a prokaryotic genome or the mouse dihydrofolate reductase gene can restore the import of a truncated precursor protein into yeast mitochondria. *Proc. Natl. Acad. Sci. U.S.A.* 84: 3117–3121.

Beasley, E.M., Muller, S., and G. Schatz (1993). The signal that sorts yeast cytochrome b_2 to the mitochondrial intermembrane space contains three distinct functional domains. *EMBO J.* 12: 2303–2311.

Berry-Lowe, S.L., and G.W. Schmidt (1991). Chloroplast protein transport. In *The Molecular Biology of Plastids* (L. Bogorad and I.K. Vasil, eds.), pp. 257–302. Academic Press, San Diego.

Bertsch, U., Soll, J., Seetharan, R., and P.V. Viitanen (1992). Identification, characterization, and DNA sequence of a functional "double" groES-like chaperonin from chloroplasts of higher plants. *Proc. Natl. Acad. Sci. U.S.A.* 89: 8696–8700.

Boutry, M., Nagy, F., Poulsen, C., Aoyagi, K., and N.-H. Chua (1987). Targeting of bacterial chloramphenicol transacetylase to mitochondria in transgenic plants. *Nature (London)* 328: 340–342.

Brandt, U., Yu, L., Chang-An, Y., and B.L. Trumpower (1993). The mitochondrial targeting presequence of the rieske iron-sulfur protein is processed in a single step after insertion into the cytochrome bc_1 complex in mammals and retained as a subunit in the complex. *J. Biol. Chem.* 268: 8387–8390.

Cheng, M.Y., Hartl, F.U., Martin, J., Pollock, R.A., Kalousek, F., Neupert, W., Hallberg, E.M., Hallberg, R., and A.L. Horwich (1989). Mitochondrial heat-shock protein hsp60 is essential for assembly of proteins imported into yeast mitochondria. *Nature (London)* 337: 620–625.

Chitnis, P.R., Nechushtai, R., and J.P. Thornber (1987). Insertion of the precursor of the light-harvesting chlorophyll a/b protein into the thylakoids requires the presence of a developmentally regulated stromal factor. *Plant Mol. Biol.* 10: 3–12.

Cline, K. (1986). Import of proteins into chloroplasts: Membrane integration of a thylakoid precursor protein reconstituted in chloroplast lysates. *J. Biol. Chem.* 261: 14804–14810.

Cline, K., Fulsom, D.R., and P.V. Viitanen (1989). An imported thylakoid protein accumulates in the stroma when insertion into thylakoids is inhibited. *J. Biol. Chem.* 264: 14225–14232.

de Boer, A.D., and P.J. Weisbeek (1991). Chloroplast protein topogenesis: Import, sorting and assembly. *Biochim. Biophys. Acta* 1071: 221–253.

Deshaies, R.J., Koch, B.D., Werner-Washburne, M., Craig, E.A., and R. Schekman (1988). A subfamily of stress proteins facilitates translocation of secretory and mitochondrial precursor polypeptides. *Nature (London)* 332: 800–805.

Douce, R., and J. Joyard (1991). Structure, organization and properties of plastid envelope membranes. In *The Molecular Biology of Plastids* (L. Bogorad and I.K. Vasil, eds.), pp. 217–256. Academic Press, San Diego.

Dumont, M.E., Schlichter, J.B., Cardillo, T.S., Hayes, M.K., Bethlendy, G., and F. Sherman (1993). CYC2 encodes a factor involved in mitochondrial import of yeast cytochrome c. *Mol. Cell. Biol.* 13: 6442–6451.

Eilers, M., Verner, K., Hwang, S., and G. Schatz (1988). Import of proteins into mitochondria. *Phil. Trans. R. Soc. London B* 319: 121–126.

Ellis, R.J. (1990). Molecular chaperones: The plant connection. *Science* 250: 954–959.

Ellis, R.J. (1993). The general concept of molecular chaperones. *Phil. Trans. R. Soc. London* B 339: 257–261.

Ellis, R.J., and S.M. van der Vies (1991). Molecular chaperones. *Annu. Rev. Biochem.* 60: 321–347.

Flugge, U.-I. (1990). On the translocation of proteins across the chloroplast envelope. *J. Bioenerget. Biomembranes* 22: 769–786.

Fujiki, M. and K. Verner (1991). Coupling of protein synthesis and mitochondrial import in a homologous yeast *in vitro* system. *J. Biol. Chem.* 266: 6841–6847.

Fujiki, M., and K. Verner (1993). Coupling of cytosolic protein synthesis and mitochondrial import in yeast: Evidence for cotranslational import *in vivo*. *J. Biol. Chem.* 268: 1914–1920

Fulson, D.R., and K. Cline (1988). A soluble protein factor is required *in vitro* for membrane insertion of the thylakoid precursor protein, pLHCP. *Plant Physiol.* 88: 1146–1153.

Gatenby, A.A. (1992). Protein folding and chaperonins. *Plant Mol. Biol.* 19: 677–687.

Geli, V., and B. Glick (1990). Mitochondrial protein import. *J. Bioenerget. Biomembranes* 22: 725–751.

Glick, B., and G. Schatz (1991). Import of proteins into mitochondria. *Annu. Rev. Genet.* 25: 21–44.

Glick, B.S., Brandt, A., Cunningham, K., Muller, S., Hallberg, R., and G. Schatz (1992a). Cytochromes

c_1 and b_2 are sorted to the intermembrane space of yeast mitochondria by a stop-transfer mechanism. *Cell* 69: 809–822.

Glick, B.S., Wachter, C., and G. Schatz (1992b). The energetics of protein import. *Biochim. Biophys. Acta* 1101: 249–251.

Gonzales, D.H., and W. Neupert (1990). Biogenesis of mitochondrial *c*-type cytochromes *J. of Bioenerget. Biomembranes* 22: 753–768.

Grivell, L.A. (1988). Protein import into mitochondria. *Int. Rev. Cytol.* 111: 107–141.

Hallberg, E.M., Shu, Y., and R.L. Hallberg (1993). Loss of mitochondrial hsp60 function: Nonequivalent effects on matrix-targeted and intermembrane-targeted proteins. *Mol. Cell. Biol.* 13: 3050–3057.

Hartl, F.-U., and W. Neupert (1990). Protein sorting to mitochondria: Evolutionary conservations of folding and assembly. *Science* 247: 930–938.

Hartl, F.-U., Ostermann, J., Guiard, B., and W. Neupert (1987). Successive translocation into and out of the mitochondrial matrix: Targeting of proteins to the intermembrane space by a bipartite signal peptide. *Cell* 51: 1027–1037.

Hartman, D.J., Hoogenraad, N.J., Contron, R., and P.B. Høj (1992). Identification of a mammalian 10-kDa heat shock protein, a mitochondrial chaperonin 10 homologue essential for assisted folding of trimeric ornithine transcarbamylase in in vitro. *Proc. Natl. Acad. Sci. USA* 89: 3394–3398.

Hartl, F.-U., Pfanner, N., Nicholson, D.W., and W. Neupert (1989). Mitochondrial protein import. *Biochim. Biophys. Acta* 988: 1–45.

Hendrick, J.P., and F.-U. Hartl (1993). Molecular chaperone functions of heat shock proteins. *Annu. Rev. Biochem.* 62: 349–384.

Hines, V., Brandt, A., Griffiths, G., Horstmann, H., Brutsch, H., and G. Schatz (1990). Protein import into yeast mitochondria is accelerated by the outer membrane protein MAS70. *EMBO J.* 9: 3191–3200.

Horst, M., Kronidou, N.G., and G. Schatz (1993). Through the mitochondrial inner membrane. *Curr. Biol.* 3: 175–177.

Horwich, A.L., and K.R. Willison (1993). Protein folding in the cell: functions of two families of molecular chaperone, hsp 60 and TF55-TCP1. *Phil. Trans. R. Soc. London* B, 339: 313–326.

Hurt, E.C., Soltinager, N., Goldschmidt-Clermont, M., Rochaix, J.-D., and G. Schatz (1986a). The cleavable presequence of an imported chloroplast protein directs attached polypeptides into yeast mitochondria. *EMBO J.* 5: 1343–1350.

Hurt, E.C., Goldschmidt-Clermont, M., Pesold-Hurt, B., Rochaix, J.-D., and G. Schatz (1986b). A mitochondrial presequence can transport a chloroplast-encoded protein into yeast mitochondria. *J. Biol. Chem.* 261: 11440–11443.

Jensen, R.E., Schmidt, S., and R.J. Mark (1992). Mutations in a 19-amino-acid hydrophobic region of the yeast cytochrome c_1 presequence prevent sorting to the mitochondrial intermembrane space. *Mol. Cell. Biol.* 12: 4677–4686.

Joyard, J., Block, M.A., and R. Douce (1991). Molecular aspects of plastid envelope biochemistry. *Eur. J. Biochem.* 199: 489–509.

Kang, P.-J., Ostermann, J., Shilling, J., Neupert, W., Craig, E.A., and N. Pfanner (1990). Hsp70 in the mitochondrial matrix is required for translocation and folding of precursor proteins. *Nature (London)* 348: 137–143.

Karlin-Neumann, G.A., and E.M. Tobin (1986). Transit peptides of nuclear-encoded chloroplast proteins share a common amino acid framework. *EMBO J.* 5: 9–13.

Keegstra, K. (1989). Transport and routing of proteins into chloroplasts. *Cell* 56: 247–253.

Keegstra, K., and C. Bauerle (1988). Targeting of proteins into chloroplasts. *BioEssays* 9: 15–19.

Keegstra, K., and G. von Heijne (1992). Transport of proteins into chloroplasts. In *Plant Gene Research: Cell Organelles* (R. G. Herrmann, ed.), pp. 353–370. Springer Verlag, Wien.

Keegstra, K., Olsen, L.J., and S.M. Theg (1989). Chloroplastic precursors and their transport across the envelope membranes. *Annu. Rev. Plant Physiol. Plant Mol. Biol.* 40: 471–501.

Keil, P., and N. Pfanner (1993). Insertion of MOM22 into the mitochondrial outer membrane depends on surface receptors. *FEEBS Lett.* 321: 197–200.

Keil, P., Weinzierl, A., Kiebler, M., Dietmeier, K., Sollner, T., and N. Pfanner (1993). Biogenesis of the mitochondrial receptor complex. Two receptors are required for binding of MOM38 to the outer membrane surface. *J. Biol. Chem.* 268: 19177–19180.

Kellems, R.E., and R.A. Butow (1972). Cytoplasmic-type 80S ribosomes associated with yeast mitochondria. I. Evidence for ribosome binding sites on yeast mitochondria. *J. Biol. Chem.* 247: 8043–8050.

Kellems, R.E., and R.A. Butow (1974). Cytoplasmic type 80S ribosomes associated with yeast mitochondria. III. Changes in the amount of bound ribosomes in response to changes in metabolic state. *J. Biol. Chem.* 249: 3304–3310.

Kellems, R.E., Allison, V.F., and R.A. Butow (1974). Cytoplasmic 80S ribosomes associated with yeast mitochondri. II. Evidence for the association of cytoplasmic ribosomes with the outer mitochondrial membrane in situ. *J. Biol. Chem.* 3297–3303.

Kiebler, M., Keil, P., van der Kiel, I.J., Pfanner, N., and W. Neupert (1993). The mitochondrial receptor complex: a central role of MOM22 in mediating preprotein transfer from receptors to the general insertion pore. *Cell* 74: 483–492.

Ko, K., Bornemisza, O., Kourtz, L., Ko, Z.W., Plaxton, W.C., and A.R. Cashmore (1992). Isolation and characterization of a cDNA clone encoding a cognate 70-kDa heat shock protein of the chloroplast envelope. *J. Biol. Chem.* 267: 2986–2993.

Kohorn, B.D., and E.M. Tobin (1989). A hydrophobic, carboxy-proximal region of a light-regulated, light-harvesting chlorophyll a/b protein is necessary for stable integration into thylakoid membranes. *Plant Cell* 1: 159–166.

Landry, S.J., and L.M. Gierasch (1991). Recognition of nascent polypeptides for targeting and folding. *Trends Biochem. Sci.* 16: 159–163.

Li, H.-m., Moore, T., and K. Keegstra (1991). Targeting of proteins to the outer envelope membrane uses a different pathway than transport into chloroplasts. *Plant Cell* 3: 709–717.

Lindquist, S., and E.A. Craig (1988). The heat shock proteins. *Annu. Rev. Genet.* 22: 631–677.

Lorimer, G.H., Todd, M.J., and P.V. Viitanen (1993). Chaperonins and protein folding: Unity and disunity of mechanisms. *Phil. Trans. R. Soc. London B* 339: 297–304.

Lubben, T.H., Gatenby, A.A., Ahlquist, P., and K. Keegstra (1989). Chloroplast import characteristics of chimeric proteins. *Plant Mol. Biol.* 12: 13–18.

Lubben, T.H., Gatenby, A.A., Donaldson, G.K., Lorimer, G.H., and P.V. Viitanen (1990). Identification of a groES-like chaperonin in mitochondria that facilitates protein folding. *Proc. Natl. Acad. Sci. U.S.A.* 87: 7683–7687.

Maarse, A.C., Blom, J., Grivell, L.A., and M. Meijer (1992). *MPI1*, an essential gene encoding a mitochondrial membrane protein is possibly involved in protein import into yeast mitochondria. *EMBO J.* 11: 3614–3628.

Marshall, J.S., and K. Keegstra (1992). Isolation and characterization of a cDNA clone encoding the major Hsp70 of the pea chloroplastic stroma. *Plant Physiol.* 100: 1048–1054.

Mishkind, M.L., Wessler, S.R., and G.W. Schmidt (1985). Functional determinants in transit sequences: Import and partial maturation by vascular plant chloroplasts of the ribulose-1,5-bisphosphate carboxylase small subunit of *Chlamydomonas. J. Cell Biol.* 100: 226–234.

Morimoto, R.I., Tissieres A., and C. Georgopoulos eds. (1990). *Stress Proteins in Biology and Medicine*. Cold Spring Harbor Laboratory Press, Cold Spring Harbor, NY.

Murakami, H., Blobel, G., and D. Pain (1990). Isolation and characterization of the gene for a yeast mitochondrial import receptor. *Nature (London)* 347: 488–491.

Murakami, H., Blobel, G., and D. Pain (1993). Signal sequence region of mitochondrial precursor proteins binds to mitochondrial import receptor. *Proc. Natl. Acad. Sci. USA* 90: 3358–3362.

Neupert, W., and N. Pfanner (1993). Roles of molecular chaperones in protein targeting to mitochondria. *Phil. Trans. R. Soc. London B* 339: 355–362.

Nicholson, D.W., Stuart, R.A., and W. Neupert (1989). Biogenesis of cytochrome c_1. Role of heme lyase and of the two proteolytic processing steps during import into mitochondria. *J. Biol. Chem.* 264: 10156–10168.

Oblong, J.E., and G.K. Lamppa (1992). Identification of two structurally related proteins involved in proteolytic processing of precursors targeted to the chloroplast. *EMBO J.* 11: 4401–4409.

Olsen, L.J., Theg, S.M., Selman, B.R., and K. Keegstra (1989). ATP is required for the binding of precursor proteins to chloroplasts. *J. Biol. Chem.* 264: 6724–6729.

Pain, D., Murakami, H., and G. Blobel (1990). Identification of a receptor for protein import into mitochondria. *Nature (London)* 347: 444–449.

Pelham, H.R.B. (1986). Speculations on the functions of the major heat shock and glucose regulated proteins. *Cell* 46: 959–961.

Pfaller, R., Steger, H.F., Rassow, J., Pfanner, N., and W. Neupert (1988). Import pathways of mitochondria: Multiple receptor sites are followed by a common membrane insertion site. *J. Cell Biol.* 107: 2483–2490.

Pfanner, N. (1992). Components and mechanisms in mitochondrial protein import. In *Plant Gene Research: Cell Organelles* (R. Herrmann, ed.), pp. 371–400. Springer-Verlag, Wien.

Pfanner, N., and W. Neupert (1990). The mitochondrial protein import apparatus. *Annu. Rev. Biochem.* 59: 331–353.

Pugsley, A.P. (1989). *Protein Targeting*. Academic Press, San Diego.

Roise, D. (1992). Interaction of a synthetic mitochondrial presequence with isolated yeast mitochondria: Mechanism of binding and kinetics of import. *Proc. Natl. Acad. Sci. U.S.A.* 89: 608–612.

Roise, D., Theiler, F., Horvath, S.J., Tomich, J.M., Richards, J.H., Allison, D.S., and G. Schatz (1988). Amphiphilicity is essential for mitochondrial presequence function. *EMBO J.* 7: 649–653.

Salomon, M., Fischer, K., Flugge, U.-I., and J. Soll (1990). Sequence analysis and protein import studies of an chloroplast envelope polypeptide. *Proc. Natl. Acad. Sci. U.S.A.* 87: 5778–5782.

Schatz, G., and R.A. Butow (1983). How are proteins imported into mitochondria? *Cell* 32: 316–318.

Schmidt, G.W., and M.L. Mishkind (1986). The transport of proteins into chloroplasts. *Annu. Rev. Biochem.* 55: 879–912.

Schwarz, E., Seytter, T., Guiard, B., and W. Neupert (1993). Targeting of cytochrome b_2 into the mitochondrial intermembrane space: specific recognition of the sorting signal. *EMBO J.* 12: 2295–2302.

Sherman, F. (1990). Studies of yeast cytochrome *c* How and why they started and why they continued. *Genetics* 125: 9–12.

Smeekens, S., van Steeg, H., Bauerle, C., Bettenbroek, H., Keegstra, K., and P. Weisbeek (1987). Import into chloroplasts of a yeast mitochondrial protein directed by ferredoxin and plastocyanin transit peptides. *Plant Mol. Biol.* 9: 377–388.

Smeekens, S., Weisbeek, P., and C. Robinson (1990). Protein transport into and within chloroplasts. *Trends Biochem. Sci.* 15: 73–76.

Soll, J., and K. Waegemann (1992). A functionally active import complex from chloroplasts. *Plant J.* 2: 253–256.

Sollner, T., Rassow, J., Wiedmann, M., Scholssman,

J., Keil, P., Neupert, W., and N. Pfanner (1992). Mapping of the protein import machinery in the mitochondrial outer membrane by crosslinking of translocation intermediates. *Nature* 355: 84–87.

Suissa, M., and G. Schatz (1982). Import of proteins into mitochondria: Translatable mRNAs for imported mitochondrial proteins are present in free as well as mitochondria-bound cytoplasmic polysomes. *J. Biol. Chem.* 257: 13048–13055.

van Loon, A.P.G.M., and G. Schatz (1987). Transport of proteins to the mitochondrial intermembrane space: The "sorting" domain of the cytochrome c_1 presequence is a stop-transfer sequence specific for the mitochondrial inner membrane. *EMBO J.* 6: 2441–2448.

Vestweber, D., and G. Schatz (1989). DNA-protein conjugates can enter mitochondria via the protein import pathway. *Nature (London)* 338: 170–172.

Verner, K. (1992). Early events in yeast mitochondrial protein targeting. *Mol. Microbiol.* 6: 1723–1728.

Verner, K. (1993). Co-translational protein import into mitochondria: an alternative view. *Trends in Biochem. Sci.* 18: 366–371.

Vierling, E. (1991). The roles of heat shock proteins in plants. *Annu. Rev. Plant Physiol. Plant Mol. Biol.* 42: 579–620.

Wienhaus, U., and W. Neupert (1992). Protein translocation across mitochondrial membranes. *BioEssays* 14: 17–23.

Wu, C., and K. Ko (1993). Identification of an uncleavable targeting signal in the 70-kilodalton spinach chloroplast outer envelope membrane protein. *J. Biol. Chem.* 268: 19384–19391.

Yalovsky, S., Paulsen, H., Michaeli, D., Chitnis, P.R., and R. Nechushtai (1992). Involvement of a chloroplast HSP70 heat shock protein in the integration of a protein (light-harvesting complex protein precursor) into the thylakoid membrane. *Proc. Natl. Acad. Sci. USA* 89: 5616–5619.

Zeilstra-Ryalls, J. *et al.* (1991). The universally conserved GroE (Hsp60) chaperonins. *Annu. Rev. Microbiol.* 45: 301–325.

16

Control of Gene Expression in Chloroplast Biogenesis

Two of the most dramatic and extreme examples of developmental change in organelles are the light-induced conversion of a colorless etioplast into a green, photosynthetically competent chloroplast and the transformation of a glucose-repressed yeast mitochondrion to a respiratory-competent organelle on transfer of cells to a nonfermentable carbon source. The parallels between the two events are striking. Each involves the synthesis and assembly of an electron transport chain and an ATP synthase in response to an environmental signal. In both cases the coordinated expression of organelle and nuclear genes and the assembly of their products are involved. The process by which this occurs must take into account the fact that organelle genes are greatly amplified with respect to their counterparts in the nucleus, that proteins encoded in the nucleus and synthesized in the cytoplasm must enter the correct organelle and the proper compartment therein, that these proteins must complex with their counterparts encoded in the chloroplast or mitochondrial genome, and that the whole assembly process must occur so that multicomponent structures result in which the individual subunits are present in a precisely defined stoichiometry. Given these constraints and the billion or so years in which chloroplasts, mitochondria, and nuclei have had to become acquainted in the comfortable aqueous milieu of the eukaryotic cell, one would expect a variety of control mechanisms to have evolved. That is the case.

One can imagine several points at which the biogenesis of a photosynthetically competent chloroplast or a respiratory sufficient yeast mitochondrion might be controlled: gene transcription; transcript processing; transcript turnover, translation, and protein processing and turnover. Examples of all of these modes of control are to be found in chloroplasts and mitochondria. Also, nuclear and organellar regulatory mechanisms differ in that control of gene expression is generally exercised at the transcriptional level in the nucleus, but is governed to a large extent by posttranscriptional and translational mechanisms in both chloroplasts and mitochondria.

Superimposed on these different possible control points for gene expression one can imagine different classes of control mechanisms (Hauser *et al.,* 1993). *Gene-specific controls* are exemplified by the many cases, discussed later in this chapter and in Chapter 17, where proteins encoded by specific nuclear genes interact with the 5' untranslated flanking regions of individual chloroplast or mitochondrial mRNAs. *Class-specific controls* discriminate between mRNAs encoding functionally related groups of proteins. For example, there is evidence in *Chlamydomonas* that expression of chloroplast-encoded mRNAs specifying photosynthetic functions and ribosomal proteins is controlled in a class-specific manner. *General controls* apply to all mRNAs irrespective of class. For instance, there is some reason to believe that the proportion of mRNAs translated within chloroplasts and mitochondria remains relatively constant in spite of fluctuations in the total mRNA pool.

There is also the question of who is the dominant partner in coordinating expression of nuclear and organellar genes. Undoubtedly, it is the nucleus, but chloroplasts and mitochondria apparently signal the nucleus as well.

Gruissem and Schuster (1993) make an important distinction between plastids and mitochondria. They remark "unlike mitochondria, chloroplasts can differentiate into nonphotosynthetic plastids that have different functions, such as starch-containing amyloplasts in root or chromoplasts in fruit, certain flower petals, and roots. Differentiation of nonphotosynthetic plastids from proplastids or chloroplasts is not a terminal process, as shown by the fact that specialized plastids maintain the capability to redifferentiate into chloroplasts." Hence, in the case of amyloplasts, chromoplasts, etc. other levels of genetic control must be interposed that signal developmental pathways distinct from that leading to mature green chloroplasts. For example, mRNA levels for specific chloroplast genes in these specialized plastids tend to be lower than in the green chloroplasts of mature leaves (Gruissem and Schuster, 1993).

This chapter describes the control of gene expression in chloroplast biogenesis and Chapter 17 does the same for mitochondria.

CHLOROPLAST BIOGENESIS

Exposure of dark-grown plants to light causes etioplasts to differentiate into chloroplasts. This process is accompanied by the synthesis and assembly of the protein components of the photosynthetic electron transport chain. The process involves several kinds of light reactions (see Hoober, 1984; Thompson and White, 1991). One is the red light-induced conversion of protochlorophyll(ide) to chlorophyll(ide) in plants that cannot form chlorophyll in the dark, following which the paracrystalline structure of the etioplast prolamellar body becomes dispersed as primary lamellar layers. This reaction is mediated by the enzyme NADPH-protochlorophyll oxidoreductase and requires NADPH in addition to light. The second involves the plant regulatory factor phytochrome (see Quail, 1991; Thompson and White, 1991 for reviews). This red-absorbing pigment can exist in two stable spectral forms, Pr and Pfr (see Chapter 1). Pr is the inactive form and is converted to the active Pfr form when it absorbs red light. The conversion is photoreversible by far red light. Pfr is required for transcription of many nuclear genes involved in chloroplast biogenesis. On the other hand, expression of certain nuclear genes seems to be repressed by Pfr (e.g., the genes encoding protochlorophyllide reductase and asparagine synthetase). Phytochrome also seems to be a photoreceptor that controls plastid gene transcription (Klein and Mullet, 1990). There are also blue/ultraviolet-A (UV-A) "cryptochrome" and ultraviolet-B (UV-B) light receptors (Thompson and White, 1991). These photoreceptors may be flavoproteins and/or carotenoids. Blue light receptors, unlike the red light receptors, are not primarily restricted to green plants, but are widely distributed in nature (see Senger, 1987; Gamble and Mullet, 1989). Excitation of the blue light receptor is known to control a wide variety of processes ranging from phototropism in the fungus *Phycomyces* to photosynthetic activity in *Acetabularia*. Blue light has also been shown to affect expression of the nuclear genes encoding the RUBISCO small subunit and the accumulation of the chloroplast-encoded *psbD-psbC* transcripts in barley (Gamble and Mullet, 1989).

General Regulation of Chloroplast Gene Expression

Chloroplast gene expression could be regulated at transcriptional, posttranscriptional, translational, and posttranslational levels (see Gruissem, 1989a,b; Gruissem and Tonkyn, 1993; Gruissem *et al.*, 1988; Link, 1988; Mullet, 1988; Mullet *et al.*, 1991; Taylor, 1989 for reviews). In fact all of these modes of regulation apply.

Transcriptional control. Accumulation of many chloroplast mRNAs increases as a function of light and the stage of chloroplast development (Gruissem, 1989a; Gruissem and Tonkyn, 1993; Herrmann *et al.*, 1992; Mullet, 1988). However, accumulation can be the result of increased mRNA synthesis, decreased mRNA stability, or both. Accumulation is relatively easy to measure, but synthesis and degradation of RNA molecules are much harder to assess since pulse–chase designs are difficult to use in land plants *in vivo*. RNA precursors (i.e., ribonucleoside triphosphates) are not readily taken up and other precursors (e.g., labeled phosphate) are far enough removed from the actual end product that pool problems can confuse the answer.

To get around this problem, run-on transcription assays have been developed for chloroplasts (see Gruissem, 1989b; Gruissem *et al.*, 1988; Mullet, 1988). In these systems, incorporation of ribonucleoside triphosphates into high-molecular-weight RNA by chloroplast lysates is measured. Since addition of exogenous cpDNA does not affect these reactions, the assumption is that the RNA being made is strictly synthesized by preinitiation transcription complexes. By comparing the accumulation of specific mRNAs with the amounts synthesized in run-on systems, synthesis and stability can be distinguished. Studies of this kind indicate that transcription activity and RNA accumulation tend to act independently for most genes examined. For example, in the course of development of a spinach plant from a 3-day-old seedling, small changes in the relative transcriptional activities of the *psbA*, *rbcL*, and *atpBE* genes are insufficient to account for the differential accumulation of their mRNAs (Gruissem *et al.*, 1988). Plastid run assays showed that transcription rates of 10 spinach chloroplast genes varied by more than 25-fold (Gruissem and Tonkyn, 1993).

Careful measurements of transcriptional activity of chloroplast genes in four-day-old dark-grown seedlings of barley have revealed dramatic variations in transcription activity (Rapp *et al.*, 1992). The 15 genes included in this study encoded components involved in photosynthesis, transcription, and translation. The genes were located in 14 different transcription units covering 50% of the barley plastid genome. Among the protein-coding genes transcription activity varied over 300-fold and transcript levels over 900-fold. Rapp *et al.* (1992) argue that their results support the hypothesis that transcription is a primary determinant of barley plastid gene expression because of the correspondence between transcription rate, mRNA level, and protein abundance.

Parallel increases in plastid gene transcription have been observed when chloroplast development is induced by light in barley, spinach, and maize (Mullet, 1988). In barley, plastid transcription activity and RNA levels are elevated during rapid chloroplast growth and decline when this phase of chloroplast development is completed (Mullet and Klein, 1987). The expression of the 15 barley chloroplast genes studied by Rapp *et al.* (1992) has been quantified as a function of developmental stage (Baumgartner *et al.*, 1993). Overall plastic transcription rates and mRNA levels were found to increase early in chloroplast development, but declined subsequently during chloroplast maturation. Furthermore, several genes of the transcription/translation apparatus were found to be transcribed preferentially early in chloroplast development while expression of chloroplast photosynthetic genes occurred at a slightly later stage.

Various explanations have been set forth to account for increases in chloroplast transcription activity during chloroplast development (see Gruissem, 1989a; Mullet, 1988 for summaries). They include variations in the number of cpDNA templates, alterations in DNA topology, and changes in specific activity of the chloroplast RNA polymerase. Bendich (1987) argued that induction of plastid transcription during chloroplast biogenesis might result from increased DNA template levels. Thus studies on wheat reveal that plastid DNA copy number can increase up to 7.5-fold during the transition from proplastids to chloroplasts (Miyamura *et al.*, 1986). However, transcription activity and DNA copy number seem not to be strictly correlated during chloroplast development arguing against a limitation of transcription activity by DNA template (Baumgartner *et al.*, 1989; Deng and Gruissem, 1987; Gruissem *et al.*, 1988). For example, in developing barley leaves Baumgartner *et al.* (1989) report that a 2-fold change in cpDNA copy number is accompanied by a 5-fold increase in transcription activity per DNA template. On the other hand, experiments with *Chlamydomonas* reveal that reductions in chloroplast genome number of over 50% induced by growth of cells in 5-fluorodeoxyuridine are matched by a parallel decline in mRNA accumulation for the three chloroplast genes examined (Hosler *et al.*, 1989). One could argue that run-on transcription assays might have revealed this parallel decline to be an artifact resulting from a decline in mRNA stability with cpDNA copy number. However, this seems unlikely.

Changes in overall plastid transcription rates might also be affected by changes in DNA conformation (Bogorad, 1991; Gruissem, 1989a; Mullet, 1988). Topological changes could influence both the overall transcription rate of the chloroplast genome as well as specific promoters. Stirdivant *et al.* (1985) conducted a detailed analysis of transcription as a function of template topology. They used purified maize chloroplast RNA polymerase and a cloned fragment from the maize chloroplast genome containing the divergently transcribed *rbcL* and *atpBE* genes. The major conclusions from this study were that (1) transcription of these genes from relaxed templates proceeded at only $\frac{1}{50}$ of the maximum rate from an appropriately negatively supercoiled template, and (2) the negative superhelical density of the template required for maximum transcription rates differed for *rbcL* and *atpBE*. These findings suggest that DNA gyrase and topoisomerase might control relative transcriptional activity of chloroplast genes *in vivo*. Consistent with this hypothesis, Lam and Chua (1987) observed that the topoisomerase II inhibitor novobiocin affects transcription from the *rbcL* and *atpB* promoters differently in chloroplast extracts from pea. Thompson and Mosig (1984) identified a chloroplast gene in *Chlamydomonas*, called A, of unknown function whose transcript level is decreased in light versus dark grown cells. The promoter of this gene seems to be regulated by torsional stress since it is stimulated by novobiocin. The gyrase that may be responsible has been identified in cell extracts from *Chlamydomonas* (Thompson and Mosig, 1985, 1987). This enzyme produces supercoils in circular DNA *in vitro* and is inhibited by novobiocin and nalidixic acid. However, Gruissem (1989a) comments that "the results currently available for chloroplast genes leave the question open as to whether topological modifications, do, in fact, have a significant role in organ-specific or developmental control of promoter activity."

RNA polymerase and promoter structure could also influence chloroplast gene expression. The major subunits of the core polymerase of flowering plant chloroplasts are encoded in the chloroplast genome, but transcription factors exist that could affect the form, function, and promoters recognized by the enzyme (Chapter 12). There is also reason to believe that there may be more than one polymerase with the second enzyme being encoded by nuclear genes (Chapter 12).

In reviewing the literature on the role of transcriptional control in chloroplast gene regulation one gets the definite impression that there is much still to be learned. Enough work has been done to raise the possibilities that DNA conformation, accessory factors, and, perhaps, even distinct polymerases originating from both the plastid and nuclear genomes could be involved in transcription of the chloroplast genome. In fact, the careful estimates of transcription rates and mRNA levels for a variety of plastid genes in barley discussed above reveal wide variations related both to developmental stage and gene function.

Posttranscriptional control. RNA processing and mRNA stabilization are potentially important mechanisms of posttranscriptional regulation in chloroplasts (Gruissem *et al.*, 1988; Gruissem, 1989a; Gruissem and Tonkyn, 1993; Mullet, 1988). The maturation of chloroplast mRNAs involves a variety of reactions that include processing of 3'- and 5'-ends, endonucleolytic cleavage events, and cis- and trans-splicing of introns (see Chapters 10 and 12). Many mRNAs are polycistronic and their processing and splicing give rise to arrays of overlapping RNAs. Most of these reactions must depend on nuclear gene products.

Detailed analyses of processing have been conducted for the *psbB* operons of maize (Barkan, 1988), spinach (Westhoff and Herrmann, 1988), and tobacco (Tanaka *et al.*, 1987) (see Fig. 12–3). These operons encode components of photosystem II *(psbB, psbH)* and the cytochrome b_6/f complex *(petB,petD)*. Two of the genes *(petB* and *petD)* contain introns. Between 18 and 20 different major transcripts of this operon accumulate and the processing pattern appears to be conserved in different plants (Gruissem *et al.*, 1988). Most of the RNA size heterogeneity results from the stable accumulation of differentially processed and spliced mRNAs, but processing of these mRNAs is not a prerequisite for translatability. In maize, polysome immunoselection experiments with antibodies specific to proteins encoded by *psbB, petB,* and *petD* show that each is translated from both monocistronic and polycistronic mRNAs (see Chapter 12 and Barkan, 1988). The maize nuclear mutant *hcf38* accumulates aberrant sets of transcripts from the *psbB* operon (Barkan *et al.*, 1986).

Mutants like *hcf38* will be important not only in establishing the genes involved in different processing events, but the pathways of processing themselves. This is particularly well illustrated in an organism such as *Chlamydomonas* where conditional mutations affecting photosynthesis are readily isolated. The *psaA* gene is organized into three exons that are transsspliced and mutants affecting this process fall into three classes depending on the transcripts they accumulate (see Fig. 10–4A). Obviously, such a system has the potential for very tight posttranscriptional regulatory control.

Complex processing patterns have also been noted for the *rps2–atpI–atpH–atpF–atpA* operons of spinach and pea (Gruissem *et al.*, 1988) and for the *psbDC* operon of tobacco (Yao *et al.*, 1989 and Fig. 12–3).

The 3'- ends of both mono- and polycistronic protein-coding transcription units as well as the individual genes themselves are followed by inverted repeat (IR) sequences with the potential to form stem–loop structures (reviewed by Gruissem and Schuster, 1993 see Chapter 12). While these IR sequences were reported not to act efficiently as transcription terminators, they were found to play an important role in transcript stabilization (Stern and Gruissem, 1987). This has been demonstrated by *in vitro* experiments with chloroplast transcription extracts from spinach employing synthetic mRNAs (Stern and Gruissem, 1987) and *in vivo* analyses of *Chlamydomonas* chloroplast transformants in which segments of the 3'-end of the *atpB* mRNA have been deleted so that all or part of the stem–loop structure has been removed (Stern *et al.*, 1991). Certain proteins bind to these sequences in a gene-specific manner (Chapter 12; Stern *et al.*, 1989). These experiments hint at the possibility that differences in protein-binding patterns seen for individual mRNAs might help to modulate the overall stability of a given mRNA. However, as discussed in Chapter 12, recent studies with *Chlamydomonas* chloroplast transformants have resurrected the possibility that the 3' IR sequences are important in transcript termination while discounting their role in transcript stabilization (Blowers *et al.*, 1993). Other experiments using chloroplast transformants of this alga further suggest that the 5' untranslated region and the gene coding region itself may play a role in stabilizing chloroplast transcripts (see Chapter 12 and Salvador *et al.*, 1993). At this point the prudent conclusion would seem to be that both the 3' and 5' untranslated regions of chloroplast mRNAs may contain sequences required for their stabilization and that variations on these themes probably exist for different mRNAs.

Light can also play an indirect role in stabilizing certain chloroplast mRNAs. In barley the adjacent *psbD* and *psbC* genes are part of a complex transcription unit that encodes several other photosynthetic genes, ORFs, and tRNAs (Berends-Sexton *et al.*, 1990). There are eight different transcripts synthesized from this gene cluster of which six are present in barley etioplasts (Berends *et al.*, 1987; Gamble *et al.*, 1988). The accumulation of two of these transcripts is light dependent (Gamble *et al.*, 1988). The photoreceptor involved seems to be activated by blue light (Gamble and Mullet, 1989). Since transcript accumulation in the light is inhibited by cycloheximide, the receptor probably activates a nuclear gene whose protein product is required for accumulation of the two transcripts. Accumulation of the two mRNAs in the light seems to involve both light-induced transcription and stabilization of these species (Mullet *et al.*, 1991). However, differential stabilization of chloroplast mRNAs barley also seems to involve light-independent signals (Kim *et al.*, 1993).

A potentially very important approach *in vivo* to the study of transcripts in flowering plants is now available because of chloroplast transformation. Thus, Staub and Maliga (1993) demonstrated the significance of the untranslated regions of the *psbA* mRNA in regulation in tobacco. They used a construct consisting of a GUS reporter gene sandwiched between the 5' and 3' *psbA* flanking sequences. Light-induced accumulation of GUS (100-fold) was accompanied by a modest increase in mRNA accumulation. When light-grown seedlings were transferred to the dark GUS synthesis was inhibited although there was no decrease in GUS mRNA. Obviously, the next step will be to determine whether both flanking regions are required for the light response, the sequence involved, etc.

Translational control. Were it not for the conflicting results mentioned above and discussed in Chapter 12 the seeming ubiquitous presence of 3' stem–loop structures in chloroplast mRNAs and their role in stabilizing these transcripts (Chapter, 12; Gruissem and Schuster, 1993) might be considered an example of a general control mechanism. Another possible general translational control mechanism identified in *Chlamydomonas* and cited earlier in this chapter involves selection for translation of mRNAs from the total mRNA pool (Hosler *et al.*, 1989). Growth of *Chlamydomonas* cells in 5-fluorodeoxyuridine causes a dramatic reduction in the number of chloroplast genomes per plastid to ca. 10–20% of normal with no effect on nuclear DNA amounts or photosynthetic growth (Fig. 16–1A; Wurtz *et al.*, 1977). Under these conditions, mRNAs for the chloroplast *rpl2, rbcL,* and *atpA* genes decline with gene copy number. However, the levels of translatable mRNA assayed *in vitro* in the rabbit reticulocyte system and the accumulation and synthesis of these proteins *in vivo* are not reduced (Fig. 16–1B; Hosler *et al.*, 1989). These results suggest that mechanisms exist to mark a subset of chloroplast mRNAs for translation. As the total mRNA pool declines, the translatable fraction increases so that gene expression remains constant at the level of the protein

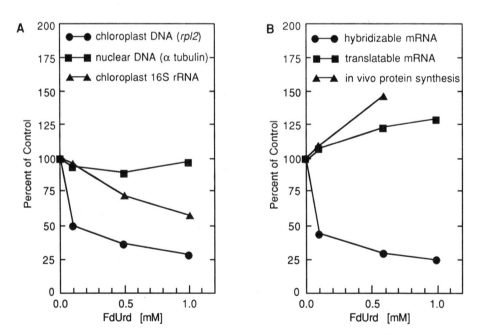

Fig. 16–1. Effects of growth of *Chlamydomonas* cells in 5-fluorodeoxyuridine for 7–8 generations on (A) levels of chloroplast and nuclear DNA and 16S chloroplast rRNA, and (B) accumulation of hybridizable and translatable mRNA and rate of synthesis for the chloroplast r-protein encoded by the *rpl2* gene. From Gillham *et al.* (1991).

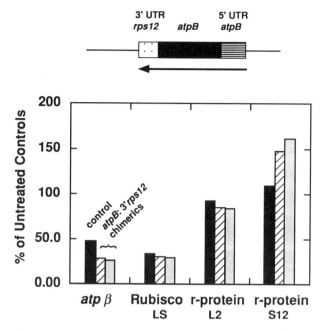

Fig. 16-2. Protein accumulation by transformants in which the *atpB:3'rps12* chimeric has replaced the endogenous *atpB* gene. Each transformant was compared to the untransformed *spr-u-1-27-3 (spr)* control strain under normal conditions and conditions of reduced chloroplast protein synthesis (acetate medium containing 40 μg/ml spectinomycin). The *spr* mutation confers low level resistance to spectinomycin on chloroplast ribosomes and in the presence of antibiotic synthesizes chloroplast encoded ribosomal (r-) proteins in preference to photosynthetic proteins. Levels of the β subunit of ATP synthase, the RUBISCO large subunit, and r-proteins L1 and S12 were determined and are given as a percentage of the untreated controls. From Hauser *et al.* (1993).

product. Such a mechanism might buffer cells against variations in organelle genome content that may arise in the course of plastid division.

Evidence for a class-specific translational control mechanism ensuring preferential synthesis of chloroplast r-proteins has also been obtained in *Chlamydomonas* (Liu *et al.*, 1989; Hauser *et al.*, 1993). When chloroplast protein synthesis is reduced, the accumulation of mRNAs for photosynthetic and r-proteins is unaffected and may even be stimulated. However, the two chloroplast r-proteins examined are made in preference to two photosynthetic proteins (Fig. 16-2). Experiments with chloroplast transformants containing a chimeric *atpB* gene with the 3' untranslated region (UTR) from the chloroplast *rps12* gene show that the *rps12* 3' UTR is not responsible for preferential expression of this mRNA under conditions of reduced chloroplast protein synthesis (Fig. 16-2; Hauser *et al.*, 1993).

Mullet and colleagues published a series of papers describing careful and systematic experiments that illustrate the caution necessary in interpreting results that seemingly demonstrate translational control. Klein and Mullet (1986, 1987) found that etioplasts of 5-day-old dark-grown barley seedlings synthesized most of the soluble and membrane proteins found in illuminated plants. These included the large subunit of RUBISCO and the α and β subunits of the chloroplast ATP synthase. However, etioplasts appeared in pulse-labeling experiments not to make the chlorophyll-binding apoproteins of PSI (P_{700} binding proteins) and PSII (CP 43 and CP 47) encoded by the chloroplast *psaA, psaB, psbB,* and *psbC* genes, respectively. The same was true of the photosystem II reaction center protein, D1, encoded by the *psbA* gene. Nevertheless, these plastid genes were found to be transcribed in etioplasts and their transcripts accumulated in the dark (Klein and Mullet, 1987; Mullet and Klein, 1987). Illumination of the dark-grown barley plants stimulated synthesis of all of the proteins.

Klein *et al.* (1988) then demonstrated that the chlorophyll apoprotein mRNAs were associated with polysomes in dark-grown plants despite the fact that their protein products were not made. On the basis of these results Klein *et al.* (1988) suggested that chlorophyll *a* activated chlorophyll apoprotein accumulation either by overcoming a block in translation elongation or by binding to and stabilizing nascent chlorophyll apoproteins. Using pulse-labeling and immunoprecipitation with specific antibodies Mullet *et al.* (1990) differentiated between these two possibilities. They found that CP43 and D1 were, after all, synthesized by etioplasts, but the proteins were unstable in the absence of chlorophyll. In chloroplasts, precursors of the 32-kDa D1 protein were converted to the mature protein, but in etioplasts they were degraded to a 23-kDa species. Similarly, newly synthesized CP43 was found to be rapidly degraded in etioplasts. In short, the earlier experiments of Klein and Mullet (1986), which had not employed immunoprecipitation, but pulse labeling in the presence of the cytoplasmic protein synthesis inhibitor cycloheximide, were misleading. They suggested minimal synthesis of these polypeptides. Subsequent experiments systematically pointed in the direction of a translational control mechanism. However, translational control proved not to be involved. Instead, chlorophyll actually stabilizes the chlorophyll apoproteins and prevents their degradation. Since photoconversion of protochlorophyllide to chlorophyll is a light-dependent process, stabilization of the chlorophyll apoproteins cannot occur in the dark.

These results illustrate that proteolysis is an important posttranslational control mechanism. This is also true for other chloroplast-encoded polypeptides that fail to assemble into functional complexes as shown mostly by experiments with *Chlamydomonas* chloroplast mutants (e.g., Mishkind and Schmidt, 1983; Leto *et al.,* 1985; Bennoun *et al.,* 1986; Jensen *et al.,* 1986; Girard-Bascou *et al.,* 1987).

Other levels of general control may exist at the levels of protein import and the availability of factors required for chloroplast protein synthesis. For instance, Dahlin and Cline (1991) analyzed developing plastids from wheat leaves for their *in vitro* import capability. They found that import capacity was high in proplastids, but declined as much as 20-fold as plastid development approached either the mature chloroplast or etioplast stage. They also found that the low import capability of mature etioplasts became transiently activated during light-mediated differentiation into green plastids. Akkaya and Breitenberger (1992) point out that work of Spremulli and colleagues in *Euglena* demonstrated the light inducibility of several chloroplast initiation and elongation factors. This prompted Akkaya and Breitenberger to see whether the same was true in flowering plants. They found that in pea seedlings germinated under continuous white or red light chloroplast EF-G specific activity reached a maximum between days 10 and 15 and then declined. Factor activity was almost undetectable in dark-grown seedlings. However, Akkaya and Breitenberger make the point that EF-G activity is probably not missing in etioplasts, but simply undetectable, possibly for technical reasons. This is important because chloroplast ribosomes still must be made as must the chloroplast RNA polymerase, both of which require chloroplast protein synthesis for the manufacture of important structural proteins. Also, Mullet *et al.* (1990), in experiments described earlier, demonstrated the synthesis of specific photosynthetic proteins in etioplasts of barley. Nevertheless, Akkaya and Breitenberger cite unpublished results that suggest a minimum 7-fold activation of EF-G by light.

The Regulation of Nuclear Genes Involved in Photosynthesis

The most extensively studied nuclear photosynthetic genes belong to the *rbcS* and *cab* families. Manipulation of these genes has led to identification of cis-acting sequences and trans-acting factors that respond to light.

Organization of the rbcS and cab gene families. In all flowering plants so far studied the *rbcS* genes are organized in multigene families with 5 to 15 members (see Dean *et al.*, 1989; Manzara and Gruissem, 1988 for reviews). In tomato the five *rbcS* genes are located on two different chromosomes mapping in three loci (Fig. 16–3). The single *rbcS1* and *rbcS2* genes are located on chromosomes 2 and 3, respectively. Three *rbcS3* genes (*rbcS3A, rbcs3B,* and *rbcS3C*) are also on chromosome 2 and arranged tandemly in a 10-kb region. The *rbcS1* and *rbcS3* genes contain two introns at conserved positions 1 and 2. The *rbcS2* gene possesses an additional intron at position 3. The exons of the *rbcS3* genes specifying the mature polypeptide differ by only 1.6% and encode identical small subunits, but the two single genes show much higher levels of nucleotide-sequence divergence, ranging from 10 to 14%.

The eight *rbcS* genes of petunia, which like tomato is a member of the family Solanaceae, are also organized into three groups (Fig. 16–3). Each of the 71 and 117 subfamilies contains single genes designated SSU301 and SSU611, respectively, while the 51 family includes six genes. Five of these are in tandem array like the *rbcS3* family in tomato. As in tomato the latter array contains introns in positions 1 and 2 as does the single SSU611 gene. The other single gene, SSU301, like the *rbcS2* gene of tomato, possesses an additional intron at position 3. In contrast, in pea (Leguminosiae) the five gene *rbcS* family maps as a cluster on chromosome 5. *Arabidopsis* has only four *rbcS* genes, consistent with the small size of its nuclear genome. These belong to two subfamilies, one of which contains three closely linked genes. Both pea and *Arabidopsis* possess introns located in the same positions they are found in the two intron-containing genes of tomato and *Petunia*. In the monocotyledons only single intron genes have so far been reported. Duckweed lacks the intron at position 1, while in wheat and maize the intron at position 2 is absent. The two *rbcS* genes of *Chlamydomonas reinhardtii* are linked in tandem and a partial pseudogene is found in the same region (Goldschmidt-Clermont, 1986; Goldschmidt-Clermont and Rahire, 1986). Each of these genes possesses three introns, but none of them corresponds exactly in location to the *rbcS* introns of flowering plants. Evolutionary arguments based on sequence comparisons suggest that the position 3 intron may be peculiar to the Solanaceae (Dean *et al.*, 1989). Furthermore, the three subfamilies of *rbcS* genes in this family arose from gene duplications at least as old as the family itself.

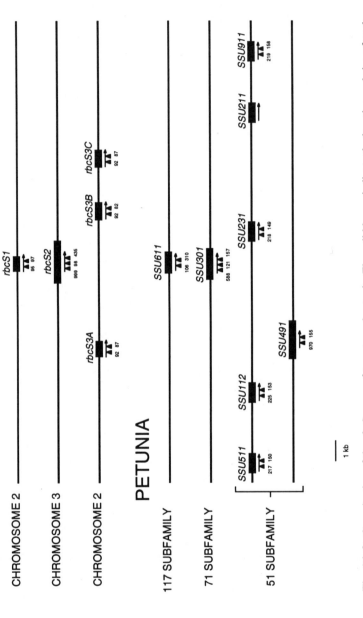

Fig. 16–3. Organization of the *rbcS* genes of tomato and petunia. The solid boxes indicate the location and size of each gene, including the intron regions. The horizontal arrows below each gene indicate the direction of transcription. The triangles represent introns, with the size of each intron indicated. From Manzara and Gruissem (1988).

The small subunit coding sequence is more conserved than the transit peptide sequence. One hexadecapeptide located in the petunia small subunit between amino acid residues 61 and 76 is completely conserved in all species so far examined (Dean et al., 1989). Since there is a relatively high level of sequence divergence within other parts of the mature small subunit, the conservation of the hexadecapeptide suggests that this region may be important in small subunit structure or function. One possibility is that this sequence is involved in binding between the small and large subunits. Six of the conserved residues are aromatic, suggesting that they are not exposed at the surface of the holoenzyme. Also relatively well conserved are the 16 amino acids located between positions 102 and 117.

The *cab* genes, like the *rbcS* genes, are also organized in families (see reviews by Chitnis and Thornber, 1988; Green et al., 1991). They number between 3 and 16 depending on plant species. Only in tomato has a virtually complete set of *cab* genes been characterized (Table 16–1). With the exception of the genes encoding LHCII Types I and II, which differ from each other by only 15%, most genes diverge substantially (60–70% at the level of protein product) suggesting that the gene duplications that gave rise to them are evolutionarily very old. Nevertheless, the sequences of the CAB proteins deduced from the *cab* gene sequences reveal that the proteins themselves seem to be composed of modules. Some of these are almost identical in sequence in all CAB proteins with others having evolved rapidly. The conserved sequences precede the first and third of the three transmembrane helices of the protein.

In most plants LHCII Type I proteins are encoded by very similar multiple gene copies within a given species (Table 16–1). In contrast, the genes specifying LHCII Type II, LHCI, CP29, and CP24 are present as single or duplicate copies in tomato and very likely other species as well (Green et al., 1991). With the exception of the LHCII Type I *cab* genes, all other members of the family contain introns. In contrast to the

Table 16–1. The Tomato Chlorophyll *a/b* Proteins and Their Genes

Complex	Role/Location	Chlorophyll a/b Ratio	Polypeptides	Gene Types	Number of	
					Gene Copies	Introns
LHCII	Major antenna PSII	1.2	2 major	Type I	8	0
				Type II	2	1
			1 minor	Type III	ND[a]	—
CP29	Core antenna PSII	4–5	2	Type I	1	5
				Type II	ND	—
CP24	Minor PSII antenna	<1	1–2	—	2	1
LHCI	PSI antenna	3–5	4	Type I	2	3
				Type II	1	4
				Type III	1	2
				Type IV	2	2

From Green et al. (1991).

[a]ND, not determined.

rbcS genes, the number and position of introns vary between different *cab* genes. Of 11 positions where introns are found, only three are common to more than one kind of *cab* gene.

Control of nuclear gene expression. The expression of nuclear genes encoding photosynthetic components has been the subject of numerous reviews (Cuozzo et al., 1988; Dean et al., 1989; Gilmartin et al., 1990; Kuhlemeier, 1992; Kuhlemeier et al., 1987; Manzara and Gruissem, 1988; Mural, 1991; Nagy et al., 1988; Silverthorne and Tobin, 1987; Taylor, 1989; Thompson and White, 1991). For the sake of convenience, this topic is most easily divided into three parts: the analysis of control as a function of development, the role of light, and cis and trans regulatory elements.

The expression of the *rbcS* genes has been used extensively to model these aspects of nuclear gene expression. The highest transcript levels for the *rbcS* genes are always found in green leaf tissue with much lower levels in other parts of the plant and in etiolated seedlings. The tissue-specific pattern of expression of different *rbcS* genes has been compared in tomato, petunia, and pea (Manzara and Gruissem, 1988). The results demonstrate that the pattern of gene expression varies between different species. In petunia a given gene tends to show about the same level of expression in all tissues whereas the tomato *rbcS* genes vary in expression in different tissues.

What are the underlying causes for variations in gene transcript level by gene family and tissue? As in the case of chloroplast genes, changes in transcription rate and transcript stability must be considered. Both are involved as illustrated by a study of Wanner and Gruissem (1991) on the expression dynamics of the tomato *rbcS* family during development. These authors used nuclear run-on transcription assays coupled with analysis of RNA accumulation to assess the relative contributions of transcriptional and posttranscriptional mechanisms to the accumulation of *rbcS* mRNAs. Their results are instructive (Fig. 16–4). Measurement of RNA abundance in etiolated, greening, and fully green seedling cotyledons reveals that transcripts of all three tomato *rbcS* families accumulate when the light is turned on (Fig. 16–4A). These experiments also show that there is some transcript accumulation in etioplasts for all *rbcS* genes except *rbcS3B* and *3C*. In run-on transcription assays the latter two genes seem not to be transcribed whereas the other three genes are. However, *rbcS3A*, whose transcript levels are highest in etioplasts, shows the lowest levels of run-on transcription in etioplasts (Fig. 16–4B). Conversely, transcripts of *rbcS1* are less abundant in etioplasts, although this gene is the most highly transcribed in etioplasts. As Wanner and Gruissem point out these "results raised the possibility that both transcriptional and post-transcriptional control mechanisms coordinate the differential expression of the tomato *rbcS* genes at other developmental stages as well."

To investigate this possibility Wanner and Gruissem (1991) examined *rbcS* mRNA stability in immature and fully developed leaves and during fruit development. The striking finding for immature and fully developed leaves is that while transcription rates for all *rbcS* genes are at least 2 times higher in mature leaves, transcript accumulation is similar or lower in mature than immature leaves. In short, *rbcS* mRNA stability declines in mature leaves and overall abundance depends on the balance between mRNA synthesis and turnover. The major effect that transcription initiation can have is illustrated in the case of fruit (Fig. 16–5A and B). Here there is no accumulation of mRNAs of the *rbcS3* family and very little synthesis as well. However, the effects of posttranscriptional controls are again evident in determining mRNA levels in the cases

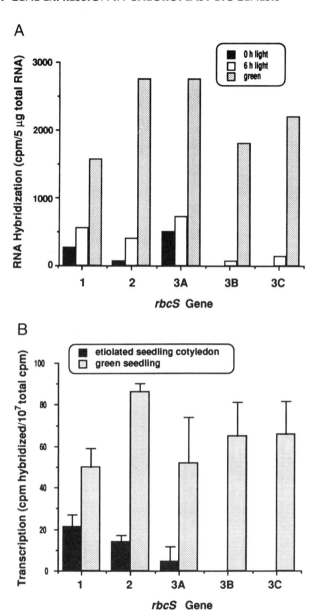

Fig. 16–4. *rbcS* mRNA in etiolated, greening, and fully green seedlings of tomato. (A) The transcript abundance for each different *rbcS* gene. (B) The relative transcription rates of each *rbcS* gene as determined from run-on transcription assays. From Wanner and Gruissem (1991).

of *rbcS1* and *2* whose mRNAs are made in developing fruit. As Wanner and Gruissem put it, ''tomato appears to have evolved precise and multiple mechanisms to coordinate the activity of the *rbcS* genes to maximize expression in different light environments and at different stages in development. These mechanisms operate at both the transcriptional and posttranscriptional levels and are specific for different members within the

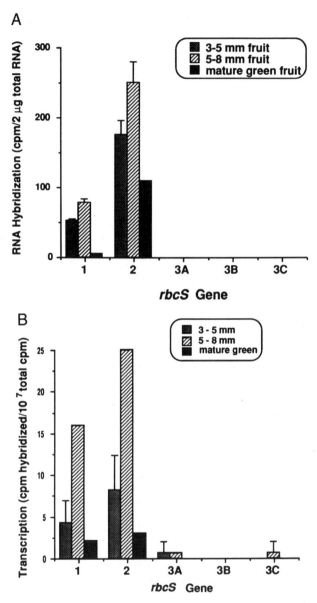

Fig. 16–5. rbcS mRNA in developing tomato fruit. (A) The transcript abundance for each different rbcS gene. (B) The relative transcription rates of each rbcS gene as determined from run-on transcription assays. From Wanner and Gruissem (1991).

rbcS gene family." In a nutshell, genetic regulation of these nuclear genes is complicated!

The role of mRNA stability in regulating levels of nuclear photosynthetic gene transcripts is becoming increasingly recognized as an important general control mechanism as also shown by studies of rbcS transcripts in *Lemna gibba* (Duckweed) and soybean (Shirley and Meagher, 1990; Shirley et al., 1990; Silverthorne and Tobin, 1990).

Light has profound effects on nuclear gene transcription in higher plants (see Mural, 1991; Quail, 1991; Thompson and White, 1991, for reviews). As mentioned earlier, plants possess several photoreceptor systems that evaluate the quantity, the wavelength (blue/UV-A, UV-B, or red/far red light), and the direction of the light source. The interaction of these different photoreceptors in expression of nuclear photosynthetic genes can be complex. For example, maximum levels of *cab* and *rbcS* gene expression require activation of all three photoreceptors; the *petE* gene, encoding plastocyanin, needs red and blue light for this purpose; and many genes are responsive to red light alone (see Thompson and White, 1991, for a summary).

The best understood photoreceptor is phytochrome. Red light converts the inactive form of phytochrome (Pr) to its active form (Pfr) and far red light reverses the process. Pfr formation initiates a transduction process that results in the altered expression of regulated genes. The phytochrome molecule consists of a polypeptide covalently linked to a linear tetrapyrrole chromophore. There are two types of phytochrome (Quail, 1991; Thompson and White, 1991). Type I predominates in etiolated tissue and is rapidly degraded in the Pfr form. Type II is low in abundance in etiolated tissue, but is found in green tissue and is stable in the Pfr form. Type I phytochrome is encoded by the *phyA* gene family while Type II may be encoded by *phyB* or perhaps *phyC*. Phytochrome regulates a variety of morphological, biochemical, and physiological responses in plants. Pfr stimulates expression of most photosynthetic genes, but certain other genes, including the phytochrome genes themselves, are repressed.

Since there is no convincing evidence that phytochrome enters the nucleus or that it has the capacity to bind directly to DNA, signal response is thought to involve a second-messenger system (Quail, 1991). A variety of transduction mechanisms, including some based on the second messenger systems of animals, have been proposed, but unequivocal evidence of a direct involvement in phytochrome signal transduction is still lacking (Quail, 1991). A powerful and direct approach to the transduction pathway has emerged via the isolation of mutants of *Arabidopsis* with aberrant light responses. These belong to two general classes: (1) mutants whose phenotypes in the light at least partially resemble dark grown seedlings; and (2) mutants that develop in darkness as if exposed to light (see Chory, 1991, 1992). The first phenotypic class of mutants defines 10 genes. Mutants in these genes are recessive and have a partially etiolated morphology when grown in white light. They are called *hy*, for long hypocotyl, because they are not subject to the normal light-induced inhibition of hypocotyl elongation. Mutations at the *hy1*, *hy2*, *hy3*, and *hy6* loci result in phytochrome deficiencies. The *hy4* mutants are blocked in the blue light response, but accumulate normal levels of phytochrome. The *hy5* mutants have normal phytochrome activity, but are defective in the red light response.

The second phenotypic class of mutants has the characteristics of light-grown plants in darkness. Mutants in three of these genes are called *det* for detiolated while the fourth gene is referred to as *COP1* (Deng et al., 1991). The DET1 and DET2 proteins seem to be required to couple the red- and blue-light signals to the downstream light-regulated developmental and gene expression responses in *Arabidopsis*. Based on the analysis of double mutant combinations, Chory (1992) constructed a simple branched pathway to describe the interactions of these gene products. The DET1, DET2, and

HY5 proteins are thought to act on a distinct signal transduction pathway to effect a downstream light-regulated response. HY5 either defines a red-light action pathway separate from phytochrome or is a transduction element on the phytochrome pathway that is influenced by DET1. The *DET1* and *DET2* gene products appear to be negative regulators of the greening (deetiolation) response. This work should lead to identification of the proteins that mediate the phytochrome response(s) of nuclear photosynthetic genes.

What are the cis-acting sequences and trans-acting factors mediating light-regulated transcription of nuclear photosynthetic genes? This question has received a lot of attention, but the answer is far from complete (see Gilmartin *et al.*, 1990; Kuhlemeier, 1992; Kuhlemeier *et al.*, 1987; Mural, 1991 for reviews). The most progress has been made in identifying cis-acting elements by modification *in vitro* of sequences upstream of specific genes (e.g., *rbcS* or *cab*). These modified genes are subsequently reintroduced into plant cells using the *Agrobacterium* Ti plasmid vectors. The recipient cells are ones that have retained the capacity to differentiate and regenerate fertile plants. Such transgenic plants provide a much better background than undifferentiated calli in which to examine gene regulation in detail. Sometimes the native, structural gene is coupled to its own modified upstream sequence whereas in other instances easily assayed reporter genes [e.g., chloramphenicol acetyltransferase (CAT) or GUS] are used. Transient expression systems have also proved useful in identifying regulatory elements of interest. In these cases chimeric constructs can be introduced directly into plant tissues of interest by techniques such as electroporation and microprojectile bombardment (see Chapter 7, Organelle Transformation).

To illustrate the results that can be obtained using these approaches a couple of examples will be instructive. To characterize the cis-acting elements in the 5' flanking regions of the *rbcS-3A* gene of pea, expression of the gene was analyzed in transgenic petunia or tobacco plants. Fluhr and Chua (1986) found that a sequence of 410 bp upstream of the *rbcS-3A* gene was sufficient to permit proper expression in transgenic petunia plants. Fluhr *et al.* (1986) then isolated *rbcS-3A* sequences from -410 to $+15$ relative to the transcription start site and fused them to a CAT reporter gene. These experiments confirmed that the 410-bp upstream region possessed the necessary sequences for light–dark regulation and for tissue-specific expression.

Bansal *et al.* (1992) used a transient system to study expression of the maize *cab-m1* and *rbcS-m3* promoter sequences of maize. The *cab-m1* gene is preferentially expressed in mesophyll cells, but not in bundle sheath cells (see Chapter 1) and is strongly photoregulated. Conversely, transcripts of the *rbcS-m3* gene are barely detectable in mesophyll cells, but are abundant in bundle sheath cells. Chimeric constructs were made between the upstream sequences of each of these genes and a GUS reporter gene. Transient expression of the intact constructs and deletion derivatives of each were then assayed following microprojectile bombardment of maize leaves. Following reaction of leaf segments with the GUS substrate, blue spots appeared in the leaf segments that could be localized to mesophyll or bundle sheath cells and quantified by counting. The results obtained with the deletion constructs showed that a 158-bp region between -1026 and -868 was required for mesophyll cell-specific expression of GUS and that a region between -359 and -39 was necessary for photoregulation of the gene. The same experiments showed that a 2.5-kb region upstream of *rbcS-m3* was responsible for part or all of light regulated expression of this gene in bundle sheath cells.

CONTROL OF GENE EXPRESSION IN CHLOROPLAST BIOGENESIS

Certain features are conserved in the DNA sequences governing eukaryotic transcriptional regulation (for reviews, see Dynan and Tjian, 1985; Kuhlemeier, 1992; Maniatis *et al.*, 1987). Thus the promoters of most eukaryotic genes contain a "TATA" box, located 25–30 bp from the transcription initiation site that plays a role in determining the transcription start site. This is the binding site for RNA polymerase II and associated proteins in human HeLa cells including TFIIA, B, D, E, and F. Only TFIIA and TFIID have so far been identified in plants (Kuhlemeier, 1992). Sequences related to "CCAAT" or "GGGCGG" motifs are also found upstream of eukaryotic transcription start sites and often serve as binding sites for promoter-specific transcriptional factors.

The extensive work done to identify light responsive cis-acting sequences upstream of nuclear photosynthetic genes and the trans-acting factors that recognize them has been reviewed (Kuhlemeier *et al.*, 1987; Gilmartin *et al.* 1990; Mural, 1991) and discussed in detail in an original paper (Manzara *et al.*, 1991). In the discussion that follows the reader should keep in mind that there may be several to many trans-acting factors (e.g., Schindler and Cashmore, 1990) and that their presence or absence may depend on the tissue examined (e.g., leaf versus fruit).

The most proximal element is the *L-box* sequence (Giuliano *et al.*, 1988 see also Table 16–2). Although this sequence is found in *rbcS* promoters from several plants, no function has been described for the L-box. In tomato the L-box is found proximal to the *rbcS1* and *3A* genes and these sequences are protected by a protein(s) from cotyledon and leaf extracts, which Manzara *et al.* (1991) term L-box binding factor.

Motif 2 of tomato (5'GGATGAGATAAGATTA; Manzara and Gruissem, 1988) is found upstream of several tomato *rbcS* genes (Manzara *et al.*, 1991). Motif 2 includes several overlapping sequences that have regulatory importance (Table 16–2). The simplest of these sequences is the conserved GATA sequence, which can be found at various positions upstream of the TATA box. GATA motifs were originally found by comparison of the upstream sequence of three *cab* genes from *Nicotiana plumbaginifolia* (Castresana *et al.*, 1987). These genes possess sequences of strong similarity between −150 and −100 relative to the translation start site with a striking conservation of the GATA motif. Further analysis using the *N. plumbaginifolia cabE* gene revealed that there were three GATA sequences between −112 and +36 and that this region was required for the light response in the context of the entire −1554 bp upstream region (Castresana *et al.*, 1988). Other experiments, summarized by Gilmartin *et al.* (1990), involving both the *N. plumbaginifolia cabE* gene and the petunia *cab22R* gene provide strong evidence that GATA sequences play a role in high-level gene expression in the light.

GATA is part of the sequence 5'GATGA*GATA*. This sequence is similar to a sequence called the *as-2* site (5'GTGGATT*GATGTGATA*TTCTCC) found in the cauliflower mosaic virus 35S promoter (CaMV 35S). The CaMV35S promoter is a very strong plant viral promoter that has been very well studied and is frequently used to drive reporter gene constructs (Kuhlemeier, 1992). The *as-2* site binds a tobacco nuclear protein factor, termed activation sequence factor 2 (ASF-2; Lam and Chua, 1989). A tetramer of the *as-2* site fused at −90 to the CaMV 35S promoter confers light-insensitive leaf-specific expression in transgenic tobacco plants suggesting that tandem GATA motifs mediate cell-type specificity, but not the light response (Gilmartin *et al.*, 1990).

Table 16–2. Conserved Sequence Motifs in *rbcS* Promoter Regions

Number	Motif	Sequence	Binding factor	Putative function
1	L-box	AATTAACCAA		Unknown
2	I-box	pyrimidine rich		Unknown
	I-box	GGATGAGATAAGA	GA-1	"Enhancer"
	GATA	GATAAG	GA-1	Unknown
3	GATA	GATGA(T)GATA[a]	ASF-2	Leaf specific
5	G-box	CACGTGG(T)C[b]	GBF	"Enhancer"
		TTAAATAGAGGGCGTAA		Unknown
8	Box II	T(G)T(A)GTGPuT(A)AAT(A)PuT(A)	GT-1	Light regulation
9		TTGT(A)AATGT(C)CAA		Unknown
10		GAG(A)CCACA		Unknown
12	CAAT box	ATCCAAC(T)		Unknown
13		GGTTAC		Unknown
15		AGATGAGG		Light regulation
16		TTTG(T)TGTCCGTTAG(A)ATG(A)	LRF-1?	Unknown
18	"LRE"	CCTTATCAT		Unknown
19	TATA	C(T)TATATAA(T)A	TFIID	Transcription initiation
	A-T rich	Varies		Unknown
	A-T rich	AATATTTTTATT	AT-1	Unknown
	A-T rich	AAATAGATAAATAAAAACATT	3AF-1	"Enhancer"

Modified from Manzara *et al.* (1991).

[a]Nucleotides in parentheses represent alternative possibilities at the previous position.

[b]This is only a segment of the canonical G-box as defined by Giuliano *et al.* (1988); see text.

Then there is the I-box (Giuliano et al., 1988). Giuliano et al. (1988) originally defined the I-box as 5'GGATGAGATAAGA for tomato *rbcS-3A,* but Donald and Cashmore (1990) subsequently defined the I-box consensus sequence as 5'GATAAG. Manzara *et al.* (1991) refer to the shorter consensus sequence as motif 2'. A similar sequence from a tobacco *cab* gene *(cab-E)* interacts with a protein GA-1 from nuclear extracts of tobacco leaves. The I-box is overlapped by the sequence 5'GATGAGATA (Table 16–2) which itself is embedded in the *as-2* site of the CaMV 35S promoter. The binding of GA-1 is competed by the *as-2* binding sequence of the CaMV 35S promoter suggesting that ASF-2 and GA-1 may be related or different proteins that target the same binding site (Schindler and Cashmore, 1990). This result is not unexpected since the two sequences overlap.

In tomato, motif 2 is found upstream of *rbcS1, 2,* and *3A* (Manzara et al., 1991). Consistent with the leaf-specific function postulated for ASF-2, motif 2 of tomato *rbcS1* and *2* is protected by nuclear extracts from leaf and cotyledon suggesting specific protein binding in these tissues, but *rbcS3A* is protected only in fruit. Relating these results to those obtained by Wanner and Gruissem (1991) is difficult because *rbcS1* and *2* are expressed in fruit, but the *rbcS3* genes are expressed poorly, if at all. To complicate matters further GATA boxes are found elsewhere in the upstream regions of several tomato *rbcS* genes. Only one of these, upstream of *rbcS3C,* seems to bind protein.

The *G-box* (Giuliano et al., 1988) with the consensus sequence 5'TCTTACA-CGTGGCAYY is 3' to motif 2 in tomato (Table 16–2). This sequence is critical for high level expression of the *Arabidopsis rbcS1A* gene attached to an *adh* reporter gene in transgenic tobacco plants (Donald and Cashmore, 1990). The G-box binds binds a protein called G-box factor (GBF) identified in nuclear extracts of *Arabidopsis* and tomato (Giuliano et al., 1988). This factor is found in leaf extracts from both light-grown and dark-adapted tomato plants, but is low in root extracts. Analysis of the upstream regions of different genes suggests a requirement for the G-box for transcriptional activity (Gilmartin et al., 1990). Although the G-box is embedded in regions that respond to light, it is not known whether the G-box acts as a regulatory component of these light-responsive elements or whether it modulates the quantitative level of expression of these sequences (Gilmartin et al., 1990). Gilmartin et al. (1990) review evidence that suggests that the G box may play a regulatory role in response to UV light. G-box like sequences have been found in several other genes encoding proteins of unrelated function (Gilmartin et al., 1990). Genes specifying several G-box binding factors have been cloned (Ueda et al., 1989; Weisshaar et al., 1991).

DNA sequences that may be necessary for light regulated *rbcS* transcription include *box II,* which binds GT-1 (tomato motif 8, Table 16–2). Comparison of several different *rbcS* promoters revealed conserved sequence motifs including box II (-151 to -138) and box III (-125 to -213) upstream of the pea *rbcS-3A* gene (Gilmartin et al., 1990). Four additional sequences further upstream, called boxes II* (-224 to -213), III* (-257 to -248), II** (-386 to -372), and III** (-360 to -347) also bind GT-1 (Green et al., 1987, 1988). A correlation between binding activity *in vitro* and transcription *in vivo* was established by showing that certain box II mutations made *in vitro* block both activities (Green et al., 1988; Kuhlemeier et al., 1988). Nuclear extracts from light- and dark-grown plants both contain GT-1 so its regulatory role does not depend on *de novo* synthesis in the light (Green et al., 1987). Nevertheless, experiments reviewed by Gilmartin *et al.* (1990) implicate GT-1 binding in light-responsive transcription, but these data also suggest that GT-1 binding is not alone sufficient for transcriptional activation. GT-1 probably has to interact with additional proteins in the formation of a stable transcription complex (Gilmartin *et al.*, 1990). A typical box II binding activity does not seem to be present in tomato nuclear extracts (Manzara *et al.*, 1991). This is consistent with the inability of a tomato leaf extract to protect the pea *rbcS-3.6* box II sequence (Giuliano *et al.*, 1988). This could mean that a GT-1 binding activity is absent from tomato or that the context of the consensus sequence is critical for GT-1 binding. Other potential light-regulated sequences include the sequences corresponding to motif 18 "LRE" (Grob and Stuber, 1987) and motif 15 in tomato (Table 16–2). The latter motif is related to an *rbcS* promoter sequence of *Lemna* that binds the LRF-1 protein. The binding activity of LRF-1 varies in relation to light (Buzby *et al.*, 1990). Motif 15 may constitute a site that participates in the light-dependent transcription of *rbcS3B* and *3C*, but not *rbcS1*. In tomato, motif 18 is closely linked to the TATA box. This sequence binds proteins from nuclear extracts from organs that transcribe *rbcS*, but binding is also observed to *rbcS* upstream sequences that are not actively transcribed (Manzara *et al.*, 1991). These findings might mean that LRE is not, after all, a light responsive element in tomato or they could reflect binding of the transcription factor TFIID to the TATA box.

As expected, TATA-box *rbcS* sequences in tomato exhibit binding activities for extracts from all organs that transcribe these genes. Certain of the A-T-rich sequences upstream of the *rbcS* genes (Table 16–2) also bind proteins.

The foregoing summary illustrates that the regulation of nuclear photosynthetic genes in plants currently seems awesomly complex. There are several reasons for this. First, cis-acting sequences and trans-acting factors are being defined by different groups using different plants and often different genes. Second, while some cis-acting promoter sequences are undoubtedly light responsive, others are required for tissue specificity or developmental response. Thus, Donald and Cashmore (1990) characterize a 196-bp promoter for the *Arabidopsis rbcS-IA* gene containing both the G- and I-boxes and find that this sequence is capable of conferring both light-regulated and tissue-specific expression on a reporter gene. Third, a swarm of trans-acting factors governs light regulation and tissue specificity. Thus, Schindler and Cashmore (1990) distinguish five different proteins that bind to the *cab-E* promoter of *Nicotiana plumbaginifolia* of which some (e.g., GBF and GA-1) are described above. What seems certain from all of this work is that these nuclear genes are transcriptionally regulated through cis-acting sequences, some of which are light responsive with others determining developmental or tissue-specific responses. These sequences interact with regulatory proteins that relate to each of these responses. However, the precise hierarchy and mechanism by which these proteins determine each of these processes is murky at present.

Dissection of the mechanisms involved in genetic control ideally requires the astute application of principles derived from three biological disciplines: biochemistry, genetics, and molecular biology. This triad embraces an arsenal of complementary technologies that can be marshalled for a complete frontal assault on the problem. What has so far been lacking in the analysis of nuclear photosynthetic gene regulation is genetics. That is why studies such as those carried being carried out with *Arabidopsis* mutants (e.g., Chory, 1991, 1992; Deng *et al.*, 1991) become so important. Through the isolation of *Arabidopsis* mutants affecting general light responses one may identify the genes involved and the proteins they encode, and unravel the light transduction pathway.

Control of Chloroplast Gene Expression by the Nucleus

An obvious question is whether control circuits exist between the chloroplast and the nucleus to coordinate gene expression. In 1977 R.J. Ellis proposed the "cytoplasmic control principle" to explain how this might be effected. The thesis underlying this principle is that cytoplasmic translation products (and nuclear genes) control transcription and translation of plastid gene products.

Analysis of nuclear mutants deficient in photosynthesis particularly in *Chlamydomonas* (see reviews by Rochaix, 1992a,b; Rochaix *et al.*, 1991) has revealed the prescience of Ellis' proposition. Not only are nuclear genes intimately involved in the general control of chloroplast genome expression, but nuclear genes control the expression of specific chloroplast genes encoding components of the photosynthetic apparatus. In *Chlamydomonas* this point has been especially well documented in the case of photosystem II. This multimolecular protein–pigment complex consists of three parts (Chapter 1). The core of the complex is embedded in the thylakoid membrane and contains at least six polypeptides all of which are encoded by the chloroplast genome (Table 16–3). The core also includes protein bound pigments (pheophytin, reaction center, and core antenna chlorophylls), lipids, electron carriers (Q_A and Q_B), and iron. The three water-soluble proteins (OEE1, OEE2, OEE3) bound to the lumen side of the thylakoid membrane form the second part of the complex. These proteins are coded by three nuclear genes and play a role in water oxidation, in particular in the stabilization of manganese and in facilitating calcium chloride binding. The third part of the complex includes the light harvesting chlorophyll protein complex (LHCP), the proteins of which are encoded by the nuclear *cab* genes discussed in the previous section. Other proteins and genes associated with PSII are enumerated in Table 1–7.

Nuclear gene products have been shown to control the expression of three of the chloroplast encoded genes listed in Table 16–3 while appropriate nuclear mutants remain to be reported for the other three (Table 16–4). All of these gene mutations except one have been shown to exert their effects at the translational level. Thus, the two allelic *nac-1* mutations exhibit a pleiotropic deficiency in polypeptides of the PSII complex associated with the thylakoid membranes, but do not alter any known structural gene in the complex (Kuchka *et al.*, 1988). Analysis of the mRNAs made by the *nac-1* mutations revealed that mRNAs for the proteins encoded by *psbA, psbB, psbC,* and *psbD* genes were present as were the mRNAs for the OEE proteins. Pulse labeling experiments showed that the protein products of all of these genes except the D2 poly-

Table 16–3. Chloroplast-Encoded Protein Components of Photosystem II

Protein	Gene
CP47 (P5)	*psbB*
CP43 (P6)	*psbC*
D1	*psbA*
D2	*psbD*
2 cyt b_{559}	*psbE*
	psbF

Table 16–4. Examples of Nuclear Gene Control of Chloroplast Gene Expression in *Chlamydomonas*

Chloroplast Gene Affected	Protein Product and Function	Effect on Chloroplast Gene Expression	Nuclear Control Gene(s)
atpA	α subunit, ATP synthase	mRNA stability translation	*nccl* F54
atpB	β subunit, ATP synthase	mRNA stability	*thm24*
psbB	CP47, PSII	Transcript accumulation	GE2.10
psbC	CP43, PSII	Translation	F34, F64
psbD	D2, PSII	Translation Transcript stability	*nac-1* *nac-2*
psaA	82kd, PSI	mRNA splicing	>14 genes

Data from Choquet *et al.* (1988); Drapier *et al.* (1992); Kuchka *et al.* (1988, 1989); Jensen *et al.*, 1986; and Rochaix *et al.* (1989)

peptide encoded by the *psbD* gene were synthesized. These results strongly suggest that the *nac-1* gene product is probably required for translation of the *psbD* mRNA.

In contrast to *nac-1*, the *nac-2* gene encodes a trans-acting factor required for *psbD* transcript stability (Kuchka *et al.*, 1989). The *psbD* mRNA is cotranscribed with exon 2 of the trans-spliced *psaA* gene (Chapter 10). When the *nac-2* mutant was coupled with a *psaA* mutant that would normally accumulate a *psbD–psaA* exon 2 splicing intermediate, the intermediate was not found. A likely explanation for this finding is that the wild-type NAC2 product protects the 5'-end of the *psbD* flanking region from nucleolytic degradation. This would be consistent with the observation that the *psbD* and *psbD–psaA* exon 2 mRNA molecules share the *psbD* 5' flanking region, but have different 3'-ends.

The nuclear gene mutation, GE2.10, blocks accumulation of the three major transcripts of the *psbB* gene encoding the CP47 (P5) core component (Jensen *et al.*, 1986; Sieburth *et al.*, 1991). The GE2.10 mutation also exhibits a partial arrest in translation of the *psbA* mRNA encoding the D1 polypeptide. Since the primary effect of the GE2.10 mutation is to block transcription of the *psbB* gene, CP47 and D1 synthesis may be coupled at the translational level. Additional support for this model is that a chloroplast mutant unable to make the *psbA* mRNA and thus the D1 polypeptide is similarly impaired in synthesis of CP47. A second mutant unable to accumulate *psbB* mRNA has also been reported, but its allelic relationship to GE2.10 is unknown (Rochaix, 1992b).

The nuclear F34 mutant blocks translation of *psbC* mRNA encoding CP43 (P6) (Rochaix *et al.*, 1989). The product of the F34 gene apparently interacts with a stem–loop sequence in the 5' untranslated leader of *psbC* mRNA (Fig. 16–6). The importance of this sequence is evident from characterization of a chloroplast mutation, FUD34, that does not synthesize CP43. FUD34 makes *psbC* mRNA, but causes insertion of two T residues opposite two unpaired A nucleotides in the stem and removal of an unpaired C 6 bp downstream (Fig. 16–6). A chloroplast suppressor of F34 (F34suI) causes a T → A change in the other half of the stem.

To explain the results Rochaix *et al.* (1989) proposed a model that supposes that the stem region may interfere with the initiation of protein synthesis. According to the model the wild-type F34 gene encodes a factor, N, that destabilizes the stem and permits translation to begin. This factor is rendered nonfunctional in the mutant. While the altered product made by the F34 mutant cannot disrupt the stem in wild type, destabilization is possible in F34suI. In F34suI a T residue in an A-T base pair is converted to an A (Fig. 16–6). This destabilizes the stem slightly since the two As cannot pair. The chloroplast FUD34 mutant has just the opposite effect. The FUD34 mutation makes possible the pairing of the two unpaired As to the inserted T residues and deletes an unpaired C in the opposite side of the stem (Fig. 16–6). This increases the stability of the stem and prevents N from disrupting or possibly even recognizing the stem.

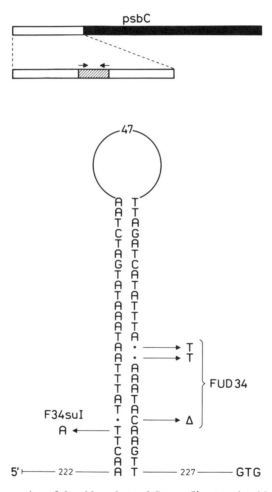

Fig. 16–6. The stem region of the chloroplast *psbC* gene 5′ untranslated leader in *Chlamydomonas*. The stem region is altered in the chloroplast FuD34 and F34suI mutants. The altered nucleotides in both mutants are indicated. Dots indicate missing nucleotides. In the upper part of the figure the two segments of the stem region are drawn as arrows within the 5′ untranslated region (white bar) of *psbC*. From Rochaix et al. (1989).

The F64 mutation, which also blocks translation of the *psbC* mRNA, belongs to a different complementation group from F34 and is not suppressed by F34suI. The nuclear mutant 6.2z is unable to accumulate *psbC* mRNA (see Rochaix, 1992b).

Mutations in three different nuclear genes have been found to control expression of the chloroplast *atpA* and *atpB* genes of *Chlamydomonas* (Drapier *et al.*, 1992). These genes map separately in the two unique sequence regions of the chloroplast genome. The F54 mutation alters a nuclear gene product that affects expression of *atpA* mRNA at the translational level so that the subunit is not made. The *ncc1* and *thm24* mutants, in contrast, cause destabilization of the *atpA* and *atpB* mRNAs, respectively. In *thm24* there is no diminution in the transcription rate of *atpB* mRNA, but no transcripts accumulate so the β subunit of the ATP synthase is not made. Similarly, in the *ncc1* mutant *atpA* mRNA synthesis is normal, but the half-life of *atpA* transcripts declined by a factor of 10. Despite this the mRNA made by *ncc1* must be efficiently translated since about half of the ATP synthase complexes present in wild type are assembled in the mutant. Consequently, the mutant grows well under photosynthetic conditions and would never have been detected in the usual screens for photosynthesis-deficient mutants in *Chlamydomonas*. In fact, *ncc1* was discovered by accident as a spontaneous mutant in the wild-type strain. Characterization of these mutants has also provided evidence for translational coupling in the synthesis of the α and β subunits of the ATP synthase even though the mRNAs encoding these subunits are not part of the same transcription unit. Thus α subunit synthesis was barely detectable in *thm24* whose primary defect is to destabilize *atpB* transcripts.

As discussed earlier, the ubiquitous 3' stem–loop structures at the ends of chloroplast mRNAs are probably important for maintenance of transcript stability and are also recognized by specific proteins some of which bind in a gene-specific manner. Obviously, certain of the *Chlamydomonas* mutants affecting transcript stability might alter such proteins. The experiments of Rochaix *et al.* (1989) detailed above show that nuclear control of the *psbC* mRNA is exerted through a stem–loop sequence in the 5' untranslated region. These experiments imply that proteins binding to the 5' untranslated regions of chloroplast mRNAs may have important regulatory functions. Danon and Mayfield (1991) have identified a 47-kDa protein complex that binds to a 36 nucleotide stem–loop located adjacent to and upstream of the ribosome binding site of the *psbA* mRNA encoding the D1 reaction center protein of photosystem II. This complex also seems to contain a 60-kDa protein. The 47-kDa protein complex binds well in light-grown cells, but not in dark-grown cells that do not exhibit high levels of *psbA* translation. *Chlamydomonas*, unlike flowering plants, synthesizes chlorophyll and a complete photosynthetic apparatus in the dark (Chapter 3). However, the nuclear *y-1* mutant does not and the chloroplasts of dark-grown *y-1* cells are comparable in many respects to etioplasts. Dark-grown *y-1* cells do not translate the *psbA* mRNA at all and are completely deficient in *psbA* mRNA-binding activity.

Yet another mechanism of posttranscriptional control of chloroplast gene expression in *Chlamydomonas* was discussed in Chapter 10. There it was noted that synthesis of the mature *psaA* mRNA, encoded by three widely separated exons and trans-spliced, requires the action of at least 14 different nuclear genes and a small RNA encoded in the chloroplast.

So far this discussion has revolved almost exclusively around *Chlamydomonas* yet large numbers of nuclear mutants in higher plants show reduced levels of photosynthetic complexes (see Gruissem and Tonkyn, 1993, for references). However, as Gruissem and Tonkyn (1993) note, there have been very few studies of chloroplast mRNA translation and turnover in these mutants. Only *hcf38* in maize, discussed earlier, alters plastid mRNAs (Barkan *et al.,* 1986). Gruissem and Tonkyn (1993) also remark that of 30 unpublished maize mutants only three have unique alterations in mRNA metabolism. They believe that the paucity of informative mutants in flowering plants may result because most chloroplast genes in these plants are parts of polycistronic transcription units. Gruissem and Tonkyn (1993) note that any nuclear mutation affecting processing or stability of a polycistronic mRNA will, by definition, be pleiotropic. While this is undoubtedly true, there is no reason why the 5' untranslated regions of these mRNAs adjacent to each coding sequence might not contain sequences that respond in a gene-specific manner to trans-acting factors encoded by nuclear genes. In *Chlamydomonas* the *rps12* gene is part of a polycistronic mRNA, yet sequences upstream of this sequence, but distal to the next coding sequence, form stem–loop structures that bind several proteins (Hauser *et al.,* 1993).

Nuclear mutants in flowering plants affecting chloroplast gene expression may also be pleiotropic for a different reason. Thus, a mutant that prevents synthesis or accumulation of a specific protein may indirectly affect accumulation of other proteins that require the affected protein for their own incorporation and stabilization into specific complexes. The properties of the nuclear viridis-115 *(vir-115)* mutant of barley can be interpreted in this light (Gamble and Mullet, 1989). Dark-grown wild-type barley plants do not accumulate the D1, D2, CP43, and CP47 polypeptides although the mRNAs for these proteins are actively transcribed since these proteins are unstable in the absence of chlorophyll (Mullet *et al.,* 1990). However, several PSII subunits do accumulate including cytochrome b_{559}, OEE1, and OEE2. Pulse-labeling experiments showed that in both wild type and the *vir-115* mutant light-induced synthesis of D1, D2, CP43, and CP47 occurs normally for the first hour, but after 16 and 72 hr synthesis of D1 and CP47 ceased in *vir-115* although further accumulation of their mRNAs occurred. Synthesis of both proteins continued in wild type. Immunoblotting experiments revealed that *vir-115* was also unable to accumulate D2 and CP43 under these conditions although the OEE1 and OEE2 proteins continued to be synthesized. Such extreme pleiotropy complicates any attempt to unravel the primary defect in a mutant like *vir-115*.

One of the major difficulties that arises in trying to understand the role of nuclear and chloroplast genes in chloroplast biogenesis is that no one body of comparable data comes from a single system. While posttranscriptional and translational control mechanisms for chloroplast gene expression have been well documented in both flowering plants and *Chlamydomonas,* virtually all work on control of nuclear genes encoding chloroplast proteins has been done on flowering plants with little genetic input. In contrast, most research on nuclear genes controlling chloroplast gene expression has been done in *Chlamydomonas* where mutants have been abundantly used. These different data sets are really not comparable. Furthermore, in the case of nuclear photosynthetic genes in flowering plants, not only does the existence of several light-responsive pathways have to be taken into consideration, but light-responsive elements and factors may overlap with developmentally regulated responses. In *Chlamydomonas* almost nothing is known about the light responses or light-responsive elements associated with nuclear, photosynthetic genes. In the future the most productive approach to dissecting these problems will likely be to concentrate on a couple of model systems (e.g., *Arabidopsis* and *Chlamydomonas*) in analyzing the genetic control of chloroplast biogenesis. This focus has been achieved for mitochondria in the case of yeast and probably explains why our current understanding of the interplay of nuclear and organelle genes in the biogenesis of yeast mitochondria seems so much easier to comprehend (Chapter 17).

Does the Chloroplast Signal the Nucleus?

Evidence for a chloroplast signal that may involve a small molecule is summarized in a review by Taylor (1989). Carotenoids protect chlorophyll from photooxidation by blocking the formation of oxygen radicals and the photodestruction of photosynthetic complexes (see Tonkyn *et al.*, 1992 for a discussion). Taylor points out that carotenoid deficiencies and ultimately photooxidation itself negatively affect transcription of the nuclear *cab* genes. Many of these effects are light conditional so that in low light chloroplasts develop the ultrastructure typical of low light growth and accumulate a normal complement of both nucleus- and plastid-encoded proteins. Very low levels of chlorophyll are also detectable and *cab* mRNA accumulates to normal levels.

Proof that photooxidative damage to the chloroplast is responsible for the *cab* mRNA deficiency came from experiments in which a maize mutant deficient in chlorophyll synthesis *(1-Blandy4)* was treated with norflurazon and compared to wild-type plants treated similarly in low and high light (Burgess and Taylor, 1988). Norflurazon is an inhibitor of carotenoid synthesis so mutant plants treated with this compound were doubly deficient in chlorophyll and carotenoids. Burgess and Taylor found that in low light norflurazon-treated wild-type and *1-Blandy4* plants both exhibited high levels of *cab* mRNA, but this was true only of the *1-Blandy4* plants at high light intensity. These experiments show that synthesis of *cab* mRNA proceeds normally when chlorophyll photooxidation is blocked.

As Taylor (1989 and references cited therein) points out, the conditional nature of carotenoid deficiencies is a useful tool and permits generalization of these observations. Light conditions permissive for chloroplast development in maize, mustard, and barley promote normal accumulation of *cab* and *rbcS* mRNAs. When these plants are transferred to higher intensity light, both mRNAs rapidly decline to very low levels. After about 6 hr at high-light intensity, 90% of the *cab* mRNA has disappeared (Mayfield and Taylor, 1987).

The nature of the chloroplast signal and its mode of action remain to be elucidated. However, Taylor (1989) reasons that the signal molecule may be destroyed by photooxidation. This molecule would be required for optimal *cab* gene transcription and presumably transcription (or translation) of other plastid and nuclear genes involved in chloroplast development as well. Support for such a model comes from the experiments of Tonkyn *et al.* (1992) who examined the effect of norflurazon treatment, on plastid and nuclear gene expression in young and mature spinach leaves. Young leaves lost essentially all pigment following norflurazon treatment which was paralleled by a decline in mRNA levels for all plastid genes tested except *psbA*. This was accompanied by a decrease in transcription rate of these genes except in the cases of *psbA* and *rbcL*. Mature leaves retained more than 60% of their chlorophyll and carotenoids following norflurazon treatment and all plastid mRNAs except one accumulated to normal or higher than normal levels. Thus, the level of plastid mRNA accumulation is related to the degree of photooxidative stess. Furthermore, Tonkyn et al. (1992) report that levels of mRNAs coded by the nuclear *rbcS* and *cab* genes were reduced although accumulation of an mRNA encoding a chloroplast RNA-binding protein was unaffected by photooxidation. These results suggest that Taylor's chloroplast signal may only be required for accumulation of a subset of nuclear-encoded photosynthetic mRNAs.

The results of Rapp and Mullet (1991) and others indicate that transcription of cpDNA is required to generate the chloroplast signal. In barley seedlings the initial activation of plastid transcription and accumulation of *rbcS* mRNA are largely light independent. If the initial increase in plastid transcription is prevented by tagetitoxin, an inhibitor of plastid RNA polymerase, *rbcS* accumulation does not occur and *cab* mRNA accumulation cannot be induced by light. Rapp and Mullet, like Tonkyn et al. (1992), also observe that illumination of seedlings grown in the presence of an inhibitor of carotenoid synthesis results in a rapid decline in plastid transcription activity and both *rbcL* and *rbcS* mRNA levels. Rapp and Mullet interpret the tagetitoxin results as showing that plastid transcription is required to produce the plastid signal. Since photooxidation interferes with plastid transcription, generation of the plastid signal is also affected.

REFERENCES

Akkaya, M.S., and C.A. Breitenberger (1992). Light regulation of protein synthesis factor EF-G in pea chloroplasts. *Plant Mol. Biol.* 20: 791–800.

Bansal, K.C., Viret, J.-F., Haley, J., Khan, B.M., Schantz, R., and L. Bogorad (1992). Transient expression from *cab-m1* and *rbcS-m3* promoter sequences is different in mesophyll and bundle sheath cells in maize leaves. *Proc. Natl. Acad. Sci. U.S.A.* 89: 3654–3658.

Barkan, A. (1988). Proteins encoded by a complex chloroplast transcription unit are each translated from both monocistronic and polycistronic mRNAs. *EMBO J.* 7:2637–2644.

Barkan, A., Miles, D., and W.C. Taylor (1986). Chloroplast gene expression in nuclear photosynthetic mutants of maize. *EMBO J.* 5: 1421–1427.

Baumgartner, B.J., Rapp, J.C., and J.E. Mullet (1989). Plastid transcription activity and DNA copy number increase early in barley chloroplast development. *Plant Physiol.* 89: 1011–1018.

Baumgartner, B.J., Rapp, J.C., and J.E. Mullet (1993). Plastid genes encoding the transcription/translation apparatus are differentially transcribed early in barley *(Hordeum vulgare)* chloroplast development. *Plant Physiol.* 101: 781–791.

Bendich, A.J. (1987). Why do chloroplasts and mitochondria contain so many copies of their genome? *BioEssays* 6: 279–282.

Bennoun, P., Spierer-Herz, M., Erickson, J., Girard-Bascou, J., Pierre, Y., Delosme, M., and J.-D. Rochaix (1986). Characterization of photosystem II mutant of *Chlamydomonas reinhardtii* lacking the *psbA* gene. *Plant Mol. Biol.* 6: 151–160.

Berends, T., Gamble, P.E., and J.E. Mullet (1987). Characterization of the barley transcription units containing *psaA-psaB* and *psbD-psbC*. *Nucl. Acids Res.* 15: 5217–5240.

Berends-Sexton, T., Jones, J.T., and J.E. Mullet (1990). Sequence and transcriptional analysis of the barley ctDNA region upstream of *psbD-psbC* encoding *trn*K(UUU), *rps*16, *trn*Q(UUG), *psb*K, *psb*I, and *trn*S(GCU). *Curr. Genet.* 17: 445–454.

Blowers, A.D., Klein, U., Ellmore, G.S., and L. Bogorad (1993). Functional in vivo analyses of the 3' flanking sequences of the *Chlamydomonas* chloroplast *rbcL* and *psaB* genes. *Mol. Gen. Genet.* 238: 339–349.

Bogorad, L. (1991). Replication and transcription of plastid DNA. In *The Molecular Biology of Plastids* (L. Bogorad and I.K. Vasil, eds.). pp. 93–124. Academic Press, San Diego.

Burgess, D.G., and W.C. Taylor (1988). The chloroplast affects the transcription of a nuclear gene family. *Mol. Gen. Genet.* 214: 89–96.

Buzby, J.S., Yamada, T., and E.M. Tobin (1990). A light-regulated DNA-binding activity interacts with a conserved region of the *Lemna gibba rbcS* promoter. *Plant Cell* 2: 805–814.

Castresana, C., Staneloni, R., Malik, V.S., and A.R. Cashmore (1987). Molecular characterization of two clusters of genes encoding the Type I CAB polypeptides of PSII in *Nicotiana plumbaginifolia*. *Plant Mol. Biol.* 10: 117–126.

Castresana, C., Garcia-Luque, I., Alonso, E., Malik, V.S., and A.R. Cashmore (1988). Both positive and negative regulatory elements mediate expression of a photoregulated CAB gene from *Nicotiana plumbaginifolia*. *EMBO J.* 7: 1929–1936.

Chitnis, P.R., and J.P. Thornber (1988). The major light-harvesting complex of Photosystem II: Aspects of its molecular and cell biology. *Photosynthesis Res.* 16: 41–63.

Choquet, Y., Goldschmidt-Clermont, M., Girard-Bascou, J., Kuck, U., Bennoun, P., and J.-D. Rochaix (1988). Mutant phenotypes support a *trans*-splicing mechanism for the expression of the tripartite *psaA* gene in the *C. reinhardtii* chloroplast. *Cell* 52: 903–913.

Chory, J. (1991). Light signals in leaf and chloroplast development: Photo-receptors and downstream responses in search of a transduction pathway. *New Biol.* 3: 538–548.

Chory, J. (1992). A genetic model for light-regulated seedling development in *Arabidopsis*. *Development* 115: 337–354.

Cuozzo, M., Kay, S.A., and N.-H. Chua (1988). Regulatory circuits of light-responsive genes. In *Temporal and Spatial Regulation of Plant Genes* (D.P.S. Verma and R.B. Goldberg, eds.), pp. 131–153, Springer-Verlag, Vienna.

Dahlin, C., and K. Cline (1991). Developmental regulation of the plastid protein import apparatus. *Plant Cell* 3: 1131–1140.

Danon, A., and S.P.Y. Mayfield (1991). Light regulated translational activators: identification of chloroplast gene specific mRNA binding proteins. *EMBO J.* 10: 3993–4001.

Dean, C., Pichersky, E., and P. Dunsmuir (1989). Structure, evolution, and regulation of *RbcS* genes in higher plants. *Annu. Rev. Plant Physiol. Plant Mol. Biol.* 40: 415–439.

Deng, X.-W., and W. Gruissem (1987). Control of plastid gene expression during development: The limited role of transcriptional regulation. *Cell* 49: 379–387.

Deng, X.-W., Casper, T., and P.H. Quail (1991). cop1: A regulatory locus involved in light-controlled development and gene expression in *Arabidopsis*. *Genes Dev.* 5: 1172–1182.

Donald, R.G.K., and A.R. Cashmore (1990). Mutation of either G box or I box sequences profoundly affects expression from the *Arabidopsis rbcS1A* promoter. *EMBO J.* 9: 1717–1726.

Drapier, D., Girard-Bascou, J., and F.-A. Wollman (1992). Evidence for nuclear control of the expression of the *atpA* and *atpB* chloroplast genes in Chlamydomonas. *Plant Cell* 4: 283–295.

Dynan, W.S., and R. Tjian (1985). Control of eukaryotic messenger RNA synthesis by sequence-specific DNA-binding proteins. *Nature (London)* 316: 774–778.

Ellis, R.J. (1977). Protein synthesis by isolated chloroplasts. *Biochim. Biophys. Acta* 463: 185–215.

Fluhr, R., and N.-H. Chua (1986). Developmental regulation of two genes encoding ribulose-bisphosphate carboxylase in pea and transgenic petunia plants: Phytochrome response and blue-light induction. *Proc. Natl. Acad. Sci. U.S.A.* 83: 2358–2362.

Fluhr, R., Kuhlemeier, C., Nagy, F., and N.-H. Chua (1986). Organ-specific and light-induced expression of plant genes. *Science* 232: 1106–1112.

Gamble, P.E., and J.E. Mullet (1989a). Blue light regulates the accumulation of two *psbD-psbC* transcripts in barley chloroplasts. *EMBO J.* 8: 2785–2794.

Gamble, P.E., and J.E. Mullet (1989). Translation and stability of proteins encoded by the plastid psbA and psbB genes are regulated by a nuclear gene during light-induced chloroplast development in barley. *J. Biol. Chem.* 264: 7236–7243.

Gamble, P.E., Berends Sexton, T., and J.E. Mullet (1988). Light-dependent changes in *psbD* and *psbC* transcripts of barley chloroplasts: Accumulation of two transcripts maintains *psbD* and *psbC* translation capability in mature chloroplasts. *EMBO J.* 7: 1289–1297.

Gillham, N.W., Harris, E.H., Randolph-Anderson, B.L., Boynton, J.E., Hauser, C.R., McElwain, K.B., and S.M. Newman (1991). Molecular genetics of chloroplast ribosomes in *Chlamydomons*. In *The Translational Apparatus of Photosynthetic Organelles* (R. Mache, E. Stutz, and A. Subramanian, eds.). NATO ASI series, Vol. H55, pp. 127–144. Springer-Verlag, Berlin.

Gilmartin, P.M., Sarokin, L., Memelink, J., and N.-H. Chua (1990). Molecular light switches for plant genes. *Plant Cell* 2: 369–378.

Girard-Bascou, J., Choquet, Y., Schneider, M., Delosme, M., and M. Dron (1987). Characterization of a chloroplast mutation in the *psaA2* gene of *Chlamydomonas reinhardtii*. *Curr. Genet.* 12: 489–495.

Giuliano, G., Pichersky, E., Malik, V.S., Timko, M.P., Scolnik, P.A., and A.R. Cashmore (1988). An evolutionarily conserved protein binding sequence upstream of a plant light-regulated gene. *Proc. Natl. Acad. Sci. U.S.A.* 85: 7089–7093.

Goldschmidt-Clermont, M. (1986). The two genes for the small subunit of RuBP carboxylase/oxygenase are closely linked in *Chlamydomonas reinhardtii*. *Plant Mol. Biol.* 6: 13–21.

Goldschmidt-Clermont, M., and M. Rahire (1986). Sequence, evolution and differential expression of the two genes encoding variant small subunits of ribulose-bisphosphate carboxylase/oxygenase in *Chlamydomonas reinhardtii*. *J. Mol. Biol.* 191: 421–432.

Green, B.R., Pichersky, E., and K. Kloppstech (1991). Chlorophyll *a/b*-binding proteins: An extended family. *Trends Biochem. Sci.* 16: 181–186.

Green, P.J., Kay, S.A., and N.-H. Chua (1987). Sequence-specific interactions of a pea nuclear factor with light-responsive elements upstream of the *rbcS-3A* gene *EMBO J.* 6: 2543–2549.

Green, P.J., Yong, M.-H., Cuozzo, M., Kano-Murakami, Y., Silverstein, P., and N.-H. Chua (1988). Binding site requirements for pea nuclear protein factor GT-1 correlate with sequences required for light-dependent transcriptional activation of the *rbcS-3A* gene. *EMBO J.* 7: 4035-4044.

Grob, U., and K. Stuber (1987). Discrimination of phytochrome dependent light-inducible from non-light-inducible plant genes. Prediction of a common light-responsive element (LRE) in phytochrome dependent light-inducible genes. *Nucl. Acids. Res.* 15: 9957-9973.

Gruissem, W. (1989a). Chloroplast gene expression: How plants turn their plastids on. *Cell* 56: 161-170.

Gruissem, W. (1989b). Chloroplast RNA: Transcription and processing. In *Biochemistry of Plants*, Vol. 15 (A. Marcus, ed.), pp. 151-191. Academic Press, San Diego.

Gruissem, W., and G. Schuster (1993). Control of mRNA degradation in organelles. In *Control of Messenger RNA Stability* (G. Braverman and J. Belasco, eds.), pp. 329-365. Academic Press, Orlando, FL.

Gruissem, W., and J.C. Tonkyn (1993). Control mechanisms of plastid gene expression. *Crit. Rev. Plant Sci.* 12: 19-55.

Gruissem, W., Barkan, A., Deng, X.-w., and D. Stern (1988). Transcriptional and posttranscriptional control of plastid mRNA levels in higher plants. *Trends Genet.* 4: 258-263.

Hauser, C.R., Randolph-Anderson, B.L., Hohl, T.M., Harris, E.H., Boynton, J.E., and N.W. Gillham (1993). Molecular genetics of chloroplast ribosomes in *Chlamydomonas reinhardtii*. In *The Translational Apparatus* (K. Nierhaus *et al.* eds.), pp. 545-554. Plenum Press, New York.

Herrmann, R.G., Westhoff, P., and G. Link (1992). Biogenesis of plastids in higher plants. In *Plant Gene Research: Cell Organelles* (R.G. Herrmann, ed.), pp. 275-349. Springer-Verlag, Wien.

Hoober, J.K. (1984). Chloroplasts. Plenum Press, New York.

Hosler, J.P., Wurtz, E.A., Harris, E.H., Gillham, N.W., and J.E. Boynton (1989). Relationship between gene dosage and gene expression in the chloroplast of *Chlamydomonas reinhardtii*. *Plant Physiol.* 91: 648-655.

Jensen, K.H., Herrin, D.L., Plumley, F.G., and G.W. Schmidt (1986). Biogenesis of photosystem II complexes: transcriptional, translational, and post-translational regulation. *J. Cell Biol.* 103: 1315-1325.

Kim, M., Christopher, D.A., and J.E. Mullet (1993). Direct evidence for selective modulation of *psbA*, *rpoA*, *rbcL*, and 16S RNA stability during barley chloroplast development. *Plant Mol. Biol.* 22: 447-463.

Klein, R.R., and J.E. Mullet (1986). Regulation of chloroplast-encoded chlorophyll-binding protein translation during higher plant chloroplast biogenesis. *J. Biol. Chem.* 261: 11138-11145.

Klein, R.R., and J.E. Mullet (1987). Control of gene expression during higher plant chloroplast biogenesis. *J. Biol. Chem.* 262: 4341-4348.

Klein, R.R., and J.E. Mullet (1990). Light-induced transcription of chloroplast genes: *psbA* transcription is differentially enhanced in illuminated barley. *J. Biol. Chem.* 265: 1895-1902.

Klein, R.R., Mason, H.S., and J.E. Mullet (1988). Light-regulated translation of chloroplast genes. I. Transcripts of *psaA-psaB*, *psbA* and *rbcL* are associated with polysomes in dark-grown and illuminated barley seedlings. *J. Cell Biol.* 106: 289-301.

Kuchka, M.R., Mayfield, S.P., and J.-D. Rochaix (1988). Nuclear mutations specifically affect the synthesis and/or degradation of the chloroplast-encoded D2 polypeptide of photosystem II in *Chlamydomonas reinhardtii*. *EMBO J.* 7: 319-324.

Kuchka, M.R., Goldschmidt-Clermont, M., van Dillewijn, J., and J.-D. Rochaix (1989). Mutation at the nuclear NAC 2 locus of *C. reinhardtii* affects the stability of the chloroplast psbD transcript encoding polypeptide D2 of photosystem II. *Cell* 58: 869-876.

Kuhlemeier, C. (1992). Transcriptional and post-transcriptional regulation of gene expression in plants. *Plant Mol. Biol.* 19: 1-14.

Kuhlemeier, C., Green, P.J., and N.-H. Chua (1987). Regulation of gene expression in higher plants. *Annu. Rev. Plant Physiol.* 38: 221-257.

Kuhlemeier, C., Cuozzo, M., Green, P., Goyvaerts, E., Ward, K., and N.-H. Chua (1988). Localization and conditional redundancy of regulatory elements in *rbcS-3A*, a pea gene encoding the small subunit of ribulose-bisphosphate carboxylase. *Proc. Natl. Acad. Sci. U.S.A.* 85: 4662-4666.

Lam, E., and N.-H. Chua (1987). Chloroplast DNA gyrase and in vitro regulation of transcription by template topology and novobiocin. *Plant Mol. Biol.* 8: 415-424.

Lam, E., and N.-H. Chua (1989). ASF-2: A factor that binds to the cauliflower mosaic virus 35S promoter and a conserved GATA motif in *Cab* promoters. *Plant Cell* 1: 1147-1156.

Leto, K.J., Bell, E., and L. McIntosh (1985). Nuclear mutation leads to an accelerated turnover of chloroplast-encoded 48 kd and 34.5 kd polypeptides in thylakoids lacking photosystem II. *EMBO J.* 4: 1645-1653.

Link, G. (1988). Photocontrol of plastid gene expression *Plant, Cell and Environ.* 11: 329-338.

Liu, X.-Q., Hosler, J.P., Boynton, J.E., and N.W. Gillham (1989). mRNAs for two ribosomal proteins are preferentially translated in the chloroplast of *Chlamydomonas reinhardtii* under conditions of reduced protein synthesis. *Plant Mol. Biol.* 12: 385-394.

Maniatis, T., Goodbourn, S., and J. Fisher (1987). Regulation of inducible tissue-specific gene expression. *Science* 236: 1237-1245.

Manzara, T., and W. Gruissem (1988). Organization and expression of the genes encoding ribulose-1,5-bisphosphate carboxylase in higher plants. *Photosynthesis Res.* 16: 117-139.

Manzara, T., Carrasco, P., and W. Gruissem (1991). Developmental and organ-specific changes in promoter DNA-protein interactions in the tomato *rbcS* gene family. *Plant Cell* 3: 1305–1316.

Mayfield, S.P., and W.C. Taylor (1987). Chloroplast photooxidation inhibits the expression of a set of nuclear genes. *Mol. Gen. Genet.* 208: 309–314.

Mishkind, M.L., and G.W. Schmidt (1983). Post-transcriptional regulation of ribulose 1,5-bisphosphate carboxylase small subunit accumulation in *Chlamydomonas reinhardtii*. *Plant Physiol.* 72: 847–854.

Miyamura, S., Nagata, T., and T. Kuroiwa (1986). Quantitative fluorescence microscopy on dynamic changes of plastid nucleoids during wheat development. *Protoplasma* 133: 66–72.

Mullet, J.E. (1988). Chloroplast development and gene expression. *Annu. Rev. Plant Physiol. Plant Mol. Biol.* 39: 475–502.

Mullet, J.E., and R.R. Klein (1987). Transcription and RNA stability are important determinants of higher plant chloroplast RNA levels. *EMBO J.* 6: 1571–1579.

Mullet, J.E., Gamble Klein, P., and R.R. Klein (1990). Chlorophyll regulates accumulation of the plastid-encoded chlorophyll apoproteins CP43 and D1 by increasing apoprotein stability. *Proc. Natl Acad. Sci. U.S.A.* 87: 4038–4042.

Mullet, J.E., Rapp, J.C., Baumgartner, B.J., Berends-Sexton, T., and D.A. Christopher (1991). Regulation of chloroplast biogenesis in barley. In *Plant Molecular Biology 2* (R.G. Herrmann and B. Larkins, eds.), pp. 439–447. Plenum Press, New York.

Mural, R.J. (1991). Fundamentals of light-regulated gene expression in plants. *Subcell. Biochem.* 17: 191–211.

Nagy, F., Kay, S.A., and N.-H. Chua (1988). Gene regulation by phytochrome. *Trends Genet.* 4: 37–42.

Quail, P.H. (1991). Phytochrome: A light-activated molecular switch that regulates plant gene expression. *Annu. Rev. Genet.* 25: 389–409.

Rapp, J.C., and J.E. Mullet (1991). Chloroplast transcription is required to express the nuclear genes *rbcS* and *cab*. Plastid copy number is regulated independently. *Plant Mol. Biol.* 17: 813–823.

Rapp, J.C., Baumgartner, B.J., and J. Mullet (1992). Quantitative analysis of transcription and RNA levels of 15 barley chloroplast genes: transcription rates and mRNA levels vary over 300-fold, predicted mRNA stabilities vary 30-fold. *J. Biol. Chem.* 267: 21404–21411.

Rochaix, J.-D. (1992a). Control of plastid gene expression in *Chlamydomonas reinhardtii*. In *Plant Gene Research: Cell Organelles* (R.G. Herrmann, ed.), pp. 249–274. Springer-Verlag, Wien.

Rochaix, J.-D. (1992b). Post-transcriptional steps in the expression of chloroplast genes. *Annu. Rev. Cell Biol.* 8: 1–28.

Rochaix, J.-D., Kuchka, M., Mayfield, S., Schirmer-Rahire, M., Girard-Bascou, J., and P. Bennoun (1989). Nuclear and chloroplast mutations affect the synthesis or stability of the chloropalst *psbC* gene product in *Chlamydomonas reinhardtii*. *EMBO J.* 8: 1013–1021.

Rochaix, J.-D., Goldschmidt-Clermont, M., Choquet, Y., Kuchka, M., and J. Girard-Bascou (1991). Nuclear and chloroplast genes involved in the expression of specific chloroplast genes of *Chlamydomonas reinhardtii*. In *Plant Molecular Biology 2* (R.G. Herrmann and B. Larkins, eds.), pp. 401–410. Plenum Press, New York.

Salvador, M.L., Klein, U., and L. Bogorad (1993). 5′ sequences are important positive and negative determinants of the longevity of *Chlamydomonas* chloroplast gene transcripts. *Proc. Natl. Acad. Sci. U.S.A.* 90: 1556–1560.

Schindler, U., and A.R. Cashmore (1990). Photoregulated gene expression may involve ubiquitous DNA binding proteins. *EMBO J.* 9: 3415–3427.

Senger, H. ed. (1987). *Blue Light Responses: Phenomena and Occurrence in Plants and Microorganisms*. Vols. 1 and 2. CRC Press, Inc. Boca Raton, Fla.

Shirley, B.W., and R.B. Meagher (1990). A potential role for RNA turnover in light regulation of plant gene expression: Ribulose-1,5-bisphosphate carboxylase small subunit in soybean. *Nucl. Acids Res.* 18: 3377–3385.

Shirley, B.W., Ham, D.P., Senecoff, J.F., Berry-Lowe, S.L., Zurfluh, L.L., Shah, D.M., and R.B. Meagher (1990). Comparison of the expression of two highly homologous members of the soybean ribulose-1,5-bisphosphate carboxylase small subunit gene family. *Plant Mol. Biol.* 14: 909–925.

Sieburth, L.E., Berry-Lowe, S., and G.W. Schmidt (1991). Chloroplast RNA stability in *Chlamydomonas*: Rapid degradation of *psbB* and *psbC* transcripts in two nuclear mutants. *Plant Cell* 3: 175–189.

Silverthorne, J., and E. Tobin (1987). Phytochrome regulation of nuclear gene expression. *Bioessays* 7: 18–23.

Silverthorne, J., and E.M. Tobin (1990). Post-transcriptional regulation of organ-specific expression of individual *rbcS* mRNAs in *Lemna gibba*. *Plant Cell* 2: 1181–1190.

Staub, J.M., and P. Maliga (1993). Accumulation of D1 polypeptide in tobacco plastids is regulated via the untranslated region of the *psbA* mRNA. *Nucl. Acids Res.* 22: 601–606.

Stern, D.B., and W. Gruissem (1987). Control of plastid gene expression: 3′ inverted repeats act as mRNA processing and stabilizing elements, but do not terminate transcription. *Cell* 51: 1145–1157.

Stern, D.B., Jones, H., and W. Gruissem (1989). Function of plastid mRNA 3′ inverted repeats: RNA stabilization and gene-specific protein binding. *J. Biol. Chem.* 264: 18742–18750.

Stern, D.B., Radwanski, E.R., and K.L. Kindle (1991). A 3′ stem/loop structure of the *Chlam-*

dyomonas chloroplast *atpB* gene regulates mRNA accumulation in vivo. *Plant Cell* 3: 285–297.

Stirdivant, S.M., Crossland, L.D., and L. Bogorad (1985). DNA supercoiling affects *in vitro* transcription of two maize chloroplast genes differently. *Proc. Natl. Acad. Sci. U.S.A.* 82: 4886–4890.

Tanaka, M., Obokata, J., Chunwongse, J., Shinozaki, K., and M. Sugiura (1987). Rapid splicing and stepwise processing of a transcript from the *psb*B operon in tobacco chloroplasts: Determination of the intron sites in *pet*B and *pet*D. *Mol. Gen. Genet.* 209: 427–431.

Taylor, W.C. (1989). Regulatory interactions between nuclear and plastid genomes. *Annu. Rev. Plant Physiol. Plant Mol. Biol.* 40: 211–233.

Thompson, R.J., and G. Mosig (1984). Light and genetic determinants in the control of specific chloroplast transcripts in *Chlamydomonas reinhardtii*. *Plant Physiol.* 76: 1–6.

Thompson, R.J., and G. Mosig (1985). An ATP-dependent supercoiling topoisomerase of *Chlamydomonas reinhardtii* affects accumulation of specific chloroplast transcripts. *Nucl. Acids Res.* 13: 873–891.

Thompson, R.J., and G. Mosig (1987). Stimulation of a *Chlamydomonas* chloroplast promoter by novobiocin in situ and in *E. coli* implies regulation by torsional stress in the chloroplast DNA. *Cell* 48: 281–287.

Thompson, W.G., and M.J. White (1991). Physiological and molecular studies of light-regulated nuclear genes in higher plants. *Annu. Rev. Plant Physiol. Plant Mol. Biol.* 42: 423–466.

Tonkyn, J.C., Deng, X.-W., and W. Gruissem (1992). Regulation of plastid gene expression during photooxidative stress. *Plant Physiol.* 99: 1406–1415.

Ueda, T., Pichersky, E., Malik, V.S., and A.R. Cashmore (1989). Level of expression of the tomato. *rbcS-3A* gene is modulated by a far upstream promoter element in a developmentally regulated manner. *Plant Cell* 1: 217–227.

Wanner, L.A., and W. Gruissem (1991). Expression dynamics of the tomato *rbcS* gene family during development. *Plant Cell* 3: 1289–1303.

Weisshaar, B., Armstrong, G.A., Block, A., da Costa e Silva, O., and K. Hahlbrock (1991). Light-inducible and constitutively expressed DNA binding proteins recognizing a plant promoter element with functional relevance in light responsiveness. *EMBO J.* 10: 1777–1786.

Westhoff, P., and R.G. Herrmann (1988). Complex RNA maturation in chloroplasts: The *psb*B operon from spinach. *Eur. J. Biochem.* 171: 551–564.

Wurtz, E.A., Boynton, J.E., and N.W. Gillham (1977). Perturbation of chloroplast DNA amounts and chloroplast gene transmission in *Chlamydomonas reinhardtii* by 5-fluorodeoxyuridine. *Proc. Natl. Acad. Sci. U.S.A.* 74: 4552–4556.

Yao, W.B., Meng, B.Y., Tanaka, M., and M. Sugiura (1989). An additional promoter within the protein-coding region of the *psb*D-*psb*C gene cluster in tobacco chloroplast DNA. *Nucl. Acids Res.* 17: 9583–9591.

17

Control of Gene Expression in Mitochondrial Biogenesis

In contrast to chloroplasts where the role of light in biogenesis has been extensively documented in a variety of plants (Chapter 16), very little is known of the signals involved in mitochondrial biogenesis except in yeast. There are two principal signals that inform a yeast cell that it needs to switch from glucose-repressed to aerobic growth. As will be discussed, they are oxygen, usually acting via heme, and a carbon source.

THE REGULATION OF MITOCHONDRIAL GENE EXPRESSION

The mechanisms involved in determining mitochondrial gene expression are discussed in several reviews (Attardi, 1985; Attardi and Schatz, 1988; Costanzo and Fox, 1990; Grivell, 1989; Nelson, 1987). Most experiments have been done with mammalian systems or yeast cells.

Regulation of mitochondrial gene expression in animal cells. Differential regulation of transcript synthesis and turnover play important roles in human mitochondrial gene expression (see Chapter 12). The balance between synthesis and degradation of different RNA molecules determines the steady-state population that is greatest for the rRNAs. The high steady-state level of rRNA transcripts reflects not only an increased rate of synthesis, but the existence of a specific termination mechanism for these transcripts (Kruse *et al.,* 1989 and Chapter 12).

There are hints that general transcriptional control mechanisms are operative in animal mitochondria. Developmental studies with sea urchins and the toad *Xenopus laevis* are instructive. They illustrate the existence of general control mechanisms for mtDNA replication and transcription that are not evident from experiments with undifferentiated cells growing in culture (reviewed by Attardi and Schatz, 1988). In both organisms, mtDNA replication occurs during oogenesis and stops at maturation. Transcription of mtDNA is activated after fertilization, long before the resumption of mtDNA synthesis.

In the fertilized egg, replication and transcription of mtDNA appear to be inhibited by the nucleus. Sea urchin eggs can be broken into two halves by centrifugation, one of which contains the nucleus and a few mitochondria while the other lacks the nucleus, but possesses most of the mitochondria (Rinaldi and Giudice, 1985). A variety of physical and chemical stimuli will cause parthenogenetic activation of unfertilized sea urchin eggs (Giudice, 1973). When enucleated halves of sea urchin eggs containing most of the mitochondria are activated parthenogenetically, a burst of mtDNA replication and transcription occurs. This does not happen in intact eggs or fertilized halves lacking the egg nucleus, but containing the male pronucleus. From these experiments Rinaldi and Giudice (1985) conclude that the nucleus in some way exerts a negative control over mtDNA transcription and replication in the fertilized egg. Similar results have also been obtained for *X. laevis* (Rinaldi and Giudice, 1985).

Mitochondrial transcription does not seem to be rate limiting for translation in mammalian cells. Thus, in a mouse cell line lacking the cytosolic thymidine kinase, 5-bromodeoxyuridine (5-BrdU) was incorporated exclusively into mtDNA with no effect on cell growth (Lansman and Clayton, 1975). When the 5-BrdU-containing mtDNA was subjected to photodamage, extensive inhibition of mitochondrial RNA synthesis occurred. However, the rate of mitochondrial protein synthesis remained essentially unchanged for at least 48 hr even though the rate of RNA synthesis had been suppressed at least 85%. Since the half-lives of mitochondrial mRNAs in HeLa cells are short (Chapter 12), this either means that the stability of these molecules has increased dramatically or that the remaining 10–20% of mtDNA transcription is sufficient to support a normal rate of mitochondrial translation (Attardi and Schatz, 1988). These results are similar to those described in Chapter 16 for *Chlamydomonas,* where it was shown that growth of cells in FdUrd, which results in a parallel reduction in cpDNA and chloroplast mRNA, does not affect the synthesis of specific chloroplast-encoded proteins *in vitro* or *in vivo.*

Other studies indicate that mitochondrial transcription is not particularly efficient (see Attardi and Schatz, 1988). In HeLa cells, the ratio of rRNA genes to rRNA molecules is more than three orders of magnitude higher in mitochondria than in the nucleocytoplasmic compartment. Despite this low transcription efficiency the high copy number of mitochondrial genomes coupled with the fact that much of the mRNA is apparently dispensable ensures that mitochondrial gene products are never rate limiting for growth.

Regulation of mitochondrial gene expression in yeast. Unlike the mammalian mitochondrial genome with its H- and L-strand promoters, yeast has at least 20 transcription initiation sites scattered around its mitochondrial genome (Chapter 12). Measurements of transcription *in vivo* and competition assays for limiting amounts of RNA polymerase *in vitro* suggest that promoter strengths vary as much as 20-fold (see Costanzo and Fox, 1990; Grivell, 1989). Promoter strength is affected both by nucleotide sequence and relative location (Biswas and Getz, 1988). Mueller and Getz (1986a) measured pulse-labeled mRNA levels from polycistronic transcription units. They found that within the same transcription unit differences as great as 17-fold could be seen between the most abundant promoter proximal transcripts and those that were promoter distal. Mueller and Getz (1986a) proposed that attenuation of RNA polymerase-catalyzed mRNA elongation might be responsible. However, selective degradation of downstream transcripts during pulse-labeling cannot be ruled out (Costanzo and Fox, 1990).

Overall levels of mitochondrial transcripts in yeast seem to be under general control. Thus, steady-state levels of most mitochondrial mRNAs are 3- to 6-fold higher in derepressed than in glucose-repressed cells (Mueller and Getz, 1986b). However, while the proportion of certain mRNAs to the total remains constant (e.g., *cox3*) under glucose-repressed and -derepressed conditions, other mRNAs increase proportionately in the absence of glucose (Baldacci and Zennaro, 1982; Zennaro et al., 1985). These steady-state increases in mitochondrial RNA levels in derepressed cells probably reflect to some extent increases in transcription rates since mitochondrial 21 S rRNA is made at a 7.2-fold faster rate in derepressed cells than in glucose-repressed cells (Mueller and Getz, 1986b). However, differences in half-lives must also be involved as rates of synthesis of various transcripts determined in pulse-labeling experiments predict 5- to 10 fold higher steady-state levels than actually seen.

Posttranscriptional processing events are also important in the regulating yeast mitochondrial gene expression. Certain of these events involve specific nuclear gene products such as the *CBP1* gene product required for stabilizing *cytb* mRNA (see Control of Mitochondrial Gene Expression by the Nucleus) or nuclear-encoded maturases required for intron splicing (see Chapter 10). Mitochondrial tRNA processing and base modification (Chapter 12) could also act to moderate overall levels of mitochondrial protein synthesis. Finally, highly specific translational control mechanisms exist whereby the products of certain nuclear genes are required for translation of certain mitochondrial mRNAs (see Control of Mitochondrial Gene Expression by the Nucleus).

THE REGULATION OF NUCLEAR GENES INVOLVED IN MITOCHONDRIAL BIOGENESIS IN YEAST

Oxygen, normally signaling via heme, and carbon source are the two principal cues determining whether a yeast cell respires (see reviews by DeWinde and Grivell, 1993; Forsburg and Guarente, 1989a; Grivell, 1989). Oxygen and heme levels are correlated since several enzymes late in heme biosynthesis are oxygenases. Under anaerobic conditions yeast cells do not make heme, but accumulate porphyrins instead (Mattoon et al., 1979). Thus, heme levels reflect oxygen levels and heme usually acts to mediate oxygen induction of respiration (Forsburg and Guarente, 1989a). Catabolite or glucose repression ensures that the cell preferentially uses glucose as a carbon source. In glucose-repressed cells, genes encoding mitochondrial proteins involved in respiratory electron transport and ATP formation are not expressed since glucose is usually fermented and not respired (for a review see Gillham, 1978). Catabolite repression of enzymes early in the heme biosynthetic pathway has also been reported (see Forsburg and Guarente, 1989a). Glucose repression in yeast involves a complex cascade of events that has been competently reviewed by De Winde and Grivell (1993).

The HAP (heme activator protein) system consisting of the *HAP 1–4* genes (Table 17–1) is subject both to heme and carbon source regulation (DeWinde and Grivell, 1993; Forsburg and Guarente, 1989a). These genes encode proteins required to activate transcription of nuclear genes involved in respiration such as *CYC1* and *CYC7*. These duplicate genes encode the two isoforms of cytochrome *c*. Similarly, *COX5A* and *COX5B*, encoding the two isoforms of cytochrome oxidase subunit V, are under control of the HAP proteins. Under aerobic growth conditions *CYC1* and *COX5A* produce the vast majority of cytochrome *c* and cytochrome oxidase subunit V, respectively. Both these genes are activated by heme via the HAP1 protein. In contrast when cells are

Table 17–1. Nuclear Regulatory and Structural Genes Known to Be Involved in Heme/Oxygen and Carbon Control of Mitochondrial Biogenesis in Yeast

Gene	Product
ANB1	Anaerobically regulated gene of unknown function
COX4	Subunit IV of cytochrome c oxidase
COX5A	Subunit V_A of cytochrome c oxidase (major isoform)
COX5B	Subunit V_B of cytochrome c oxidase (minor isoform)
CYC1	Iso-1-cytochrome c
CYC7	Iso-2-cytochrome c
CYT1	Cytochrome $c1$
HAP1	Heme activator protein; positive transcriptional regulator of UAS1 of *CYC1* and other respiratory genes
HAP2,3,4	Heme activator protein; positive transcriptional regulators of UAS2 of *CYC1* and other respiratory genes
HEM1	δ-Aminolevulinate synthetase; a heme biosynthetic enzyme
REO1	Represses *COX5B* under aerobic conditions
ROX1	Regulation by oxygen; a negative transcriptional regulator of *ANB1* and *CYC7*

Adapted from Forsburg and Guarente (1989a).

anaerobically grown so that heme does not accumulate, expression of these two genes is turned off while *CYC7* and *COX5B* continue to be expressed.

There are two upstream activator sequences (UAS1 and UAS2) proximal to the *CYC1* structural gene (Fig. 17-1A). UAS1 is the target sequence for the HAP1 protein while UAS2 binds a heteromeric complex containing the HAP2, HAP3, and HAP4 proteins. The HAP1 protein binds to UAS1 in a heme-dependent manner consistent with the hypothesis that heme binding is required for HAP1 to activate *CYC1* transcription (Pfeifer *et al.*, 1987a,b). The N-terminal domain (residues 1-148) is responsible for sequence-specific DNA binding while a second domain (residues 245-445) responds to heme, and heme binding seems to be required for HAP1 binding to DNA (De Winde and Grivell, 1993; Forsburg and Guarante, 1990; Pfeifer *et al.*, 1989). In fact, the heme-binding domain possesses seven repeats of a peptide sequence that bear similarity to a metal or heme-binding site (De Winde and Grivell, 1993). HAP1 binding sits similar to UAS1 have been detected upstream of a number of genes encoding components of the electron transport chain, catalase, and genes involved in heme, sterol, and fatty acid biosynthesis (De Winde and Grivell, 1993).

HAP1 also binds to a sequence upstream of the *CYC7* gene with no obvious similarity to UAS1 with an affinity similar to its affinity for UAS1 (Pfeifer *et al.*, 1987a,b). By examining mutant derivatives of HAP1 Pfeifer *et al.* (1989) showed that the same 148 amino acid domain in HAP1 recognizes UAS1 and the *CYC7* binding sites. However, within this single domain, nonidentical sets of amino acids were found to recognize the two motifs. HAP1 bound to UAS1 stimulates transcription more than 300-fold with most of this increase being dependent on heme. Stimulation of *CYC7* transcription is at least 10-fold lower with the heme dependence being marginal (Pfeifer *et al.*, 1987a). Thus, the HAP1 protein appears to be down-regulated when bound to

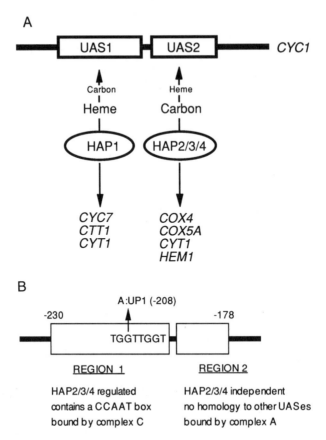

Fig. 17–1. The heme- and carbon-responsive (HAP) proteins of *S. cerevisiae* and their binding sites. (A) Organization of the upstream region of the yeast *CYC1* gene, encoding isocytochrome c. *CYC1* is regulated by two independent systems. The HAP1 system responds principally to heme, and has been shown to regulate *CYC7, CTT1,* and *CYT1* expression. The HAP2/3/4 system responds to a carbon source, such that cells deficient in any one of the three are petite, have reduced levels of all cytochromes and are unable to grow on a nonfermentable carbon source. HAP2/3/4 have also been shown to regulate *COX4, COX5A, CYT1,* and *HEM1* expression. From Forsburg and Guarante (1989a). (B) Organization of the UAS2 binding region. UAS2 consists of two regions. Region 1 is bound by complex C, known to contain HAP2 and HAP3. Region 1 extends from −230 to −200, relative to the upstream-most RNA start site. It contains the sequence TGGTTGGT, which possesses a G → A transition at −208 in the mutant UAS2UP1, increasing the homology to the HAP2 and HAP3 consensus TNATTGGT. This sequence encodes a CCAAT box on the opposite strand. Region 2, from −192 to −178, has little activity by itself, but substantially enhances the activity of region 1. Region 2 is bound by a HAP2/3-independent factor. From Forsburg and Guarante (1989b).

the *CYC7* site explaining why this gene is expressed less actively than *CYC1* under aerobic conditions. Forsburg and Guarante (1989a) propose a model in which the DNA sequence of *CYC7* recognized by HAP1 behaves like an allosteric effector, which causes the DNA binding domain of HAP1 to mask its activation domain.

The major form of control at the UAS2 sequence upstream of *CYC1* is a 50-fold induction when cells are shifted from glucose to lactate (Guarente et al., 1984). The UAS2 site spans about 65 nucleotides and includes two important regions (Forsburg and Guarente, 1988; Fig. 17–1B). Region 1 contains the sequence CCAAC, which is similar to the CCAAT box found in many eukaryotic promoters. Conversion of this sequence to a true CCAAT box by a GC → AT base pair change (Fig. 17–1B) increases the activity of UAS2 10-fold (Guarente et al., 1984). A similar consensus binding sequence (5′ACCAATNA3′) has been identified in the UAS regions of many nuclear genes encoding mitochondrial proteins (De Winde and Grivell, 1993). Region 2 is separated by 7 bp from region 1 and has no obvious similarity to other UASs. This region has little activity by itself, but substantially enhances the activity of region 1 (Forsburg and Guarente, 1988). Region 1 is the binding site for the HAP2/3/4 complex while region 2 is bound by a second protein complex that remains to be characterized.

The *HAP2* and *HAP3* genes are constitutively expressed, but *HAP4* transcription is induced when cells are shifted from glucose to lactate suggesting that HAP4 is probably the regulatory subunit of the complex (see Forsburg and Guarente, 1989b). HAP2 and HAP3 are required for sequence-specific DNA binding of the complex while HAP4 possesses the primary transcriptional activation domain. *HAP4* may be negatively regulated by the MIG1 receptor (De Winde and Grivell, 1993). the *MIG1* gene seems to be at the end of the regulatory cascade involved in glucose repression and negative regulation of the *HAP4* gene by the MIG1 repressor would permit glucose repression to act indirectly on many nuclear genes encoding mitochondrial proteins.

Since $hap2^-$ and $hap3^-$ mutants utilize ammonia poorly as a nitrogen source, Forsburg and Guarente (1989a) imagine that HAP2 and HAP3 are necessary to bind to an element common to UAS sequences upstream of catabolite repressible and ammonia utilization genes. However, HAP4 would be required for recognition of a second sequence element upstream of the former genes and an as yet uncharacterized equivalent of HAP4 would recognize a different element upstream of the ammonia utilization genes.

The four genes encoding the HAP proteins are not the only regulatory genes that respond to heme and carbon source. Transcription of the *ROX1* repressor gene is activated by heme (Lowry and Zitomer, 1988). Its product acts as a negative transcriptional regulator of the *CYC7* gene and *ANB1*. The latter gene is anaerobically regulated and encodes a protein that is homologous to a translational initiation factor, eIF4d (Lowry and Lieber, 1986; Lowry and Zitomer, 1984). Similarly, the *REO1* gene product represses expression of the *COX5B* gene under aerobic conditions (Trueblood et al., 1988). Hence, the anaerobically expressed *CYC7* and *COX5B* genes are repressible under aerobic conditions by the products of these regulatory genes. This is also true of the *HEM13* gene encoding coproporphyrinogen oxidase (Zagorec and Labbe-Bois, 1986). Lastly, multifunctional DNA binding proteins such as ABF1 and CPF1 play a role in regulating expression of nuclear genes encoding mitochondrial proteins (De Winde and Grivell, 1992; Trawick et al., 1992). These proteins each bind to different consensus sequences, but since these consensus sequences are associated with different genes the proteins are multifunctional. Thus, ABF1 binds to the ARS1B element, the silent mating type loci, and is involved in the initiation of DNA replication and transcriptional activation in yeast. ABF-1 is also required for efficient transcription of the yeast *QCR8* gene, which encodes an 11-kDa subunit of the ubiquinol–cytochrome *c* reductase complex.

CONTROL OF MITOCHONDRIAL GENE EXPRESSION BY THE NUCLEUS

Striking parallels exist between the ways in which nuclear genes control the expression of specific chloroplast and mitochondrial genes. Demonstration of these interactions has in both cases required the analyses to be performed in experimental systems where genetic analysis is readily done (i.e., *Chlamydomonas* and yeast). The relevant work on yeast has been discussed in several reviews (Costanzo and Fox, 1990; Fox, 1986; Fox et al., 1988; Grivell, 1989).

Posttranscriptional control of mitochondrial gene expression in yeast takes place at both the translational and mRNA processing levels and has been particularly well characterized in the case of the *cytb, cox2,* and *cox3* genes (Table 17–2). In the case of the *cytb* gene opportunities for regulation by nuclear genes also exist at the levels of mRNA splicing (see Chapter 10), mRNA processing, and translation.

The *CBP1* gene is the best characterized example of a nuclear gene whose product is involved in mRNA processing. The protein encoded by this gene is required for stabilization of the *cytb* mRNA. The *cytb* gene is cotranscribed with an upstream tRNA^{Glu} gene with the region between the two genes encompassing 1100 nucleotides (Christianson et al., 1983). The mature 5'-end of the *cytb* mRNA is formed by a cleavage of 145 nucleotides downstream of the tRNA^{Glu}, leaving a long (954 nucleotide) untranslated leader attached to the mature *cytb* mRNA (Dieckmann et al., 1984). In the *cbp1* mutants the tRNA^{Glu} and *cytb* genes are cotranscribed and the tRNA is processed out of the precursor. However, the remaining downstream sequence containing the *cytb* coding region is degraded (Dieckmann et al., 1982).

A mitochondrial suppressor of *cbp1* was isolated in which a chimeric mitochon-

Table 17–2. Nuclear Genes Controlling Processing and Translation of Specific Yeast Mitochondrial mRNAs

Target Mitochondrial mRNA for Nuclear Gene Product	Nuclear Gene	
	Designation	Putative Action of Gene Product
cytb	*CBP1*	5'-Processing of mRNA
	CBP3	Translation
	CBP6	Translation
	CBS1 (MK2)	Translation
	CBS2	Translation
	CBP2	Splicing (bI5)
cox1	*MSS18*	Splicing (aI5B)[a]
	MSS51	Splicing (aI1, aI2, aI4, aI5)
cox2	*PET111*	Translation
	PET112	Translation?
	SCO1	Translation?
cox3	*PET54*	Translation
	PET122	Translation
	PET494	Translation

See Attardi and Schatz (1988) and Costanzo and Fox (1990) for original papers.

[a]Table 10–2 also lists nuclear genes whose products are required for splicing, including those splicing introns in more than one gene.

drial gene fusion attached the 5' leader sequence from *atp9* upstream of *cytb* (Dieckmann et. al., 1984). The *atp9* leader of the resulting mRNA does not require 5' processing, as the transcription start site forms the mature 5'-end of the mRNA. Since the resulting chimeric mRNA was stable, the 5' leader of the *cytb* mRNA must be the target of the *CBP1* gene product.

Such mitochondrial suppressor mutations have proved invaluable in targeting the region with which the nuclear gene product interacts not only in the case of the *CBP1* gene, but also in the case of the *cox* genes. In each case the suppressor mutation results from a deletion that fuses the mRNA leader sequence of another mitochondrial gene (e.g., *atp9*) to the mitochondrial gene that is not expressed in the nuclear mutant (e.g., *cytb*). This permits expression of the mitochondrial gene making the suppressed stock respiratory competent. However, respiratory competence also requires that the suppressed strain be maintained as a heteroplasmon since some mitochondrial genomes in each cell must retain an intact copy of the gene (e.g., *atp9*) whose leader sequence has been deleted.

Subsequently, Dieckmann and Mittelmeier (1987) narrowed the site of action of the *CBP1* gene product to the 5' third of the *cytb* leader. They obtained a strain containing a tribrid gene in which only the 3' half of the leader region of the *cytb* gene was replaced by leader sequence from *atp9*. This *cytb–atp9–cytb* tribrid gene produced a stable mRNA that was translated in a *CBP1* strain, but not in a *cbp1* mutant. These results demonstrate that the *CBP1* gene product must interact with the 5' half of the leader region.

The *CBP1* gene is itself transcriptionally regulated (Mayer and Dieckmann, 1989). Two different transcripts of the gene are synthesized that share common 5'-ends. The 2.2-kb transcript is required for respiration and encodes the entire protein, but the 1.3-kb transcript terminates in the middle of the *CBP1* coding sequence. Paradoxically, the former transcript *decreases* in amount on glucose derepression. Mayer and Dieckmann (1989) suppose that the functional 2.2-kb transcript is required in highest amount at the moment derepression begins because cytochrome *b* synthesis is maximal at this time. The transcript is needed in lower amounts later as a steady-state level of cytochrome *b* is reached.

Several other nuclear gene products are also required for translation of the *cytb* mRNA (Table 17–2). One of these is the product of the *CBS1* gene (Roedel, 1986; Roedel *et al.*, 1985). Although the *cbs1* mutant strain accumulates unspliced cytochrome *b* mRNA, the primary defect is at the level of translation since mitochondrial suppressor mutations fuse the *atp9* 5' leader sequence to the *cytb* structural gene. Interestingly, a mitochondrial suppressor mutation of the nuclear *pet494* mutant that is blocked in translation of *cox3* fuses the *cytb* 5' leader sequence to the *cox3* structural gene (Roedel and Fox, 1987). Translation of the chimeric *cox3* transcript is now under the control of *CBS1*. Both these results indicate that the *CBS1* gene product, like the *CBP1* gene product, acts on the 5' untranslated leader of the *cytb* gene. The accumulation of the unspliced cytochrome *b* precursor in the *cbs1* mutant is probably a secondary effect resulting because the intron-encoded maturases, required for *cytb* mRNA splicing, also cannot be translated. *CBS2*, like *CBS1*, is also required for translation of *cytb* mRNA as demonstrated by chimeric gene experiments (Roedel, 1986). CBS1 seems to be a membrane component, while CBS2 appears to recognize a protein of the small subunit of the mitochondrial ribosome (Michaelis and Roedel, 1990; Michaelis *et al.*, 1991). A model has been proposed which supposes that the 5' untranslated region of *cytb* mRNA associates with CBS1 at the mitochondrial inner membrane surface while CBS2 serves as a kind of bridge between the 5' untranslated region and the small subunit of the mitochondrial ribosome (Michaelis et al., 1991). A third nuclear gene, *CBP6*, also encodes a protein required posttranscriptionally for normal cytochrome *b* accumulation (Dieckmann and Tzagoloff, 1985).

Three nuclear genes (*PET54, PET122,* and *PET494*) have been identified that regulate translation of *cox3* mRNA (Costanzo and Fox, 1990; Table 17–2). Mutations in all three genes block accumulation of cytochrome oxidase subunit III, but none of them affects the steady-state level of *cox3* mRNA. A mitochondrial suppressor of a mutation in one of the genes *(PET494)* also suppressed mutations in the other two genes indicating that all three gene products targeted the 5'-untranslated leader of the *cox3* mRNA. The site(s) at which the *PET494, PET54,* and *PET122* gene products act has been mapped to the 5' two thirds (about 400 bases) of the *cox3* mRNA (Costanzo and Fox, 1990).

The foregoing observations led Costanzo and Fox (1990) to speculate that the proteins encoded by the three aforementioned genes might "mediate an interaction between the *cox3* mRNA leader and a component of the mitochondrial translation system." To test this hypothesis Fox and colleagues set out to isolate allele-specific suppressors of mutations in the *PET494, PET54,* and *PET122* genes. This search initially yielded suppressors of *pet122* mutations that affected two different proteins of the small subunit of the mitochondrial ribosome, one of which has been identified as MRP1 (Haffter et al., 1990, 1991; McMullin et al., 1990). The other mutations mapped in a newly discovered gene, *PET123*, which also encodes a small subunit r-protein. Haffter et al. (1991) speculated that the PET122 protein, together with the PET54 and PET494 proteins, may interact with the two r-proteins to activate translation of the *cox3* mRNA. Further suppressor analysis has allowed Haffter and Fox (1992) to extend this model to yet another small subunit r-protein, MRP17, plus a protein encoded by the *PET127* gene, which could be a unique component of the mitochondrial translation system that promotes accuracy of translational initiation.

Direct biochemical evidence on the localization of the three-protein complex required for *cox3* translation has been obtained using antisera specific for each protein (McMullin and Fox, 1993). These experiments revealed that PET122 and PET 494 are present in small quantities and associated with the inner membrane, but that only half of the PET 54 protein is membrane-bound, with the rest being found in the soluble phase of the mitochondrion. Since PET 54 is a bifunctional protein with two distinct functional domains, one of which is required to splice the *cox1* aI5B intron (Chapter 10), the soluble form of the protein may be required for intron-splicing, while the membrane-bound form is part of the *cox3* translational activation complex. Thus, the picture that emerges is similar to the one proposed above for the *cytb* mRNA. A membrane-bound complex of nuclear-encoded proteins seems to be required both for mitochondrial ribosome binding and to guide the 5' end of *cox3* mRNA to its correct destination in the inner membrane. In this panoply of interactions the role of PET122 seems to be to provide the link between the small subunit of the mitochondrial ribosome and the *cox3* 5' untranslated leader the results in correct translation initiation (Costanzo and Fox, 1993).

Translational control of *cox2* mRNA by nuclear gene products has also been investigated (Costanzo and Fox, 1990). The products of three nuclear genes are known to be required posttranscriptionally for expression of *cox2* mRNA. The role of *PET111* is best understood. Cytochrome oxidase subunit II is undetectable in *pet111* mutants although the corresponding mRNA is present in somewhat reduced amounts (Poutre and Fox, 1987). Mulero and Fox (1993) constructed chimeric genes in which the 5' untranslated leader of the *cox3* gene is fused precisely to the *cox2* structural gene and showed that *cox2* no longer requires the *PET111* gene product for translation. This demonstrates unequivocally that the PET111 protein facilitates translation of *cox2* mRNA by interacting with 5' untranslated leader sequence. Other experiments involving *in vitro*-generated mutations in the 5' untranslated leader sequence and mitochondrial transformation have defined more precisely the sequences important for translation (Mulero and Fox, 1993a). Furthermore, suppressor analysis has revealed that a mutation in the *PET111* gene will suppress a one base pair deletion in the 5' untranslated region of the *cox2* mRNA. These results provide additional evidence that the PET111 protein interacts productively with the 5' untranslated region of this mRNA. Like the three proteins required for *cox3* mRNA expression, PET111 also seems to be membrane-bound, further supporting the model that these nuclear-encoded translational activators are required both for ribosome-binding and to dock the mitochondrial mRNA at the appropriate berth on the inner membrane.

Yet another mitochondrial gene transcript should soon be well enough characterized to add further breadth to the emerging regulatory picture sketched above. The *AEP1* and *AEP2* gene products seem to be necessary respectively for the translation and maturation/stabilization of the *atp9* transcript (Finnegan *et al.*, 1991; Payne, *et al.*, 1991). The product of another nuclear gene, *NCA1*, also seems to be required for stabilization of *atp9* mRNA (Ziaja *et al.*, 1993). However, the deduced protein product of this gene bears strong resemblence to *AEP1*, except that the AEP1 protein has a longer C-terminal extension, raising the question of how distinct AEP1 and NCA1 really are.

How are these nuclear genes themselves regulated? One presumes that heme acting through the HAP system is often involved, but the *PET494* gene seems to be regulated directly by oxygen and not heme (Marykwas and Fox, 1989).

DO MITOCHONDRIA SIGNAL THE NUCLEUS?

At least two classes of mechanisms can be envisioned by which an organelle might signal the nucleus. First, there could be signals that tell the nucleus whether the organelle is dividing or else inform the nucleus about organelle genome copy number. Second, there might be signals from the organelle that relate to its state of development.

In yeast, an intriguing pathway has been identified from the mitochondrion to the nucleus by Butow and colleagues (Butow *et al.*, 1988; Parikh *et al.*, 1987) which has been dubbed retrograde communication (Liao and Butow, 1993; Shyjan and Butow, 1993). The initial experiments involved comparisons between wild-type cells, a mit^- mutant and ρ^0 and ρ^- mitochondrial petites. These experiments identified a class of nuclear transcripts that were more abundant in petites than in the mit^- mutant or wild-type (Butow *et al.*, 1988; Parikh *et al.*, 1987). The best characterized of these transcripts mapped to the nontranscribed spacer region of the nuclear rDNA repeat. More recently retrograde regulation of peroxisomal citrate synthase has been reported (Liao and Butow, 1993; Shyjan and Butow, 1993). In yeast the *CIT1* and *CIT2* genes encode the mitochondrial (CS1) and peroxisomal (CS2) forms of citrate lyase respectively. In ρ^o petites transcription of *CIT2* as well as enzyme activity is elevated by 6- to 30-fold compared to the isochromosomal ρ^+ strain. A cis-acting upstream activation site named $UAS_{r(retrograde)}$ has been identified. Furthermore, two previously unknown nuclear genes, *RTG1* and *RTG2*, whose products are required for retrograde *CIT2* expression have been discovered. One *(RTG1)* encodes a protein with similarity to a transcription factor while the other *(RTG2)* specifies a protein of unknown function. Shyjan and Butow (1993) argue that one function of retrograde regulation is to compensate for a mitochondrial deficiency by changes in nuclear gene expression. They point out that in yeast the mitochondrial tricarboxylic acid cycle and the peroxisomal glyoxalate cycle are linked, since common metabolites can be exchanged. Thus, disruption of *CIT1* leads to an increase in *CIT2* transcription and CS2 activity, which can partially compensate for the CS1 deficiency. Shyjan and Butow (1993) regard the upregulation of CS2 in ρ^o cells an extreme case in which "the attempt to compensate for mitochondrial deficiencies is a futile process."

Shyjan and Butow (1993) note that while interactions between chloroplasts (Chapter 16) and mitochondria can now be viewed as a dialogue, the nature of the signals transmitted by the two organelles and the means by which they modulate gene expression remain obscure. They do, however, point out that metabolites may be involved. One obvious example, of course, is heme (Forsburg and Guarente, 1989a). *HEM1* mutant strains cannot make heme and show a 100-fold reduction in expression of a *CYC1-lacZ* chimeric gene construct (Guarante and Mason, 1983). Exogenously added heme induces transcription of this gene. While we have seen that HAP1 is the major activator of genes that respond to heme, the mechanism by which heme exits the mitochondrion is unknown.

REFERENCES

Attardi, G. (1985). Animal mitochondrial DNA: An extreme example of genetic economy. *Int. Rev. Cytol.* 93: 93–145.

Attardi, G., and G. Schatz (1988). Biogenesis of mitochondria. *Annu. Rev. Cell Biol.* 4: 289–333.

Baldacci, G., and E. Zennaro (1982). Mitochondrial transcripts in glucose repressed cells of *Saccharomyces cerevisiae*. *Eur. J. Biochem.* 127: 411–416.

Biswas, T., and G.S. Getz (1988). Promoter-promoter interactions influencing transcription of the yeast mitochondrial gene, Oli1, coding for ATPase subunit 9: *cis* and *trans* effects. *J. Biol. Chem.* 263: 4844–4851.

Butow, R.A., Docherty, R., and V.S. Parikh (1988). A path from mitochondria to the yeast nucleus. *Phil. Trans. R. Soc. London B* 319: 127–133.

Christianson, T., Edwards, J.C., Mueller, D.M., and M. Rabinowitz (1983). Identification of a single transcriptional initiation site for the glutamic tRNA and *COB* genes in yeast mitochondria. *Proc. Natl. Acad. Sci. U.S.A.* 80: 5564–5568.

Costanzo, M.C., and T.D. Fox (1990). Control of mitochondrial gene expression in *Saccharomyces cerevisiae*. *Annu. Rev. Genet.* 24: 91–113.

Costanzo, M.C., and T.D. Fox (1993). Suppression of a defect in the 5' untranslated leader of mitochondrial *COX3* mRNA by a mutation affecting an mRNA-specific translational activator protein. *Mol. Cell. Biol.* 13: 4806–4813.

De Winde, J.H., and L.A. Grivell (1992). Global regulation of mitochondrial biogenesis in *Saccharomyces cerevisiae*: ABF1 and CPF1 play opposite roles in regulating expression of the *QCR8* gene, which encodes subunit VIII of the mitochondrial ubiquinol-cytochrome *c* oxidoreductase. *Mol. Cell. Biol.* 12: 2872–2883.

De Winde, J.H., and L.A. Grivell (1993). Global regulation of mitochondrial biogenesis in *Saccharomyces cerevisiae*. *Prog. in Nucl. Acids Res and Mol. Biol.* 46: 51–91.

Dieckmann, C.L., and T.M. Mittelmeier (1987). Nuclearly-encoded CBP1 interacts with the 5' end of mitochondrial cytochrome *b* pre-mRNA. *Curr. Genet.* 12: 391–397.

Dieckmann, C.L., and A. Tzagoloff (1985). Assembly of the mitochondrial membrane system: *CBP6*, a yeast nuclear gene necessary for the synthesis of cytochrome b. *J. Biol. Chem.* 260: 1513–1520.

Dieckmann, C.L., Pape, L.K., and A. Tzagoloff (1982). Identification and cloning of a yeast nuclear gene *(CBP1)* inolved in expression of mitochondrial cytochrome b. *Proc. Natl. Acad. Sci. U.S.A.* 79: 1805–1809.

Dieckmann, C.L., Koerner, T.J., and A. Tzagoloff (1984). Assembly of the mitochondrial membrane system: *CBP1*, a yeast nuclear gene necessary for synthesis of cytochrome b. *J. Biol. Chem.* 259: 4722–4731.

Finnegan, P.M., Payne, M.J., Keramidaris, E., and H.B. Lukins (1991). Characterization of a yeast nuclear gene, *AEP2*, required for accumulation of mitochondrial mRNA encoding subunit 9 of the ATP synthase. *Curr. Genet.* 20: 53–61.

Forsburg, S.L., and L. Guarente (1988). Mutational analysis of upstream activation site 2 of the *Saccharomyces cerevisiae CYC1* gene: A *HAP2-HAP3* responsive site. *Mol. Cell. Biol.* 8: 647–654.

Forsburg, S.L., and L. Guarente (1989a). Communication between mitochondria and the nucleus in regulation of cytochrome genes in the yeast *Saccharomyces cerevisiae*. *Annu. Rev. Cell. Biol.* 5: 153–180.

Forsburg, S.L., and L. Guarente (1989b). Identification and characterization of HAP4: A third component of the CCAAT-bound HAP2/HAP3 heteromer. *Genes Dev.* 3: 1166–1178.

Fox, T.D. (1986). Nuclear gene products required for translation of specific mitochondrially encoded mRNAs in yeast. *Trends Genet.* 2: 97–100.

Fox, T.D., Costanzo, M.C., Strick, C.A., Marykwas, D.L., Seaver, E.C., and J.K. Rosenthal (1988). Translational regulation of mitochondrial gene expression by nuclear genes of *Saccharomyces cerevisiae*. *Phil. Trans. R. Soc. London B* 319: 97–105.

Gillham, N.W. (1978). *Organelle Heredity*. Raven Press, New York.

Giudice, G. (1973). *Developmental Biology of Sea Urchin Embryos*. Academic Press, New York.

Grivell, L.A. (1989). Nucleo-mitochondrial interactions in yeast mitochondrial biogenesis. *Eur. J. Biochem.* 182: 477–493.

Guarente, L., and T. Mason (1983). Heme regulates transcription of the *CYC1* genein *S. cerevisiae* via an upstream activation site. *Cell* 32: 1279–1286.

Guarente, L., Lalonde, B., Gifford, P., and E. Alani (1984). Distinctly regulated tandem upstream activation sites mediate catabolite repression of the *CYC1* gene of *S. cerevisiae*. *Cell* 36: 503–511.

Haffter, P., and T.D. Fox (1992). Suppression of carboxy-terminal truncations of the yeast mitochondrial mRNA-specific translational activator PET122 by mutations in two new genes *MRP17* and *PET127*. *Mol. Gen. Genet.* 235: 64–73.

Haffter, P., McMullin, T.W., and T.D. Fox (1990). A genetic link between an mRNA-specific translational activator and the translation system in yeast mitochondria. *Genetics* 125: 495–503.

Haffter, P., McMullin, T.W., and T.D. Fox (1991). Functional interactions among two yeast mitochondrial ribosomal proteins and an mRNA-specific translational activator. *Genetics* 127: 319–326.

Kruse, B., Narasimhan, N., and G. Attardi (1989). Termination of transcription in human mitochondria: Identification and purification of a DNA binding protein factor that promotes termination. *Cell* 58: 391–397.

Lansman, R.A., and D.A. Clayton (1975). Mitochondrial protein synthesis in mouse L-cells: Effect of selective nicking of mitochondrial DNA. *J. Mol. Biol.* 99: 777–793.

Liao, X., and R.A. Butow (1993). *RTG1* and *RTG2*: two yeast genes required for a novel path of communication from mitochondria to the nucleus. *Cell* 72: 61–71.

Lowry, C.V., and R.H. Lieber (1986). Negative regulation of the *Saccharomyces cerevisiae ANB1* gene by heme as mediated by *ROX1* gene product. *Mol. Cell. Biol.* 6: 4145–4148.

Lowry, C.V., and R.S. Zitomer (1984). Oxygen regulation of anaerobic and aerobic genes mediated by a common factor in yeast. *Proc. Natl. Acad. U.S.A.* 81: 6129–6133.

Lowry, C.V., and R.S. Zitomer (1988). *ROX1* encodes a heme-induced repression factor regulating *ANB1* and *CYC7* of *Saccharomyces cerevisiae*. *Mol. Cell. Biol.* 8: 4651–4658.

Marykwas, D.L., and T.D. Fox (1989). Control of the *Saccharomyces cerevisiae* regulatory gene *PET494*: Transcriptional repression by glucose and translational induction by oxygen. *Mol. Cell. Biol.* 9: 484–491.

Mattoon, J.R., Lancashire, W.E., Sanders, H.K., Carvajal, E., Malamud, D.R. et al. (1979). Oxygen and catabolite regulation in yeast. In *Biochemical and Clinical Aspects of Oxygen* (W.J. Caughey, ed.), pp. 421–435. Academic Press, New York.

Mayer, S.A., and C.L. Dieckmann (1989). The yeast *CBP1* gene produces two differentially regulated transcripts by alternative 3'-end formation. *Mol. Cell. Biol.* 9: 4161–4169.

Michaelis, U., and G. Roedel (1990). Identification of CBS2 as a mitochondrial protein in *Saccharomyces cerevisiae*. *Mol. Gen. Genet.* 223: 394–400.

Michaelis, U., Korte, A., and G. Roedel (1991). Association of cytochrome *b* translational activator proteins with the mitochondrial membrane: implications for cytochrome *b* expression in yeast. *Mol. Gen. Genet.* 230: 177–185.

McMullin, T.W., Haffter, P., and T.D. Fox (1990). A novel small subunit ribosomal protein interacts functionally with an mRNA-specific translational activator. *Mol. Cell. Biol.* 10: 4590–4595.

Mueller, D.M., and G.S. Getz (1986a). Transcriptional regulation of the mitochondrial genome of yeast *Saccharomyces cerevisiae*. *J. Biol. Chem.* 261: 11756–11764.

Mueller, D.M., and G.S. Getz (1986b). Steady state analysis of mitochondrial RNA after growth of yeast *Saccharomyces cerevisiae* under catabolite repression and derepression. *J. Biol. Chem.* 261: 11816–11822.

Mulero, J.J., and T.D. Fox (1993). *PET111* acts in the 5' leader of the *Saccharomyces cerevisiae* mitochondrial *COX2* mRNA to promote its translation. *Genetics* 133: 509–516.

Mulero, J.J. and T.D. Fox (1993a). Alteration of the *Saccharomyces cerevisiae COX2* mRNA 5'-untranslated leader by mitochondrial gene replacement and functional interaction with translational activator protein PET111. *Mol. Biol. of the Cell*, 4: 1327–1335.

Nelson, B.D. (1987). Biogenesis of mammalian mitochondria. *Curr. Topics Bioenerget.* 15: 222–272.

Payne, M.J., Schweizer, E., and H.B. Lukins (1991). Properties of two nuclear *pet* mutants affecting expression of the mitochondrial *oliI* gene of *Saccharomyces cerevisiae*. *Curr. Genet.* 19: 343–351.

Parikh, V.S., Morgan, M.M., Scott, R., Clements, L.S., and R.A. Butow (1987). The mitochondrial genotype can influence nuclear gene expression in yeast. *Science* 235: 576–580.

Pfeifer, K., Arcangioli, B., and L. Guarente (1987a). Yeast HAP1 activator competes with the factor RC2 for binding to the upstream activation site UAS1 of the *CYC1* gene. *Cell* 49: 9–18.

Pfeifer, K., Prezant, T., and L. Guarente (1987b). Yeast HAP1 activator binds to two upstream sites of different sequences. *Cell* 49: 19–27.

Pfeifer, K., Kim, K.-S., Kogan, S., and L. Guarente (1989). Functional dissection and sequence of the yeast HAP1 activator. *Cell* 56: 291–301.

Poutre, C.P., and T.D. Fox (1987). PET111, a *Saccharomyces cerevisiae* nuclear gene required for translation of the mitochondrial mRNA encoding cytochrome *c* oxidase subunit II. *Genetics* 115: 637–647.

Rinaldi, A.M., and G. Giudice (1985). Nuclear-cytoplasmic interactions in early development. In *Biology of Fertilization*, Vol. 3 (C.B. Metz and A. Monroy, eds.), pp. 367–377. Academic Press, Orlando, FL.

Roedel, G. (1986). Two yeast nuclear genes, *CBS1* and *CBS2*, are required for translation of mitochondrial transcripts bearing the 5'-untranslated *COB* leader. *Curr. Genet.* 11: 41–45.

Roedel, G., and T.D. Fox (1987). The yeast nuclear gene *CBS1* is required for translation of mitochondrial mRNAs bearing the *cob* 5'-untranslated leader. *Mol. Gen. Genet.* 206: 45–50.

Roedel, G., Korte, A., and F. Kaudewitz (1985). Mitochondrial suppression of a yeast nuclear mutation which affects the translation of the mitochondrial apocytochrome *b* transcript. *Curr. Genet.* 9: 641–648.

Shyjan, A., and R.A. Butow (1993). Intracellular Dialogue. *Curr. Biol.* 3: 398–400.

Trawick, J.D., Kraut, N., Simon, F.R., and R.O. Poyton (1992). Regulation of yeast *COX6* by the general transcription factor ABF1 and separate HAP2-and heme-responsive elements. *Mol. Cell. Biol.* 12: 2302–2314.

Trueblood, C.E., Wright, R.M., and R.O. Poyton (1988). Differential regulation of the two genes encoding *Saccharomyces cerevisiae* cytochrome *c* oxidase subunit V by heme and *HAP1* and *REO1* genes. *Mol. Cell. Biol.* 8: 4537–4540.

Zagorec, M., and R. Labbe-Bois (1986). Negative control of yeast coproporphyrinogen oxidase synthesis by heme and oxygen. *J. Biol. Chem.* 261: 2506–2509.

Zennaro, E., Grimaldi, L., Baldacci, G., and L. Frontali (1985). Mitochondrial transcription and processing of transcripts during release from glucose repression in "resting cells" of *Saccharomyces cerevisiae*. *Eur. J. Biochem.* 147: 191–196.

Ziaja, K., Michaelis, G., and T. Lisowsky (1993). Nuclear control of messenger RNA expression for mitochondrial ATPase subunit 9 in a new yeast mutant. *J. Mol. Biol.* 229: 909–916.

Index

Italic letter *t* denotes a table.

aadA bacterial gene, use in chloroplast transformation, 173
ABF1, encodes DNA binding protein, yeast, 401
ABF2, involvement in yeast mtDNA transcription, 273
Acanthamoeba, RNA editing, 325, 326
accD, land plant chloroplast genomes, 76, 79
Acetabularia
    blue light effects on photosynthesis, 366
    nucleoid transmission, 183
Achlya, mitochondrial genome, 70
acpA, algal chloroplast genomes, 81
Adiantum, chloroplast rRNA genes, 296
AEP genes, regulated *atp9* expression, yeast, 405
Agaricus, mitochondrial genome, 70
Agrobacterium
    rRNA phylogeny, 40
    Ti-mediated transformation, 172
Allomyces, mtDNA transmission, 185
Alternative oxidase
    Effects of fungal mitochondrial gene deletions, 243, 247
    Function, genetics, structure, 14–15
Alveolates, phylogeny, 39
Amaranthus, transfer of *psbA* to nucleus, 171–172
Amia, mtDNA evolution, 133–135
Ammodramus, mtDNA evolution, 136–137
ANB1, control of expression, yeast, 401
Anodonta, mtDNA transmission, 184
Anacystis, tRNA^{Ile} gene, 307
Antimycin A
    resistant mutants, yeast, 157*t*, 214
    respiratory inhibitor, 11
Apis, mtDNA and hybridization, 138
Arabidopsis
    cis-acting factors and nuclear gene expression, 383
    nuclear mutants affecting photocontrol, 379–380
    rbcS gene family, 373
    RecA gene and protein, 103, 106
Archaebacteria, phylogeny, 38, 39

Archezoa, 39
Ascaris. See Nematoda
Ascomycetes, filamentous. See also *Aspergillus; Neurospora; Podospora*
    mitochondrial genomes, 68–70
Aspergillus
    absence of mitochondrial *var-1* gene, 69
    mitochondrial genetics, 151–152, 168
    mitochondrial genome structure, 69–70
    mitochondrial rRNA phylogeny, 40
    nuclear location of *atp9*, 94
Astasia, plastic genome, 83
Asteridae, chloroplast DNA evolution, 140
atp synthase genes
    antibiotic resistant mutants, 156, 157*t*
    chloroplast genomes, 80, 100–101
    E. coli, 80
    flowering plant mitochondria, 65
atpA
    Chlamydomonas chloroplast, 80
    mRNA processing, 369
    translational control, 370
atpB
    Chlamydomonas chloroplast, 80
    Cuscuta chloroplast, 84
    initiation codon, 293
    phylogenetic studies, 140
    promoter, 282, 283
    transcript stability, 283
    transcription, 366–367
    transformation, 172, 224–225, 282–283
    translational control, 371
atpE
    Chlamydomonas chloroplast, 80
    Cuscuta chloroplast, 84
    mixed operon, 301
    phylogenetic studies, 140
    promoter, 282
    transcription, 366–367
atpF
    Chlamydomonas chloroplast, 80
    mRNA processing, 369

atpH
   Chlamydomonas chloroplast, 80
   mRNA processing, 369
atpI
   Chlamydomonas chloroplast, 80
   mixed operon, 301
   mRNA processing, 369
atp6
   cytoplasmic male sterility, 259, 262
   human mitochondrial NARP syndrome, 250
   location in different taxa, 58, 66, 72, 74, 101
   splicing in *Neurospora,* 218
atp8
   artificial transfer to nucleus, yeast, 171
   location in different taxa, 58, 66, 72
atp9
   involvement in cytoplasmic male sterility, 262–263
   location in different taxa, 66, 72, 94, 101
   regulation of expression, 405

Bacillariophyta, chloroplast structure, 16, 17*t*
Bacteriophage
   phage analogy model, 159, 202–203
   phage T4 introns, 210, 219, 225
Barley. *See Hordeum*
Beechdrops. *See Epifagus*
Bipolaris maydis toxin and CMS, 259, 262
Bos
   mitochondrial ribosomes, 301–302
   mtDNA segregation in heteroplasmons, 201–202
Bottlenecks and lineage extinction, mtDNA, 137–138
Bovines. *See Bos*
Bowfin. *See Amia*
Brassica. See also Raphanus
   chloroplast nucleoids, 113
   cytoplasmic male sterility, 263
   mitochondrial genome, 62
   mitochondrial transcripts, 280
Brettanomyces. See Dekkera/Brettanomyces
Broadbean. See *Vicia*
Brown algae. *See* Phaeophyta
Bryopsis, chloroplast nucleoids, 113, 183
5-bromodeoxyuridine, incorporation into mtDNA, 397

cab genes, 24*t*, 26*t*, 375*t*
   expression, 380–381, 383–384, 390–391
   nuclear gene families, 27, 375–376
Cabbage. *See Brassica*
Caenorhabditis. See Nematoda
Candida glabrata
   master mtDNA molecule, 198–199
   mitochondrial gene order, 101
   mitochondrial genome size and structure, 67, 68–69
   petite mutants, 156
   replication of mtDNA, 120
   tRNA processing, 310
   var1 gene, 69
CBP genes and *cytb* expression, 216, 398, 402–403
CBS genes and *cytb* expression, 403–404
ccsA (chlorophyll biosynthesis) gene, *Euglena* chloroplast genome, 81
Centrurus, mitochondrial partitioning, 199
Chaperones, in protein import, 342–347, 354, 356–357
Chara, chloroplast nucleoid transmission, 183
Charales, location of *tufA* gene, 94
Charophyta
   land plant ancestors, 93–94
   location of *tufA* gene, 94
Chicken. *See Gallus*
Chlamydomonas, rRNA phylogeny, 98
Chlamydomonas eugametos/moewusii
   chloroplast rRNA genes, 296
   introns in chloroplast rRNA genes, 98, 225
   mitochondrial genome, 65
   transmission of cpDNA in crosses, 164
Chlamydomonas monoica, control of chloroplast genome transmission, 188
Chlamydomonas reinhardtii/smithii. See also Genetics, chloroplast, *Chlamydomonas;* Genetics, mitochondrial *Chlamydomonas*
   chloroplast DNA deletion mutants, 85
   chloroplast DNA duplication mutations, 96
   chloroplast DNA molecules, number 51, 115*t*
   chloroplast DNA replication, 124
   chloroplast DNA, short dispersed repeats, 64, 96
   chloroplast heat shock protein, 338
   chloroplast nucleoid variation, 116
   chloroplast photoreactivating enzyme, 106
   chloroplast protein synthesis, essential, 85
   chloroplast ribosomal proteins, 299–301
   chloroplast rRNA genes, 296–298
   chloroplast sigma factors, 275
   chlororespiration, 27, 83
   discovery of cpDNA, 50
   fluorodeoxyuridine, as chloroplast mutagen, 152, 165
   fluorodeoxyuridine, reduction of chloroplast genome number, 367, 397
   introns in organelle genes, 65, 98–99, 210, 224–225
   life cycle, 161–163
   ndh genes absent in chloroplast, 83
   mitochondrial genome, 65
   mitochondrial rRNA genes, 65, 299
   mitochondrial tRNA genes, 65, 311

INDEX

proteolysis, 372
pyrenoid, 17
rbcS genes, 373
RUBISCO, small subunit import, 334
rRNA phylogeny of organelles, 47–48
topoisomerase, 367–368
transformation, chloroplasts, 172–174, 282–285, 369, 371
transformation, mitochondria, 175
trans-splicing of *psaA,* 80, 212, 219–221
tufA gene, 295
Chloramphenicol
    inhibitor experiments, 271–272
    mitochondrial resistance mutants, 159, 169, 222–223, 303, 157*t*
Chorella
    chloroplast rRNA gene organization, 296
    mitochondrial genome, 66
Chlorobiaceae, 15–16
Chlorophyll
    absorption shifts, 19
    antenna, 22
    chloroplast genes for synthesis in dark, 76–79
    protochlorophyllide, 19, 365
    reaction center, 22, 23, 26, 44
    taxonomic variation, 43*t,* 44–45
Chlorophyll binding (CAB) proteins, 22, 25, 27, 340, 359. *See also cab* genes.
Chloroplast envelopes
    number, 43*t,* 44–47
    protein sorting, 352
    protein targeting, 340
    receptors, 348
    structure, function, composition 15, 20–21
Chloroplast signal, 390–391
Chloroplast stroma, 15, 27–29
Chloroplast stroma lamellae, 15
Chloroplast thylakoids
    protein targeting, 340, 359
    protein targeting to lumen, 340, 357–358
    structure, 15, 21–27
    thylakoid bound polysomes, 304–305
Chloroplasts
    biogenesis, 17, 365–366
    chloroplast endoplasmic reticulum, 16
    classification. *See individual entries,* e.g., Chromoplast etc.)
    classification, plastid types, 16*t*
    grana, 15–19
    origin, 37–48
    partitioning, 199
    pigment composition, 43*t,* 44–45
    purification, 20
    structure/function, 15–29
Chloroplasts, algal
    bounding membranes, 43*t,* 45–47
    chloroplast endoplasmic reticulum, 16
    girdle bands, 16
    nucleomorph, 16, 47, 82
    pigment composition, 43*t,* 44–45
    pyrenoid, 17
    structural characteristics, 16–17, 17*t*
Chloroplasts, land plants
    development and structure, 16*t,* 17–19
    division, 114
Chlorophyta
    chloroplast pigments, 43*t*
    chloroplast structure, 16, 43*t*
Chlororespiration, 27, 76, 83
Chlorosome, 16
Chromatiaceae, 16
Chromatophore, 16
Chromista, 17
Chromophytes, chloroplast genes, 101
Chromoplast, 15, 16*t*
Chrysophyta
    chloroplast structure 16, 17*t,* 43*t*
    pigments, 43*t*
CIT genes, retrograde regulation, 405–406
ClpP, 76, 84
Cnemidophorus, mtDNA, 58
Cnidaria
    mitochondrial genome structure, 53, 58
    tRNA import, 311
Code (see Genetic Code)
Coleochaetales, location of *tufA* gene, 94
Conophilus, chloroplast genome, 84
Convolvulaceae. *See Cuscuta*
COP genes, photocontrol, 379
Coprinus, mtDNA transmission, 185
Coriander, chloroplast genome, 96
Coupling factors, 29–31, 30*t*
Cow. *See Bos*
cox genes, mtDNA of diverse groups, 54, 59, 65, 66, 72, 74
cox1
    editing, 326
    initiator codon, *Drosophila* mtDNA, 322
    introns, *Podospora,* 244–247
    intron splicing, *Saccharomyces,* 213–217, 225
    mitochondrial transformation, 174–175
    Saccharomyces oxi3, 157
    mtDNA of diverse groups, 66, 72, 74, 97
cox2
    absence from *Chlamydomonas* mitochondrial genome, 65
    cytoplasmic male sterility, chimeric involvement, 262–263
    editing, 326
    expression, nuclear gene control, 404–405
    mitochondrial transformation, 93, 175
    mRNA leader, 294

cox2 (*continued*)
    mtDNA of diverse groups, 72
    NCS mutational alteration, 257
    Paramecium, mtDNA, 72
    rate of evolution, 106
    Saccharomyces, *oxi1*, 157
    transfer to nucleus in certain plants, 94–95
cox 3
    absence from certain mitochondrial genomes, 65, 72
    expression, control by nuclear genes, 404
    Plasmodium mitochondrial genome, 74
    rate of evolution, 104
    Saccharomyces oxi2, 157
COX4, 263, 341, 355
COX5A, 398
COX5B, 398–399
CPF1, encodes mtDNA binding protein, 401
Crithidia
    kDNA, 73
    RNA editing, 328
    rRNA gene fragmentation, 299
Cryptomonas
    chloroplast gene content, 82
    evolution, 47, 82
    absence of chloroplast *ndh* genes, 83
    nucleomorph, 16, 47, 82
    rRNA phylogeny, 47
Cryptophyta (Cryptomonads)
    chloroplast structure, 16, 43*t*
    pigments, 43*t*
Cucumber. *See* Cucurbitaceae
Cucurbitaceae
    mitochondrial genomes, 61–62
    mitochondrial transcripts, 280
Cuscuta, chloroplast genome, 84–85
Cyanobacteria
    evolution of plastids, 37, 40
    photosynthetic apparatus, 16
    photosynthetic pigments, 16, 44
    tRNA^{Leu} intron, 97, 99
Cyanelle
    absence of *ndh* genes, 83
    genome, 60*t*, 81–82, 101
    ribosomal protein genes, 60*t*, 300
    tRNA^{Leu} intron, 97–98
    pigments, 44
    structure, 45–47
Cyanidium, chloroplast nucleoids, 112–114
Cyanophora. *See also* Cyanelle
Cybrids, 169, 190–191
CYC genes, 13, 359–360, 398–401
Cychloheximide
    import experiments, 334
    inhibitor of cytoplasmic protein synthesis, 271–272, 303*t*
Cyt4, *Neurospora*, 218

Cyt18, intron splicing in *Neurospora*, 218–219
Cyt19, *Neurospora*, 218
cytb
    control of expression of *Saccharomyces*, 398, 402–404
    deletion in *Chlamydomonas dum-1*, 168
    human LHON syndrome, 249–250
    initiation codon, maize, 294
    intron splicing, 213–217, 218, 225
    in various mitochondrial genomes, 54, 59, 65, 66, 72, 74, 97
    mitochondrial transformation, *Chlamydomonas*, 175
    mRNA leader, 294
    mutations in yeast, 156, 157*t*
    rate of evolution, 106
Cytochrome *b*, 11, 12*t*
Cytochrome b_2, sorting, 352–357
Cytochrome b_{559}, 23–24
Cytochrome b_6, 12, 25
Cytochrome *c*, 11, 13. *See also CYC* genes
Cytochrome c_1
    complex III, 11, 12*t*
    sorting, 352–357
Cytochrome oxidase, 13–14, 13*t*. *See also cox* genes
Cytochrome *f*
    function, 12, 25
    sorting, 358
Cytoplasmic male sterility
    maize, 231–233, 258–262
    other plants, 263
    Petunia, 262–263

Dekkera/Brettanomyces
    mitochondrial AUA codon, 323
    mitochondrial genome size and organization, 53, 68, 69, 101
Desulfovibrio, protein sorting, 358
DET genes, photocontrol, 379–380
Diatom. *See* Bacillariophyta
Dictyosphaeria, nucleoid transmission, 183
Dinoflagellata
    membranes, 320*t*
    pigments, 320*t*
    relationship to *Plasmodium*, 39
Dionaea, chloroplast DNA evolution, 140
Diplomonads, 39
DNA, chloroplast, *See also* Genomes, chloroplast
    buoyant density, 50
    discovery, 50
    fossils, 141–143
    membrane attachment, 115
    molecules per nucleoid, 115*t*
    phylogeny, 139–141
    molecules per plastid, 51, 115*t*
    replication, 123–124

rates of evolution, 103t, 106
size and conformation, 51, 77t
DNA, mitochondria. *See also* Genomes, mitochondrial
dead and fossilized material, 141–142
discovery, 53
diseases involving. *See* Mitochondrial disease
Eve hypothesis, 128–132
human migration studies, 137
evolutionary studies, 128–139
kinetoplast DNA. *See* Genomes, kinetoplast
molecules per mitochondrion, 11, 115t
molecules per nucleoid, 115t
rates of evolution, 103–106, 103t
repair, 102–103
replication, 116–123
size and conformation, 50–51, 52t–53t, 54t
DNA, nuclear, 5-methylcytosine in plants, 50
dnaK
Cryptomonas chloroplast genome, 82
encodes a chaperone for import, 344–345
Dryopteris, chloroplast rRNA gene organization, 296–297
Drosophila
mitochondrial heteroplasmons, 201
mitochondrial initiation and termination codons, 278, 322
rate of evolution of mtDNA, 104, 106
replication of mtDNA, 120

Ectocarpus, chloroplast nucleoid, 113
Editing. *See* RNA editing
Electron transport. *See* Mitochondrial and Photosynthetic electron transport
Endonucleases
chloroplast mRNA processing, 283
intron encoded, 65, 68, 221, 223
Envelope. *See* Chloroplast envelopes
Epifagus
chloroplast genome structure and gene content, 76, 83–84
chloroplast gene promoters, 275
infA present in chloroplast genome, 292
RNA polymerase genes, absence in chloroplast genome, 275
rpl22 absent from chloroplast genome, 301
Epilobium, plastic segregation, 150
Equus, mtDNA phylogeny, 141
Erythromycin
chloroplast mutations, to resistance, 303, 304t
chloroplast ribosomal proteins, 301
mitochondrial mutations to resistance, 157t, 159, 169, 222–223, 303
Escherichia coli
atp synthase genes, 101
chaperones, 342–347
C1p protease, 76

ribosomal protein, genes, 59–61, 60t, 79–80, 295, 299–301
RNase P, 278
RNA polymerase and *rpo* gene organization, 273–276
rRNA genes, 296
rRNA phylogeny, 40
Etioplast, 16t, 18–19
Euglena
bleached mutants, 86
chloroplast genome structure and gene content, 51, 81
chloroplast DNA introns, 81, 97, 99, 212–213
chloroplast DNA replication, 124
chloroplast ribosomal proteins and genes, 60t, 81, 299
chloroplast rRNA genes, 81, 296
cytoplasmic rRNA gene fragmentation, 299
factors involved in chloroplast protein synthesis, 292–293, 295, 372–373
infA absence, chloroplast genome, 292
TAC polymerase, 274
tRNA deficiency in chloroplasts, 321
Euglenophyta
chloroplast structure, 16, 17t, 43t
pigments, 43t, 44
Euglenozoa, 39
Eve hypothesis, 128–132
Evening primrose. *See Oenothera*
Exonuclease, chloroplast mRNA processing, 283
ezy-1, and chloroplast gene transmission, *Chlamydomonas,* 189

F34, F64, *psbC* expression, *Chlamydomonas,* 386–388, 386t
F54, *atpA* expression, *Chlamydomonas,* 386t, 388
Ferredoxin, 26
Fireweed. *See Epilobium*
5-fluorodeoxyuridine
chloroplast DNA reduction, *Chlamydomonas,* 367, 370, 397
chloroplast gene mutagen, *Chlamydomonas,* 152, 165
Fossils, organelle DNA from, 141–143
Four-O'clock. *See Mirabilis*
French bean. *See Phaseolus*
Fruitfly. *See Drosophila*
FUM1, dual function, *Saccharomyces,* 311

Gallus, mtDNA sequence, 58
Gamophyta
membranes, 43t, 44
pigments, 43t
GE2.10, *psbB* expression, *Chlamydomonas,* 386
Gene expression, chloroplasts
control by nuclear genes, 385–390, 386t
posttranscriptional control, 283–288, 368–370

Gene expression, chloroplasts (*continued*)
  role of proteolysis, 372
  translational control, 370–372
  transcriptional control, 366–368
Gene expression, mitochondrion
  control by nuclear genes, 402–405, 402*t*
  posttranscriptional control, 278, 282
  regulation in animal cells, 396–397
  regulation in *Saccharomyces*, 397–398
Gene expression, nucleus
  chloroplast signal, 390–391
  genes encoding chloroplast proteins, 373–384, 282*t*
  genes encoding mitochondrial proteins, 398–402, 399*t*
  mitochondrial signal, 405–406
Gene expression, organelles, control mechanisms, 364–365
Gene transfers
  mitochondrion to nucleus, 94–95
  nucleus to mitochondrion, 95
  plastid to mitochondrion, 95
  transfer to nucleus by gene manipulation, 171–172
  plastic to nucleus, 93–94
Genes, bifunctional, 217–219, 310–311, 313, 404
Genetic code, 320*t*
  chloroplasts, 321
  mitochondria, 322–323, 322*t*
  families and nonfamilies, 319, 321
  two-out-of-three, 319
  wobble, classical and four way, 319, 321
Genetics, chloroplasts
  chloroplast gene transmission, 149–150, 170, 181–184, 182*t*, 186–189
  chloroplast gene transmission in somatic fusions, 170–171, 190–191
  chloroplast mutations, 164–165, 170
  chloroplast-nuclear gene compatibility, 190–191
  heteroplasmon defined, 149
  homoplasmon defined, 149
  chloroplast gene recombination, 102–103, 167
  somatic (vegetative) segregation, defined, 149
  transformation, 172–174, 282–285, 369, 371
Genetics, chloroplasts, *Chlamydomonas*
  chloroplast gene complementation, 152, 165
  chloroplast gene mapping, 165–167
  chloroplast gene mutations, 164–165, 303–304, 304*t*
  chloroplast gene transformation, 172–174, 282–285, 369, 371
  chloroplast gene transmission in crosses, 152, 163–164, 186–189
  fluorodeoxyuridine as chloroplast gene mutagen, 152, 165
  intron mobility, 224–225

nucleoid segregation model, 198
persistent heteroplasmy, 200
phage analogy model, 202–203
trans-splicing, 212, 219–221
Genetics, mitochondrial, gene transmission, 184–186
Genetics, mitochondrial, *Aspergillus*, 151–152, 168, 247
Genetics, mitochondrial, *Chlamydomonas*, 164, 167–168, 175, 186–188
Genetics, mitochondrial, human, 169–170
  mitochondrial injection experiments, 169–170, 252
  mutations and disease, 169, 249–257, 251*t*, 276–277
  threshold effect, 169, 252
Genetics, mitochondrial, land plants
  cytoplasmic male sterility, 258–262
  mitochondrial gene recombination, 61–65, 170
  mitochondrial gene transmission in crosses, 183
  mitochondrial gene transmission in somatic hybrids, 170–171
  trans-splicing of mitochondrial genes, 221
Genetics, mitochondrial, mammals (except humans), 169
Genetics, mitochondrial, *Neurospora*
  intron splicing, 217–219
  mutations and disease, 151, 168, 240–243, 247
  plasmids, 228–231, 247–248
Genetics, mitochondrial, *Paramecium*, 168, 190
Genetics, mitochondrial, *Podospora*
  senescence, 152, 243–247
Genetics, mitochondrial, *Saccharomyces*. See also *Saccharomyces cerevisiae*
  complementation, 151, 160–161
  control of mitochondrial gene expression, 398, 402–405
  control of nuclear genes involved in mitochondrial biogenesis, 398–402
  crosses, 157–159
  G+C cluster mobility, 222, 227–228
  intron mobility, 221–223, 223*t*
  intron splicing, 213–217, 214*t*, 223*t*
  mutant induction with Mn, 151, 157
  nucleoid segregation, 198
  origin of mitochondrial mutations, 195–196
  phage analogy model, 158–159, 202–203
  physical mapping of mitochondrial genes, 159–160, 159*t*
  polarity, 159
  retrograde regulation, 405–406
  transformation, 151, 174–175
  var-1 mobility, 222, 226–227
Genetics, mitochondrial, *Saccharomyces*, mutants except petites

INDEX

antibiotic resistant, 150–151, 156, 157t, 159, 214, 222, 303–304
box, 157t, 213–217
cob, 157t, 214
mit, 156, 159
other, 157t, 215, 223
syn, 156, 159
Genetics, mitochondrial, *Saccharomyces,* petites
classification, 150, 154–156
deletion mapping, 160
marker retention analysis, 151, 160
mtDNA organization, 151, 154–156
petite negative yeast, 165
Genetics, organelles
compatibility of organelle and nuclear genes, 190–191
heteroplasmy, 199–202
hypothesis or intragenomic conflict, 189–190
master molecules, 120, 198–199
origin of mutations, 195–196
partitioning of organelles, 199
phage analogy model, 158–159, 202–203
population genetics, 203–204
relaxed control, 197
segregating unit, 196–199
Genomes, chloroplast
fossils, 142–143
phylogenetic uses, 139–141
rates of evolution, 106, 103t
recombination, 102–103, 167
replication, 124
structure and function, 60t, 75–86, 77t, 95–96, 97–99, 100–101
Genomes, chloroplast, algal. *See also individual taxa*
gene content, 60t, 80–83
size and structure, 75–80, 77t
Genomes, chloroplast, colorless plants, 83–86
Genomes, chloroplast, land plants, *See individual taxa*
copy correction, 76, 106
gene content, 60t, 76–80
inverted repeat 76
rates of evolution, 103t, 106
recombination, 102–103
size and structure, 75–80, 77t
Genomes, kinetoplast
genes and structure, 73–74
replication, 121–123
Genomes, mitochondrial
conformation and size, 52t–53t
dead material and fossils, 141–142
gene content, 55t, 56t
rates of evolution, 103–106, 103t
replication, 116–123
Genomes, mitochondrial, algae, 65–66, 97

Genomes, mitochondrial, animals
copy number, 115t, 116
evolution, 128–139
gene content, 54–58, 97, 99
recombination, lack of, 102
repair, 104–105
replication, 116–120
size and structure, 50, 51–59, 52t–53t, 99, 101
Genomes, mitochondrial, Ascomycetes
recombination, 101, 102–103
replication, 120–121
size and structure, 54t, 66–70, 97, 99
Genomes, mitochondrial, Basidiomycetes, 70–71
Genomes, mitochondrial, Hypochytridiomycetes, 70
Genomes, mitochondrial, land plants
rates of evolution, 103t, 106
recombination, 62–64, 102
size and structure, 52t–53t, 59–65, 96, 97, 99, 101
Genomes, mitochondrial, Oomycetes, 70, 103, 120
Genomes, mitochondrial, protists (also Genomes, kinetoplast)
replication, 121–123
size and structure, 72–74
Genomes, organelle
changes in genome size and composition, 95–100
changes in genome organization, 100–102
rates of evolution, 103–106, 103t
recombination, 102–103
Geranium. *See* Pelargonium
Giardia, 39
Glycine, 281, 325
Green algae. *See* Chlorophyta
GroE operon, 346
Ground sloth. *See* Mylodon
Gryllus, mtDNA, 58
Guide RNAs, 325, 328–330
GUS (B-glucoronidase. *See uidA*
Gymnosperms
chlorophyll synthesis in the dark, 76–77
chloroplast genomic rearrangements, 100
short direct repeats in chloroplast genome, 96, 102
transmission of organelle genomes in crosses, 183

Hadrurus, mitochondrial partitioning, 199
HAP genes, 398–401
Heliobacterium, 44–45
Heliothis, no paternal mtDNA transmission, 184
HEM genes, 401, 406
Heme, role in regulating yeast mitochondrial biogenesis, 398–401, 406
Herpetomonas, RNA editing, 330

Heteroplasmons, 149, 199–202
hisH, Cyanophora chloroplast genome, 81
hlpA, Cryptomonas chloroplast genome, 82
Hominoids
 Eve hypothesis, 128–132
 mitochondrial diseases, 249–257
 mitochondrial DNA evolution, 128–132
 mitochondrial DNA and migration, 137
 mitochondrial DNA from fossils, 141–142
 mitochondrial DNA transcription, 272–273
Homoplasmon, 149
Honeybee. *See Apis*
Hordeum
 chloroplast gene transcription, 366, 369
 chloroplast genes, translational control, 371–372
 chloroplast signal, 391
 nuclear control of chloroplast gene expression, 389
 ribosome-deficient plastids, 274–275
Horse. *See Equus*
HSP60, protein import into organelles, 345–347, 356–357
HSP70, 82, 344–345
HTS1, tRNA synthetase, 313
HY genes, photocontrol, 380
Hybridization, analysis using mitochondrial DNA, 138–139
Hydra, mitochondrial genome, 53
Hypochytridiomycetes, mitochondrial genome, 70

Import. *See Protein import*
ilv B, Cryptomonas chloroplast genome, 82
inf genes and chloroplast genomes, 81, 292
Intragenomic conflict, 189–190
Introns
 distribution in organelle genomes, 66–68, 80, 81, 97–99, 214*t*
 early versus late debate, 97–99
 involvement in senescence, 244–247
 mobile introns, 221–226, 223*t*, 228
 proteins that facilitate splicing, 213–219, 216*t*
 twintrons, 212–213
Introns, group I
 conservation of tRNA^{Leu} in different chloroplast genomes, 97, 99
 in different organelle genomes, 58, 61, 65, 80, 98–99, 213–219, 214*t*, 216*t*, 223*t*
 mobility, 221–226, 223*t*
 reverse transcriptase-like genes and proteins, 229–230
 splicing, 210
 structure, 208–209
Introns, group II
 comparison with nuclear pre-mRNA introns, 211–212
 in different organelle genomes, 61, 80, 98–99
 mobility, 228
 in purple bacteria, 99
 reverse transcriptase-like genes and proteins, 211, 216, 228, 244–246
 splicing, 211
 structure, 210–211
 trans-splicing, 80, 212, 219–221
Introns, group III, 212–213
Introns, nuclear pre-mRNA, 207–208, 211–212
Introns, nuclear tRNA genes, 208

Kinetoplast. *See Genomes, kinetoplast.*
 structure, 73
Kloeckeria, mitochondrial genome, 68–69
Kluyveromyces
 group II intron mobility, 228
 petite negative, 169

Leguminoseae, chloroplast genome
 absence of inverted repeat in Fabaceae, 76
 chloroplast genomic inversions, 76, 100
 rpl22 gene transfer to nucleus, 94, 301
Leguminoseae, mitochondrial genome
 transfer of *cox2* to nucleus, 94–95
Leishmania
 kDNA, 73
 RNA editing, 326
Lemna, gene expression, 373, 378, 384
LEU4, dual function, 311
Leucoplast, 15, 16*t*
Lilium, organelle nucleoid transmission, 182–183
Lincomycin, resistant mutants, 85, 303*t*, 304*t*
Lipids of organelles, 5, 7, 20–21
Liverwort. *See Marchantia*
Lupus, mitochondrial DNA and hybridization, 138–139
Lycopersicon
 control of expression of *rbcS* genes, 376–378, 380–384
 nuclear gene *rbcS* and *cab* families, 373–375, 375*t*
 plastid transmission in crosses, 181–182
Lycopsida, chloroplast genome inversions, 100

Magnolia, chloroplast DNA from fossil species, 142–143
Marchantia
 chlorophyll synthesis in dark, 76–77
 chloroplast genome, gene content, organization, structure, 75–79, 96, 100, 216, 296, 319–320
 mitochondrial genome, gene content, organization, structure, 59–61, 303
 organelle introns, 61, 216
 organelle ribosomal protein gene organization, 60*t*
Maize. *See Zea mays*

INDEX

Marsiliea, loss of male plastids in spermatazoids, 183
MAS genes, mitochondrial peptide processing, 349
Maturases. *See* Introns, proteins that facilitate splicing
mbp genes, chloroplast genome of *Marchantia,* 76
Meloidogyne. See Nematoda
Melons. *See* Cucurbitaceae
Membranes. *See also* Chloroplast envelopes; Chloroplast thylakoids
Metridium
  mitochondrial genome, 58
  mitochondrial tRNA genes, 278
Microsporidia, 39
MIF4 encodes yeast chaperonin, 60, 346, 356–357
MIG1 and glucose repression, 401
MIP1
  encodes subunit of yeast mitochondrial DNA polymerase, 121
  yeast mitochondrial import gene, 349
Mirabilis, plastid transmission, 149
Mitochondria
  editing in. *See* RNA editing
  evolution, 38–48
  nucleoids. *See* Nucleoids, mitochondrial
  partitioning, 199
  structure and function, 6–15, 10*t,* 12*t,* 13*t,* 14*t*
Mitochondrial diseases
  flowering plants, 257–263
  fungi, 240–248
  humans 249–257, 251*t,* 276–277
Mitochondrial inner membrane
  membrane bound organelle ribosomes, 304
  protein targeting, 338, 340, 358–359
  structure and function, 7–15
  translocation contact sites, 348
Mitochondrial intermembrane space
  enzymes, 7
  protein sorting, 352–357
  protein targeting, 338
Mitochondrial matrix
  enzymes, 4–5, 15
  protein targeting, 340
Mitochondrial outer membrane
  presequence binding, 337
  protein sorting, 351–352
  receptors, 347–348
  structure and function, 7
  translocation contact sites, 348
Mobile elements in organelles, 221–228, 223*t*
MOD5, mitochondrial tRNA modification, 310
Mouse. *See Mus*
MRP genes, mitochondrial ribosomal proteins, 302–303, 404
MRS genes, intron splicing, 216*t,* 217
MSS genes, intron splicing, 216–217, 216*t*

MTF1, RNA polymerase specificity factor, 272
MTF2, mitochondrial IF2, 293
Mucidin, 157*t*
Mus
  mitochondrial genome, 51, 59
  mitochondrial rRNA phylogeny, 40
  paternal leakage of mitochondrial DNA, 184
Mussels. *See Anodonta; Mytilis*
Mustard. *See Sinapis*
Mutation, origin of organelle mutations, 195–196
Mycoplasma, UGA equals tryptophan, 322
Mylodon, mitochondrial DNA, 141
Mytilis, mitochondrial DNA transmission, 184
Myxothiazol, 11

nac genes, *psbD* expression, 386, 386*t*
NADH-ubiquinone reductase (Complex I), structure and function, 8–10, 10*t*
Nalidixic acid, 368
NAM1, dual function, 217
NAM2, dual function, 217
NAM9, mitochondrial ribosomal protein, 302
NCA, expression of *atp9,* 405
ncc1, atpA transcript stability, 388
ndh genes
  chloroplasts, 27, 76
  absence from certain chloroplast genomes, 81, 83
  editing in organelle genomes, 325–326
  mitochondria, 10*t*
  mitochondrial genomes of various taxa, 59, 65, 66, 72, 74
ndh1
  editing in maize, 325
  human LHON disease, 249–250
  trans-splicing in plant mitochondria, 221
  various mitochondrial genomes, 59, 101
ndh2
  deletion in stopper mutants, 242–243
  human LHON disease, 249
  maize mitochondria, 95
  trans-splicing in plant mitochondria, 221
ndh3
  adjacent of *pcf* in *Petunia,* 263
  deletion in stopper mutants, 242–243
  editing in *Petunia,* 325–326
  flowering plant mitochondria, 101
ndh4
  animal mitochondria, 58, 59
  deletion and death in *Chlamydomonas,* 168
  human LHON disease, 249–250
  NCS2 alteration in maize, 257
ndh4L, animal mitochondria, 58, 59
ndh5
  animal mitochondria, 58, 59
  human LHON disease, 249

ndh6
  human heteroplasmy, 202
  mammalian mitochondria, 58
ndh7
  editing, 327
  protozoan mitochondria, 72, 74
ndh8, Trypanosome mitochondria, 74
ndh11, *Paramecium* mitochondria, 72
Nematoda, mitochondrial genome structure, 58–59
  tRNA genes, 308–309
Nepenthes, chloroplast DNA evolution, 140
Neurospora
  membrane bound ribosomes in mitochondria, 304
  mitochondrial DNA inheritance, 185
  mitochondrial genome, 69–70
  mitochondrial genetics. *See* Genetics, mitochondrial, *Neurospora*
  mitochondrial intron mobility, 225
  mitochondrial intron splicing, 210, 217–219
  mitochondrial plasmids, 229–231, 247–248
  mitochondrial receptors, 347
  mitochondrial rRNA gene organization, 298–299
  mitochondrial transcript processing, 280
  mitochondrial tRNA synthetases, 313
  nuclear location of *atp9,* 94
  outer membrane targeting in mitochondria, 352
  protein transport components, Table 344*t*
  protein translocation channels, 349
  PstI palindromes, 280
  ribosomal protein encoded in mitochondria, 302
Nicotiana
  chloroplast genetics, 150, 170
  chloroplast genome, structure and gene content, 27, 75–76, 95–96
  chloroplast mutations to antibiotic resistance, 85, 303–304, 304*t*
  chloroplast nucleoid transmission, 182
  chloroplast protein synthesis, 85
  chloroplast *rpo* genes, 274
  chloroplast transcript processing, 368
  chloroplast transformation, 173–174, 369–370
  genomic compatibility in intergeneric crosses, 190–191
  nuclear genes encoding chloroplast proteins, 381, 384
  recombination of mitochondrial genome, 170
Nitella, mitochondrial nucleoid, 111
Norflurazon, effects on plants, 390–391
Novobiocin, topoisomerase inhibition, 368
Nucleoids
  DNA molecules per organelle and nucleoid, 115*t*
  organization and classification, 110–116
  segregation model, 198
  transmission in crosses, 182–183, 187–189
Nucleomorph, 16, 47, 82

Ochromonas, mitochondrial genome, 66
Oenothera
  biparental transmission of chloroplast genes, 170, 182
  gene transfer from nucleus to mitochondrion, 95
  mitochondrial promoter, 281
  plastome-genome compatibility, 150, 191
  trans-splicing of mitochondrial introns, 221
Olisthodiscus
  chloroplast genome, 83
  chloroplast partitioning, 199
OmpR, Cryptomonas chloroplast genome, 82
Oomycetes, mitochondrial genome, 70, 103, 120
Open reading frame (ORF or URF)
  cytoplasmic male sterility, 258–263
  various organelle genomes, 59, 65, 66, 73–74, 79, 84
Opisthacanthus, mitochondrial partitioning, 199
Organelle origins, 37–48
Orobranchaceae, chloroplast genomes, 83–84
Oryza
  chloroplast DNA deletions, 85–86
  chloroplast genome structure and gene content, 75–77, 95–96
  chloroplast *ndh* genes, 27
  chloroplast *rpoC2* gene, 274
Oxidative phosphorylation, 29–30

Pandorina, mitochondrial genome, 65
Paracoccus, 40, 357
Paramecium
  inheritance of mitochondrial DNA, 185
  intracellular selection of mitochondria, 196
  mitochondrial genome structure and gene content, 72–73
  mitochondrial genetics. *See* Genetics, mitochondrial, *Paramecium*
  mitochondrial DNA replication, 121
  mitochondrial-nuclear compatibility, 190
  mitochondrial nucleoids, 110–111
  mitochondrial ribosomal protein genes, 303
  mitochondrial rRNA phylogeny, 40
  mitochondrial tRNA gene deficiency, 72, 311
pcf, cytoplasmic male sterility, *Petunia,* 263
PCR. *See* Polymerase chain reaction
Pea. *See Pisum*
Pelargonium
  chloroplast genome structure, 76, 96, 100
  chloroplast genome transmission, 149, 182–183
  chloroplast rRNA gene organization, 298
PET genes, 154
PET54, functions in splicing and *cox3* expression, 217, 404
PET111, functions in *cox2* expression, 405
PET122, functions in *cox3* expression, 404

INDEX

PET123, mitochondrial ribosomal protein gene, 302, 404
PET494, functions in *cox3* expression, 404
petA
   absence from *Euglena* chloroplast genome, 81
   encodes cytochrome *f*, 25
petB
   encodes cytochrome b_6, 25
   polycistronic transcript, 285, 288
petC, encodes iron sulfur protein, 25
petD
   absence from *Euglena* chloroplast genome, 81
   encodes subunit IV of cytochrome b_6f complex, 25
   initiation codon, 293
   transcript processing, 283–285
petE, encodes subunit V of cytochrome b_6f complex, 25
petF, encodes subunit VI of cytochrome b_6f complex, 25
petG, encodes subunit VII of cytochrome b_6f complex, 25
Petite mutants, *See* Genetics, mitochondrial; *Saccharomyces*
Petunia
   cytoplasmic male sterility, 262–263
   genome-plastome incompatibility in crosses, 190–191
   plastid transmission in crosses, 182
   rbcS families, 375
   RNA editing, 325–326
   trans-splicing of mitochondrial introns, 221
Phaeophyta
   chloroplast genome, 82
   chloroplast structure, 16
   membranes and pigments, 43*t*
Phage analogy model, 158–159, 202–203
Phaseolus, mitochondrial tRNAs, 309
Photophosphorylation, 29–31
Photoreactivation, 105, 106
Photoreceptors, 19, 365–366, 379–380
Photosynthesis
   C_3 and C_4 pathways, 19–20
   electron transport, 21–27
Photosynthetic bacteria. 15–16. *See also* Chlorobiaceae; Chromatiaceae; Cyanobacteria; Rhodospirillaceae
Photosystem I
   genes, 26*t*
   structure and function, 26–27
Photosystem II
   genes, 24*t*, 385*t*
   structure and function, 22–25
Phycobilisome, 16
Phycobiliproteins, 16, 81, 82
Phycomyces, 366

Phyllosticta maydis, 258–259
Phylogeography, 132–137
Phylogeny
   chloroplast DNA, 139–141
   fossil DNA in phyogeny, 141–143
   hominoid mtDNA, 128–132
   multigene, 40, 47
   rRNA phylogenies, 38–40
Physarum
   intron splicing, 210
   mitochondrial nucleoids, 110–111, 115
   mtDNA transmission in crosses, 186
   RNA editing, 330
Phytochrome, 19, 365, 379–380
Phytophora, mitochondrial genome, 70
PIF1, encodes mitochondrial helicase, 121
Pisum
   chloroplast outer membrane receptor complex, 348
   chloroplast transcript processing, 369
   cpDNA replication, 123–124
   mitochondrial genome, 61
   rbcS genes, 373
   RecA protein, 103
Pitcher plant. *See* Nepenthes
Plasmids, mitochondrial
   Maize, 231–233, 232*t*
   Neurospora, 229–231, 247–248
   Podospora, 244–247
Plasmodium
   organelle genomes, 74–75
   phylogeny, 39
Plastocyanin, 25
Plastome defined, 149
Pneumocystis, self-splicing intron, 210
Podospora
   mitochondrial genetics. *See* Genetics, mitochondrial; *Podospora*
   mitochondrial intron splicing by tRNA synthetase, 219
   senescence, 152, 243–247
Polymerase chain reaction (PCR)
   evolutionary studies, 128
   fossil DNA, 141–143
Polymerase, DNA
   chloroplast, 124
   mitochondrial, 104, 120
Polymerase, RNA
   chloroplast, 273–276, 368
   mitochondrial, 272–273
Polysphondylium
   mobility of nuclear rDNA intron, 225
   mtDNA transmission, 185
Polysomes. *See* Ribosomes
Polytoma, chloroplast genome, 83

Porphyra
  absence of chloroplast *ndh* genes, 83
  chloroplast genome and genes, 82, 116, 292
  chloroplast r-protein genes, 60t, 82, 300
  intron absence in chloroplast genome, 98–99
Potato. *See Solanum*
Presequences
  cleavage, 349–350
  evolution, 92–93
  structure and function, 334–337
Prochlorophytes, 44–45
Proplastid, 17, 113–114, 372
Protein import
  chaperone involvement. *See* Chaperones
  chloroplasts and proplastids compared, 372
  cotranslational versus posttranslational, 333–334
  cytochrome *c*, 359–360
  energetics, 350–351
  mechanisms, 342–351
  mitochondrial contact site involvement, 356
  mitochondrial import components, 344t
  organelle specificity, 341–342
  presequences. *See* Presequences
  protein translocation channels, 349
  receptors, 347–348
  routing and sorting sequences, 338–340
  sorting to different compartments, 351–359
  targeting signals, 334–335
Protein synthesis, organelles
  capping absence, mitochondria, 293
  elongation, 294–295, 372–373
  initiation, 292–294, 372
  essential in chloroplasts? 83–86
  Shine-Dalgarno sequences, 293–294
  termination, 295
Proteus vulgaris, rRNA phylogeny, 40
Protochloroplast, 19
Prototheca, mitochondrial genome, 66
psa genes, function 26t, 27
psaA
  operon organization, 301
  polysomes, 305
  translational control, 372
  trans-splicing in *Chlamydomonas*, 80, 212, 219–221
psaB
  mapping in *Chlamydomonas* chloroplast, 165, 166
  operon organization, 301
  polysomes, 305
  translational control, 371
psaI, absence from *Euglena* chloroplast genome, 81
psaM, presence in *Euglena* chloroplast genome, 81
psb genes
  function, 23–25, 24t
  in polycistronic transcripts, 285–288

psbA
  chloroplast transformation, 172
  expression, 369–370, 388, 390
  gene structure various species, 84, 98–99, 282
  herbicide resistance mutations, 166, 170
  phylogenetic reconstructions, 44, 47, 140
  polysomes, 305
  transfer to nucleus, 171–172
psbB, expression, 368, 372, 386
psbC, expression, 366, 372, 386–388
psbD, expression, 366, 369, 385–386
psbF, *Euglena* twintron, 211
psbJ, operon in *Chlamydomonas*, 301
psbM, absence in *Euglena* chloroplast genome, 81
psCL, nuclear genes specifying chloroplast r-proteins, 301
Psilopsida, chloroplast genome inversions, 100
Pteris, nucleoid transmission in crosses, 183
Puromycin, 303t, 334
Purple bacteria
  group II intron, 99
  origin of mitochondria, 40
Pylaiella, chloroplast genome, 82
Pyrenoid, 17

QCR8, transcriptional control, 401
Quinones, 8, 10–11, 22–23, 25, 26

Radish. *See Raphanus*
Raphanus. See also *Brassica*
  mitochondrial genome, 62
rbcL
  chloroplast genomes, various species, 76, 83, 84
  heteroplasmy, 200
  mapping in *Chlamydomonas*, 165, 166
  maize mitochondria, 95
  mutations, 170
  phylogeny, 47, 139–140, 142–143
  polysomes, 305
  transcription, 367
  translational control, 370
rbcS
  expression, 376–378, 380–384, 382t, 390–391
  gene location in different plants and algae, 83
  nuclear gene families, 373–376
  phylogeny, 47
RecA, in chloroplasts, 103, 106
Recombination
  chloroplasts, 76, 167, 170, 191
  land plant mitochondria, 62–64, 170–171
  Neurospora stopper formation, 243
  role in organelle genome evolution, 102–103
Red algae. *See Rhodophyta*
Red Wolf. *See Lupus*
Repression, catabolite in mitochondria, 398–401
Retrograde regulation, 405–406
Reverse transcription

INDEX

Chlamydomonas chloroplast rRNA introns, 98
  mitochondrial group II introns, 211, 216, 229, 244, 246
  mitochondrial plasmids, 229–231
  plant mitochondrial ORFs, 65
Rhizobacteria, 40
Rhizopogon, mitochondrial genome, 71
Rhodobacter
  chlorophyll synthesis in dark, 79
  cytochromes, 357, 359
Rhodophyta
  chloroplast membranes and pigments, 43t, 44
  chloroplast structure, 16, 17t
Rhodopseudomonas, ribulose bisphosphate carboxylase, 28
Rhodospirillaceae, 16
Rhodospirillum, ribulose bisphosphate carboxylase, 28
Ribosomes, chloroplasts, 296t
  antibiotic sensitivity and resistance, 303–304, 303t
  conditional lethal mutants, *Chlamydomonas*, 85
  discovery, 271
  polysomes, 288, 304–305, 368
  ribosomal proteins, 299–301, 303t
  ribosome-deficient plastids, 274–275
  rRNAs, 296–298
Ribosomes, cytoplasmic, 296t
  cotranslational import, 333–334
  inhibitor sensitivity, 303t
Ribosomes, mitochondrial, 296t
  antibiotic sensitivity and resistance, 303–304, 303t
  discovery, 271
  polysomes, 304
  ribosomal proteins, 301–303
  rRNAs, 298–299
Ribosomal protein genes
  algal chloroplast genomes, 60t, 81–82
  chloroplast ribosomal proteins and genes, 299–301
  Epifagus chloroplast genome, 83
  flowering plant mitochondria, 65
  land plant chloroplasts, 60t, 76
  Marchantia mitochondrial genome, 60t
  mitochondrial ribosomal proteins and genes, 301–303
  operon organization in chloroplast genomes, 60t, 79–80
  Paramecium mitochondria, 72
Ribulose bisphosphate carboxylase. *See also rbcL* and *rbcS*
  structure and functions, 27–28
Rice. *See Oryza*
Rickettsia, 40
RIM1, mitochondrial single stranded DNA binding protein, 121

Roridula, chloroplast genome evolution, 140
ROX1, catabolite repression, 401
rpo genes
  transcription in bleached plants, 275
  various plastid genomes, 74–75, 76, 81, 84, 273, 274
rpoB, initiator codon, 293
rpoC₁, *Euglena* twintron, 212
RPO41, mitochondrial core RNA polymerase, 272
rpl2
  absence from *Cuscuta* chloroplast genome, 84
  intron distribution in land plant chloroplast genomes, 98
  Paramecium mitochondrial genome, 72, 303
  translational control, 370
rpl4, *Paramecium* mitochondrial genome, 72
rpl5, *Euglena* chloroplast genome, 81
rpl12, *Euglena* chloroplast genome, 81
rpl14
  Oenothera chloroplast genome, 191
  Paramecium mitochondrial genome, 303
rpl15, *Euglena* chloroplast genome, 81
rpl16
  Euglena twintron, 212
  NCS3 mutant of maize, 257
  Oenothera chloroplast genome, 191
rpl22, location in land plants, 94, 301
rpl23, absence from *Cuscuta* chloroplast genome, 84
RPM genes, mitochondrial tRNA processing, yeast, 310
rps2, polycistronic transcript, chloroplasts, 369
rps3
  Euglena twintron, 212
  NCS3 mutant of maize, 257
rps7
  land plant mitochondrial genomes, 59, 95
  organization and structure, chloroplasts, 80, 301
rps8, *Oenothera* chloroplast genome, 191
rps12
  antibiotic resistant mutants, 85, 166, 304
  chloroplast genome organization and structure, 90, 221, 301
  chloroplast transformation, 173
  mitochondrial genomes, 59, 72, 95, 101, 263, 303
  translational regulation, 371
rps13, flowering plant mitochondrial genomes, 101, 103
rps14
  mitochondrial genomes, 72, 303
  operon in chloroplasts, 301
rps15, transcription in heat bleached plants, 275
rps16
  initiation codon, chloroplasts, 293
  promoter, chloroplasts, 282
  transcription, chloroplasts, 283

rps18, Euglena twintron, 212
rps19, flowering plant chloroplasts, 303
RNA editing, 323–330, 324*t,* 326*t,* 327*t*
rRNA
  anti Shine-Dalgarno sequence, 293
  cytoplasmic 18S in mitochondria, 95
  mitochondrial intron splicing, *Neurospora,* 217–219
  editing in *Physarum,* 330
  plant mitochondrial 5S rRNA, 59, 65
  phylogeny, 37, 40, 47–48, 72–73, 82
  secondary structure conservation, 299*t*
rRNA genes
  antibiotic resistant mutants, 156, 159, 165, 166, 168–169, 173, 222–223
  chloroplast promoter, 283
  cytoplasmic male sterility, 259
  deletion in *poky* mutant, 241
  interrupted genes, 65, 74, 299
  mitochondria of various taxa, 54, 59, 66, 74, 95, 101
  mobile introns, 221–223
  plastid genomes of various taxa, 72, 75, 76, 81, 83, 101
  rates of evolution in mitochondria, 104, 128
  structure and organization in organelles, 296–299
  transcription in heat bleached plastids, 275
  transcription by TAC polymerase, 274
  transformation, 173
RTG genes, retrograde regulation, 405–406
Rye. *See Secale*

Saccharomyces capensis, mobile G+C elements, 227
Saccharomyces carlsbergensis
  cytb gene introns, 213
  mtDNA, 51
Saccharomyces cerevisiae. See also Genetics, mitochondrial, *Saccharomyces*
  chaperones, 344–346
  expression of genes involved in mitochondrial biogenesis, 398–405, 399*t,* 402*t*
  G+C clusters, 68, 97, 227–228
  introns, 66–68, 97, 210, 213–217, 214*t,* 216*t,* 221–223, 223*t*
  intron splicing proteins, 213–217, 216*t*
  life cycle and growth, 153–154
  mitochondrial gene transfer to nucleus, 93, 94, 171
  mitochondrial genome replication, 120–121
  mitochondrial genome structure, 51, 66–68, 99–100
  mitochondrial location of *atp9,* 94
  mitochondrial nucleoids, 111–112
  mitochondrial ribosomal proteins and genes, 302–303

mitochondrial rRNA gene organization, 298–299
mitochondrial tRNA import, 311
mitochondrial tRNA processing, 309–311
mitochondrial tRNA synthetases, 313
mobile elements, 221–223, 226–228
ori/rep elements, 97, 156
protein import into mitochondria, 333–334, 347–357, 359–360
protein transport components, 344*t*
recombination of mitochondrial genome, 103
retrograde regulation, 405–406
sequence evolution of mitochondrial genome, 106
site clusters, 68
spacers, 68, 99
transcription of mitochondrial DNA, 272–273, 279–280
translation in mitochondria, 293, 294–295
Saccharomyces exiguus, mitochondrial gene order, 101
Saccharomycopsis, mitochondrial tRNA processing, 310
Scenedesmus, mitochondrial genome, 66
Schizophyllum, mitochondrial UGA termination codon, 322
Schizosaccharomyces
  absence of *var1* gene, 69
  mitochondrial genetics, 168–169
  mitochondrial genome structure, 53, 68
  mitochondrial rRNA gene organization, 69, 299
  petites, 169
  mitochondrial transcripts, 280
Scorpions. *See Centrurus; Hadrurus; Opisthacanthus*
Seaside Sparrow. *See Ammodramus*
SecA, Cryptomonas chloroplast genome, 82
Secale, chloroplast ribosome deficiency, 274–275
SecY, Cryptomonas chloroplast genome, 82
Sea anemone. *See Metridium*
Sea urchin, mitochondrial DNA transcription, 396–397
Selaginella, chloroplast DNA, 50
Senescence in fungi, 152, 243–247
Signal peptide. *See* Presequence
Sinapis, chloroplast sigma factors, 275–276
Solanaceae, 190–191, 373
Solanum
  mitochondrial tRNAs, 309
  plastid transmission, 182
  somatic hybrids, 170
Somatic (vegetative) segregation, defined, 149
Soybean. *See Glycine*
Spectinomycin
  chloroplast rRNA resistance mutations, 173–174, 303–304
  detoxification by *aadA* gene product, 173
  inhibition of organelle protein synthesis, 303*t*

INDEX

Spinacia (Spinach)
   chloroplast transcript processing, 368–369
   editing in chloroplasts, 325
   norflurazon induced bleaching, 390–391
SSA genes, yeast HSP70 chaperones, 344
SSC, yeast HSP70 mitochondrial chaperone, 345
Stramenopiles, 39
Streptomycin
   callus bleaching, 85
   chloroplast resistance mutations, 85, 165, 173–174, 303–304
   inhibition of organelle protein synthesis, 303*t*
Stroma. *See* Chloroplast stroma
Subclover, short dispersed repeats in chloroplast genome, 96
Succinate-ubiquinone oxidoreductase (Complex II), 10–11
Suillus, mitochondrial genome variation, 70–71
Sunfish, mitochondrial genome evolution, 136

Tagetitoxin, chloroplast transcriptional inhibitor, 391
Taxodium, chloroplast DNA from fossils, 143
Tetrahymena
   mitochondrial genome, 72
   mitochondrial nucleoids, 111
   mitochondrial DNA replication, 121
   mitochondrial rRNA gene organization, 299
   tRNA import into mitochondria, 311
thm24, controls stability of *atpB* mRNA, 388
Thylakoids. *See* Chloroplast thylakoids
Tobacco. *See* Nicotiana
Tomato. *See* Lycopersicon
Topoisomerase
   chloroplast DNA transcription, 367–368
   kDNA replication, 123
Transcription, chloroplasts, 273–276, 282–288
Transcription, mitochondria, 251, 272–273, 276–282
Transformation, *Agrobacterium,* 172
Transformation, chloroplasts
   methods, 152, 172–174
   uses, 282–285, 369–371
Transformation, mitochondria
   Chlamydomonas, 174
   Saccharomyces, methods, 151, 174–175
   uses, 93, 405
Transit peptide. *See* Presequence
Trans-splicing. *See* Introns, group II, trans-splicing
Triticum
   chloroplast division, 114
   chloroplast transmission in crosses, 182
   mitochondrial promoter, 281
   mitochondrial RNA editing, 324, 326
   mitochondrial rRNA phylogeny, 40
   nucleoids, 113–114

TRM genes, mitochondrial tRNA modification, 310, 313
trn genes
   promoters, 282
   transcript processing, 283
tRNAs
   chloroplast tRNA^{Glu} and porphyrin synthesis, 85–86, 312
   chloroplasts, 311–312
   editing, 326, 330
   import, 58, 65, 72, 74, 84, 305, 311
   mitochondria, 308–309, 311
   mitochondrial transcript processing, 58, 276–278, 280, 307
   processing and modification, 309–310
tRNA^{Ala}
   Anacystis, 307
   chloroplast intron, 97
tRNA^{Arg}, *Zea mays* mitochondrial genome, 95, 259
tRNA^{Asp}, *Zea mays* mitochondrial genome, 95
tRNA^{Cys}, *Zea mays* mitochondrial genome, 95
tRNA^{Fmet}, 321
tRNA^{Glu}
   porphyrin biosynthesis, plastids, 85–86
   three species in plastids, 312
tRNA^{His}, *Drosophila* mitochondria, 59
tRNA^{Ile}
   Anacystis, 307
   chloroplast intron, 97
   codon recognition, chloroplasts, 312
   Zea mays mitochondrial genome, 95
tRNA^{Leu}
   ancient intron, chloroplasts and cyanobacteria, 82, 97–99
   human MELAS syndrome, 250, 276–277
   mitochondrial transcript processing, 276–277
   Zea mays mitochondrial genome, 95
tRNA^{Lys}
   human MERRF syndrome, 250
   import in yeast, 311
tRNA^{Met}
   initiation animal mitochondria, 321
   Neurospora mitochondrial recombination repeat, 243
tRNA^{Ser}, structure in animal mitochondria, 308–309
tRNA^{Trp}, nonsense mutation, *Chlamydomonas* chloroplast, 200
tRNA^{Val}
   Drosophila mitochondria, 59
   maize mitochondria, 95
tRNA genes
   chloroplast promoters, 282
   hominoid evolution, 128
   various mitochondrial genomes, 54, 59, 65, 66, 69, 72, 74, 95
   various plastid genomes, 74, 76, 84–86

tRNA synthetases
  characteristics of organelle enzymes, 313
  role in mitochondrial intron splicing, 217–219, 313
trpD, Cyanophora chloroplast genome, 81
Trypanosoma
  kDNA, 73–74
  RNA editing, 326–330
Trypanosomes
  kDNA replication, 121–123
  mitochondrial genome 73–74
  mitochondrial rRNA gene organization, 298
  RNA editing, 326–330
  tRNA import into mitochondria, 311
tscA, RNA required for trans-splicing in *Chlamydomonas,* 212, 221
tufA, in different plants and algae, 47, 81, 83, 93–94, 295
T-ruf13, cytoplasmic male sterility, *Zea mays,* 259–262
T-ruf25, Zea mays, 259
Turnip. *See Brassica*
Twintrons, 212–213

Ubiquinol-cytochrome *c* reductase (Complex III), 12*t*, 411–413
URA3
  cotransformation, 174
  transfer from mitochondrion to nucleus, 93, 174–175
urfS, Petunia cytoplasmic male sterility, 263
uidA, expression in plastids, 174, 369–370

Vairimorpha, 39

var1
  in mitochondria, 66, 69
  mobility, 226–228
VAS1, valyl tRNA synthetase, 313
Venus' Flytrap. *See Dionaea*
Vicia, chloroplast polysomes, 305
vir-115, chloroplast gene expression, 389

Wheat. *See Triticum*

Xenopus
  mitochondrial DNA, 58
  mitochondrial DNA replication and transcription, 273, 396

Yeast, *See Saccharomyces*

Zea mays
  chloroplast DNA replication, 123–124
  chloroplast polysomes, 288, 368
  chloroplast signal, 390
  chloroplast DNA transcription, 273–274, 367
  chloroplast DNA variation, 141
  cytoplasmic male sterility, 232–233, 258–262
  expression of genes in mesophyll and bundle sheath plastids, 380
  mitochondrial genome structure and function 61, 62–63
  mitochondrial plasmids, 231–233, 232*t*
  mitochondrial transcription, 280–281
  nonchromosomal striped mutants, 257–258
  RNA editing, 324–325
  transfer of genes to mitochondria, 95
Zebra. *See* Equus
Zygnematales, location of *tufA* gene, 94